AF361673

Optimization for Decision Making

Optimization for Decision Making

Editors

Víctor Yepes
José María Moreno-Jiménez

MDPI • Basel • Beijing • Wuhan • Barcelona • Belgrade • Manchester • Tokyo • Cluj • Tianjin

Editors
Víctor Yepes
ICITECH, Universitat Politècnica
de València
Spain

José María Moreno-Jiménez
Universidad de Zaragoza
Spain

Editorial Office
MDPI
St. Alban-Anlage 66
4052 Basel, Switzerland

This is a reprint of articles from the Special Issue published online in the open access journal *Mathematics* (ISSN 2227-7390) (available at: https://www.mdpi.com/journal/mathematics/special_issues/Optimization_Decision_Making).

For citation purposes, cite each article independently as indicated on the article page online and as indicated below:

LastName, A.A.; LastName, B.B.; LastName, C.C. Article Title. *Journal Name* **Year**, *Article Number*, Page Range.

ISBN 978-3-03943-220-2 (Hbk)
ISBN 978-3-03943-221-9 (PDF)

Contents

About the Editors

Víctor Yepes Full Professor of Construction Engineering; he holds a Ph.D. degree in civil engineering. He serves at the Department of Construction Engineering, Universitat Politecnica de Valencia, Valencia, Spain. He has been the Academic Director of the M.S. studies in concrete materials and structures since 2007 and a Member of the Concrete Science and Technology Institute (ICITECH). He is currently involved in several projects related to the optimization and life-cycle assessment of concrete structures as well as optimization models for infrastructure asset management. He is currently teaching courses in construction methods, innovation, and quality management. He authored more than 250 journals and conference papers including more than 100 published in the journal quoted in JCR. He acted as an Expert for project proposals evaluation for the Spanish Ministry of Technology and Science, and he is the Main Researcher in many projects. He currently serves as the Editor-in-Chief of the *International Journal of Construction Engineering and Management* and a member of the editorial board of 12 international journals (*Structure & Infrastructure Engineering, Structural Engineering and Mechanics, Mathematics, Sustainability, Revista de la Construcción, Advances in Civil Engineering*, and *Advances in Concrete Construction*, among others).

José María Moreno-Jiménez Full Professor of Operations Research and Multicriteria Decision Making, received the degrees in mathematics and economics as well as a Ph.D. degree in applied mathematics from the University of Zaragoza, Spain; where he is teaching from the course 1980–1981. He is the Head of the Quantitative Methods Area in the Faculty of Economics and Business of the University of Zaragoza from 1997, the Chair of the Zaragoza Multicriteria Decision Making Group from 1996, a member of the Advisory Board of the Euro Working Group on Decision Support Systems from 2017, and an Honorary Member of the International Society on Applied Economics ASEPELT from 2019. He has also been the President of this international scientific society (2014–2018) and the Coordinator of the Spanish Multicriteria Decision Making Group (2012–2015). His research interests are in the general area of Operations Research theory and practice, with an emphasis on multicriteria decision making, electronic democracy/cognocracy, performance analysis, and industrial and technological diversification. He has published more than 250 papers in scientific books and journals in the most prestigious editorials and is a member of the Editorial Board of several national and international journals.

Preface to "Optimization for Decision Making"

Decision making is one of the distinctive activities of the human being; it is an indication of the degree of evolution, cognition, and freedom of the species. In the past, until the end of the 20th century, scientific decision-making was based on the paradigms of substantive rationality (normative approach) and procedural rationality (descriptive approach). Since the beginning of the 21st century and the advent of the Knowledge Society, decision-making has been enriched with new constructivist, evolutionary, and cognitive paradigms that aim to respond to new challenges and needs; especially the integration into formal models of the intangible, subjective, and emotional aspects associated with the human factor, and the participation in decision-making processes of spatially distributed multiple actors that intervene in a synchronous or asynchronous manner.

To help address and resolve these types of questions, this book comprises 13 chapters that present a series of decision models, methods, and techniques and their practical applications in the fields of economics, engineering, and social sciences. The chapters collect the papers included in the "Optimization for Decision Making" Special Issue of the *Mathematics* journal, 2019, 7(3), first decile of the JCR 2019 in the Mathematics category.

We would like to thank both the MDPI publishing editorial team, for their excellent work, and the 47 authors who have collaborated in its preparation. The papers cover a wide spectrum of issues related to the scientific resolution of problems; in particular, related to decision making, optimization, metaheuristics, simulation, and multi-criteria decision-making.

We hope that the papers, with their undoubted mathematical content, can be of use to academics and professionals from the many branches of knowledge (philosophy, psychology, economics, mathematics, decision science, computer science, artificial intelligence, neuroscience, and more) that have, from such diverse perspectives, approached the study of decision-making, an essential aspect of human life and development.

Víctor Yepes, José María Moreno-Jiménez
Editors

 *mathematics* 

Article

Algorithm for Probabilistic Dual Hesitant Fuzzy Multi-Criteria Decision-Making Based on Aggregation Operators with New Distance Measures

Harish Garg * and Gagandeep Kaur

School of Mathematics, Thapar Institute of Engineering & Technology (Deemed University) Patiala, Punjab 147004, India; gdeep01@ymail.com
* Correspondence: harishg58iitr@gmail.com or harish.garg@thapar.edu; Tel.: +91-86990-31147

Received: 2 November 2018; Accepted: 21 November 2018; Published: 25 November 2018

Abstract: Probabilistic dual hesitant fuzzy set (PDHFS) is an enhanced version of a dual hesitant fuzzy set (DHFS) in which each membership and non-membership hesitant value is considered along with its occurrence probability. These assigned probabilities give more details about the level of agreeness or disagreeness. By emphasizing the advantages of the PDHFS and the aggregation operators, in this manuscript, we have proposed several weighted and ordered weighted averaging and geometric aggregation operators by using Einstein norm operations, where the preferences related to each object is taken in terms of probabilistic dual hesitant fuzzy elements. Several desirable properties and relations are also investigated in details. Also, we have proposed two distance measures and its based maximum deviation method to compute the weight vector of the different criteria. Finally, a multi-criteria group decision-making approach is constructed based on proposed operators and the presented algorithm is explained with the help of the numerical example. The reliability of the presented decision-making method is explored with the help of testing criteria and by comparing the results of the example with several prevailing studies.

Keywords: probabilistic dual hesitant fuzzy sets; distance measures; aggregation operators; consumer behavior; multi-criteria decision-making; maximum deviation method

---

1. Introduction

With growing advancements in economic, socio-cultural as well as technical aspects of the world, uncertainties have started playing a dominant part in decision-making (DM) processes. The nature of DM problems is becoming more and more complex and the data available for the evaluation of these problems is increasingly having uncertain pieces of unprocessed information [1,2]. Such data content leads to inaccurate results and increase the risks by many folds. To decrease the risks and to reach the accurate results, decision-making has attained the attention of a large number of researchers. In the complex decision-making systems, often large cost and computational efforts are required to address the information, to evaluate it to form accurate results. In such situations, the major aim of the decision makers remain to decrease the computational overheads and to reach the desired objective in less space of time.

Time-to-time such DM techniques are framed which captures the uncertain information in an efficient way and results are calculated in such a manner that they comply easily to the real-life situations. From the crisp set theory, an analysis was shifted towards the fuzzy sets (FSs) and further Atanassov [3] extended the FS theory given by Zadeh [4] to Intuitionistic FSs (IFSs) by acknowledging the measures of disagreeness along with measures of agreeness. Afterward, Atanassov and Gargov [5] extended the IFS to the Interval-valued intuitionistic fuzzy sets (IVIFSs) which contain the degrees of agreeness and disagreeness as interval values instead of single digits. As it is quite a common phenomenon that different attributes play a vital part during the selection of best alternative among the available ones, so suitable aggregation operators to evaluate the data are to be chosen carefully by the experts to address the nature of the DM problem. In these approaches, preferences are given as falsity and truth membership values in the crisp or interval number respectively such that the corresponding degrees altogether sum to be less than or equal to one. In above-stated environments, various researchers have constructed their methodologies for solving the DM problems focussing on information measures, aggregation operators etc. For instance, Xu [6] presented some weighted averaging aggregation operators (AOs) for intuitionistic fuzzy numbers (IFNs). Wang et al. [7] presented some AOs to aggregate various interval-valued intuitionistic fuzzy (IVIF) numbers (IVIFNs). Garg [8,9] presented some improved interactive AOs for IFNs. Wang and Liu [10] gave interval-valued intuitionistic fuzzy hybrid weighted AOs based on Einstein operations. Wang and Liu [11] presented some hybrid weighted AOs using Einstein norm operations. Garg [12] presented a generalized AOs using Einstein norm operations for Pythagorean fuzzy sets. Garg and Kumar [13] presented some new similarity measures for IVIFNs based on the connection number of the set pair analysis theory. However, apart from these, a comprehensive overview of the different approaches under the IFSs and/or IVIFSs to solve MCDM problems are summarized in [14–24]. In the above theories, it is difficult to capture cases where the preferences related to different objects are given in the form of the multiple numbers of possible membership entities. To handle it, Torra [25] came up with the idea of hesitant fuzzy sets (HFSs). Zhu et al. [26] enhanced it to the dual hesitant fuzzy sets (DHFSs) by assigning equal importance to the possible non-membership values as that of possible membership values in the HFSs. In the field of AOs, Xia and Xu [27] established different operators to aggregated their values. Garg and Arora [28] presented some AOs under the dual hesitant fuzzy soft set environment and applied them to solve the MCDM problems. Wei and Zhao [29] presented some induced hesitant AOs for IVIFNs. Apart from these, some other kinds of the algorithms for solving the decision-making problems are investigated by the authors [30–38] under the hesitant fuzzy environments.

Although, these approaches are able to capture the uncertainties in an efficient way, yet these works are unable to model the situations in which the refusal of an expert in providing the decision plays a dominant role. For example, suppose a panel of 6 experts is approached to select the best candidate during the recruitment process and 2 of them refused to provide any decision. While evaluating the informational data using the existing approaches, the number of decision makers is considered to be 4 instead of 6 i.e., the refusal providing experts are completely ignored and the decision is framed using the preferences given by the 4 decision-providing experts only. This cause a significant loss of information and may lead to inadequate results. In order to address such refusal-oriented cases, Zhu and Xu [39] corroborated probabilistic hesitant fuzzy sets (PHFSs). Wu et al. [40] gave the notion of AOs on interval-valued PHFSs (IVPHFSs) whereas Zhang et al. [41] worked on preference relations based on IVPHFSs and accessed the findings by applying to real life decision scenarios. Hao et al. [42] corroborated the concept of PDHFSs. Later on, Li et al. [43] presented the concept of dominance degrees and presents a DM approach based on the best-worst method under the PHFFSs. Li and Wang [44] comprehensively expressed way to address their vague and uncertain information. Lin and Xu [45] determined various probabilistic linguistic distance measures. Apart from them, several researchers [46–52] have shown a keen interest in applying probabilistic hesitant fuzzy set environments to different decision making approaches. Based on these existing studies, the primary motivation of this paper is summarized as below:

(i)   In the existing DHFSs, each and every membership value has equal probability. For instance, suppose a person has to buy a commodity X, and he is confused that either he is 10% sure or 20% sure to buy it, and is uncertain about 30% or 40% in not buying it. Thus, under DHFS environment, this information is captured as $(\{0.10, 0.20\}, \{0.30, 0.40\})$. Here, in dual hesitant fuzzy set, each hesitant value is assumed to have probability 0.5. So, mentioning the same probability value repeatedly is omitted in DHFSs. But, if the buyer is more confident about 10% agreeness than that of 20% i.e., suppose he is certain that his agreeness towards buying the commodity is 70% towards 10% and 30% towards 20% and similarly, for the non-membership case, he is 60% favoring to the 40% rejection level and 40% favoring the 30% rejection level. Thus, probabilistic dual hesitant fuzzy set is formulated as $(\{0.10|0.70, 0.20|0.30\},\ \{0.30|0.4, 0.40|0.6\})$. So, to address such cases, in which even the hesitation has a some preference over the another hesitant value, DHFS acts as an efficient tool to model them.

(ii)   In the multi-expert DM problems, there may often arise conflicts in the preferences given by different experts. These issues can easily be resolved using DHFSs. For example, let A and B be two experts giving their opinion about buying a commodity X. Suppose opinion provided by A is noted in form of DHFS as $(\{0.20, 0.30\}, \{0.10, 0.15\})$ and similarly B gave opinion as $(\{0.20, 0.25\}, \{0.10\})$. Now, both the experts are providing different opinions regarding the same commodity X. This is a common problem that arises in the real life DM scenarios. To address this case, the information is combined into PDHFS by analyzing the probabilities of decision given by both the experts. The PDHFS, thus formed, is given as $\left(\left\{0.20\big|\frac{0.5+0.5}{2}, 0.30\big|\frac{0.5}{2}, 0.25\big|\frac{0.5}{2}\right\},\ \left\{0.10\big|\frac{0.5+1}{2}, 0.15\big|\frac{0.5}{2}\right\}\right)$. In simple form, it is $(\{0.20|0.5, 0.30|0.25, 0.25|0.25\},\ \{0.10|0.75, 0.15|0.25\})$. Thus, this paper is motivated by the need of capturing the more favorable values among the hesitant values.

(iii)   The existing decision-making approaches based on DHFS environment are numerically more complex and time consuming because of redundancy of the membership (non-membership) values to match the length of one set to another. This manuscript is motivated by the fact of reducing this data redundancy and making the DM approach more time-efficient.

    Motivated by the aforementioned points regarding shortcomings in the existing approaches, this paper focusses on eradicating them by developing a series of AOs. In order to do so, the supreme objectives are listed below:

(i)   To consider the PDHFS environment to capture the information.
(ii)   To propose two novel distance measures on PDHFSs.
(iii)   To capture some weighted information regarding the available information by solving a non-linear mathematical model.
(iv)   To develop average and geometric Einstein AOs based on the PDHFS environment.
(v)   To propose a DM approach relying on the developed operators.
(vi)   To check numerical applicability of the approach to a real-life case and to compare the outcomes with prevailing approaches.

    To achieve the first objective and to provide more degrees of freedom to practitioners, in this article, we consider PDHFS environment to extract data. For achieving the second objective, two distance measures are proposed; one in which the size of two PDHFSs should be the same whereas in the second one the size may vary. For achieving the third objective, a non-linear model is solved to capture the weighted information. For achieving fourth objective average and geometric Einstein AOs are proposed. To attain the fifth and sixth objective a real-life based case-study is conducted and its comparative analysis with the prevailing environments is carried out.

    The rest of this paper is organized as follows: Section 2 highlights the basic definitions related to DHFSs, PHFSs, and PDHFSs. Section 3 introduces the two distance measures for PDHFSs along with their desirable properties. Section 4 introduces some Einstein operational laws on PDHFSs with the investigation of some properties. In Section 5, some averaging and geometric weighted Einstein AOs are proposed. A non-linear programming model for weights determination is elicited in Section 6. In Section 7, an approach is constructed to address the DM problems and includes the real-life marketing problem including a comparative analysis with the existing ones. Finally, concluding remarks are given in Section 8.

2. Preliminaries

This section emphasizes on basic definitions regarding the DHFSs, PHFSs and PDHFSs.

Definition 1. *On the universal set X, Zhu et al. [26] defined dual hesitant fuzzy set as:*

$$\alpha = \{(x, h(x), g(x)) \mid x \in X\} \tag{1}$$

where the sets $h(x)$ and $g(x)$ have values in $[0,1]$, which signifies possible membership and non-membership degrees for $x \in X$. Also,

$$0 \leq \gamma, \eta \leq 1; 0 \leq \gamma^+ + \eta^+ \leq 1 \tag{2}$$

in which, $\gamma \in h(x)$; $\eta \in g(x)$; $\gamma^+ \in h^+(x) = \bigcup\limits_{\gamma \in h(x)} \max\{\gamma\}$ and $\eta^+ \in g^+(x) = \bigcup\limits_{\eta \in g(x)} \max\{\eta\}$

Definition 2. *Let X be a reference set, then a probabilistic hesitant fuzzy set (PHFS) [39] P on X is given as*

$$P = \{\langle x, h_x(p_x) \rangle \mid x \in X\} \tag{3}$$

Here, the set h_x contains several values in $[0,1]$, and described by the probability distribution p_x. Also, h_x denotes membership degree of x in X. For simplicity, $h_x(p_x)$ is called a probabilistic hesitant fuzzy element (PHFE), denoted as $h(p)$ and is given as

$$h(p) = \{\gamma_i(p_i) \mid i = 1, 2, \ldots, \#H\},$$

where p_i satisfying $\sum\limits_{i=1}^{\#H} p_i \leq 1$, is the probability of the possible value γ_i and $\#H$ is the number of all $\gamma_i(p_i)$.

Definition 3 ([49]). *A probabilistic dual hesitant fuzzy set (PDHFS) on X is defined as:*

$$\alpha = \{(x, h(x)|p(x), g(x)|q(x)) \mid x \in X\} \tag{4}$$

Here, the sets $h(x)|p(x)$ and $g(x)|q(x)$ contains possible elements where $h(x)$ and $g(x)$ represent the hesitant fuzzy membership and non-membership degrees $x \in X$, respectively. Also, $p(x)$ and $q(x)$ are their associated probabilistic information. Moreover,

$$0 \leq \gamma, \eta \leq 1; 0 \leq \gamma^+ + \eta^+ \leq 1 \tag{5}$$

and

$$p_i \in [0,1], q_j \in [0,1], \sum_{i=1}^{\#h} p_i = 1, \sum_{j=1}^{\#g} q_j = 1 \tag{6}$$

where $\gamma \in h(x)$; $\eta \in g(x)$; $\gamma^+ \in h^+(x) = \bigcup\limits_{\gamma \in h(x)} \max\{\gamma\}$; $\eta^+ \in g^+(x) = \bigcup\limits_{\eta \in g(x)} \max\{\eta\}$. The symbols $\#h$ and $\#g$ are total values in $(h(x)|p(x))$ and $(g(x)|q(x))$ respectively. For sake of convenience, we shall denote it as $(h|p, g|q)$ and name it as probabilistic dual hesitant fuzzy element (PDHFE).

Definition 4 ([49]). *For a PDHFE α, defined over a universal set X, the complement is defined as*

$$
\alpha^c = \begin{cases}
\bigcup\limits_{\gamma \in h, \eta \in g} (\{\eta \mid q_\eta\}, \{\gamma \mid p_\gamma\}), & \text{if } g \neq \phi \text{ and } h \neq \phi \\
\bigcup\limits_{\gamma \in h} (\{1 - \gamma\}, \{\phi\}), & \text{if } g = \phi \text{ and } h \neq \phi \\
\bigcup\limits_{\eta \in g} (\{\phi\}, \{1 - \eta\}), & \text{if } h = \phi \text{ and } g \neq \phi
\end{cases}
\tag{7}
$$

Definition 5 ([49]). *Let $\alpha = (h\mid p, g\mid q)$ be a PDHFE, then the score function is defined as:*

$$
S(\alpha) = \sum_{i=1}^{\#h} \gamma_i \cdot p_i - \sum_{j=1}^{\#g} \eta_j \cdot q_j
\tag{8}
$$

where $\#h$ and $\#g$ are total numbers of elements in the components $(h\mid p)$ and $(g\mid q)$ respectively and $\gamma \in h$, $\eta \in g$. For two PDHFEs α_1 and α_2, if $S(\alpha_1) > S(\alpha_2)$, then the PDHFE α_1 is regarded more superior to α_2 and is denoted as $\alpha_1 \succ \alpha_2$.

3. Proposed Distance Measures for PDHFEs

In this section, we propose some measures to calculate the distance between two PDHFEs defined over a universal set $X = \{x_1, x_2, \ldots, x_n\}$. Throughout this paper, the main notations used are listed below:

Notations	Meaning	Notations	Meaning
n	number of elements in the universal set	N_A	number of elements in g_A
h_A	hesitant membership values of set A	p_A	probability for hesitant membership of set A
g_A	hesitant non-membership values of set A	q_A	probability for hesitant non-membership of set A
M_A	number of elements in h_A	ω	weight vector

Let $A = \left\{\left(x, h_{A_i}(x)\mid p_{A_i}(x), g_{A_j}(x)\mid q_{A_j}(x)\right) \mid x \in X\right\}$ and $B = \left\{\left(x, h_{B_{i'}}(x)\mid p_{B_{i'}}(x), g_{B_{j'}}(x)\mid q_{B_{j'}}(x)\right) \mid x \in X\right\}$ where $i = 1, 2, \ldots, M_A$; $j = 1, 2, \ldots, N_A$; $i' = 1, 2, \ldots, M_B$ and $j' = 1, 2, \ldots, N_B$, be two PDHFSs. Also, let $M = \max\{M_A, M_B\}$, $N = \max\{N_A, N_B\}$, be two real numbers, then for a real-number $\lambda > 0$, we define distance between A and B as:

$$
d_1(A, B) = \left(\sum_{k=1}^{n} \frac{1}{n} \left(\frac{1}{M+N} \left(\begin{array}{l} \sum\limits_{i=1}^{M} \left| \gamma_{A_i}(x_k) p_{A_i}(x_k) - \gamma_{B_i}(x_k) p_{B_i}(x_k) \right|^\lambda \\ + \sum\limits_{j=1}^{N} \left| \eta_{A_j}(x_k) q_{A_j}(x_k) - \eta_{B_j}(x_k) q_{B_j}(x_k) \right|^\lambda \end{array} \right) \right) \right)^{\frac{1}{\lambda}}
\tag{9}
$$

where $\gamma_{A_i} \in h_{A_i}$, $\gamma_{B_i} \in h_{B_{i'}}$, $\eta_{A_i} \in g_{A_i}$, $\eta_{B_i} \in g_{B_{j'}}$. It is noticeable that, there may arise the cases in which $M_A \neq M_B$ as well as $N_A \neq N_B$. Under such situations, for operating distance d_1, the lengths of these elements should be equal to each other. To achieve this, under the hesitant environments, the experts repeat the least or the greatest values among all the hesitant values, in the smaller set, till the length of both A and B becomes equal. In other words, if $M_A > M_B$, then repeat the smallest value in set h_B till M_B becomes equal to M_A and if $M_A < M_B$, then repeat the smallest value in set h_A till M_A becomes equal to M_B. Alike the smallest values, the largest values may also be repeated. This choice of the smallest or largest value's repetition entirely depends on decision-makers optimistic or pessimistic approach. If the expert opts for the optimistic approach then he will expect the highest membership values and thus will repeat the largest values. However, if the expert chooses to follow the pessimistic approach, then he will expect the least favoring values and will go with repeating the smallest values till the same length is achieved. But sometimes, length of A and B cannot be matched by increasing the numbers of elements, then in such cases, the distance d_1 can be unappropriate for the data evaluations. To handle such cases, we propose another distance measure d_2 in which there is no need to repeat the values for matching the length of the elements under consideration. This distance d_2 is calculated as:

$$
d_2(A,B) = \left(\sum_{k=1}^{n} \frac{1}{n} \left(\frac{\left| \frac{1}{M_A} \sum_{i=1}^{M_A} \left(\gamma_{A_i}(x_k) p_{A_i}(x_k) \right) - \frac{1}{M_B} \sum_{i'=1}^{M_B} \left(\gamma_{B_i'}(x_k) p_{B_i'}(x_k) \right) \right|^\lambda}{2} + \frac{\left| \frac{1}{N_A} \sum_{j=1}^{N_A} \left(\eta_{A_j}(x_k) q_{A_j}(x_k) \right) - \frac{1}{N_B} \sum_{j'=1}^{N_B} \left(\eta_{B_j'}(x_k) q_{B_j'}(x_k) \right) \right|^\lambda}{2} \right) \right)^{\frac{1}{\lambda}}
\tag{10}
$$

The distance measures proposed above satisfy the axiomatic statement given below:

Theorem 1. *Let A and B be two PDHFSs, then the distance measure d_1 satisfies the following conditions:*

(P1) $0 \leq d_1(A,B) \leq 1$;

(P2) $d_1(A,B) = d_1(B,A)$;

(P3) $d_1(A,B) = 0$ *if* $A = B$;

(P4) *If* $A \subseteq B \subseteq C$, *then* $d_1(A,B) \leq d_1(A,C)$ *and* $d_1(B,C) \leq d_1(A,C)$.

Proof. Let $X = \{x_1, x_2, \ldots, x_n\}$ be the universal set and A, B be two PDHFSs defined over X. Then for each $x_k, k = 1, 2, \ldots, n$, we have

(P1) Since, $0 \leq \gamma_{A_i}(x_k) \leq 1$ and $0 \leq p_{A_i}(x_k) \leq 1$, for all $i = 1, 2, \ldots, M$, this implies that $0 \leq \gamma_{A_i}(x_k) p_{A_i}(x_k) \leq 1$ and $0 \leq \gamma_{B_i}(x_k) p_{B_i}(x_k) \leq 1$. Thus, for any $\lambda > 0$, we have $0 \leq |\gamma_{A_i}(x_k) p_{A_i}(x_k) - \gamma_{B_i}(x_k) p_{B_i}(x_k)|^\lambda \leq 1$. Further, $\sum_{i=1}^{M} 0 \leq \sum_{i=1}^{M} |\gamma_{A_i}(x_k) p_{A_i}(x_k) - \gamma_{B_i}(x_k) p_{B_i}(x_k)|^\lambda \leq \sum_{i=1}^{M} 1$ which leads to $0 \leq \sum_{i=1}^{M} |\gamma_{A_i}(x_k) p_{A_i}(x_k) - \gamma_{B_i}(x_k) p_{B_i}(x_k)|^\lambda \leq M$. Similarly, for $j = 1, 2, \ldots, N$,

$$
0 \leq \sum_{j=1}^{N} \left| \eta_{A_j}(x_k) q_{A_j}(x_k) - \eta_{B_j}(x_k) q_{B_j}(x_k) \right|^\lambda \leq N \text{ which yields}
$$

$$
0 \leq \sum_{i=1}^{M} \left| \gamma_{A_i}(x_k) p_{A_i}(x_k) - \gamma_{B_i}(x_k) p_{B_i}(x_k) \right|^\lambda + \sum_{j=1}^{N} \left| \eta_{A_j}(x_k) q_{A_j}(x_k) - \eta_{B_j}(x_k) q_{B_j}(x_k) \right|^\lambda \leq M + N.
$$

Thus,

$$
0 \leq \left(\sum_{k=1}^{n} \frac{1}{n} \left(\frac{1}{M+N} \left(\sum_{i=1}^{M} \left| \gamma_{A_i}(x_k) p_{A_i}(x_k) - \gamma_{B_i}(x_k) p_{B_i}(x_k) \right|^\lambda + \sum_{j=1}^{N} \left| \eta_{A_j}(x_k) q_{A_j}(x_k) - \eta_{B_j}(x_k) q_{B_j}(x_k) \right|^\lambda \right) \right) \right)^{\frac{1}{\lambda}} \leq 1,
$$

which clearly implies that $0 \leq d_1(A,B) \leq 1$.

(P2) Since

$$
\begin{aligned}
d_1(A,B) &= \left(\sum_{k=1}^{n} \frac{1}{n} \left(\frac{1}{M+N} \left(\begin{array}{l} \sum_{i=1}^{M} \left| \gamma_{A_i}(x_k) p_{A_i}(x_k) - \gamma_{B_i}(x_k) p_{B_i}(x_k) \right|^{\lambda} \\ + \sum_{j=1}^{N} \left| \eta_{A_j}(x_k) q_{A_j}(x_k) - \eta_{B_j}(x_k) q_{B_j}(x_k) \right|^{\lambda} \end{array} \right) \right) \right)^{\frac{1}{\lambda}} \\[2mm]
&= \left(\sum_{k=1}^{n} \frac{1}{n} \left(\frac{1}{M+N} \left(\begin{array}{l} \sum_{i=1}^{M} \left| \gamma_{B_i}(x_k) p_{B_i}(x_k) - \gamma_{A_i}(x_k) p_{A_i}(x_k) \right|^{\lambda} \\ + \sum_{j=1}^{N} \left| \eta_{B_j}(x_k) q_{B_j}(x_k) - \eta_{A_j}(x_k) q_{A_j}(x_k) \right|^{\lambda} \end{array} \right) \right) \right)^{\frac{1}{\lambda}} \\[2mm]
&= d_1(B,A)
\end{aligned}
$$

Hence, the distance measure d_1 possess a symmetric nature.

(P3) For $A = B$, we have $\gamma_{A_i}(x_k) = \gamma_{B_i}(x_k)$ and $p_{A_i}(x_k) = p_{B_i}(x_k)$. Also, $\eta_{A_j}(x_k) = \eta_{B_j}(x_k)$ and $q_{A_j}(x_k) = q_{B_j}(x_k)$. Thus, we have $\left| \gamma_{A_i}(x_k) p_{A_i}(x_k) - \gamma_{A_i}(x_k) p_{A_i}(x_k) \right|^{\lambda} = 0$ and $\left| \eta_{A_j}(x_k) q_{A_j}(x_k) - \eta_{A_j}(x_k) q_{A_j}(x_k) \right|^{\lambda} = 0$. Hence, it implies that

$$
\left(\sum_{k=1}^{n} \frac{1}{n} \left(\frac{1}{M+N} \left(\begin{array}{l} \sum_{i=1}^{M} \left| \gamma_{A_i}(x_k) p_{A_i}(x_k) - \gamma_{B_i}(x_k) p_{B_i}(x_k) \right|^{\lambda} \\ + \sum_{j=1}^{N} \left| \eta_{A_j}(x_k) q_{A_j}(x_k) - \eta_{B_j}(x_k) q_{B_j}(x_k) \right|^{\lambda} \end{array} \right) \right) \right)^{\frac{1}{\lambda}} = 0
$$

$$
\Rightarrow \quad d_1(A,B) = 0.
$$

(P4) Since, $A \subseteq B \subseteq C$, then $\gamma_{A_i}(x_k) p_{A_i}(x_k) \leq \gamma_{B_i}(x_k) p_{B_i}(x_k) \leq \gamma_{C_i}(x_k) p_{C_i}(x_k)$ and $\eta_{A_j}(x_k) q_{A_j}(x_k) \geq \eta_{B_j}(x_k) q_{B_j}(x_k) \geq \eta_{C_j}(x_k) q_{C_j}(x_k)$. Further, $\left| \gamma_{A_i}(x_k) p_{A_i}(x_k) - \gamma_{B_i}(x_k) q_{B_i}(x_k) \right|^{\lambda} \leq \left| \gamma_{A_i}(x_k) p_{A_i}(x_k) - \gamma_{C_i}(x_k) q_{C_i}(x_k) \right|^{\lambda}$ and $\left| \eta_{A_j}(x_k) q_{A_j}(x_k) - \eta_{B_j}(x_k) q_{B_j}(x_k) \right|^{\lambda} \geq \left| \eta_{A_j}(x_k) q_{A_j}(x_k) - \eta_{C_j}(x_k) q_{C_j}(x_k) \right|^{\lambda}$. Therefore, $d_1(A,B) \leq d_1(A,C)$ and $d_1(B,C) \leq d_1(A,C)$.

$\square$

Theorem 2. *Let A and B be two PDHFSs, then the distance measure d_2 satisfies the following conditions:*

(P1) $0 \leq d_2(A,B) \leq 1$;

(P2) $d_2(A,B) = d_2(B,A)$;

(P3) $d_2(A,B) = 0$ if $A = B$;

(P4) *If $A \subseteq B \subseteq C$, then $d_2(A,B) \leq d_2(A,C)$ and $d_2(B,C) \leq d_2(A,C)$.*

Proof. The proof is similar to Theorem 1, so we omit it here. $\square$

4. Einstein Aggregation Operational laws for PDHFSs

In this section, we propose some operational laws and the investigate some of their properties associated with PDHFEs.

Definition 6. *Let α, α_1 and α_2 be three PDHFEs such that $\alpha = (h|p_h, g|q_g)$, $\alpha_1 = (h_1|p_{h_1}, g_1|q_{g_1})$ and $\alpha_2 = (h_2|p_{h_2}, g_2|q_{g_2})$. Then, for $\lambda > 0$, we define the Einstein operational laws for them as follows:*

(i) $\alpha_1 \oplus \alpha_2 = \bigcup\limits_{\substack{\gamma_1 \in h_1, \eta_1 \in g_1 \\ \gamma_2 \in h_2, \eta_2 \in g_2}} \left(\left\{ \dfrac{\gamma_1 + \gamma_2}{1 + \gamma_1 \gamma_2} \,\middle|\, p_{\gamma_1} p_{\gamma_2} \right\}, \left\{ \dfrac{\eta_1 \eta_2}{1 + (1 - \eta_1)(1 - \eta_2)} \,\middle|\, q_{\eta_1} q_{\eta_2} \right\} \right);$

(ii) $\alpha_1 \otimes \alpha_2 = \bigcup\limits_{\substack{\gamma_1 \in h_1, \eta_1 \in g_1 \\ \gamma_2 \in h_2, \eta_2 \in g_2}} \left(\left\{ \dfrac{\gamma_1 \gamma_2}{1 + (1 - \gamma_1)(1 - \gamma_2)} \,\middle|\, p_{\gamma_1} p_{\gamma_2} \right\}, \left\{ \dfrac{\eta_1 + \eta_2}{1 + \eta_1 \eta_2} \,\middle|\, q_{\eta_1} q_{\eta_2} \right\} \right);$

(iii) $\lambda \alpha = \bigcup\limits_{\substack{\gamma \in h, \\ \eta \in g}} \left(\left\{ \dfrac{(1+\gamma)^\lambda - (1-\gamma)^\lambda}{(1+\gamma)^\lambda + (1-\gamma)^\lambda} \,\middle|\, p_\gamma \right\}, \left\{ \dfrac{2(\eta)^\lambda}{(2-\eta)^\lambda + (\eta)^\lambda} \,\middle|\, q_\eta \right\} \right);$

(iv) $\alpha^\lambda = \bigcup\limits_{\substack{\gamma \in h, \\ \eta \in g}} \left(\left\{ \dfrac{2(\gamma)^\lambda}{(2-\gamma)^\lambda + (\gamma)^\lambda} \,\middle|\, p_\gamma \right\}, \left\{ \dfrac{(1+\eta)^\lambda - (1-\eta)^\lambda}{(1+\eta)^\lambda + (1-\eta)^\lambda} \,\middle|\, q_\eta \right\} \right)$

Theorem 3. *For real value $\lambda > 0$, the operational laws for PDHFEs given in Definition 6 that is $\alpha_1 \oplus \alpha_2$, $\alpha_1 \otimes \alpha_2$, $\lambda \alpha$, and α^λ are also PDHFEs.*

Proof. For two PDHFEs α_1 and α_2, we have

$$\alpha_1 \oplus \alpha_2 = \bigcup\limits_{\substack{\gamma_1 \in h_1, \eta_1 \in g_1 \\ \gamma_2 \in h_2, \eta_2 \in g_2}} \left(\left\{ \dfrac{\gamma_1 + \gamma_2}{1 + \gamma_1 \gamma_2} \,\middle|\, p_{\gamma_1} p_{\gamma_2} \right\}, \left\{ \dfrac{\eta_1 \eta_2}{1 + (1 - \eta_1)(1 - \eta_2)} \,\middle|\, q_{\eta_1} q_{\eta_2} \right\} \right)$$

As $0 \le \gamma_1, \gamma_2, \eta_1, \eta_2 \le 1$, thus it is evident that $0 \le \gamma_1 + \gamma_2 \le 2$ and $1 \le 1 + \gamma_1 \gamma_2 \le 2$, thus it follows that $0 \le \frac{\gamma_1 + \gamma_2}{1 + \gamma_1 \gamma_2} \le 1$. On the other hand, $0 \le \eta_1 \eta_2 \le 1$ and $1 \le 1 + (1 - \eta_1)(1 - \eta_2) \le 2$. Thus, $0 \le \frac{\eta_1 \eta_2}{1 + (1 - \eta_1)(1 - \eta_2)} \le 1$ Also, since $0 \le p_{\gamma_1}, p_{\gamma_2}, q_{\eta_1}, q_{\eta_2} \le 1$, thus $0 \le p_{\gamma_1} p_{\gamma_2} \le 1$ and $0 \le q_{\eta_1} q_{\eta_2} \le 1$. Similarly, $\alpha_1 \otimes \alpha_2$, $\lambda \alpha$ and α^λ are also PDHFEs. $\square$

Theorem 4. *Let $\alpha_1, \alpha_2, \alpha_3$ be three PDHFEs and $\lambda, \lambda_1, \lambda_2 > 0$ be three real numbers, then following results hold:*

(i) $\alpha_1 \oplus \alpha_2 = \alpha_2 \oplus \alpha_1$;

(ii) $\alpha_1 \otimes \alpha_2 = \alpha_2 \otimes \alpha_1$;

(iii) $(\alpha_1 \oplus \alpha_2) \oplus \alpha_3 = \alpha_1 \oplus (\alpha_2 \oplus \alpha_3)$;

(iv) $(\alpha_1 \otimes \alpha_2) \otimes \alpha_3 = \alpha_1 \otimes (\alpha_2 \otimes \alpha_3)$;

(v) $\lambda(\alpha_1 \oplus \alpha_2) = \lambda \alpha_1 \oplus \lambda \alpha_2$;

(vi) $\alpha_1^\lambda \otimes \alpha_1^\lambda = (\alpha_1 \otimes \alpha_2)^\lambda$.

Proof. Let $\alpha_1 = \left(h_1 | p_{h_1}, g_1 | q_{g_1} \right)$, $\alpha_2 = \left(h_2 | p_{h_2}, g_2 | q_{g_2} \right)$, $\alpha_3 = \left(h_3 | p_{h_3}, g_3 | q_{g_3} \right)$ be three PDHFEs. Then, we have

(i) For two PDHFEs α_1 and α_2, from Definition 6, we have

$$\alpha_1 \oplus \alpha_2 = \bigcup\limits_{\substack{\gamma_1 \in h_1, \eta_1 \in g_1 \\ \gamma_2 \in h_2, \eta_2 \in g_2}} \left(\left\{ \dfrac{\gamma_1 + \gamma_2}{1 + \gamma_1 \gamma_2} \,\middle|\, p_{\gamma_1} p_{\gamma_2} \right\}, \left\{ \dfrac{\eta_1 \eta_2}{1 + (1 - \eta_1)(1 - \eta_2)} \,\middle|\, q_{\eta_1} q_{\eta_2} \right\} \right)$$

$$= \bigcup\limits_{\substack{\gamma_1 \in h_1, \eta_1 \in g_1 \\ \gamma_2 \in h_2, \eta_2 \in g_2}} \left(\left\{ \dfrac{\gamma_2 + \gamma_1}{1 + \gamma_2 \gamma_1} \,\middle|\, p_{\gamma_2} p_{\gamma_1} \right\}, \left\{ \dfrac{\eta_2 \eta_1}{1 + (1 - \eta_2)(1 - \eta_1)} \,\middle|\, q_{\eta_2} q_{\eta_1} \right\} \right)$$

$$= \alpha_2 \oplus \alpha_1$$

(ii) Proof is obvious so we omit it here.

(iii) For three PDHFEs α_1, α_2 and α_3, consider L.H.S. i.e.,

$$(\alpha_1 \oplus \alpha_2) \oplus \alpha_3$$

$$= \left(\bigcup_{\substack{\gamma_1 \in h_1, \eta_1 \in g_1 \\ \gamma_2 \in h_2, \eta_2 \in g_2}} \left(\left\{ \frac{\gamma_1 + \gamma_2}{1 + \gamma_1 \gamma_2} \,\Big|\, p_{\gamma_1} p_{\gamma_2} \right\}, \left\{ \frac{\eta_1 \eta_2}{1 + (1 - \eta_1)(1 - \eta_2)} \,\Big|\, q_{\eta_1} q_{\eta_2} \right\} \right) \right) \oplus \alpha_3$$

$$= \bigcup_{\substack{\gamma_1 \in h_1, \eta_1 \in g_1 \\ \gamma_2 \in h_2, \eta_2 \in g_2 \\ \gamma_3 \in h_3, \eta_3 \in g_3}} \left(\left\{ \frac{\gamma_1 + \gamma_2 + \gamma_3 + \gamma_1 \gamma_2 \gamma_3}{1 + \gamma_1 \gamma_2 + \gamma_2 \gamma_3 + \gamma_3 \gamma_1} \,\Big|\, p_{\gamma_1} p_{\gamma_2} p_{\gamma_3} \right\}, \left\{ \frac{\eta_1 \eta_2 \eta_3}{4 - 2\eta_1 - 2\eta_2 - 2\eta_3 + \eta_1 \eta_2 + \eta_2 \eta_3 + \eta_1 \eta_3} \,\Big|\, q_{\eta_1} q_{\eta_2} q_{\eta_3} \right\} \right) \tag{11}$$

Also, on considering the R.H.S., we have

$$\alpha_1 \oplus (\alpha_2 \oplus \alpha_3)$$

$$= \alpha_1 \oplus \left(\bigcup_{\substack{\gamma_2 \in h_2, \eta_2 \in g_2 \\ \gamma_3 \in h_3, \eta_3 \in g_3}} \left(\left\{ \frac{\gamma_2 + \gamma_3}{1 + \gamma_2 \gamma_3} \,\Big|\, p_{\gamma_2} p_{\gamma_3} \right\}, \left\{ \frac{\eta_2 \eta_3}{1 + (1 - \eta_2)(1 - \eta_3)} \,\Big|\, q_{\eta_2} q_{\eta_3} \right\} \right) \right)$$

$$= \bigcup_{\substack{\gamma_1 \in h_1, \eta_1 \in g_1 \\ \gamma_2 \in h_2, \eta_2 \in g_2 \\ \gamma_3 \in h_3, \eta_3 \in g_3}} \left(\left\{ \frac{\gamma_1 + \gamma_2 + \gamma_3 + \gamma_1 \gamma_2 \gamma_3}{1 + \gamma_1 \gamma_2 + \gamma_2 \gamma_3 + \gamma_3 \gamma_1} \,\Big|\, p_{\gamma_1} p_{\gamma_2} p_{\gamma_3} \right\}, \left\{ \frac{\eta_1 \eta_2 \eta_3}{4 - 2\eta_1 - 2\eta_2 - 2\eta_3 + \eta_1 \eta_2 + \eta_2 \eta_3 + \eta_1 \eta_3} \,\Big|\, q_{\eta_1} q_{\eta_2} q_{\eta_3} \right\} \right) \tag{12}$$

From Equations (11) and (12), the required result is obtained.

(iv) Proof is obvious so we omit it here.

(v) For $\lambda > 0$, consider

$$\lambda(\alpha_1 \oplus \alpha_2) = \lambda \left(\bigcup_{\substack{\gamma_1 \in h_1, \eta_1 \in g_1 \\ \gamma_2 \in h_2, \eta_2 \in g_2}} \left(\left\{ \frac{(1 + \gamma_1)(1 + \gamma_2) - (1 - \gamma_1)(1 - \gamma_2)}{(1 + \gamma_1)(1 + \gamma_2) + (1 - \gamma_1)(1 - \gamma_2)} \,\Big|\, p_{\gamma_1} p_{\gamma_2} \right\}, \left\{ \frac{2\eta_1 \eta_2}{(2 - \eta_1)(2 - \eta_2) + \eta_1 \eta_2} \,\Big|\, q_{\eta_1} q_{\eta_2} \right\} \right) \right)$$

For sake of convenience, put $(1 + \gamma_1)(1 + \gamma_2) = a$; $(1 - \gamma_1)(1 - \gamma_2) = b$; $\eta_1 \eta_2 = c$ and $(2 - \eta_1)(2 - \eta_2) = d$. This implies

$$\lambda(\alpha_1 \oplus \alpha_2) = \lambda \bigcup_{\substack{\gamma_1 \in h_1, \eta_1 \in g_1 \\ \gamma_2 \in h_2, \eta_2 \in g_2}} \left(\left\{ \frac{a - b}{a + b} \,\Big|\, p_{\gamma_1} p_{\gamma_2}, \right\} \left\{ \frac{2c}{d + c} \,\Big|\, q_{\eta_1} q_{\eta_2} \right\} \right)$$

$$= \bigcup_{\substack{\gamma_1 \in h_1, \eta_1 \in g_1 \\ \gamma_2 \in h_2, \eta_2 \in g_2}} \left(\left\{ \frac{\left(1 + \frac{a - b}{a + b}\right)^{\lambda} - \left(1 - \frac{a - b}{a + b}\right)^{\lambda}}{\left(1 + \frac{a - b}{a + b}\right)^{\lambda} + \left(1 - \frac{a - b}{a + b}\right)^{\lambda}} \,\Big|\, p_{\gamma_1} p_{\gamma_2} \right\}, \left\{ \frac{2\left(\frac{2c}{d + c}\right)^{\lambda}}{\left(2 - \frac{2c}{d + c}\right)^{\lambda} + \left(\frac{2c}{d + c}\right)^{\lambda}} \,\Big|\, q_{\eta_1} q_{\eta_2} \right\} \right)$$

$$= \bigcup_{\substack{\gamma_1 \in h_1, \eta_1 \in g_1 \\ \gamma_2 \in h_2, \eta_2 \in g_2}} \left(\left\{ \frac{\left(\frac{2a}{a + b}\right)^{\lambda} - \left(\frac{2b}{a + b}\right)^{\lambda}}{\left(\frac{2a}{a + b}\right)^{\lambda} + \left(\frac{2b}{a + b}\right)^{\lambda}} \,\Big|\, p_{\gamma_1} p_{\gamma_2} \right\}, \left\{ \frac{2\left(\frac{2c}{d + c}\right)^{\lambda}}{\left(\frac{2d}{d + c}\right)^{\lambda} + \left(\frac{2a}{d + c}\right)^{\lambda}} \,\Big|\, q_{\eta_1} q_{\eta_2} \right\} \right)$$

$$= \bigcup_{\substack{\gamma_1 \in h_1, \eta_1 \in g_1 \\ \gamma_2 \in h_2, \eta_2 \in g_2}} \left(\left\{ \frac{(a^{\lambda} - b^{\lambda})}{(a^{\lambda} + b^{\lambda})} \,\Big|\, p_{\gamma_1} p_{\gamma_2} \right\}, \left\{ \frac{2c^{\lambda}}{d^{\lambda} + c^{\lambda}} \,\Big|\, q_{\eta_1} q_{\eta_2} \right\} \right)$$

Re-substituting a, b, c and d we have

$$= \bigcup_{\substack{\gamma_1 \in h_1, \eta_1 \in g_1 \\ \gamma_2 \in h_2, \eta_2 \in g_2}} \left(\left\{ \frac{(1 + \gamma_1)^{\lambda} (1 + \gamma_2)^{\lambda} - (1 - \gamma_1)^{\lambda} (1 - \gamma_2)^{\lambda}}{(1 + \gamma_1)^{\lambda} (1 + \gamma_2)^{\lambda} + (1 - \gamma_1)^{\lambda} (1 - \gamma_2)^{\lambda}} \,\Big|\, p_{\gamma_1} p_{\gamma_2} \right\}, \left\{ \frac{2 (\eta_1 \eta_2)^{\lambda}}{(2 - \eta_1)^{\lambda} (2 - \eta_2)^{\lambda} + \eta_1 \eta_2} \,\Big|\, q_{\eta_1} q_{\eta_2} \right\} \right)$$

$$= \lambda \alpha_1 \oplus \lambda \alpha_2$$

(vi)   For $\lambda > 0$,

$$(\alpha_1 \otimes \alpha_2)^\lambda \;=\; \left(\bigcup_{\substack{\gamma_1 \in h_1, \eta_1 \in g_1 \\ \gamma_2 \in h_2, \eta_2 \in g_2}} \left(\left\{ \frac{2\gamma_1\gamma_2}{1+(1-\gamma_1)(1-\gamma_2)} \;\Big|\; p_{\gamma_1}p_{\gamma_2} \right\}, \left\{ \frac{(1+\eta_1)(1+\eta_2)-(1-\eta_1)(1-\eta_2)}{(1+\eta_1)(1+\eta_2)+(1-\eta_1)(1-\eta_2)} \;\Big|\; q_{\eta_1}q_{\eta_2} \right\} \right) \right)^\lambda$$

For sake of convenience, put

$$\gamma_1\gamma_2 = a;\; (2-\gamma_1)(2-\gamma_2) = b;\; (1+\eta_1)(1+\eta_2) = c \text{ and } (1-\eta_1)(1-\eta_2) = d$$

So we obtain

$$(\alpha_1 \otimes \alpha_2)^\lambda \;=\; \left(\bigcup_{\substack{\gamma_1 \in h_1, \eta_1 \in g_1 \\ \gamma_2 \in h_2, \eta_2 \in g_2}} \left(\left\{ \frac{2a}{b+a} \;\Big|\; p_{\gamma_1}p_{\gamma_2} \right\}, \left\{ \frac{c-d}{c+d} \;\Big|\; q_{\eta_1}q_{\eta_2} \right\} \right) \right)^\lambda$$

$$= \bigcup_{\substack{\gamma_1 \in h_1, \eta_1 \in g_1 \\ \gamma_2 \in h_2, \eta_2 \in g_2}} \left(\left\{ \frac{2\left(\frac{2a}{b+a}\right)^\lambda}{\left(2-\frac{2a}{b+a}\right)^\lambda + \left(\frac{2a}{b+a}\right)^\lambda} \;\Big|\; p_{\gamma_1}p_{\gamma_2} \right\}, \left\{ \frac{\left(1+\frac{c-d}{c+d}\right)^\lambda - \left(1-\frac{c-d}{c+d}\right)^\lambda}{\left(1+\frac{c-d}{c+d}\right)^\lambda + \left(1-\frac{c-d}{c+d}\right)^\lambda} \;\Big|\; q_{\eta_1}q_{\eta_2} \right\} \right)$$

$$= \bigcup_{\substack{\gamma_1 \in h_1, \eta_1 \in g_1 \\ \gamma_2 \in h_2, \eta_2 \in g_2}} \left(\left\{ \frac{2\left(\frac{2a}{b+a}\right)^\lambda}{\left(\frac{2b}{b+a}\right)^\lambda + \left(\frac{2a}{b+a}\right)^\lambda} \;\Big|\; p_{\gamma_1}\gamma_2 \right\}, \left\{ \frac{\left(\frac{2c}{c+d}\right)^\lambda - \left(\frac{2d}{c+d}\right)^\lambda}{\left(\frac{2c}{c+d}\right)^\lambda + \left(\frac{2d}{c+d}\right)^\lambda} \;\Big|\; q_{\eta_1}q_{\eta_2} \right\} \right)$$

$$= \bigcup_{\substack{\gamma_1 \in h_1, \eta_1 \in g_1 \\ \gamma_2 \in h_2, \eta_2 \in g_2}} \left(\left\{ \frac{2a^\lambda}{b^\lambda + a^\lambda} \;\Big|\; p_{\gamma_1}p_{\gamma_2} \right\}, \left\{ \frac{c^\lambda - d^\lambda}{c^\lambda + d^\lambda} \;\Big|\; q_{\eta_1}q_{\eta_2} \right\} \right)$$

Re-substituting values of a, b, c and d we get

$$= \bigcup_{\substack{\gamma_1 \in h_1, \eta_1 \in g_1 \\ \gamma_2 \in h_2, \eta_2 \in g_2}} \left(\left\{ \frac{2(\gamma_1\gamma_2)^\lambda}{(2-\gamma_1)^\lambda(2-\gamma_2)^\lambda + (\gamma_1\gamma_2)^\lambda} \;\Big|\; p_{\gamma_1}p_{\gamma_2} \right\}, \left\{ \frac{(1+\eta_1)^\lambda(1+\eta_2)^\lambda - (1-\eta_1)^\lambda(1-\eta_2)^\lambda}{(1+\eta_1)^\lambda(1+\eta_2)^\lambda + (1-\eta_1)^\lambda(1-\eta_2)^\lambda} \;\Big|\; q_{\eta_1}q_{\eta_2} \right\} \right)$$

$$= \alpha_1^\lambda \otimes \alpha_2^\lambda$$

$\square$

Theorem 5. *Let* $\alpha = \left(h|p_h, g|q_g\right)$ $\alpha_1 = \left(h_1|p_{h_1}, g_1|q_{g_1}\right)$, *and* $\alpha_2 = \left(h_2|p_{h_2}, g_2|q_{g_2}\right)$ *be three PDHFEs, and* $\lambda > 0$ *be a real number, then*

(i) $(\alpha^c)^\lambda = \lambda\alpha^c;$

(ii) $\lambda(\alpha^c) = (\alpha^\lambda)^c;$

(iii) $\alpha_1^c \oplus \alpha_2^c = (\alpha_1 \otimes \alpha_2)^c;$

(iv) $\alpha_1^c \otimes \alpha_2^c = (\alpha_1 \oplus \alpha_2)^c.$

Proof.   (i)   Let $\alpha = \left(h|p_h, g|q_g\right)$ be a PDHFE, then using Definition 4, the proof for the three possible cases is given as:

(Case 1)   If $h \neq \phi;\, g \neq \phi$ then for a PDHFE $\alpha = \left(h|p_h, g|q_g\right)$, from Equation (7) we have

$$(\alpha^c)^\lambda \;=\; \left(\bigcup_{\substack{\gamma \in h \\ \eta \in g}} \left(\left\{ \eta \;\Big|\; q_\eta \right\}, \left\{ \gamma \;\Big|\; p_\gamma \right\} \right) \right)^\lambda$$

$$= \bigcup_{\substack{\gamma \in h \\ \eta \in g}} \left(\left\{ \frac{2(\eta)^\lambda}{(2-\eta)^\lambda + (\eta)^\lambda} \;\Big|\; q_{\eta'} \right\}, \left\{ \frac{(1+\gamma)^\lambda - (1-\gamma)^\lambda}{(1+\gamma)^\lambda + (1-\gamma)^\lambda} \;\Big|\; p_\gamma \right\} \right)$$

$$= \left(\bigcup_{\substack{\gamma \in h \\ \eta \in g}} \left(\left\{ \frac{(1+\gamma)^\lambda - (1-\gamma)^\lambda}{(1+\gamma)^\lambda + (1-\gamma)^\lambda} \, \Big| \, p_\gamma \right\}, \left\{ \frac{2(\eta)^\lambda}{(2-\eta)^\lambda + (\eta)^\lambda} \, \Big| \, q_\eta \right\} \right) \right)^c$$

$$= \left(\lambda \left(\bigcup_{\substack{\gamma \in p \\ \eta \in q}} \left\{ \gamma \, \big| \, p_\gamma \right\}, \left\{ \eta \, \big| \, q_\eta \right\} \right) \right)^c = (\lambda \alpha)^c$$

(Case 2)   If $g = \phi, h \neq \phi$, then

$$(\alpha^c)^\lambda = \left(\bigcup_{\gamma \in h} \left(\left\{ 1 - \gamma \, \big| \, p_\gamma \right\}, \{\phi\} \right) \right)^\lambda$$

$$= \bigcup_{\gamma \in h} \left(\left\{ \frac{2(1-\gamma)^\lambda}{(2 - (1-\gamma))^\lambda + (1-\gamma)^\lambda} \, \Big| \, p_\gamma \right\}, \{\phi\} \right)$$

$$= (\lambda \alpha)^c$$

(Case 3)   If $h = \phi, g = \phi$, then

$$(\alpha^c)^\lambda = \left(\bigcup_{\eta \in g} \left(\{\phi\}, \left\{ 1 - \eta \, \big| \, q_\eta \right\} \right) \right)^\lambda$$

$$= \bigcup_{\eta \in g} \left(\{\phi\}, \left\{ \frac{(1 + (1-\eta))^\lambda - (1 - (1-\eta))^\lambda}{(1 + (1-\eta))^\lambda + (1 - (1-\eta))^\lambda} \, \Big| \, q_\eta \right\} \right)$$

$$= \left(\bigcup_{\eta \in g} \left(\left\{ \frac{(2-\eta)^\lambda - (\eta)^\lambda}{(2-\eta)^\lambda + (\eta)^\lambda} \, \Big| \, q_\eta \right\}, \{\phi\} \right) \right)^c$$

$$= \left(\lambda \bigcup_{\eta \in g} \left\{ (1 - \eta) \, \big| \, q_\eta \right\}, \{\phi\} \right)^c = (\lambda \alpha)^c$$

(ii)   Similar to above, so it is omitted.

(iii)   For two PDHFEs α_1, α_2 and a real number $\lambda > 0$, using Definitions 4 and 6 we have,

(Case 1)   If $h_1 \neq \phi, g_1 \neq \phi, h_2 \neq \phi$ and $g_2 \neq \phi$

$$\alpha_1^c \oplus \alpha_2^c$$

$$= \bigcup_{\substack{\gamma_1 \in h_1 \\ \eta_1 \in g_1}} \left(\left\{ \eta_1 \, \big| \, q_{\eta_1} \right\}, \left\{ \gamma_1 \, \big| \, p_{\gamma_1} \right\} \right) \oplus \bigcup_{\substack{\gamma_2 \in h_2 \\ \eta_2 \in g_2}} \left(\left\{ \eta_2 \, \big| \, q_{\eta_2} \right\}, \left\{ \gamma_2 \, \big| \, p_{\gamma_2} \right\} \right)$$

$$= \bigcup_{\substack{\gamma_1 \in h_1, \eta_1 \in g_1 \\ \gamma_2 \in h_2, \eta_2 \in g_2}} \left(\left\{ \frac{\eta_1 + \eta_2}{1 + \eta_1 \eta_2} \, \Big| \, q_{\eta_1} q_{\eta_2} \right\}, \left\{ \frac{\gamma_1 \gamma_2}{1 + (1 - \gamma_1)(1 - \gamma_2)} \, \Big| \, p_{\gamma_1} p_{\gamma_2} \right\} \right)$$

$$= \bigcup_{\substack{\gamma_1 \in h_1, \eta_1 \in g_1 \\ \gamma_2 \in h_2, \eta_2 \in g_2}} \left(\left\{ \frac{\gamma_1 \gamma_2}{1 + (1 - \gamma_1)(1 - \gamma_2)} \, \Big| \, p_{\gamma_1} p_{\gamma_2} \right\}, \left\{ \frac{\eta_1 + \eta_2}{1 + \eta_1 \eta_2} \, \Big| \, q_{\eta_1} q_{\eta_2} \right\} \right)^c$$

$$= (\alpha_1 \otimes \alpha_2)^c$$

(Case 2)   If $h_1 \neq \phi, g_1 = \phi, h_2 \neq \phi$ and $g_2 = \phi$, then

$$
\begin{aligned}
\alpha_1^c \oplus \alpha_2^c &= \bigcup_{\substack{\gamma_1 \in h_1, \\ \eta_1 \in g_1}} \left(\left\{ 1 - \gamma_1 \mid p_{\gamma_1} \right\}, \{\phi\} \right) \oplus \bigcup_{\substack{\gamma_2 \in h_2, \\ \eta_2 \in g_2}} \left(\left\{ 1 - \gamma_2 \mid p_{\gamma_2} \right\}, \{\phi\} \right) \\
&= \bigcup_{\substack{\gamma_1 \in h_1, \eta_1 \in g_1 \\ \gamma_2 \in h_2, \eta_2 \in g_2}} \left(\left\{ \frac{(1 - \gamma_1) + (1 - \gamma_2)}{1 + (1 - \gamma_1)(1 - \gamma_2)} \mid p_{\gamma_1} p_{\gamma_2} \right\}, \{\phi\} \right) \\
&= (\alpha_1 \otimes \alpha_2)^c
\end{aligned}
$$

(Case 3)   If $h_1 = \phi, g_1 \neq \phi, h_2 = \phi, g_2 \neq \phi$

$$
\begin{aligned}
\alpha_1^c \oplus \alpha_2^c &= \bigcup_{\substack{\gamma_1 \in h_1 \\ \eta_1 \in g_1}} \left(\{\phi\}, \left\{ 1 - \eta_1 \mid q_{\eta_1} \right\} \right) \oplus \bigcup_{\substack{\gamma_2 \in h_2 \\ \eta_2 \in g_2}} \left(\{\phi\}, \left\{ 1 - \eta_2 \mid q_{\eta_2} \right\} \right) \\
&= \bigcup_{\substack{\gamma_1 \in h_1, \eta_1 \in g_1 \\ \gamma_2 \in h_2, \eta_2 \in g_2}} \left(\{\phi\}, \left\{ \frac{(1 - \eta_1)(1 - \eta_2)}{1 + \eta_1 \eta_2} \mid q_{\eta_1} q_{\eta_2} \right\} \right) \\
&= (\alpha_1 \otimes \alpha_2)^c
\end{aligned}
$$

(iv)   Similar, so we omit it here.

□

5. Probabilistic Dual Hesitant Weighted Einstein AOs

In this section, we have defined some weighted aggregation operators by using aforementioned laws for a collection of PDHFEs. For it, let Ω be the family of PDHFEs.

Definition 7. *Let Ω be the family of PDHFEs α_i $(i = 1, 2, \ldots, n)$ with the corresponding weights $\omega = (\omega_1, \omega_2, \ldots, \omega_n)^T$, such that $\omega_i > 0$ and $\sum_{i=1}^{n} \omega_i = 1$. If PDHFWEA: $\Omega^n \to \Omega$, is a mapping defined by*

$$
PDHFWEA(\alpha_1, \alpha_2, \ldots, \alpha_n) = \omega_1 \alpha_1 \oplus \omega_2 \alpha_2 \oplus \ldots \oplus \omega_n \alpha_n \tag{13}
$$

then, PDHFWEA is called probabilistic dual hesitant fuzzy weighted Einstein average operator.

Theorem 6. *For a family of PDHFEs $\alpha_i = \left(h_i \mid p_{h_i}, g_i \mid q_{g_i} \right), (i = 1, 2, \ldots, n)$, the aggregated value obtained by using PDHFWEA operator is still a PDHFE and is given as*

$$
PDHFWEA(\alpha_1, \alpha_2, \ldots, \alpha_n) = \bigcup_{\substack{\gamma_i \in h_i \\ \eta_i \in g_i}} \left(\begin{array}{c} \left\{ \dfrac{\prod_{i=1}^{n} (1 + \gamma_i)^{\omega_i} - \prod_{i=1}^{n} (1 - \gamma_i)^{\omega_i}}{\prod_{i=1}^{n} (1 + \gamma_i)^{\omega_i} + \prod_{i=1}^{n} (1 - \gamma_i)^{\omega_i}} \mid \prod_{i=1}^{n} p_{\gamma_i} \right\}, \\[4ex] \left\{ \dfrac{2 \prod_{i=1}^{n} (\eta_i)^{\omega_i}}{\prod_{i=1}^{n} (2 - \eta_i)^{\omega_i} + \prod_{i=1}^{n} (\eta_i)^{\omega_i}} \mid \prod_{i=1}^{n} q_{\eta_i} \right\} \end{array} \right) \tag{14}
$$

where $\omega = (\omega_1, \omega_2, \ldots, \omega_n)^T$ is a weight vector such that $\sum_{i=1}^{n} \omega_i = 1$ where $0 < \omega_i < 1$.

Proof. We will prove the Equation (14) by following the steps mathematical induction on n, and the proof is executed as below:

(Step 1)   For $n = 2$, we have $\alpha_1 = \left(h_1 \mid p_{h_1}, g_1 \mid q_{g_1} \right)$ and $\alpha_2 = \left(h_2 \mid p_{h_2}, g_2 \mid q_{g_2} \right)$. Using operational laws on PDHFEs as stated in Definition 6 we get

$$
\omega_1 \alpha_1 \;=\; \bigcup_{\gamma_1 \in h_1, \eta_1 \in g_1} \left(\begin{array}{l} \left\{ \dfrac{(1+\gamma_1)^{\omega_1} - (1-\gamma_1)^{\omega_1}}{(1+\gamma_1)^{\omega_1} + (1-\gamma_1)^{\omega_1}} \mid p_{\gamma_1} \right\}, \\[4mm] \left\{ \dfrac{2(\eta_1)^{\omega_1}}{(2-\eta_1)^{\omega_1} + (\eta_1)^{\omega_1}} \mid q_{\eta_1} \right\} \end{array} \right)
$$

$$
\text{and}\quad \omega_2 \alpha_2 \;=\; \bigcup_{\gamma_2 \in h_2, \eta_2 \in g_2} \left(\begin{array}{l} \left\{ \dfrac{(1+\gamma_2)^{\omega_2} - (1-\gamma_2)^{\omega_2}}{(1+\gamma_2)^{\omega_2} + (1-\gamma_2)^{\omega_2}} \mid p_{\gamma_2} \right\}, \\[4mm] \left\{ \dfrac{2(\eta_2)^{\omega_2}}{(2-\eta_2)^{\omega_2} + (\eta_2)^{\omega_2}} \mid q_{\eta_2} \right\} \end{array} \right)
$$

Hence, by addition of PDHFEs, we get

$$
\text{PDHFWEA}(\alpha_1, \alpha_2) = \omega_1 \alpha_1 \oplus \omega_2 \alpha_2
$$

$$
= \bigcup_{\substack{\gamma_1 \in h_1, \eta_1 \in g_1 \\ \gamma_2 \in h_2, \eta_2 \in g_2}} \left(\begin{array}{l} \left\{ \dfrac{\prod\limits_{i=1}^{2}(1+\gamma_i)^{\omega_i} - \prod\limits_{i=1}^{2}(1-\gamma_i)^{\omega_i}}{\prod\limits_{i=1}^{2}(1+\gamma_i)^{\omega_i} + \prod\limits_{i=1}^{2}(1-\gamma_i)^{\omega_i}} \;\middle|\; \prod\limits_{i=1}^{2} p_{\gamma_i} \right\}, \\[8mm] \left\{ \dfrac{2\prod\limits_{i=1}^{2}(\eta_i)^{\omega_i}}{\prod\limits_{i=1}^{2}(2-\eta_i)^{\omega_i} + \prod\limits_{i=1}^{2}(\eta_i)^{\omega_i}} \;\middle|\; \prod\limits_{i=1}^{2} q_{\eta_i} \right\} \end{array} \right)
$$

Thus, the result holds for $n = 2$.

(Step 2)   If Equation (14) holds for $n = k$, then for $n = k + 1$, we have

$$
\text{PDHFWEA}(\alpha_1, \alpha_2, \ldots, \alpha_{k+1}) = \left(\bigoplus_{i=1}^{k} \omega_i \alpha_i \right) \oplus (\omega_{k+1} \alpha_{k+1})
$$

$$
= \bigcup_{\gamma_i \in h_i, \eta_i \in g_i} \left(\left\{ \dfrac{\prod\limits_{i=1}^{k}(1+\gamma_i)^{\omega_i} - \prod\limits_{i=1}^{k}(1-\gamma_i)^{\omega_i}}{\prod\limits_{i=1}^{k}(1+\gamma_i)^{\omega_i} + \prod\limits_{i=1}^{k}(1-\gamma_i)^{\omega_i}} \;\middle|\; \prod\limits_{i=1}^{k} p_{\gamma_i} \right\} \left\{ \dfrac{2\prod\limits_{i=1}^{k}(\eta_i)^{\omega_i}}{\prod\limits_{i=1}^{k}(2-\eta_i)^{\omega_i} + \prod\limits_{i=1}^{k}(\eta_i)^{\omega_i}} \;\middle|\; \prod\limits_{i=1}^{k} q_{\eta_i} \right\} \right)
$$

$$
\oplus \bigcup_{\substack{\gamma_{k+1} \in h_{k+1}, \\ \eta_{k+1} \in g_{k+1}}} \left(\left\{ \dfrac{(1+\gamma_{k+1})^{\omega_{k+1}} - (1-\gamma_{k+1})^{\omega_{k+1}}}{(1+\gamma_{k+1})^{\omega_{k+1}} + (1-\gamma_{k+1})^{\omega_{k+1}}} \;\middle|\; p_{\gamma_{k+1}} \right\} \left\{ \dfrac{2(\eta_{k+1})^{\omega_{k+1}}}{(2-\eta_{k+1})^{\omega_{k+1}} + (\eta_{k+1})^{\omega_{k+1}}} \;\middle|\; q_{\eta_{k+1}} \right\} \right)
$$

$$
= \bigcup_{\gamma_i \in h_i, \eta_i \in g_i} \left(\left\{ \dfrac{\prod\limits_{i=1}^{k+1}(1+\gamma_i)^{\omega_i} - \prod\limits_{i=1}^{k+1}(1-\gamma_i)^{\omega_i}}{\prod\limits_{i=1}^{k+1}(1+\gamma_i)^{\omega_i} + \prod\limits_{i=1}^{k+1}(1-\gamma_i)^{\omega_i}} \;\middle|\; \prod\limits_{i=1}^{k+1} p_{\gamma_i} \right\} \left\{ \dfrac{2\prod\limits_{i=1}^{k+1}(\eta_i)^{\omega_i}}{\prod\limits_{i=1}^{k+1}(2-\eta_i)^{\omega_i} + \prod\limits_{i=1}^{k+1}(\eta_i)^{\omega_i}} \;\middle|\; \prod\limits_{i=1}^{k+1} q_{\eta_i} \right\} \right)
$$

Thus,

$$
\text{PDHFWEA}(\alpha_1, \alpha_2, \ldots, \alpha_n)
$$

$$
= \bigcup_{\substack{\gamma_i \in h_i \\ \eta_i \in g_i}} \left(\begin{array}{l} \left\{ \dfrac{\prod\limits_{i=1}^{n}(1+\gamma_i)^{\omega_i} - \prod\limits_{i=1}^{n}(1-\gamma_i)^{\omega_i}}{\prod\limits_{i=1}^{n}(1+\gamma_i)^{\omega_i} + \prod\limits_{i=1}^{n}(1-\gamma_i)^{\omega_i}} \;\middle|\; \prod\limits_{i=1}^{n} p_{\gamma_i} \right\}, \\[8mm] \left\{ \dfrac{2\prod\limits_{i=1}^{n}(\eta_i)^{\omega_i}}{\prod\limits_{i=1}^{n}(2-\eta_i)^{\omega_i} + \prod\limits_{i=1}^{n}(\eta_i)^{\omega_i}} \;\middle|\; \prod\limits_{i=1}^{n} q_{\eta_i} \right\} \end{array} \right)
$$

which completes the proof. $\square$

Further, it is observed that the proposed PDHFWEA operator satisfies the properties of boundedness and monotonicity, for a family of PDHFEs α_i, $(i = 1, 2, \ldots, n)$ which can be demonstrated as follows:

Property 1. *(Boundedness) For* $\alpha_i = \left(h_i \mid p_{h_i}, g_i \mid q_{g_i} \right)$ *where* $i = (1, 2, \ldots, n)$, *let* $\alpha^- = \left(\min(h_i) \mid \min(p_{h_i}), \max(g_i) \mid \max(q_{g_i}) \right) = \left(\left\{ \gamma_{\min} \mid p_{\min} \right\}, \left\{ \eta_{\max} \mid q_{\max} \right\} \right)$ *and* $\alpha^+ = \left(\max(h_i) \mid \max(p_{h_i}), \min(g_i) \mid \min(q_{g_i}) \right) = \left(\left\{ \gamma_{\max} \mid p_{\max} \right\}, \left\{ \eta_{\min} \mid q_{\min} \right\} \right)$ *be PDHFEs, then* $\alpha^- \leq PDHFWEA(\alpha_1, \alpha_2, \ldots, \alpha_n) \leq \alpha^+$.

Proof. Since each α_i is a PDHFE, it is obvious that $\min(h_i) \leq h_i \leq \max(h_i), \min(g_i) \leq g_i \leq \max(g_i), p_{\min} \leq p_i \leq p_{\max}$ and $q_{\min} \leq q_i \leq q_{\max}$. Let $f(x) = \frac{1-x}{1+x}, x \in [0, 1]$, $f'(x) = \frac{-2}{(1+x)^2} < 0$ i.e., $f(x)$ is a decreasing function. Since, $\gamma_{\min} \leq \gamma_i \leq \gamma_{\max}$, for all i, then $f(\gamma_{\max}) \leq f(\gamma_i) \leq f(\gamma_{\min})$ i.e., $\frac{1-\gamma_{\max}}{1+\gamma_{\max}} \leq \frac{1-\gamma_i}{1+\gamma_i} \leq \frac{1-\gamma_{\max}}{1+\gamma_{\max}}$. Let $\omega = (\omega_1, \omega_2, \ldots, \omega_n)^T$ be the weight vector of $(\alpha_1, \alpha_2, \ldots, \alpha_n)$ such that each $\omega_i \in (0, 1)$ and $\sum_{i=1}^{n} \omega_i = 1$, then we have

$$\left(\frac{1 - \gamma_{\max}}{1 + \gamma_{\max}} \right)^{\omega_i} \leq \left(\frac{1 - \gamma_i}{1 + \gamma_i} \right)^{\omega_i} \leq \left(\frac{1 - \gamma_{\min}}{1 + \gamma_{\min}} \right)^{\omega_i}$$

Thus, we get

$$1 + \left(\frac{1 - \gamma_{\max}}{1 + \gamma_{\max}} \right) \leq 1 + \prod_{i=1}^{n} \left(\frac{1 - \gamma_i}{1 + \gamma_i} \right)^{\omega_i} \leq 1 + \left(\frac{1 - \gamma_{\min}}{1 + \gamma_{\min}} \right)$$

$$\Rightarrow \frac{2}{1 + \gamma_{\max}} \leq \frac{\prod_{i=1}^{n}(1 + \gamma_i)^{\omega_i} + \prod_{i=1}^{n}(1 - \gamma_i)^{\omega_i}}{\prod_{i=1}^{n}(1 + \gamma_i)^{\omega_i}} \leq \frac{2}{1 + \gamma_{\min}}$$

$$\Rightarrow \gamma_{\min} \leq \frac{1 - \prod_{i=1}^{n} \left(\frac{1-\gamma_i}{1+\gamma_i} \right)^{\omega_i}}{1 + \prod_{i=1}^{n} \left(\frac{1-\gamma_i}{1+\gamma_i} \right)^{\omega_i}} \leq \gamma_{\max}$$

$$\Rightarrow \gamma_{\min} \leq \frac{\prod_{i=1}^{n}(1 + \gamma_i)^{\omega_i} - \prod_{i=1}^{n}(1 - \gamma_i)^{\omega_i}}{\prod_{i=1}^{n}(1 + \gamma_i)^{\omega_i} + \prod_{i=1}^{n}(1 - \gamma_i)^{\omega_i}} \leq \gamma_{\max}$$

Hence, we obtain the required result for membership values.

Now, for non-membership, let $c(y) = \frac{2-y}{y}, y \in (0, 1]$, then $c'(y) < 0$ i.e., $c(y)$ is the decreasing function. Since, $\eta_{\min} \leq \eta_i \leq \eta_{\max}$, then for all i, we have $c(\eta_{\max}) \leq c(\eta_i) \leq c(\eta_{\min})$, that is $\frac{2-\eta_{\max}}{\eta_{\max}} \leq \frac{2-\eta_i}{\eta_i} \leq \frac{2-\eta_{\min}}{\eta_{\min}}$. Let $\omega = (\omega_1, \omega_2, \ldots, \omega_n)^T$ be the weight vector of $(\alpha_1, \alpha_2, \ldots, \alpha_n)$ such that $\omega_i \in (0, 1)$ and $\sum_{i=1}^{n} \omega_i = 1$, then

$$\left(\frac{2-\eta_{\max}}{\eta_{\max}}\right)^{\omega_i} \leq \left(\frac{2-\eta_i}{\eta_i}\right)^{\omega_i} \leq \left(\frac{2-\eta_{\min}}{\eta_{\min}}\right)^{\omega_i}$$

Thus,
$$\prod_{i=1}^{n}\left(\frac{2-\eta_{\max}}{\eta_{\max}}\right)^{\omega_i} \leq \prod_{i=1}^{n}\left(\frac{2-\eta_i}{\eta_i}\right)^{\omega_i} \leq \prod_{i=1}^{n}\left(\frac{2-\eta_{\min}}{\eta_{\min}}\right)^{\omega_i}$$

$$\Rightarrow \frac{2}{\eta_{\min}} \leq \frac{1}{1+\prod\limits_{i=1}^{n}\left(\frac{2-\eta_i}{\eta_i}\right)^{\omega_i}} \leq \frac{2}{\eta_{\max}}$$

$$\Rightarrow \eta_{\min} \leq \frac{2\prod\limits_{i=1}^{n}(\eta_i)^{\omega_i}}{\prod\limits_{i=1}^{n}(\eta_i)^{\omega_i} + \prod\limits_{i=1}^{n}(2-\eta_i)^{\omega_i}} \leq \eta_{\max}$$

Hence, the required for non-membership values is obtained.

Now, for probabilities, since $p_{\min} \leq p_i \leq p_{\max}$ and $q_{\min} \leq q_i \leq q_{\max}$ this implies that $\prod\limits_{i=1}^{n} p_{\min} \leq \prod\limits_{i=1}^{n} p_i \leq \prod\limits_{i=1}^{n} p_{\max}$ and $\prod\limits_{i=1}^{n} q_{\min} \leq \prod\limits_{i=1}^{n} q_i \leq \prod\limits_{i=1}^{n} q_{\max}$. According to the score function, as defined in Definition 5, we obtain $S(\alpha^-) \leq S(\alpha) \leq S(\alpha^+)$. Hence, from all the above notions, $\alpha^- \leq \text{PDHFWEA}(\alpha_1, \alpha_2, \ldots, \alpha_n) \leq \alpha^+$. $\square$

Property 2. (Monotonicity) Let $\alpha_i = \left(h_i \mid p_{h_i}, g_i \mid q_{g_i}\right)$ and $\alpha_i^* = \left(h_i^* \mid p_{h_i^*}, g_i^* \mid q_{g_i^*}\right)$, for all $i = (1, 2, \ldots, n)$ be two families of PDHFEs where for each element in α_i and α_i^*, there are $\gamma_{\alpha_i} \leq \gamma_{\alpha_i^*}$ and $\eta_{\alpha_i} \geq \eta_{\alpha_i^*}$ while the probabilities remain the same i.e., $p_{h_i} = p_{h_i^*}, q_{g_i} = q_{g_i^*}$ then $\text{PDHFWEA}(\alpha_1, \alpha_2, \ldots, \alpha_n) \leq \text{PDHFWEA}(\alpha_1^*, \alpha_2^*, \ldots, \alpha_n^*)$.

Proof. Similar to that of Property 1, so we omit it here. $\square$

However, the PDHFWEA operator does not satisfy the idempotency. To illustrate this, we give the following example:

Example 1. *Let $\alpha_1 = \alpha_2 = \left(\left\{0.3|0.25, 0.4|0.75\right\}, \left\{0.2|0.4, 0.3|0.6\right\}\right)$ be two PDHFEs and $\omega = (0.2, 0.8)^T$ be the weight vector, then for $(i = 1, 2)$ the aggregated value using PDHFWEA operator is obtained as*

$$\text{PDHFWEA}(\alpha_1, \alpha_2) = \bigcup_{\substack{\gamma_i \in h_i \\ \eta_i \in g_i}} \left(\left\{ \frac{\prod\limits_{i=1}^{2}(1+\gamma_i)^{\omega_i} - \prod\limits_{i=1}^{2}(1-\gamma_i)^{\omega_i}}{\prod\limits_{i=1}^{n}(1+\gamma_i)^{\omega_i} + \prod\limits_{i=1}^{2}(1-\gamma_i)^{\omega_i}} \;\middle|\; \prod_{i=1}^{n} p_{\gamma_i} \right\}, \left\{ \frac{2\prod\limits_{i=1}^{2}(\eta_i)^{\omega_i}}{\prod\limits_{i=1}^{2}(2-\eta_i)^{\omega_i} + \prod\limits_{i=1}^{2}(\eta_i)^{\omega_i}} \;\middle|\; \prod_{i=1}^{2} q_{\eta_i} \right\} \right)$$

$$= \left(\begin{array}{l} \left\{0.3|0.625, 0.3807|0.1875, 0.3206|0.1875, 0.4|0.5625\right\}, \\ \left\{0.2|0.16, 0.2772|0.24, 0.2173|0.24, 0.30|0.36\right\} \end{array} \right)$$

which clearly shows that $\text{PDHFWEA}(\alpha_1, \alpha_1) \neq \alpha_1$. Thus, it does not satisfy idempotency.

Definition 8. *Let α_i $(i = 1, 2, \ldots, n)$ be the collection of PDHFEs, and PDHFOWEA: $\Omega^n \to \Omega$, if*

$$\text{PDHFOWEA}(\alpha_1, \alpha_2, \ldots, \alpha_n) = \omega_1 \alpha_{\sigma(1)} \oplus \omega_2 \alpha_{\sigma(2)} \oplus \ldots \oplus \omega_n \alpha_{\sigma(n)} \tag{15}$$

where Ω is the set of PDHFEs and $\omega = (\omega_1, \omega_2, \ldots, \omega_n)^T$ is the weight vector of α_i such that $\omega_i > 0$ and $\sum_{i=1}^{n} \omega_i = 1$. $(\sigma(1), \sigma(2), \ldots, \sigma(n))$ is a permutation of $(1, 2, \ldots, n)$ such that $\alpha_{\sigma(i-1)} \geq \alpha_{\sigma(i)}$ for $(i = 2, 3, \ldots, n)$, then PDHFOWEA is called probabilistic dual hesitant fuzzy ordered weighted Einstein AO.

Theorem 7. *For a family of PDHFEs $\alpha_i = \left(h_i \mid p_{h_i}, g_i \mid q_{g_i} \right), (i = 1, 2, \ldots, n)$, the combined value obtained by using PDHFOWEA operator is given as*

$$
PDHFOWEA(\alpha_1, \alpha_2, \ldots, \alpha_n) = \bigcup_{\substack{\gamma_{\sigma(i)} \in h_{\sigma(i)}, \\ \eta_{\sigma(i)} \in g_{\sigma(i)}}} \left(\left\{ \frac{\prod_{i=1}^{n}(1+\gamma_{\sigma(i)})^{\omega_{\sigma(i)}} - \prod_{i=1}^{n}(1-\gamma_{\sigma(i)})^{\omega_{\sigma(i)}}}{\prod_{i=1}^{n}(1+\gamma_{\sigma(i)})^{\omega_{\sigma(i)}} + \prod_{i=1}^{n}(1-\gamma_{\sigma(i)})^{\omega_{\sigma(i)}}} \, \middle| \, \prod_{i=1}^{n} p_{\gamma_{\sigma(i)}} \right\}, \right.
$$
$$
\left. \left\{ \frac{2 \prod_{i=1}^{n}(\eta_{\sigma(i)})^{\omega_{\sigma(i)}}}{\prod_{i=1}^{n}(2-\eta_{\sigma(i)})^{\omega_{\sigma(i)}} + \prod_{i=1}^{n}(\eta_{\sigma(i)})^{\omega_{\sigma(i)}}} \, \middle| \, \prod_{i=1}^{n} q_{\eta_{\sigma(i)}} \right\} \right)
$$

$$(16)$$

where $\omega = (\omega_1, \omega_2, \ldots, \omega_n)^T$ is a weight vector such that $\sum_{i=1}^{n} \omega_i = 1$ where $0 < \omega_i < 1$.

Proof. Similar to Theorem 6. $\square$

Property 3. *For all PDHFEs, $\alpha_i = \left(h_i \mid p_{h_i}, g_i \mid q_{g_i} \right)$ where $i = (1, 2, \ldots, n)$ and for an associated weight vector $\omega = (\omega_1, \omega_2, \ldots, \omega_n)^T$, such that each $\omega_i > 0$ and $\sum_{i=1}^{n} \omega_i = 1$, we have*

- (P1) *(Boundedness) For $\alpha_i = \left(h_i \mid p_{h_i}, g_i \mid q_{g_i} \right)$ where $i = (1, 2, \ldots, n)$, let $\alpha^- = \left(\min(h_i) \mid \min(p_{h_i}), \max(g_i) \mid \max(q_{g_i}) \right) = \left(\left\{ \gamma_{\min} \mid p_{\min} \right\}, \left\{ \eta_{\max} \mid q_{\max} \right\} \right)$ and $\alpha^+ = \left(\max(h_i) \mid \max(p_{h_i}), \min(g_i) \mid \min(q_{g_i}) \right) = \left(\left\{ \gamma_{\max} \mid p_{\max} \right\}, \left\{ \eta_{\min} \mid q_{\min} \right\} \right)$ be PDHFEs, then $\alpha^- \leq PDHFOWEA(\alpha_1, \alpha_2, \ldots, \alpha_n) \leq \alpha^+$.*
- (P2) *(Monotonicity) Let $\alpha_i = \left(h_i \mid p_{h_i}, g_i \mid q_{g_i} \right)$ and $\alpha_i^* = \left(h_i^* \mid p_{h_i^*}, g_i^* \mid q_{g_i^*} \right)$, for all $i = (1, 2, \ldots, n)$ be two families of PDHFEs where for each element in α_i and α_i^*, there are $\gamma_{\alpha_i} \leq \gamma_{\alpha_i^*}$ and $\eta_{\alpha_i} \geq \eta_{\alpha_i^*}$ while the probabilities remain the same i.e., $p_{h_i} = p_{h_i^*}, q_{g_i} = q_{g_i^*}$ then $PDHFOWEA(\alpha_1, \alpha_2, \ldots, \alpha_n) \leq PDHFOWEA(\alpha_1^*, \alpha_2^*, \ldots, \alpha_n^*)$.*

Proof. Similar to Properties 1 and 2. $\square$

Definition 9. *Let Ω be a family of all PDHFEs α_i $(i = 1, 2, \ldots, n)$ with the corresponding weights $\omega = (\omega_1, \omega_2, \ldots, \omega_n)^T$, such that $\omega_i > 0$ and $\sum_{i=1}^{n} \omega_i = 1$. If PDHFWEG: $\Omega^n \to \Omega$, is a mapping defined by*

$$
PDHFWEG(\alpha_1, \alpha_2, \ldots, \alpha_n) = \alpha_1^{\omega_1} \otimes \alpha_2^{\omega_2} \otimes \ldots \otimes \alpha_n^{\omega_n} \tag{17}
$$

then, PDHFWEG is called probabilistic dual hesitant fuzzy weighted Einstein geometric operator.

Theorem 8. *For a collection of PDHFEs $\alpha_i = \left(h_i \mid p_{h_i}, g_i \mid q_{g_i} \right), (i = 1, 2, \ldots, n)$, the combined value obtained by using PDHFWEG operator is still a PDHFE and is given as*

$$PDHFWEG(\alpha_1, \alpha_2, \ldots, \alpha_n)$$

$$= \bigcup_{\gamma_i \in h_i, \eta_i \in g_i} \left(\left\{ \frac{2 \prod\limits_{i=1}^{n} (\gamma_i)^{\omega_i}}{\prod\limits_{i=1}^{n} (2 - \gamma_i)^{\omega_i} + \prod\limits_{i=1}^{n} (\gamma_i)^{\omega_i}} \middle| \prod_{i=1}^{n} p_{\gamma_i} \right\}, \left\{ \frac{\prod\limits_{i=1}^{n} (1 + \eta_i)^{\omega_i} - \prod\limits_{i=1}^{n} (1 - \eta_i)^{\omega_i}}{\prod\limits_{i=1}^{n} (1 + \eta_i)^{\omega_i} + \prod\limits_{i=1}^{n} (1 - \eta_i)^{\omega_i}} \middle| \prod_{i=1}^{n} q_{\eta_i} \right\} \right) \tag{18}$$

where $\omega = (\omega_1, \omega_2, \ldots, \omega_n)^T$ *is a weight vector such that* $\sum\limits_{i=1}^{n} \omega_i = 1$ *where* $0 < \omega_i < 1$.

Proof. Same as Theorem 6. $\square$

Also, it has been seen that the PDHFWEG operator satisfies the properties of boundedness and monotonicity.

Definition 10. *Let* α_i $(i = 1, 2, \ldots, n)$ *be the family of PDHFEs, and PDHFOWEG:* $\Omega^n \to \Omega$, *if*

$$PDHFOWEG(\alpha_1, \alpha_2, \ldots, \alpha_n) = \alpha_{\sigma(1)}^{\omega_1} \oplus \alpha_{\sigma(2)}^{\omega_2} \cdots \oplus \alpha_{\sigma(n)}^{\omega_n} \tag{19}$$

where Ω *is the set of PDHFEs and* $\omega = (\omega_1, \omega_2, \ldots, \omega_n)^T$ *is the weight vector of* α_i *such that* $\omega_i > 0$ *and* $\sum\limits_{i=1}^{n} \omega_i = 1$. $(\sigma(1), \sigma(2), \ldots, \sigma(n))$ *is a permutation of* $(1, 2, \ldots, n)$ *such that* $\alpha_{\sigma(i-1)} \geq \alpha_{\sigma(i)}$ *for* $(i = 2, 3, \ldots, n)$, *then PDHFOWEG is called probabilistic dual hesitant fuzzy ordered weighted Einstein geometric operator.*

Theorem 9. *For a family of PDHFEs* $\alpha_i = \left(h_i \middle| p_{h_i}, g_i \middle| q_{g_i} \right), (i = 1, 2, \ldots, n)$, *the combined value obtained by using PDHFOWEG operator is given as*

$$PDHFOWEG(\alpha_1, \alpha_2, \ldots, \alpha_n)$$

$$\bigcup_{\substack{\gamma_{\sigma(i)} \in h_{\sigma(i)}, \\ \eta_{\sigma(i)} \in g_{\sigma(i)}}} \left(\left\{ \frac{2 \prod\limits_{i=1}^{n} (\gamma_{\sigma(i)})^{\omega_{\sigma(i)}}}{\prod\limits_{i=1}^{n} (2 - \gamma_{\sigma(i)})^{\omega_{\sigma(i)}} + \prod\limits_{i=1}^{n} (\gamma_{\sigma(i)})^{\omega_{\sigma(i)}}} \middle| \prod_{i=1}^{n} p_{\gamma_{\sigma(i)}} \right\}, \left\{ \frac{\prod\limits_{i=1}^{n} (1 + \eta_{\sigma(i)})^{\omega_{\sigma(i)}} - \prod\limits_{i=1}^{n} (1 - \eta_{\sigma(i)})^{\omega_{\sigma(i)}}}{\prod\limits_{i=1}^{n} (1 + \eta_{\sigma(i)})^{\omega_{\sigma(i)}} + \prod\limits_{i=1}^{n} (1 - \eta_{\sigma(i)})^{\omega_{\sigma(i)}}} \middle| \prod_{i=1}^{n} q_{\eta_{\sigma(i)}} \right\} \right) \tag{20}$$

where $\omega = (\omega_1, \omega_2, \ldots, \omega_n)^T$ *is a weight vector such that* $\sum\limits_{i=1}^{n} \omega_i = 1$ *where* $0 < \omega_i < 1$.

Proof. Similar to Theorem 6. $\square$

Also, it has been seen that the PDHFOWEG operator satisfies the properties of boundedness and monotonicity.

6. Maximum Deviation Method for Determination the Weights

The choice of weights directly affects the performance of weighted aggregation operators. For this purpose, in this subsection, the effective maximizing deviation method is adapted to calculate the weights in MCDM when the weights are unknown or partially known.

Given the set of alternatives $A = \{A_1, A_2, \ldots, A_m\}$ and the set of criteria $C = \{C_1, C_2, \ldots, C_t\}$ which is being evaluated by a decision maker under the PDHFS environment over the universal set $X = \{x_1, x_2, \ldots, x_n\}$. Assume that the rating values corresponding to each alternative is expressed in terms of PDHFEs as

$$A_r = \left\{ (C_1, s_{r1}), (C_2, s_{r2}), \ldots, (C_t, s_{rv}) \right\}, \tag{21}$$

where $s_{rv} = (h_{rv}(x_k) | p_{rv}(x_k), g_{rv}(x_k) | q_{rv}(x_k))$, where $r = 1, 2, \ldots, m; v = 1, 2, \ldots, t, k = 1, 2, \ldots, n$. Assume that the importance of each criterion are given in the form of weights as $(\omega_1, \omega_2, \ldots, \omega_t)$ respectively such that $0 < \omega_v \leq 1$ and $\sum_{v=1}^{t} \omega_v = 1$. Now, by using the proposed distances d_1 in Equation (9) or d_2 in (10) ; the deviation measure between the alternative A_r and all other alternatives with respect to the criteria C_v is given as:

$$D_{rv}(\omega) = \sum_{b=1}^{m} \omega_v D(s_{rv}, s_{bv}) \quad r = 1, 2, \ldots, m; v = 1, 2, \ldots, t \tag{22}$$

In accordance to the notion of maximizing deviation method, if the distance between the alternatives is smaller for a criteria, then it should have smaller weight. This one shows that the alternatives are homologous to the criterion. Contrarily, it should have larger weights. Let,

$$D_v(\omega) = \sum_{r=1}^{m} D_{rv}(\omega) = \sum_{r=1}^{m} \sum_{b=1}^{m} \omega_v D(s_{rv}, s_{bv}), \quad v = 1, 2, \ldots, t \tag{23}$$

Here $D_v(\omega)$ represents the distance of all the alternatives to the other alternatives under the criteria $C_v \in C$. Moreover, 'D' represents either distance d_1 or d_2 as given in Equations (9) and (10) respectively. Based on the concept of maximum deviation, we have to choose a weight vector 'ω' to maximize all the deviations measures for the criteria. For this, we construct a non-linear programming model as given below:

$$\begin{cases} \max\ D(\omega) = \sum_{v=1}^{t} \sum_{r=1}^{m} D_{rv}(\omega) = \sum_{v=1}^{t} \sum_{r=1}^{m} \sum_{b=1}^{m} D(s_{rv}, s_{bv}) \omega_v \\ \text{s.t.}\quad \omega_v > 0; \quad \sum_{v=1}^{t} \omega_v = 1; \quad v = 1, 2, \ldots, t \end{cases} \tag{24}$$

where 'D' can be either d_1 or d_2.

If $D = d_1$, then for $\lambda > 0$, we have

$$D(\omega) = \sum_{v=1}^{t} \sum_{r=1}^{m} \sum_{b=1}^{m} \omega_v \left(\sum_{k=1}^{n} \frac{1}{n} \left(\frac{1}{M+N} \left(\begin{array}{c} \sum_{i=1}^{M} \left| (\gamma_{A_i}(x_k) p_{A_i}(x_k))(x_{rv}) - (\gamma_{B_i}(x_k) p_{B_i}(x_k))(x_{bv}) \right|^{\lambda} \\ + \sum_{j=1}^{N} \left| (\eta_{A_j}(x_k) q_{A_j}(x_k))(x_{rv}) - (\eta_{B_j}(x_k) q_{B_j}(x_k))(x_{bv}) \right|^{\lambda} \end{array} \right) \right) \right)^{\frac{1}{\lambda}};$$

and if $D = d_2$, then

$$D(\omega) = \sum_{v=1}^{t} \sum_{r=1}^{m} \sum_{b=1}^{m} \omega_v \left(\sum_{k=1}^{n} \frac{1}{n} \left(\frac{\left| \frac{1}{M_A} \sum_{i=1}^{M_A} (\gamma_{A_i}(x_k) p_{A_i}(x_k))(x_{rv}) - \frac{1}{M_B} \sum_{i'=1}^{M_B} (\gamma_{B_{i'}}(x_k) p_{B_{i'}}(x_k))(x_{bv}) \right|^{\lambda}}{2} \right. \right.$$
$$\left. \left. + \frac{\left| \frac{1}{N_A} \sum_{j=1}^{N_A} (\eta_{A_j}(x_k) q_{A_j}(x_k))(x_{rv}) - \frac{1}{N_B} \sum_{j'=1}^{N_B} (\eta_{B_{j'}}(x_k) q_{B_{j'}}(x_k))(x_{bv}) \right|^{\lambda}}{2} \right) \right)^{\frac{1}{\lambda}}$$

If the information about criteria weights is completely unknown, then another programming method can be established as:

$$\begin{cases} \max\ D(\omega) = \sum\limits_{v=1}^{t} \sum\limits_{r=1}^{m} D_{rv}(\omega) = \sum\limits_{v=1}^{t} \sum\limits_{r=1}^{m} \sum\limits_{b=1}^{m} D(s_{rv}, s_{bv})\omega_v \\ \text{s.t.}\quad \omega_v \geq 0;\quad \sum\limits_{v=1}^{n} \omega_v^2 = 1;\quad v = 1,2,\ldots,t \end{cases} \tag{25}$$

To solve this, a Lagrange's function is constructed as

$$L(\omega,\zeta) = \sum_{v=1}^{t} \sum_{r=1}^{m} \sum_{b=1}^{m} D(s_{rv}, s_{bv})\omega_v + \frac{\zeta}{2}\left(\sum_{v=1}^{t} \omega_v^2 - 1\right) \tag{26}$$

where ζ is the Lagrange's parameter. Computing the partial derivatives of Lagrange's function w.r.t ω_v as well as ζ and letting them equal to zero.

$$\begin{cases} \frac{\partial L}{\partial \omega_v} = \sum\limits_{r=1}^{m} \sum\limits_{b=1}^{m} D(s_{rv}, s_{bv}) + \zeta\omega_v = 0;\quad v = 1, 2, \ldots, t \\ \frac{\partial L}{\partial \zeta} = \sum\limits_{v=1}^{t} \omega_v^2 - 1 = 0 \end{cases} \tag{27}$$

Solving, Equation (27) we can obtain,

$$\omega_v = \frac{\sum\limits_{r=1}^{m} \sum\limits_{b=1}^{m} D(s_{rv}, s_{bv})}{\sqrt{\sum\limits_{v=1}^{t} \left(\sum\limits_{r=1}^{m} \sum\limits_{b=1}^{m} D(s_{rv}, s_{bv})\right)^2}};\quad v = 1, 2, \ldots, t \tag{28}$$

Normalizing Equation (28) we get

$$\omega_v = \frac{\sum\limits_{r=1}^{m} \sum\limits_{b=1}^{m} D(s_{rv}, s_{bv})}{\sum\limits_{v=1}^{t} \sum\limits_{r=1}^{m} \sum\limits_{b=1}^{m} D(s_{rv}, s_{bv})} \tag{29}$$

In DM process, the data values for evaluation are available as DHFSs or PDHFSs which are integrated to form the PDHFSs. In order to gather the information, the probability values are assigned to each possible membership or non-membership value. An algorithm followed for this information fusion is outlined in Algorithm 1.

Algorithm 1 Aggregating probabilities for more than one Probabilistic fuzzy sets.

Input: $\alpha^{(1)}, \alpha^{(2)}, \ldots, \alpha^{(d)}$ where $\alpha^{(d)} = \left(h^{(d)} \big| p^{(d)}\right)$ where $d = 1, 2, \ldots, D$ such that D is the total number of elements to be fused together.

  Output: $\alpha^{(out)} = \left(h^{(out)} \big| p^{(out)}\right)$

1: Let $u = \frac{1}{D}$, be the normalized unit.
2: List all the probabilistic membership values in a set and represent it as $M = \{m_l \big| s_l\}$, where $m_l \big| s_l = h^{(d)} \big| p^{(d)}, \forall d = 1, 2, \ldots, D$, and $l = 1, 2, \ldots, \#L$, such that $\#L$ is the total number of probabilistic membership values of all the considered elements.
3: Set $i = 1$
4: Set $m_e = m_i$
5: $f^{(l)}_{(mem)} = \begin{cases} 1, & \text{if } m_e = m_l \\ 0, & \text{if } m_e \neq m_l \end{cases}$
6: Set $l = l + 1$ and repeat 5, until $l = \#L$
7: Set $h^{(out)} = \bigcup_i m_e$
8: $p^{(out)} = \left(\sum_l \left(f^{(l)}_{(mem)} \cdot s_l \right) \cdot u \right)$
9: Set $i = i + 1$ and goto 4, until $i = \#L$

To demonstrate the working of aforementioned algorithm, an example is given below.

Example 2. *Let* $\alpha^{(1)} = (\{0.1|0.1, 0.2|0.5, 0.3|0.4\}, \{0.5|1\})$; $\alpha^{(2)} = (\{0.2|0.4, 0.3|0.6\}, \{0.5|0.2, 0.6|0.8\})$ *and* $\alpha^{(3)} = (\{0.1|0.4, 0.2|0.4, 0.6|0.2\}, \{0.1|1\})$ *be three PDHFEs to be fused together. Since,* $\left(h^{(1)}, p^{(1)}\right) = (\{0.1|0.1, 0.2|0.5, 0.3|0.4\})$, $\left(h^{(2)}, p^{(2)}\right) = (\{0.2|0.4, 0.3|0.6\})$ *and* $\left(h^{(3)}, p^{(3)}\right) = (\{0.1|0.4, \ 0.2|0.4, 0.6|0.2\})$, *so we get* $M = \{0.1|0.1, 0.2|0.5, 0.3|0.4, 0.2|0.4, 0.3|0.6, 0.1|0.4, 0.2|0.4, 0.6|0.2\}$ *where* $\#L = 8$ *and thus* $l = 1, 2, \ldots, 8$. *Clearly, here* $D = 3$. *Now, by following Algorithm 1 for both membership and non-membership degrees, we obtained the final PDHFE as:*

$$\alpha^{(out)} = (\{0.1|0.1667, 0.2|0.4333, 0.3|0.3333, 0.6|0.066\}, \{0.5|0.4, 0.6|0.2666, 0.1|0.3333\})$$

7. Decision Making Approach Using the Proposed Operators

In this section, a DM approach based on proposed AOs is given followed by a numerical example.

7.1. Approach Based on the Proposed Operators

Consider a set of m alternatives $A = \{A_1, A_2, \ldots, A_m\}$ which are evaluated by the experts classified under criteria information $C = \{C_1, C_2, \ldots, C_t\}$. The ratings for each alternative in PDHFEs are given as:

$$A_r = \left\{ (C_1, \alpha_{r1}), (C_2, \alpha_{r2}), \ldots, (C_t, \alpha_{rv}) \right\}, \tag{30}$$

where $\alpha_{rv} = \left(h_{rv} \big| p_{rv}, g_{rv} \big| q_{rv}\right)$, where $r = 1, 2, \ldots, m$; $v = 1, 2, \ldots, t$. In order to get the best alternative(s) for a problem, DM approach is summarized in the following steps by utilizing proposed AOs as:

Step 1: Construct decision matrices $R^{(d)}$ for 'd' number of decision makers in form of PDHFEs as:

$$R^{(d)} = \begin{array}{c} \\ A_1 \\ A_2 \\ \vdots \\ A_m \end{array} \begin{pmatrix} \left(h_{11}^{(d)}\Big|p_{11}^{(d)}, g_{11}^{(d)}\Big|q_{11}^{(d)}\right) & \left(h_{12}^{(d)}\Big|p_{12}^{(d)}, g_{12}^{(d)}\Big|q_{12}^{(d)}\right) & \cdots & \left(h_{1t}^{(d)}\Big|p_{1t}^{(d)}, g_{1t}^{(d)}\Big|q_{1t}^{(d)}\right) \\ \left(h_{21}^{(d)}\Big|p_{21}^{(d)}, g_{21}^{(d)}\Big|q_{21}^{(d)}\right) & \left(h_{22}^{(d)}\Big|p_{22}^{(d)}, g_{22}^{(d)}\Big|q_{22}^{(d)}\right) & \cdots & \left(h_{2t}^{(d)}\Big|p_{2t}^{(d)}, g_{2t}^{(d)}\Big|q_{2t}^{(d)}\right) \\ \vdots & \vdots & \ddots & \vdots \\ \left(h_{m1}^{(d)}\Big|p_{m1}^{(d)}, g_{m1}^{(d)}\Big|q_{m1}^{(d)}\right) & \left(h_{m2}^{(d)}\Big|p_{m2}^{(d)}, g_{m2}^{(d)}\Big|q_{m2}^{(d)}\right) & \cdots & \left(h_{mt}^{(d)}\Big|p_{mt}^{(d)}, g_{mt}^{(d)}\Big|q_{mt}^{(d)}\right) \end{pmatrix}$$

where $\left(h_{rv}^{(d)}\Big|p_{rv}^{(d)}, g_{rv}^{(d)}\Big|q_{rv}^{(d)}\right) = \left(\{\gamma_{rv}^{(d)}|p_{rv}^{(d)}\}, \{\eta_{rv}^{(d)}|q_{rv}^{(d)}\}\right)$, such that $r = 1, 2, \ldots, m$ and $v = 1, 2, \ldots, t$.

Step 2: If $d = 1$, then $\left(h_{rv}^{(d)}\Big|p_{rv}^{(d)}, g_{rv}^{(d)}\Big|q_{rv}^{(d)}\right)$ is equal to $(h_{rv}|p_{rv}, g_{rv}|q_{rv})$, where $(h_{rv}|p_{rv}, g_{rv}|q_{rv}) = (\{\gamma_{rv}|p_{rv}\}, \{\eta_{rv}|q_{rv}\})$; such that $r = 1, 2, \ldots, m$ and $v = 1, 2, \ldots, t$ and goto Section 7.1 Step 3. If $d \geq 2$, then a matrix is formed by combining the probabilities in accordance to the Algorithm 1. The comprehensive matrix so obtained is given as:

$$R = \begin{array}{c} \\ A_1 \\ A_2 \\ \vdots \\ A_m \end{array} \begin{pmatrix} (h_{11}|p_{11}, g_{11}|q_{11}) & (h_{12}|p_{12}, g_{12}|q_{12}) & \cdots & (h_{1t}|p_{1t}, g_{1t}|q_{1t}) \\ (h_{21}|p_{21}, g_{21}|q_{21}) & (h_{22}|p_{22}, g_{22}|q_{22}) & \cdots & (h_{2t}|p_{2t}, g_{2t}|q_{2t}) \\ \vdots & \vdots & \ddots & \vdots \\ (h_{m1}|p_{m1}, g_{m1}|q_{m1}) & (h_{m2}|p_{m2}, g_{m2}|q_{m2}) & \cdots & (h_{mt}|p_{mt}, g_{mt}|q_{mt}) \end{pmatrix}$$

where $(h_{rv}|p_{rv}, g_{rv}|q_{rv}) = (\{\gamma_{rv}|p_{rv}\}, \{\eta_{rv}|q_{rv}\})$, where $r = 1, 2, \ldots, m$ and $v = 1, 2, \ldots, t$.

Step 3: Choose the appropriate distance measure among d_1 or d_2 as given in Equations (9) and (10), on the basis of need the expert. If the repeated values of the largest or smallest dual-hesitant probabilistic values can be repeated according to the optimistic or pessimistic behavior of the expert then choose measure d_1 otherwise choose measure d_2 and determine the weights of different criteria using Equation (29).

Step 4: Compute the overall aggregated assessment 'Q_r' of alternatives using PDHFWEA or PDHFOWEA or PDHFWEG or PDHFOWEG operators as given below in Equations (31)–(34) respectively.

$$\begin{aligned} Q_r &= \text{PDHFWEA}(\alpha_{r1}, \alpha_{r2}, \ldots, \alpha_{rv}) \\ &= \bigcup_{\substack{\gamma_{rv}\in h_{rv} \\ \eta_{rv}\in g_{rv}}} \left(\left\{ \frac{\prod\limits_{v=1}^{t}(1+\gamma_{rv})^{\omega_v} - \prod\limits_{v=1}^{t}(1-\gamma_{rv})^{\omega_v}}{\prod\limits_{v=1}^{t}(1+\gamma_{rv})^{\omega_v} + \prod\limits_{v=1}^{t}(1-\gamma_{rv})^{\omega_v}} \; \Bigg| \prod\limits_{v=1}^{t} p_{\gamma_{rv}} \right\}, \right. \\ &\qquad\qquad \left. \left\{ \frac{2\prod\limits_{v=1}^{t}(\eta_{rv})^{\omega_v}}{\prod\limits_{v=1}^{t}(2-\eta_{rv})^{\omega_v} + \prod\limits_{v=1}^{t}(\eta_{rv})^{\omega_v}} \; \Bigg| \prod\limits_{v=1}^{t} q_{\eta_{rv}} \right\} \right) \end{aligned} \tag{31}$$

or

$$Q_r = \text{PDHFOWEA}(\alpha_{r1}, \alpha_{r2}, \ldots, \alpha_{rv})$$

$$= \bigcup_{\substack{\gamma_{\sigma(rv)} \in h_{\sigma(rv)} \\ \eta_{\sigma(rv)} \in g_{\sigma(rv)}}} \left(\left\{ \left. \frac{\prod_{v=1}^{t}(1+\gamma_{\sigma(rv)})^{\omega_{\sigma(v)}} - \prod_{v=1}^{t}(1-\gamma_{\sigma(rv)})^{\omega_{\sigma(v)}}}{\prod_{v=1}^{t}(1+\gamma_{\sigma(rv)})^{\omega_{\sigma(v)}} + \prod_{v=1}^{t}(1-\gamma_{\sigma(rv)})^{\omega_{\sigma(v)}}} \;\right|\; \prod_{v=1}^{t} p_{\gamma_{\sigma(rv)}} \right\}, \left\{ \left. \frac{2\prod_{v=1}^{t}(\eta_{\sigma(rv)})^{\omega_{\sigma(v)}}}{\prod_{v=1}^{t}(2-\eta_{\sigma(rv)})^{\omega_{\sigma(v)}} + \prod_{v=1}^{t}(\eta_{\sigma(rv)})^{\omega_{\sigma(v)}}} \;\right|\; \prod_{v=1}^{t} q_{\eta_{\sigma(rv)}} \right\} \right) \tag{32}$$

or

$$Q_r = \text{PDHFWEG}(\alpha_{r1}, \alpha_{r2}, \ldots, \alpha_{rv})$$

$$= \bigcup_{\substack{\gamma_{rv} \in h_{rv} \\ \eta_{rv} \in g_{rv}}} \left(\left\{ \left. \frac{2\prod_{v=1}^{t}(\gamma_{rv})^{\omega_v}}{\prod_{v=1}^{t}(2-\gamma_{rv})^{\omega_v} + \prod_{v=1}^{t}(\gamma_{rv})^{\omega_v}} \;\right|\; \prod_{v=1}^{t} p_{\gamma_{rv}} \right\}, \left\{ \left. \frac{\prod_{v=1}^{t}(1+\eta_{rv})^{\omega_v} - \prod_{v=1}^{t}(1-\eta_{rv})^{\omega_v}}{\prod_{v=1}^{t}(1+\eta_{rv})^{\omega_v} + \prod_{v=1}^{t}(1-\eta_{rv})^{\omega_v}} \;\right|\; \prod_{v=1}^{t} q_{\eta_{rv}} \right\} \right) \tag{33}$$

or

$$Q_r = \text{PDHFOWEG}(\alpha_{r1}, \alpha_{r2}, \ldots, \alpha_{rv})$$

$$= \bigcup_{\substack{\gamma_{\sigma(rv)} \in h_{\sigma(rv)} \\ \eta_{\sigma(rv)} \in g_{\sigma(rv)}}} \left(\left\{ \left. \frac{2\prod_{v=1}^{t}(\gamma_{\sigma(rv)})^{\omega_{\sigma(v)}}}{\prod_{v=1}^{t}(2-\gamma_{\sigma(rv)})^{\omega_{\sigma(v)}} + \prod_{v=1}^{t}(\gamma_{\sigma(rv)})^{\omega_{\sigma(v)}}} \;\right|\; \prod_{v=1}^{t} p_{\gamma_{\sigma(rv)}} \right\}, \left\{ \left. \frac{\prod_{v=1}^{t}(1+\eta_{\sigma(rv)})^{\omega_{\sigma(v)}} - \prod_{v=1}^{t}(1-\eta_{\sigma(rv)})^{\omega_{\sigma(v)}}}{\prod_{v=1}^{t}(1+\eta_{\sigma(rv)})^{\omega_{\sigma(v)}} + \prod_{v=1}^{t}(1-\eta_{\sigma(rv)})^{\omega_{\sigma(v)}}} \;\right|\; \prod_{v=1}^{t} q_{\eta_{\sigma(rv)}} \right\} \right) \tag{34}$$

Step 5: Utilize Definition 5 to rank the overall aggregated values and select the most desirable alternative(s).

7.2. Illustrative Example

An illustrative example (based on consumer's buying behavior) for eliciting the numerical applicability of our proposed approach is given below:

In a company's production oriented decision-making processes, consumers or buyers play a vital role. In order to increase sales and to be in good books of every customer, every production company pays a great attention to customer's buying behavior. This consumer behavior is the main driving force behind the change of trends, need of updation in the products etc., to which the production company must remain in contact to have a great mutual relationship with the customers and to maintain a strong position in the competitive market environment.

Suppose a multi-national company wants to launch the new products on the basis of different consumers in different countries. For that, they have delegated works to the company heads of three different countries viz. India, Canada, and Australia. The company heads of these countries have to analyze the customer's buying behavior and for that, they have information available in the form of PDHFEs. Each expert ($d = 1, 2, 3$) from the three different countries accessed the available information oriented to four company products A_i's where ($i = 1, 2, 3, 4$) classified under four criteria determining the customer's buying behavior namely C_1 : 'Suitability to cultural environment'; C_2 : 'Global trend accordance'; C_3 : 'Suitability to weather conditions' ; C_4 : 'Good quality after-sale services'. The aim of the company is to access the main criteria which affect the customer's buying behavior so as to figure out which product among A_i's ($i = 1, 2, 3, 4$) has to be launched first. Following steps are adopted to find the most suitable product for the first launch.

Step 1: The preference information corresponding to three decision-makers (d = 1; 2; 3) is given in Tables 1–3.

Table 1. Preference values provided by decision-maker 1.

	C_1	C_2	C_3	C_4
A_1	$\{0.2\|0.4, 0.3\|0.6\}$ $\{0.4\|1\}$	$\{0.45\|0.42, 0.60\|0.58\}$ $\{0.2\|0.4, 0.3\|0.6\}$	$\{0.9\|1\}$ $\{0.1\|1\}$	$\{0.6\|1\}$ $\{0.3\|1\}$
A_2	$\{0.8\|0.9, 0.6\|0.1\}$ $\{0.1\|1\}$	$\{0.30\|1\}$ $\{0.6\|1\}$	$\{0.6\|1\}$ $\{0.2\|0.5, 0.1\|0.5\}$	$\{0.2\|1\}$ $\{0.8\|1\}$
A_3	$\{0.05\|0.7, 0.2\|0.3\}$ $\{0.5\|1\}$	$\{0.50\|1\}$ $\{0.5\|1\}$	$\{0.8\|0.6, 0.6\|0.4\}$ $\{0.15\|1\}$	$\{0.12\|1\}$ $\{0.7\|0.9, 0.6\|0.1\}$
A_4	$\{0.4\|1\}$ $\{0.3\|0.5, 0.2\|0.5\}$	$\{0.50\|1\}$ $\{0.2\|0.3, 0.4\|0.7\}$	$\{0.3\|1\}$ $\{0.65\|1\}$	$\{0.5\|1\}$ $\{0.2\|0.3, 0.4\|0.7\}$

Table 2. Preference values provided by decision-maker 2.

	C_1	C_2	C_3	C_4
A_1	$\{0.3\|0.5, 0.5\|0.5\}$ $\{0.4\|1\}$	$\{0.20\|1\}$ $\{0.7\|0.1\}$	$\{0.2\|1\}$ $\{0.4\|0.8, 0.6\|0.2\}$	$\{0.6\|0.7, 0.7\|0.3\}$ $\{0.25\|1\}$
A_2	$\{0.2\|1\}$ $\{0.7\|1\}$	$\{0.30\|0.5, 0.2\|0.5\}$ $\{0.20\|0.5, 0.15\|0.5\}$	$\{0.2\|1\}$ $\{0.6\|1\}$	$\{0.2\|0.3, 0.3\|0.7\}$ $\{0.6\|1\}$
A_3	$\{0.4\|0.4, 0.5\|0.6\}$ $\{0.5\|1\}$	$\{0.45\|1\}$ $\{0.5\|1\}$	$\{0.8\|0.4, 0.6\|0.6\}$ $\{0.2\|0.7, 0.1\|0.3\}$	$\{0.1\|1\}$ $\{0.6\|0.6, 0.8\|0.4\}$
A_4	$\{0.4\|0.2, 0.5\|0.8\}$ $\{0.3\|1\}$	$\{0.2\|0.4, 0.5\|0.6\}$ $\{0.4\|0.2, 0.3\|0.8\}$	$\{0.4\|0.1, 0.5\|0.9\}$ $\{0.3\|1\}$	$\{0.4\|1\}$ $\{0.6\|1\}$

Table 3. Preference values provided by decision-maker 3.

	C_1	C_2	C_3	C_4
A_1	$\{0.75\|1\}$ $\{0.2\|1\}$	$\{0.50\|1\}$ $\{0.2\|0.5, 0.5\|0.5\}$	$\{0.3\|1\}$ $\{0.6\|1\}$	$\{0.6\|1\}$ $\{0.3\|1\}$
A_2	$\{0.6\|0.6, 0.8\|0.4\}$ $\{0.1\|1\}$	$\{0.20\|1\}$ $\{0.7\|1\}$	$\{0.9\|1\}$ $\{0.1\|1\}$	$\{0.3\|1\}$ $\{0.5\|0.4, 0.6\|0.6\}$
A_3	$\{0.9\|1\}$ $\{0.1\|1\}$	$\{0.6\|1\}$ $\{0.25\|0.5, 0.1\|0.5\}$	$\{0.8\|1\}$ $\{0.2\|1\}$	$\{0.2\|1\}$ $\{0.8\|1\}$
A_4	$\{0.3\|0.7, 0.5\|0.3\}$ $\{0.4\|0.6, 0.5\|0.4\}$	$\{0.1\|1\}$ $\{0.8\|1\}$	$\{0.3\|1\}$ $\{0.3\|1\}$	$\{0.35\|1\}$ $\{0.6\|1\}$

Step 2: Since number of decision makers i.e., $d \geq 2$, therefore, using Algorithm 1, the comprehensive matrix obtained after integrating all the preferences given by the panel of experts is given in Table 4.

Table 4. Comprehensive matrix.

	C_1	C_2	C_3	C_4
A_1	$\left(\{0.2\|0.1333, 0.3\|0.3667, 0.5\|0.1667, 0.75\|0.3333\}, \{0.4\|0.6667, 0.2\|0.3333\}\right)$	$\left(\{0.45\|0.14, 0.6\|0.1934, 0.2\|0.3333, 0.5\|0.3333\}, \{0.2\|0.3, 0.3\|0.2, 0.7\|0.3333, 0.5\|0.1667\}\right)$	$\left(\{0.9\|0.3333, 0.2\|0.3333, 0.3\|0.3334\}, \{0.1\|0.3333, 0.4\|0.2667, 0.6\|0.4\}\right)$	$\left(\{0.6\|0.9, 0.7\|0.1\}, \{0.3\|0.6667, 0.25\|0.3333\}\right)$
A_2	$\left(\{0.8\|0.4333, 0.6\|0.2334, 0.2\|0.3333\}, \{0.1\|0.6667, 0.7\|0.3333\}\right)$	$\left(\{0.30\|0.75, 0.2\|0.5\}, \{0.6\|0.3333, 0.2\|0.1667, 0.15\|0.1667, 0.7\|0.3333\}\right)$	$\left(\{0.6\|0.3333, 0.2\|0.3334, 0.9\|0.3333\}, \{0.2\|0.1667, 0.1\|0.6667, 0.6\|0.1666\}\right)$	$\left(\{0.2\|0.4333, 0.3\|0.5667\}, \{0.8\|0.3333, 0.6\|0.3333, 0.5\|0.1333\}\right)$
A_3	$\left(\{0.05\|0.2334, 0.2\|0.1, 0.4\|0.1333, 0.5\|0.2, 0.9\|0.3333\}, \{0.5\|0.6667, 0.1\|0.3333\}\right)$	$\left(\{0.5\|0.3333, 0.45\|0.3333, 0.6\|0.3334\}, \{0.5\|0.6667, 0.2\|0.1667, 0.1\|0.1666\}\right)$	$\left(\{0.8\|0.6667, 0.6\|0.3333\}, \{0.15\|0.3333, 0.2\|0.5666, 0.1\|0.1\}\right)$	$\left(\{0.12\|0.3333, 0.1\|0.3333, 0.2\|0.3334\}, \{0.7\|0.3, 0.6\|0.2333, 0.8\|0.4667\}\right)$
A_4	$\left(\{0.4\|0.4, 0.5\|0.3667, 0.3\|0.2333\}, \{0.3\|0.5, 0.2\|0.1667, 0.4\|0.2, 0.5\|0.1333\}\right)$	$\left(\{0.5\|0.5333, 0.2\|0.1333, 0.1\|0.3334\}, \{0.2\|0.1, 0.4\|0.3, 0.3\|0.2667, 0.8\|0.3333\}\right)$	$\left(\{0.30\|0.6667, 0.4\|0.0333, 0.5\|0.3\}, \{0.65\|0.3333, 0.3\|0.6667\}\right)$	$\left(\{0.5\|0.3333, 0.4\|0.3333, 0.35\|0.3334\}, \{0.2\|0.1, 0.4\|0.2334, 0.6\|0.6666\}\right)$

Step 3: The experts chose to have an optimistic behavior towards the analysis and thus utilizing distance d_1 in Equation (29), the weights are determined as $\omega = (0.4385, 0.1986, 0.1815, 0.1814)^T$.

Step 4: The aggregated values for each alternative $A_i, i = (1, 2, 3, 4)$ by using PDHFWEA operator as given in Equation (31) are :

$$Q_1 = \left(\begin{Bmatrix} 0.5213|0.0056, 0.5439|0.0006, \\ 0.5546|0.0154, 0.5760|0.0017, \\ \dots\dots\dots, 0.6347|0.0037 \end{Bmatrix}, \begin{Bmatrix} 0.2617|0.0444, 0.2531|0.0222, \\ 0.1909|0.0222, 0.1844|0.0111, \\ \dots\dots\dots, 0.3120|0.0074 \end{Bmatrix}\right)$$

$$Q_2 = \left(\begin{Bmatrix} 0.6080|0.0469, 0.6201|0.0614, \\ 0.4838|0.0253, 0.4985|0.0331, \\ \dots\dots\dots, 0.4240|0.0157 \end{Bmatrix}, \begin{Bmatrix} 0.2531|0.0123, 0.2359|0.0198, \\ 0.2266|0.0049, 0.5372|0.0062, \\ \dots\dots\dots, 0.6427|0.0025 \end{Bmatrix}\right)$$

$$Q_3 = \left(\begin{Bmatrix} 0.3384|0.0173, 0.3352|0.0173, \\ 0.3515|0.0173, 0.3963|0.0074, \\ \dots\dots\dots, 0.7379|0.0123 \end{Bmatrix}, \begin{Bmatrix} 0.4391|0.0444, 0.4251|0.0346, \\ 0.4256|0.0691, 0.2226|0.0222, \\ \dots\dots\dots, 0.1540|0.0026 \end{Bmatrix}\right)$$

$$Q_4 = \left(\begin{Bmatrix} 0.4225|0.0474, 0.4036|0.0474, \\ 0.3947|0.0474, 0.4667|0.0435, \\ \dots\dots\dots, 0.3110|0.0078 \end{Bmatrix}, \begin{Bmatrix} 0.3016|0.0017, 0.3413|0.0039, \\ 0.3698|0.0111, 0.2533|0.0006, \\ \dots\dots\dots, 0.5259|0.0197 \end{Bmatrix}\right)$$

Step 5: The score values are obtained as $S(Q_1) = 0.1810, S(Q_2) = 0.1799, S(Q_3) = 0.1739$ and $S(Q_4) = -0.0002$

Step 6: Since, the ranking order is $S(Q_1) > S(Q_2) > S(Q_3) > S(Q_4)$, thus the ranking is obtained as $A_1 \succ A_2 \succ A_3 \succ A_4$.

Thus, it is clear that according to the experts product A_1 should be launched first.

However, on the other hand, if we utilize the PDHFWEG operator instead of PDHFWEA operator to aggregate the different preferences, then the following steps of the proposed approach are executed to reach the optimal alternative(s) as.

Step 1: Similar as above Section 7.2 Step 1.

Step 2: Similar as above Section 7.2 Step 2.

Step 3: Similar as above Section 7.2 Step 3.

Step 4: The aggregated values for each alternative $A_i, i = (1, 2, 3, 4)$ by using PDHFWEG operator as given in Equation (33) are :

$$Q_1 = \left(\left\{ \begin{array}{l} 0.3959|0.0056, 0.4092|0.0006, \\ 0.4642|0.0154, 0.4792|0.0017, \\ \ldots\ldots\ldots\ldots, 0.5908|0.0037 \end{array} \right\}, \left\{ \begin{array}{l} 0.2917|0.0444, 0.2827|0.0222, \\ 0.2008|0.0222, 0.1913|0.0111, \\ \ldots\ldots\ldots\ldots, 0.3541|0.0074 \end{array} \right\} \right)$$

$$Q_2 = \left(\left\{ \begin{array}{l} 0.5090|0.0469, 0.5415|0.0614, \\ 0.4391|0.0253, 0.4685|0.0331, \\ \ldots\ldots\ldots\ldots, 0.2959|0.0157 \end{array} \right\}, \left\{ \begin{array}{l} 0.3950|0.0123, 0.3312|0.0198, \\ 0.3078|0.0049, 0.6376|0.0062, \\ \ldots\ldots\ldots\ldots, 0.6516|0.0025 \end{array} \right\} \right)$$

$$Q_3 = \left(\left\{ \begin{array}{l} 0.1667|0.0173, 0.1615|0.0173, \\ 0.1828|0.0173, 0.2950|0.0074, \\ \ldots\ldots\ldots\ldots, 0.6164|0.0123 \end{array} \right\}, \left\{ \begin{array}{l} 0.4890|0.0444, 0.4646|0.0346, \\ 0.5203|0.0691, 0.3256|0.0222, \\ \ldots\ldots\ldots\ldots, 0.2742|0.0026 \end{array} \right\} \right)$$

$$Q_4 = \left(\left\{ \begin{array}{l} 0.4150|0.0474, 0.3981|0.0474, \\ 0.3886|0.0474, 0.4580|0.0435, \\ \ldots\ldots\ldots\ldots, 0.2774|0.0078 \end{array} \right\}, \left\{ \begin{array}{l} 0.3395|0.0017, 0.3744|0.0039, \\ 0.4157|0.0111, 0.2974|0.0006, \\ \ldots\ldots\ldots\ldots, 0.5656|0.0197 \end{array} \right\} \right)$$

Step 5: The score values are obtained as $S(Q_1) = 0.0937, S(Q_2) = -0.0073, S(Q_3) = -0.0202$ and $S(Q_4) = -0.0545$

Step 6: Since, the ranking order is $S(Q_1) > S(Q_3) > S(Q_2) > S(Q_4)$, thus the ranking is obtained as $A_1 \succ A_2 \succ A_3 \succ A_4$.

The most desirable alternative is A_1.

If we analyze the impact of the all the proposed operators along with the distance d_1 and d_2 onto the final ranking order of the alternative, we perform an experiment where the steps of the proposed algorithms are executed. The final score values of each alternative A_i ($i = 1, 2, 3, 4$), are obtained and are summarized in Table 5. It is seen that utilizing different distance measures i.e., d_1 and d_2 do not affect the best alternative A_1 in most of the cases. Moreover, the score values obtained by the proposed operators namely: PDHFWEA, PDHFWEG, and PDHFOWEG represent the same alternative A_1 as the best alternative which is to be launched first while the operator PDHOWEA represents the alternative A_3 as the best one. However, it can be seen that corresponding average PDHFWEA, PDHFOWEA score values are greater than that of PDHFWEG, PDHFOWEG aggregation operators showing that the average aggregation operators offer the decision maker more optimistic score-values as compared to the geometric ones. Also, it can be seen that both the distances, despite providing, a huge variation in numerical evaluation and data processing flexibility lead to the same result as A_1 as the best choice in most of the cases among the alternatives to be launched first.

Table 5. Score values of proposed approach.

	Operator	A_1	A_2	A_3	A_4	Ranking
Distance d_1	PDHFWEA	0.1810	0.1799	0.1739	−0.0002	$A_1 \succ A_2 \succ A_3 \succ A_4$
	PDHFOWEA	0.2293	0.2239	0.2940	0.0013	$A_3 \succ A_1 \succ A_2 \succ A_4$
	PDHFWEG	0.0937	−0.0073	−0.0202	−0.0545	$A_1 \succ A_2 \succ A_3 \succ A_4$
	PDHFOWEG	0.1458	0.0283	0.0856	−0.0515	$A_1 \succ A_3 \succ A_2 \succ A_4$
Distance d_2	PDHFWEA	0.1968	0.0754	0.1213	−0.0459	$A_1 \succ A_3 \succ A_2 \succ A_4$
	PDHFOWEA	0.1684	0.0832	0.0971	−0.0472	$A_1 \succ A_3 \succ A_2 \succ A_4$
	PDHFWEG	0.1006	−0.1189	−0.1072	−0.1056	$A_1 \succ A_4 \succ A_3 \succ A_2$
	PDHFOWEG	0.0691	−0.1118	−0.1268	−0.1091	$A_1 \succ A_4 \succ A_2 \succ A_3$

7.3. Comparative Studies

In order to analyze the alignment of the proposed approach's results with the existing theories and to validate our proposed results, the score values corresponding to different operators are given in Table 6. The operators in the considered existing theories are: probabilistic dual hesitant fuzzy weighted average (PDHFWA) by Hao et al. [42], hesitant probabilistic fuzzy Einstein weighted average and Einstein weighted geometric (HPFEWEA, HPFEWEG) by Park et al. [50] and hesitant probabilistic fuzzy weighted average (HPFWA), hesitant probabilistic fuzzy weighted geometric (HPFWG), hesitant probabilistic fuzzy ordered weighted average (HPFOWA) , hesitant probabilistic fuzzy ordered weighted geometric (HPFOWG) aggregation operators by Xu and Zhou [48]. Noticeably, the approach outlined by Hao et al. [42] by utilizing PDHFWA operator figures out A_2 as the best alternative and the least preferred alternative A_4 remains same as that of our proposed approach. However, if we consider only the probabilistic hesitant fuzzy information and ignores the non-membership probabilistic hesitant values, then the best alternative starts fluctuating among A_1 and A_3 by varying the different aggregation operators and the least preferred alternative remains same as A_4, which coincides the outcomes of our proposed approach. This variation is due to the negligence of the non-membership values and their corresponding probabilities. Thus, the proposed approach is advantageous among the traditional approaches because it remains firm on the same output ranking for different operators. Moreover, the best alternative chosen by the proposed approach remains the same as that with that of the existing approaches signifies that the proposed approach is the valid one.

Further, a deep insight into the comparison of our method with the existing ones is given by comparing the characteristics of all the approaches with the proposed one. In Table 7, it can be seen that the approaches put-forth by Hao et al. [42] and Xu and Zhou [48] considers multiple experts in analysis process whereas Park et al. [50] does not consider the multi-expert problems. All the existing approaches are the probabilistic approaches so they consider probabilities corresponding to their considered membership or non-membership values. Moreover, it is analyzed that the method proposed by [42] considers the non-membership probabilistic information but the rest two only considers the hesitant values and their probabilities. In all the three existing approaches, the weights are not derived by using any non-linear technique such as maximum deviation method for determination of weights but the weights corresponding to two different distance measures are considered in the proposed methodology.

In addition to above comparison studies, we elicit some characteristic comparison of our approach with existing DM methods proposed in [42,48,50] which are tabulated in Table 7.

In Table 7, the symbol '√' describes that the corresponding DM approach considers more than one decision maker, handles probabilities, accounts for non-membership entities and has weights derived by the non-linear approach, whereas the symbol '×' means that the associated method fails. The symbols tabulated in Table 7 depicts that the MCDM mentioned in [42] as well as [48] consider multiple multiple decision-makers whereas the approach utilized by [50] consists of preference evaluations through single expert. It is seen that all the three considered approaches considers the probabilities along with their respective fuzzy environments whereas only [42] considers only the non-membership values along with the membership ones while the other two considers only the membership value ratings. On the other hand, none of the existing approach among the specified ones, adopt a non-linear weight determination technique. Thus, it is analyzed that our proposed approach consists of all the four said characteristics and thus it deals with the real life situations, more efficiently as compared to the existing approaches [42,48,50].

Table 6. Comparison of overall rating values and ranking order of alternatives.

Existing Approaches	Operators	Score Values				
		A_1	A_2	A_3	A_4	Ranking
Hao et al. [42]	PDHFWA	0.1985	0.2135	0.2061	0.0098	$A_2 \succ A_3 \succ A_1 \succ A_4$
Park et al. [50]	HPFEWA	0.5131	0.4915	0.5243	0.3917	$A_3 \succ A_1 \succ A_2 \succ A_4$
	HPFEWG	0.4569	0.4094	0.4056	0.3723	$A_1 \succ A_2 \succ A_3 \succ A_4$
Xu and Zhou [48]	HPFWA	0.5253	0.5091	0.5445	0.3953	$A_3 \succ A_1 \succ A_2 \succ A_4$
	HPFWG	0.4457	0.3937	0.3837	0.3685	$A_1 \succ A_2 \succ A_3 \succ A_4$
	HPFOWA	0.5585	0.5215	0.6078	0.3957	$A_3 \succ A_1 \succ A_2 \succ A_4$
	HPFOWG	0.4826	0.3998	0.4385	0.3699	$A_1 \succ A_3 \succ A_2 \succ A_4$

Table 7. Characteristic comparison of the proposed approach with different methods.

Methods	Whether Consider More Than One Decision Maker	Whether Considers Probabilities	Whether Considers Non-Membership	Weights Derived By Non-Linear Approach
Hao et al. [42]	✓	✓	✓	×
Park et al. [50]	×	✓	×	×
Xu and Zhou [48]	✓	✓	×	×
Our proposed approach	✓	✓	✓	✓

8. Conclusions

In this manuscript, we have utilized the concept of PDHFS to handle the uncertainty in the data so as to capture the information with some more degree of freedom. For it, we have defined some new distance measures based on the size of two PDHFSs. Further, by focussing on the advantages of the aggregation operators into the decision-making process, we propose some series of weighted averaging and geometric aggregation operators by using Einstein norm operations. The major advantages of the proposed operators are that it considers the probability information to each dual hesitant membership degrees which give more information and help for the decision maker to take a decision more clearly. Further, since the decision makers are more sensitive to the loss and their bounded rationality, so there is a need for the probabilistic information into the analysis to solve the related MCDM problems. Also, its prominent characteristic is that it can consider the decision makers psychological behavior. The primary contribution of this paper is summarized as follows:

(1) To introduce the two new distance measures between the pairs of the PDFHEs and explore their properties. Further, some basic operational laws for this proposed structure are discussed and explore the various relationships among them using Einstein norm operations.

(2) To obtain the optimal selection in the group decision making (GDM) under the probabilistic dual hesitant fuzzy environment, we have proposed a maximum deviation method (MDM) algorithm and developed several weighted aggregation operators. In this case, the MDM method has been used to determine the optimal weight of each criterion.

(3) Four new aggregation operators, namely, the PDHFWEA, PDHFOWEA, PDHFWEG, and PDHFOWEG operators have been developed to aggregate the PDHFE information. In addition to it, on a comprehensive scrutiny of DHFSs and PDHFSs, we have devised an algorithm to formulate PDHFSs from the given probabilistic fuzzy information. Based on the decision maker preferences in order to optimize their desired goals, the person can choose the required proposed distance measures and/or aggregation operators.

(4)   Finally, the presented group decision-making approach is explained with the help of numerical example and an extensive comparative analysis has been conducted with the existing decision making theories [42,48,50] to show the advantages of the proposed approach.

Thus, we can conclude that the proposed notion about the PDHFSs is widely used in the different scenarios such as when a person provides the information about the fact that 'how much he/she sure about the uncertain information evaluated by him/her?'; in the situations, when the evaluators have no knowledge of the importance of their decision as well the considered criteria. Thus, the proposed concepts are efficaciously applicable to the situation under uncertainties and expected to have wide applications in complex DM problems. In the future, there is a scope of extending the proposed method to some different environment and its application in the various fields related to decision-theory [53–63].

Author Contributions: Conceptualization, H.G. and G.K.; Methodology, G.K.; Validation, H.G.; Formal analysis, H.G. and G.K.; Investigation, H.G. and G.K.; Writing-original draft preparation, H.G. and G.K.; Writing-review and editing, H.G.; Visualization, H.G.

Funding: This research received no external funding.

Acknowledgments: The authors wish to thank the anonymous reviewers for their valuable suggestions.

Conflicts of Interest: The authors declare no conflict of interest.

References

1.   Kaur, G.; Garg, H. Multi-Attribute Decision—Making Based on Bonferroni Mean Operators under Cubic Intuitionistic Fuzzy Set Environment. *Entropy* **2018**, *20*, 65. [CrossRef]
2.   Kaur, G.; Garg, H. Generalized cubic intuitionistic fuzzy aggregation operators using t-norm operations and their applications to group decision-making process. *Arab. J. Sci. Eng.* **2018**, 1–20. [CrossRef]
3.   Atanassov, K.T. Intuitionistic fuzzy sets. *Fuzzy Sets Syst.* **1986**, *20*, 87–96. [CrossRef]
4.   Zadeh, L.A. Fuzzy sets. *Inf. Control* **1965**, *8*, 338–353. [CrossRef]
5.   Atanassov, K.; Gargov, G. Interval-valued intuitionistic fuzzy sets. *Fuzzy Sets Syst.* **1989**, *31*, 343–349. [CrossRef]
6.   Xu, Z.S. Intuitionistic fuzzy aggregation operators. *IEEE Trans. Fuzzy Syst.* **2007**, *15*, 1179–1187.
7.   Wang, W.; Liu, X.; Qin, Y. Interval-valued intuitionistic fuzzy aggregation operators. *J. Syst. Eng. Electron.* **2012**, *23*, 574–580. [CrossRef]
8.   Garg, H. Generalized intuitionistic fuzzy interactive geometric interaction operators using Einstein t-norm and t-conorm and their application to decision making. *Comput. Ind. Eng.* **2016**, *101*, 53–69. [CrossRef]
9.   Garg, H. Novel intuitionistic fuzzy decision making method based on an improved operation laws and its application. *Eng. Appl. Artif. Intell.* **2017**, *60*, 164–174. [CrossRef]
10.   Wang, W.; Liu, X. Interval-valued intuitionistic fuzzy hybrid weighted averaging operator based on Einstein operation and its application to decision making. *J. Intell. Fuzzy Syst.* **2013**, *25*, 279–290.
11.   Wang, W.; Liu, X. The multi-attribute decision making method based on interval-valued intuitionistic fuzzy Einstein hybrid weighted geometric operator. *Comput. Math. Appl.* **2013**, *66*, 1845–1856. [CrossRef]
12.   Garg, H. A New Generalized Pythagorean Fuzzy Information Aggregation Using Einstein Operations and Its Application to Decision Making. *Int. J. Intell. Syst.* **2016**, *31*, 886–920. [CrossRef]
13.   Garg, H.; Kumar, K. An advanced study on the similarity measures of intuitionistic fuzzy sets based on the set pair analysis theory and their application in decision making. *Soft Comput.* **2018**, *22*, 4959–4970. [CrossRef]
14.   Wei, G.W.; Xu, X.R.; Deng, D.X. Interval-valued dual hesitant fuzzy linguistic geometric aggregation operators in multiple attribute decision making. *Int. J. Knowl.-Based Intell. Eng. Syst.* **2016**, *20*, 189–196. [CrossRef]
15.   Peng, X.; Dai, J.; Garg, H. Exponential operation and aggregation operator for q-rung orthopair fuzzy set and their decision-making method with a new score function. *Int. J. Intell. Syst.* **2018**, *33*, 2255–2282. [CrossRef]
16.   Garg, H. Some robust improved geometric aggregation operators under interval-valued intuitionistic fuzzy environment for multi-criteria decision -making process. *J. Ind. Manag. Optim.* **2018**, *14*, 283–308. [CrossRef]

17. Liu, P. Some Frank Aggregation Operators for Interval-valued Intuitionistic Fuzzy Numbers and their Application to Group Decision Making. *J. Mult.-Valued Log. Soft Comput.* **2017**, *29*, 183–223.

18. Chen, S.M.; Cheng, S.H.; Tsai, W.H. Multiple attribute group decision making based on interval-valued intuitionistic fuzzy aggregation operators and transformation techniques of interval-valued intuitionistic fuzzy values. *Inf. Sci.* **2016**, *367–368*, 418–442. [CrossRef]

19. Garg, H. Some arithmetic operations on the generalized sigmoidal fuzzy numbers and its application. *Granul. Comput.* **2018**, *3*, 9–25. [CrossRef]

20. Chen, S.M.; Cheng, S.H.; Chiou, C.H. Fuzzy multiattribute group decision making based on intuitionistic fuzzy sets and evidential reasoning methodology. *Inf. Fus.* **2016**, *27*, 215 –227. [CrossRef]

21. Garg, H. A new generalized improved score function of interval-valued intuitionistic fuzzy sets and applications in expert systems. *Appl. Soft Comput.* **2016**, *38*, 988–999. [CrossRef]

22. Wei, G.W. Interval Valued Hesitant Fuzzy Uncertain Linguistic Aggregation Operators in Multiple Attribute Decision Making. *Int. J. Mach. Learn. Cybern.* **2016**, *7*, 1093– 1114. [CrossRef]

23. Kumar, K.; Garg, H. Connection number of set pair analysis based TOPSIS method on intuitionistic fuzzy sets and their application to decision making. *Appl. Intell.* **2018**, *48*, 2112–2119. [CrossRef]

24. Kumar, K.; Garg, H. Prioritized Linguistic Interval-Valued Aggregation Operators and Their Applications in Group Decision-Making Problems. *Mathematics* **2018**, *6*, 209. [CrossRef]

25. Torra, V. Hesitant fuzzy sets. *Int. J. Intell. Syst.* **2010**, *25*, 529–539. [CrossRef]

26. Zhu, B.; Xu, Z.; Xia, M. Dual Hesitant Fuzzy Sets. *J. Appl. Math.* **2012**, *2012*, 13. [CrossRef]

27. Xia, M.; Xu, Z.S. Hesitant fuzzy information aggregation in decision-making. *Int. J. Approx. Reason.* **2011**, *52*, 395–407. [CrossRef]

28. Garg, H.; Arora, R. Dual hesitant fuzzy soft aggregation operators and their application in decision making. *Cognit. Comput.* **2018**, *10*, 769–789. [CrossRef]

29. Wei, G.; Zhao, X. Induced hesitant interval-valued fuzzy Einstein aggregation operators and their application to multiple attribute decision making. *J. Intell. Fuzzy Syst.* **2013**, *24*, 789–803.

30. Meng, F.; Chen, X. Correlation Coefficients of Hesitant Fuzzy Sets and Their Application Based on Fuzzy Measures. *Cognit. Comput.* **2015**, *7*, 445–463. [CrossRef]

31. Garg, H. Hesitant Pythagorean fuzzy Maclaurin symmetric mean operators and its applications to multiattribute decision making process. *Int. J. Intell. Syst.* **2018**, 1–26. [CrossRef]

32. Zhao, N.; Xu, Z.; Liu, F. Group decision making with dual hesitant fuzzy preference relations. *Cognit. Comput.* **2016**, *8*, 1119–1143. [CrossRef]

33. Farhadinia, B.; Xu, Z. Distance and aggregation-based methodologies for hesitant fuzzy decision making. *Cognit. Comput.* **2017**, *9*, 81–94. [CrossRef]

34. Arora, R.; Garg, H. A robust correlation coefficient measure of dual hesitant fuzzy soft sets and their application in decision making. *Eng. Appl. Artif. Intell.* **2018**, *72*, 80–92. [CrossRef]

35. Garg, H.; Arora, R. Distance and similarity measures for Dual hesitant fuzzy soft sets and their applications in multi criteria decision-making problem. *Int. J. Uncertain. Quantif.* **2017**, *7*, 229–248. [CrossRef]

36. Wei, G.W.; Alsaadi, F.E.; Hayat, T.; Alsaedi, A. Hesitant fuzzy linguistic arithmetic aggregation operators in multiple attribute decision making. *Irani. J. Fuzzy Syst.* **2016**, *13*, 1–16.

37. Garg, H. Hesitant Pythagorean fuzzy sets and their aggregation operators in multiple attribute decision making. *Int. J. Uncertain. Quantif.* **2018**, *8*, 267–289. [CrossRef]

38. Garg, H; Kumar, K. Group decision making approach based on possibility degree measure under linguistic interval-valued intuitionistic fuzzy set environment *J. Ind. Manag. Optim.* **2018**, 1–23. [CrossRef]

39. Zhu, B.; Xu, Z.S. Probability-hesitant fuzzy sets and the representation of preference relations. *Technol. Econ. Dev. Econ.* **2018**, *24*, 1029–1040. [CrossRef]

40. Wu, W.; Li, Y.; Ni, Z.; Jin, F.; Zhu, X. Probabilistic Interval-Valued Hesitant Fuzzy Information Aggregation Operators and Their Application to Multi-Attribute Decision Making. *Algorithms* **2018**, *11*, 120. [CrossRef]

41. Zhang, S.; Xu, Z.; Wu, H. Fusions and preference relations based on probabilistic interval-valued hesitant fuzzy information in group decision making. *Soft Comput.* **2018**, 1–16. [CrossRef]

42. Hao, Z.; Xu, Z.; Zhao, H.; Su, Z. Probabilistic dual hesitant fuzzy set and its application in risk evaluation. *Knowl.-Based Syst.* **2017**, *127*, 16–28. [CrossRef]

43. Li, J.; Wang, J.Q.; Hu, J.H. Multi-criteria decision-making method based on dominance degree and BWM with probabilistic hesitant fuzzy information. *Int. J. Mach. Learn. Cybern.* **2018**, 1–15. [CrossRef]

44. Li, J.; Wang, J.Q. Multi-criteria outranking methods with hesitant probabilistic fuzzy sets. *Cognit. Comput.* **2017**, *9*, 611–625. [CrossRef]

45. Lin, M.; Xu, Z. Probabilistic Linguistic Distance Measures and Their Applications in Multi-criteria Group Decision Making. In *Soft Computing Applications for Group Decision-Making and Consensus Modeling*; Springer: Berlin, Germany, 2018; pp. 411–440.

46. Xu, Z.; He, Y.; Wang, X. An overview of probabilistic-based expressions for qualitative decision-making: Techniques, comparisons and developments. *Int. J. Mach. Learn. Cybern.* **2018**, 1–16. [CrossRef]

47. Song, C.; Xu, Z.; Zhao, H. A Novel Comparison of Probabilistic Hesitant Fuzzy Elements in Multi-Criteria Decision Making. *Symmetry* **2018**, *10*, 177. [CrossRef]

48. Xu, Z.; Zhou, W. Consensus building with a group of decision makers under the hesitant probabilistic fuzzy environment. *Fuzzy Optim. Decis. Mak.* **2017**, *16*, 481–503. [CrossRef]

49. Zhou, W.; Xu, Z. Probability calculation and element optimization of probabilistic hesitant fuzzy preference relations based on expected consistency. *IEEE Trans. Fuzzy Syst.* **2018**, *26*, 1367–1378. [CrossRef]

50. Park, J.; Park, Y.; Son, M. Hesitant Probabilistic Fuzzy Information Aggregation Using Einstein Operations. *Information* **2018**, *9*, 226. [CrossRef]

51. Wang, Z.X.; Li, J. Correlation coefficients of probabilistic hesitant fuzzy elements and their applications to evaluation of the alternatives. *Symmetry* **2017**, *9*, 259. [CrossRef]

52. Zhou, W.; Xu, Z. Group consistency and group decision making under uncertain probabilistic hesitant fuzzy preference environment. *Inf. Sci.* **2017**, *414*, 276–288. [CrossRef]

53. Garg, H. Linguistic Pythagorean fuzzy sets and its applications in multiattribute decision-making process. *Int. J. Intell. Syst.* **2018**, *33*, 1234–1263. [CrossRef]

54. Garg, H. New Logarithmic operational laws and their aggregation operators for Pythagorean fuzzy set and their applications. *Int. J. Intell. Syst.* **2019**, *34*, 82–106. [CrossRef]

55. Garg, H.; Arora, R. Generalized and Group-based Generalized intuitionistic fuzzy soft sets with applications in decision-making. *Appl. Intell.* **2018**, *48*, 343–356. [CrossRef]

56. Garg, H.; Nancy. New Logarithmic operational laws and their applications to multiattribute decision making for single-valued neutrosophic numbers. *Cognit. Syst. Res.* **2018**, *52*, 931–946. [CrossRef]

57. Rani, D.; Garg, H. Complex intuitionistic fuzzy power aggregation operators and their applications in multi-criteria decision-making. *Expert Syst.* **2018**, e12325. [CrossRef]

58. Garg, H.; Rani, D. Complex Interval- valued Intuitionistic Fuzzy Sets and their Aggregation Operators. *Fund. Inf.* **2019**, *164*, 61–101.

59. Liu, X.; Kim, H.; Feng, F.; Alcantud, J. Centroid Transformations of Intuitionistic Fuzzy Values Based on Aggregation Operators. *Mathematics* **2018**, *6*. [CrossRef]

60. Wang, J.; Wei, G.; Gao, H. Approaches to Multiple Attribute Decision Making with Interval-Valued 2-Tuple Linguistic Pythagorean Fuzzy Information. *Mathematics* **2018**, *6*. [CrossRef]

61. Garg, H.; Kaur, J. A Novel (R, S)-Norm Entropy Measure of Intuitionistic Fuzzy Sets and Its Applications in Multi-Attribute Decision-Making. *Mathematics* **2018**, *6*. [CrossRef]

62. Joshi, D.K.; Beg, I.; Kumar, S. Hesitant Probabilistic Fuzzy Linguistic Sets with Applications in Multi-Criteria Group Decision Making Problems. *Mathematics* **2018**, *6*. [CrossRef]

63. Garg, H.; Nancy. Linguistic single-valued neutrosophic prioritized aggregation operators and their applications to multiple-attribute group decision-making. *J. Ambient Intell. Hum. Comput.* **2018**, *9*, 1975–1997. [CrossRef]

 *mathematics*

Article

An Application of Multicriteria Decision-making for the Evaluation of Alternative Monorail Routes

Mustafa Hamurcu and Tamer Eren *

Department of Industrial Engineering, Faculty of Engineering, Kirikkale University, 71451 Kirikkale, Turkey; hamurcu.mustafa.55@gmail.com
* Correspondence: tamereren@gmail.com; Tel.: +90-3183-574-242

Received: 13 November 2018; Accepted: 20 December 2018; Published: 24 December 2018

Abstract: Urban transportation planning is important for a metropolitan city. Route selection, which is among the decisions of urban transportation planning, is also important in terms of developing the urban transportation. This study contains the route selection for the planned monorail transport system that is a new system in Ankara. The most suitable monorail route was selected among the determined eight alternative monorail routes. In this decision process, we used the Analytic Network Process (ANP) and Technique for Order Preference by Similarity to Ideal Solution (TOPSIS) method, which is one of the multi-criteria decision-making methods. Finally, we provided the most suitable ranking and planning with the selection process for the development of urban transportation.

Keywords: Monorail; urban transportation planning; Analytic Network Process; Technique for Order Preference by Similarity to Ideal Solution

1. Introduction

Transport systems are complex socio-technical systems that affect the social, economic, and environmental dimensions of a community [1]. In this context, transport planning is typically a decision-making process that based on rationality, aimed at defining and implementing transport system operations [2]. Transportation planning is now a fundamental support to a rational and sustainable development of the territorial system due to the increase of environmental issues and constraints, the worldwide financial crisis, and the numerous interactions of the transportation system with the social and economic contexts [3]. Strategic planning involves decisions on long-term nearly 10–20 years, capital investment programs for the realization of new infrastructures such as roads, railways, and ports, and the acquisition of vehicles and technologies [4]. So, the route selection problem is important in the metropolitan city for urban transportation planning processes, involving decisions on a medium- or long-term basis.

On the other hand, transportation development plays an essential role in a society's economy and has long-lasting effects on the financial, social, and political life of individuals and the community. It is essential to develop a transportation network that best suits the public's needs, to build a contemporary city [5]. Public transportation is one of the most important systems in transportation, especially in metropolis cities. So, evaluation of public transportation systems is a strategic decision-making problem for urban area [6]. At the same time, public transport is an essential element of urban life since it reduces car traffic and gives mobility to city residents. In addition, more use of public transport reduces emissions such as carbon dioxide. This feature has become more important due to the Kyoto Protocol came into effect [7]. It is important to consider the multifactorial evaluation of transportation projects due to these reasons.

The assessment of projects, meant here as capital investments that create transport infrastructure, supports the activity of decision makers. The assessment is deal with achieving social objectives, such as improvement of economic efficiency, reduction of the damage on the environment, improvement of safety. In the case of public decision makers, the assessment is used as a tool to assist the process of planning transport infrastructure. Multicriteria analysis are widely used due to the simplicity it's in taking into account nonmarketable effects and qualitative criteria for these aims [8]. Multicriteria decision-making methods (MCDM) are widely used in transport planning to include in a comparative assessment of alternative projects their contributions to different evaluation criteria [9]. MCDM has gained importance as an evaluation method for transport projects and use of these methods increase day by day to evaluating transport projects such as passenger and freight transport, infrastructure investments, location decisions, etc. [10,11]. MCDM methodologies are rapidly growing in the various transportation problems [12–14]. At the same time, there are some studies using multi-objective optimization about transportation subjects [15] and solving multicriteria transportation-location problems [16]. Besides, these methods also have been applied in various area such as supply chain management and supply chain performance measurement [17,18].

Route selection is one of the most important activity for the planning of the urban traffic that needs MCDM process. Because constructing a new structure or installing new systems are big investments and require large budget, there should be good planning. Briefly, route selection is a process in which selection or ranking are carried out among the alternative routes. At the same time, the route selection is named by some names such as "investment project selection" [19], "project selection" [20], "transportation planning" [21], "infrastructure projects selection" [22] or "corridor selection" [23]. The aim of route selection is to provide maximum benefit for traffic and the developing urban transportation. So, it will provide livable urban environment and city center. These investments need big resources such as large budget. Planned investments should be addressed in a wide range by the executives. Otherwise, it will be inevitable that the investment will become a waste. Therefore, this process is dependent on lot of criteria such as social effect, environmental effect, cost, demand level etc.

The monorail, which is one of the rail system investments, is also one of the major investment projects. The monorail is one of the urban public transportation systems that acts on its own line. This new system for Turkey is used in various countries such as Japan and China. But Turkey does not have this technology yet. However, this system is planned for various cities in Turkey and studies on this subject are still ongoing. Monorail has a lot of advantages such as to be independent of vehicle traffic, to be safe, to be fast, to be comfortable, to use low area, to be environmentally friendly and to have its own road among the other rail systems. Therefore, monorails have been becoming common day by day in the urban transportation worldwide. In terms of environment, it is environmentally friendly because of quietness and usage of energy. This system is alternative to the other rail systems and public mass transportation vehicles due to all these reasons. But it also has some negative aspects, such as high initial investment costs and the electric is not free. This system has high visual impact. This situation can be developed with high construction cost. It is important to select this technology, but the planning process is the most important of all. Therefore, selecting the best or the most suitable route is needed as the first step. This process is difficult due to the effects of many factors.

Selecting a route and a new system are complex problems which involve and effect the development of urban areas, use of land, future of the city or various other criteria and sub-criteria. There are various transportation types used for urban transportation such as bus, metro, private vehicle, taxi, subway, tramway and monorail etc. Monorail has been being applied in the European countries, the USA, in Asia (especially Japan and Chine) and Middle East countries such as Saudi Arabia and United Arab Emirates (UAE).

There are various studies about monorail in the literature. Kuwabara et al. [24] mentioned that monorail is an effective vehicle for urban transportation due to the short construction time and low-cost advantages. Wang [25] also talked about the short construction processes of monorail projects, the cost and the quality of the transportation. Kato et al. [26] talked about the advantages of a saddle-type monorail system and pointed out that in the coming years, driverless monorail systems would be used more and more and system costs would be even lower. With simulation application, Sadatugu et al. [27] talked about alternative policies and scenarios for monorail. Sekitani et al. [28] mentioned a thrust-type monorail system for the solution of rugged roads, traffic congestion and air pollution, and they also described the technical characteristics of the line. Considering the rapid transportation of monorail, Kennedy [29] defined and mentioned their types and features. Kimijima et al. [30] gave information about the monorail by mentioning its active use in the place where the monorail was installed. Ghafooripour et al. [31] examined the countries with metro and monorail applications for developing countries and evaluated them in terms of cost-effectiveness. By evaluating its effectiveness in terms of user satisfaction, Das et al. [32] offered suggestions for the monorail transportation systems. Marathe and Hajian [33] pointed out that the monorail was ideal for the use in urban transportation in terms of economy, security and environmental sensitivity. Parekh et al. [34] discussed the features of the monorails which are popular in urban areas. Liu et al. [35] compared the conventional rail transport systems with the monorail system and discussed the advantages and disadvantages of monorail systems. Hussien [36] made a comparison between the monorail system and other public transportation vehicles. Li et al. [37] made a technical feasibility of suspended monorail type by analyzing the urban adaptation, capacity, specifications and construction costs. Timan [38] emphasized that monorail systems would be a suitable solution for the traffic problems in metropolitan cities. In his study, He [39] mentioned about the features of straddle-type monorail and noted the increase in its popularity day by day.

In the literature, related to this subject, there are a lot of studies focusing on route planning, route selection, local selection, station site selection, project selection and transportation planning. These studies have been carried out in various area and they examined different vehicle types. At the same time, authors of this research have conducted some studies related to this subject and they have contributed the literature with those studies. Hamurcu and Eren [40] proposed the monorail mass transportation for Turkey as first. Hamurcu and Eren [41] used multicriteria decision-making methods for monorail route selection in Ankara. Hamurcu et al. [42] used analytic hierarchy process (AHP) and 0-1 goal programming (GP) in the monorail project selection under the capacity constraints. Gür et al. [20] carried out monorail project selection for different route alternatives by using AHP and goal programming methods. Hamurcu and Eren [43], in their conference paper, used Analytic Network Process (ANP) and Similarity to Ideal Solution (TOPSIS) in order to carry out the monorail route selection in Ankara. Besides, selection of monorail technology [44], rail system projects selection in Istanbul [45], prioritization of high-speed rail projects [46], transportation planning [21] and decision-making for rail systems projects with MCDM and GP [47] are some of the studies of the authors of this article. So, multicriteria decision-making methods are today widely used in transport project studies commissioned by public bodies and city ad transportation planners.

Decision-making processes in transportation can be grouped different topic in terms of subject. Some of them and study areas are route planning for tramway [48], high-speed rail [49,50], railway [51], for highway [52]; route selection for light rail system [53–55], metro line [56,57], and bicycle [14]; location selection for metro [58]; station location selection for rail system network [59]; project selection for rail systems network [11]; transportation planning for transport network [60–65].

These studies show that transportation planning decisions are very important processes for planners and managers, are need analytic methods. Transportation planning is the process of identifying and incorporating stakeholder concerns, needs and values in the transport decision-making process. MCDM makes it possible to incorporate, account and quantify human opinion and preferences; solve decision problems taking into account tangible and intangible aspects; provide a methodology to calibrate the numeric scale for the measurement of quantitative as well as qualitative performances. In this study, using the analytic network process and TOPSIS from MCDMs, the challenges faced by planners in route design these decision processes were eliminated in this study. Use of ANP and TOPSIS hybrid from multi-criteria decision-making methods which are effective in terms of analysis, selection and ranking, are effective tools for quantitatively considering qualitative concepts.

In this study, we focus on the selection of monorail route. Sections of this study are as follows: In Section 2, research methodology is shown. In Section 3, the multi-criteria decision-making methods used in this study are explained. In Section 4, application of the route selection in Ankara is presented. Finally, the ranking of the best route selection is shown in Section 5.

2. Materials and Methods

Multicriteria decision analysis has seen frequently used the last several decades. Its application in different areas has increased significantly, especially as new methods develop and as old methods improve.it has allowed for more complex decision analysis methods with technology advancement over the past couple of decades to be developed in addition to applying single MCDM methods to real-world decisions. This process with hybrid multi-criteria decision-making methods and their application has provided a whole new approach to decision analysis [66].

In this study, the research was carried out on eight monorail routes in Ankara, the capital city of Turkey. This study involved two methods related to the multi-criteria decision-making. These methods were ANP and TOPSIS which were used for the determination of the criteria and alternative routes for urban public transportation in Ankara. Ankara hasn't got a monorail technology for urban transportation yet. Considered monorail projects were selected from expert opinions for urban transport planning. In the implementation of this research, there were four main parts (Figure 1), which were;

- Identification of the goal and criteria
- Use of the multi-criteria decision-making methods
- Determination of alternative eight route
- Selection of the best route and evaluation at the end of the selection

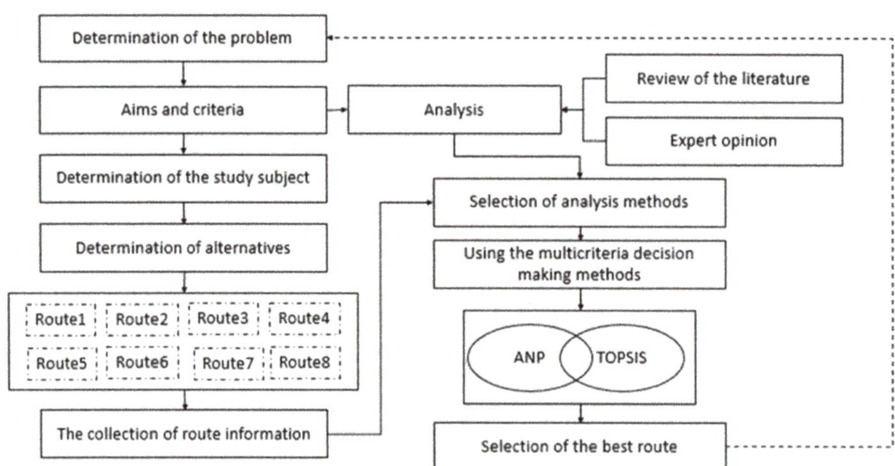

Figure 1. Research methodology.

This process was used in order to select the best monorail route. The alternative route characteristics were taken from Ankara metropolitan municipality and the criteria were determined by expert opinions and literature research.

The contribution of this study to the state-of-the-art can be summarized as follows: This study presents new example application and proposes a comprehensive multicriteria decision-making model for the route selection problem, which accounts for the criterion components reported in the literature. The proposed work is one of the first few works to investigate application of ANP and TOPSIS for evaluation of monorail projects under various criteria. Besides, this new system for Ankara will be first. Thus, selection of the best alternative route by using MCDM play an important role for sustainability and public transportation in Ankara metropolitan area.

2.1. Analytic Network Process (ANP)

There ANP can improve communication and resolve conflicts, help diffusion of responsibility, and assist decision makers in understanding other members' viewpoints. These characteristics are attractive when a good decision calls for actions that may not be well-liked, such as outsourcing. The ANP can evaluate a wide range of criteria including tangible and intangible factors related to the outcome. Because ANP allows for complex interactions and influences among the various components of the decision problem, it can be seen as a better choice for studying more complex decision problems [67]. ANP brings all the decision objectives, criteria, alternatives and actors (such as decision makers, stakeholder, and influencers) into a single unified framework, and it facilitates interaction and feedback of elements (alternatives, criteria and actors) within groups (inner dependence) and between groups (outer dependence) [68]. Briefly, ANP is more concerned with network structure. In terms of advantages, it allows for dependence and includes independence and has the ability to prioritize groups or clusters of elements. Besides, it can support complex, networked decision-making with various intangible criteria [69]. ANP is often utilized in project selection, product planning, green supply chain management, and optimal scheduling problems and also transportation.

Most of the complex real-world decision problems have numerous inter-dependent elements that can be captured and processed by utilizing the feedback and interaction capabilities of an ANP model. In this regard, the ANP method was used directly or indirectly by Lee and Kim [70] in the information system project, by Meade and Presley [71] for the selection of research and development projects, by Ravi et al. [72] for the selection of the reverse logistics projects, by Büyüközkan and Öztürkcan [73] in six sigma projects selection, by Wey and Wu [74] for the selection of projects among transportation systems, by Begičević et al. [75] for the selection of projects at higher education institutions, by El-Abbasy et al. [76] for the selection of highway projects and by Tuzkaya and Yolver [77] for the selection of research and development projects.

To derive the global priorities of the criteria by using ANP, it is necessary first to carry out the pairwise comparison of the criteria with respect to the node representing their category and to all other criteria with which they interact or on which they have effect. Next, the principal right eigenvector of each comparison matrix is computed to obtain the local priority of every criterion [78]. In the last step, a super-matrix consisting of all the local-limiting matrices is formed for overall criteria prioritization and alternative ranking. The weighted super matrix is taken to the limit for the results.

2.2. TOPSIS

This technique, developed by Hwang and Yoon [79], is based on the selection of the shortest distance from the positive ideal solution and the longest distance alternative from the negative ideal solution. The positive-ideal solution is the best possible combination of the criteria. The negative ideal solution consists of the worst criterion values that can be reached. The only assumption in this method is the assumption that each measure is either a monotone increasing or monotonously decreasing one-way benefit. The steps of the TOPSIS method will be shown on the handling problem.

TOPSIS is an approach for identifying an alternative that is closest to the ideal solution and farthest to the negative ideal solution [80]. It has numerous advantages such as a simple process and easy to use and programmable. The number of steps remains the same regardless of the number of alternatives and criteria [81]. This method has a wide range of application areas such as multi-criteria inventory planning [82], freight transport selection [83], selection of the scholarship with the AHP [84], selection of the service providers [85], performance evaluation [86], personnel selection [87], reverse logistics supplier selection [88].

In addition, ANP and TOPSIS methods have been used together in some studies. Ersoz and his colleagues determined the weights of the criteria that were effective in the course selection of graduate students by the ANP method and alternative courses were ranked by using the TOPSIS method [89]. ANP and TOPSIS methods have been used together to evaluate the supplier's selection process [90] and to rank strategies in the mining industry [91].

3. Using the ANP and TOPSIS Approach for Route Selection

Firstly, the criteria and alternatives were identified for selection. Then, the interdependence between criteria, sub-criteria and alternatives was determined. The pairwise comparisons were carried out between these criteria and sub-criteria by using Super Decision program. The pairwise comparisons were made by expert opinions. By this way, weights of the criteria were found for TOPSIS method. In the other step, TOPSIS method were applied by using ANP weights. Then the negative and positive ideal solution and separation were calculated in the TOPSIS steps. At the end of the solution process, the best ranking was created among the alternative routes. This process:

Step_1. Identify criteria and alternatives
Step_2. Determination interdependence relationship between criteria and sub-criteria. Then finding the criteria weight with ANP.
Step_3. Using the ANP weights for TOPSIS method
Step_4. Calculate the negative-positive ideal solution and separation.
Step_5. The best ranking for among the alternative route

4. An Application in ANKARA

In this study, a route selection was applied for Ankara. Monorail is a new urban mass transportation system. It will be the first example of this system in urban transportation in Turkey with its implementation in Ankara. Ankara is a region covered with plains formed by confined the Kızılırmak and Sakarya rivers in the north-western part of Central Anatolia. The population of Ankara is 5,045,083 according to the results of the 2013 census using Address-Based Population Registration System. The largest districts of Ankara in terms of population are Cankaya, Keciören, Yenimahalle, Mamak, Sincan, Etimesgut, Altındag, Pursaklar and Polatli. The largest district in terms of surface area is Polatli. The main determinant of Ankara's socio-economic structure is the fact that the city of Ankara is the administrative center of the country at the same time. For this reason, the public service sector has an important place in Ankara's economic life. Economical, technological and political developments have initiated the population migration to Ankara from other settlements.

Due to the increasing population and migration, public transportation systems have to be used for urban transportation in Ankara. Public transportation services are provided municipal buses, private buses, minibuses, subways and suburban in this city. Efforts are continuing to establish the monorail system in Ankara.

4.1. Determination of the Alternatives

Ankara is a big city and its population density is very high. For this reason, it has traffic problems. Therefore, municipal administrators have been producing projects for the solution of traffic problems. The first of their projects is urban mass transportation projects. Therefore, monorail technology, one of the types of public transportation, was considered. And 8 alternative routes were identified within the scope of this study.

Table 1 shows characteristic of the routes in terms of distance, number of stations, number of vehicles, number of series, total number of vehicles and approximate total cost of the routes. In this study eight monorail routes were used to determine the best route. These routes and their pictures are shown in Figure 2.

Table 1. Characteristic features of the routes.

Route	Distance (m)	Number of Stations	Number of Vehicles	Number of Series	Total Number of Vehicles	Approximate Total Cost ($)
Route_1	11140	11	4	20	80	412,180,000
Route_2	5020	5	4	10	40	185,740,000
Route_3	8076	8	4	15	60	298,812,000
Route_4	7763	7	4	15	60	287,231,000
Route_5	11596	10	4	22	88	429,052,000
Route_6	11426	10	4	22	88	422,762,000
Route_7	4069	4	4	8	32	150,553,000
Route_8	19168	18	4	36	144	709,216,000

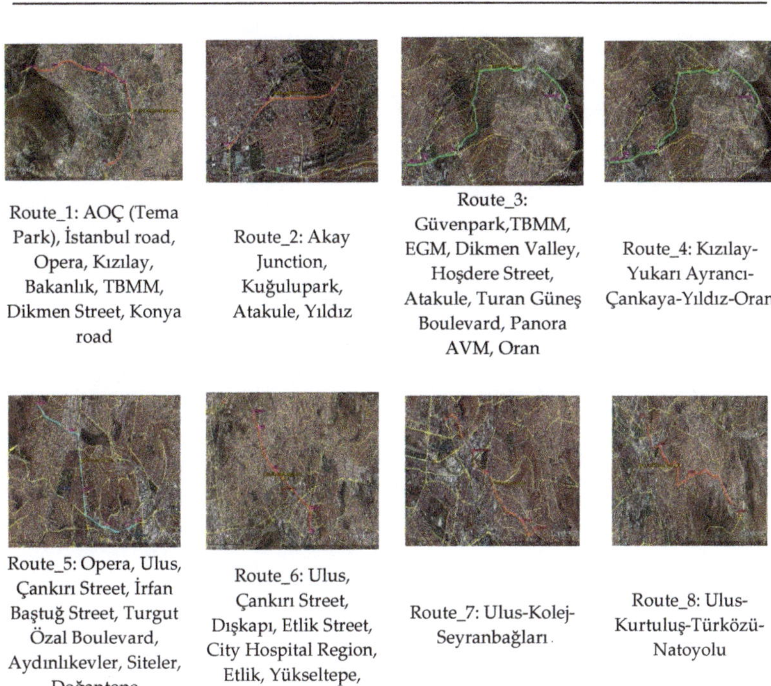

Figure 2. Alternative routes and their pictures in the map.

4.2. Determination of the Criteria

Criteria and sub-criteria were determined by taking the expert opinions and as a result of the literature review. Some of the experts were personnel of Ankara Metropolitan Municipality and they were working in urban planning and traffic planning sections. Other experts were academicians studying in the related field. The determined criteria, sub-criteria and their explanations are shown in Table 2.

Economic: Refers to the use of monetary resources. This criterion deal with construction costs, infrastructure investment, fuel costs. Social impact: This type of criteria refers to both benefits and negative impacts on society because of decisions made regarding the transport system such as the access to shopping-employment-resistant. In addition, the criterion deal with mobility, population density and visual impact for urban area. Engineering: These criteria are related with issues technical of transportation planning such as travel time, demand, accessibility, traffic capacity, ability to develop and to improve, the integration of transport. Environmental impact: This set of criteria is associated to the impacts on the natural environment and historical-cultural area. In this category, we find the sensitive areas and use of land. The attribute of criteria K14 influences criteria K15, the attribute of criteria K9 influences criteria K10, K11, K12 and K13, and criteria K3 influences criteria K4, K5, K7, K1, K2, K14, K15 and sub-criteria of engineering.

4.3. Determination of the Weights of the Criteria by ANP Technique

One of the most important parts is to determine the criteria and measuring indicators in decision-making models. To be determined criteria and their interdependence for this purpose that the important aspects and characteristics of alternatives being measured. Therefore, the design for decision-making model has a direct impact on model efficiency. The criteria and sub-criteria affecting on selection processes differ based on objectives, in this study, we used expert opinion (academic and engineer planners) in order to identify criteria, with regard to municipality strategic goals. For evaluation of the monorail projects, we need quantitative data on environmental impact, engineering, economic and social impact main criteria. Since these projects are new and implemented for the first time in Ankara context, there is very limited quantitative data available, thereby making the evaluation process difficult. At the same time, these projects have yet been considered and have in the process of being planned. To address this situation, a decision-making committee comprising of subject matter experts (4 academic researchers from industrial engineering and civil engineering, 2 transportation experts as rail system planner and transportation planner from Ankara Metropolitan Municipality) made qualitative ratings by using Saaty's 1–9 importance scale for assessing the alternatives and the criteria. In the TOPSIS method, the criterion values of the alternatives were found by using ANP with this scale according to expert opinions.

In this research, to be able to identify the relationship and degree of interdependency among the criteria, opinions of the experts from academia and from metropolitan municipality staff were consulted. Those experts were working and studying in the urban planning and traffic planning area. The relationship having interdependence among the four essential criteria and fifteen sub-criteria taken in this research is shown in Figure 3. There is an interdependence relationship among these criteria in the route selection problem. For example, population density criterion would result in an increase in public mobility and increase the demand level for the selection of alternative routes. All the criteria are linearly related to each other under the engineering criterion. And these criteria also related to the other essential 3 criteria. Likewise, sensitive areas increase the construction cost and these areas affect the current situation such as traffic capacity, ability to develop and to improve, the total travel time, the integration of transport. Thus, there is an interdependency among these criteria and sub-criteria, economy, environmental impact, social impact and engineering.

Comparing the structure of AHP hierarchy, there is the same level relationship among factors in the solution of ANP. At the same time, "Super Decision" program was used in this study.

In order to create pairwise comparisons in the direction of determining the relationships between criteria and alternatives and present them to the user, this program was used. In Table 3, the pairwise comparison of sub-criteria under the engineering factor is shown. This process was carried out for other criteria and sub-criteria. The interrelated criteria is made with ANP using 1–9 Saaty'scale to compare two alterative with respect to attribute.

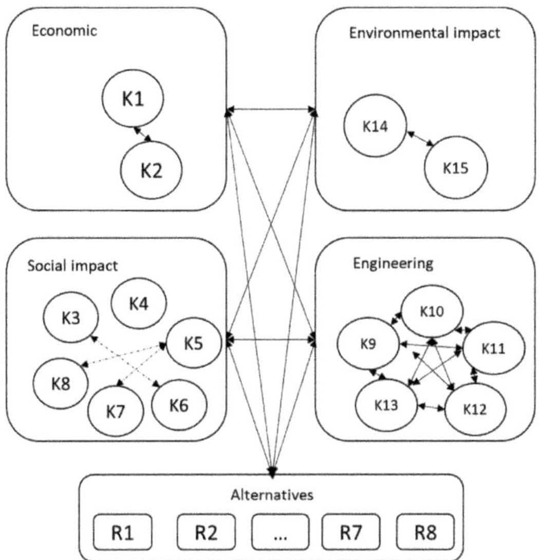

Figure 3. Strategic decision framework.

Table 2. Criteria for the route selection.

Criteria	Sub-Criteria	Explanation
Economic	(K1) Construction cost	Refers to the use of monetary resources such as implementation costs, infrastructure investments, operational costs, maintenance costs, infrastructure and others
	(K2) Expropriation	
Social impact	(K3) Access to shopping and residential areas	This type of criteria refers to the effects on the society because of decisions made regarding the transportation system.
	(K4) Access to employment and education	
	(K5) Aesthetic and visual impact	
	(K6) Population density	
	(K7) Public mobility	
	(K8) Accessibility	
Engineering	(K9) Ability to develop and to improve	These criteria are related to the issues regarding technical requirements and prudential situations of the city. They were not explicitly computed like monetary functions and they are possible predictions about the city.
	(K10) The total travel time	
	(K11) The integration of transport	
	(K12) Traffic capacity	
	(K13) Demand level	
Environmental impact	(K14) Sensitive area	This set of criteria is associated to the impact(s) on the natural environment and sensitive areas such as cultural and historical or other.
	(K15) Land structure	

This decision process is done for every pair of among the each other as shown in Table 3. A basic questionnaire has been prepared and feedback has been taken from academics and planner experts to find out the relative importance of the selected criteria. The pairwise comparison for population density is also shown in this table.

Table 3. Comparisons between population density and social impact.

	Comparisons wrt "Population density" Node in "Social impact" Cluster																	
1	Access to employment and educ …	9	8	7	6	5	4	3	2	**1**	2	3	4	5	6	7	8	9	Access to shopping and res …
2	Access to employment and educ …	9	8	7	6	5	4	3	2	1	**2**	3	4	5	6	7	8	9	Accessibility
3	Access to employment and educ …	9	8	7	6	5	4	3	2	**1**	2	3	4	5	6	7	8	9	Public mobility
4	Access to shopping and residen …	9	8	7	6	5	4	3	2	1	2	3	**4**	5	6	7	8	9	Accessibility
5	Access to shopping and residen …	9	8	7	6	5	4	3	2	**1**	2	3	4	5	6	7	8	9	Public mobility
6	Accessibility	9	8	7	6	5	4	3	**2**	1	2	3	4	5	6	7	8	9	Public mobility

Then, for each criterion, comparisons were carried out with each alternative. Prioritization of the weight of the criteria is ranked in Table 4. In the table, weights of the criteria found with ANP are seen. The criteria having the highest weight values are construction cost, sensitive area, land structure, population density, and ability to develop and to improve, respectively.

Table 4. Weights of the criteria.

Name of the Criteria	Grap	Normalized by Cluster	Limiting
(K1) Construction cost		0.89436	0.06841
(K2) Expropriation		0.10564	0.00808
(K3) Access to shopping and residential areas		0.06914	0.02289
(K4) Access to employment and education		0.15248	0.05056
(K5) Aesthetic and visual impact		0.15663	0.05194
(K6) Population density		0.3877	0.12855
(K7) Public mobility		0.12976	0.04303
(K8) Accessibility		0.10438	0.03462
(K9) Ability to develop and to improve		0.37311	0.10872
(K10) The total travel time		0.07499	0.02185
(K11) The integration of transport		0.27847	0.08114
(K12) Traffic capacity		0.04108	0.01197
(K13) Demand level		0.23235	0.0677
(K14) Sensitive area		0.53773	0.16162
(K15) Land structure		0.46227	0.13894

4.4. Ranking Monorail Route Alternatives by Using TOPSIS

The In this step, TOPSIS technique played role for ranking the routes. The weights were obtained by the ANP technique using Equations (1) and (2). Table 5 shows the normalized weighted matrix by using Equation (1).

Table 5. Normalized weighted matrix.

Alternatives	K1	K2	K3	K4	...	K12	K13	K14	K15
R1	0.014	0.184	0.224	0.426	...	0.031	0.331	0.26	0.042
R2	0.363	0.092	0.107	0.018	...	0.155	0.03	0.127	0.153
R3	0.037	0.148	0.216	0.056	...	0.144	0.11	0.132	0.141
R4	0.036	0.101	0.228	0.11	...	0.132	0.157	0.17	0.134
R5	0.109	0.139	0.072	0.059	...	0.119	0.071	0.066	0.139
R6	0.126	0.102	0.059	0.144	...	0.188	0.114	0.066	0.149
R7	0.252	0.121	0.042	0.114	...	0.091	0.046	0.068	0.155
R8	0.041	0.114	0.053	0.073	...	0.142	0.141	0.113	0.086

TOPSIS method was applied by using weights of the criteria that are results of the ANP method in the ANP-TOPSIS combine model. The used standard decision matrix in TOPSIS is found at the end of the comparisons of the alternatives with each criterion.

The weights of the evaluation criteria using ANP is shown in Table 6. In addition, the weighted normalized decision matrix by TOPSIS is shown in Table 7. The vector normalization technique is used for computing the element (a_{ij}) of the normalized decision matrix, which is given as:

$$a_{ij} = \frac{r_{ij}}{\sqrt{\sum_{i=1}^{m} r_{ij}}} \tag{1}$$

The weighted normalized decision matrix can be calculated by multiplying each row (r_{ij}) of the normalized decision matrix with its associated attribute weight N_D. The weighted normalized value v_{ij} is calculated as below:

$$V_{ij} = N_D * r_{ij} \quad \text{where } j = 1, 2, \ldots, n; \; i = 1, 2, \ldots, m. \tag{2}$$

Table 6. The weights of the evaluation criteria using ANP.

	Criteria							
	K1	K2	K3	K4	K5	K6	K7	K8
Important Weight	0.8944	0.1056	0.0691	0.1525	0.1566	0.3877	0.1298	0.1044
	K9	K10	K11	K12	K13	K14	K15	
	0.3731	0.75	0.2785	0.0411	0.2324	0.5377	0.4623	

Table 7. The weighted normalized decision matrix.

Alternatives	K1	K2	K3	K4	...	K12	K13	K14	K15
R1	0.013	0.019	0.015	0.065	...	0.001	0.077	0.14	0.02
R2	0.325	0.01	0.007	0.003	...	0.006	0.007	0.068	0.071
R3	0.033	0.016	0.015	0.009	...	0.006	0.026	0.071	0.065
R4	0.032	0.011	0.016	0.017	...	0.005	0.036	0.091	0.062
R5	0.117	0.015	0.005	0.009	...	0.005	0.016	0.035	0.064
R6	0.113	0.011	0.004	0.022	...	0.008	0.026	0.035	0.069
R7	0.225	0.013	0.003	0.017	...	0.004	0.011	0.036	0.072
R8	0.037	0.012	0.004	0.011	...	0.006	0.033	0.061	0.04

Then, using Equations (3) and (4) positive and negative ideal solutions were obtained. The obtained results are shown in Table 8. Compute the positive ideal solution (PIS/A$^+$) and the negative ideal solution (NIS/A$^-$) for each criterion:

$$A^+ = \left\{ \left(max\, X_{ij} | j \in J^* \right), \left(min\, X_{ij} | j \in J^- \right) \right\} = \left\{ X_1^+, X_2^+, \ldots, X_n^+ \right\} \tag{3}$$

$$A^- = \left\{ \left(min\, X_{ij} | j \in J^* \right), \left(max\, X_{ij} | j \in J^- \right) \right\} = \left\{ X_1^-, X_2^-, \ldots, X_n^- \right\} \tag{4}$$

where J^* is the set of benefit attributes and J^- is the set of cost attributes.

Table 8. The ideal solution and negative solution.

Criteria	K1	K2	K3	K4	K5	K6	K7	K8
A^+	0.013	0.01	0.016	0.065	0.027	0.057	0.031	0.043
A^-	0.325	0.019	0.003	0.003	0.013	0.035	0.007	0.002

Criteria	K9	K10	K11	K12	K13	K14	K15
A^+	0.129	0.002	0.069	0.008	0.077	0.035	0.02
A^-	0.015	0.025	0.011	0.001	0.007	0.14	0.072

The ideal solution which maximizes the benefit criteria (criteria of K3, K4, K5, K6, K7, K8, K9, K11 and K12 in this study) and minimizes the cost criteria (criteria of K1, K2, K10, K14 and K15 in this study), whereas the negative ideal solution criteria in this study maximizes the cost criteria/attributes and minimizes the benefit criteria/attributes. The negative ideal solution consists of the worst performance values whereas the best alternative is the one that is nearest to the ideal solution.

The next step of TOPSIS technique is to calculate the Euclidean distance of each alternative. For the positive and negative ideals, the Euclidean distance of each alternative was calculated by using Equations (5) and (6). The distance (di^+, di^-) of each weighted alternative $i = 1, 2 \ldots , m$ from the PIS(d_i^+) and the NIS(d_i^-) is computed as follows:

$$d_i^+ = \left\{ \sum_{j=1}^{n} \left(X_{ij} - X_j^+ \right)^2 \right\}^{0.5} i = 1, 2, \ldots, m; \; j = 1, 2, \ldots, n \tag{5}$$

$$d_i^- = \left\{ \sum_{j=1}^{n} \left(X_{ij} - X_j^- \right)^2 \right\}^{0.5} \; i = 1, 2, \ldots, m; \; j = 1, 2, \ldots, n \tag{6}$$

In the final stage, relative closeness of suppliers to ideal solution was obtained by using Equation (7) and the results were ranked in terms of relative approximately descending order of routes. Table 9 presents the ranking of alternative routes based on combination of ANP and TOPSIS techniques. The closeness coefficient CL_i^+ represents the distances to the positive ideal solution (A^+) and the negative ideal solution (A^-) simultaneously. The closeness coefficient of each alternative is calculated as:

$$CL_i^+ = \frac{d_i^-}{\left(d_i^+ + d_i^- \right)}, \qquad 0 \leq CL_i^+ \leq 1, \; i = 1, 2, \ldots, m \tag{7}$$

Table 9. Final ranking in two-phase ANP-TOPSIS approaches.

Route	A^+	A^-	Ci(S-/S-+S*)	Ranking
R1	0.1061	0.3585	0.7716	1
R2	0.3476	0.0897	0.2052	8
R3	0.1438	0.305	0.6796	4
R4	0.1366	0.3035	0.6897	3
R5	0.1885	0.2353	0.5553	6
R6	0.1744	0.2404	0.5795	5
R7	0.2713	0.1454	0.349	7
R8	0.1302	0.3059	0.7014	2

As it is shown in Table 9, route R1 can give the best score among all alternative routes. The order of alternative monorail routes according to the obtained closeness coefficients is R1 > R8 > R4 > R3 > R6 > R5 > R7 > R2. According to this ranking, the best monorail route is "Route_1(R1): AOÇ (Tema Park), İstanbul road, Opera, Kızılay" which has the highest closeness coefficient. The results are also shown graphically in Figure 4.

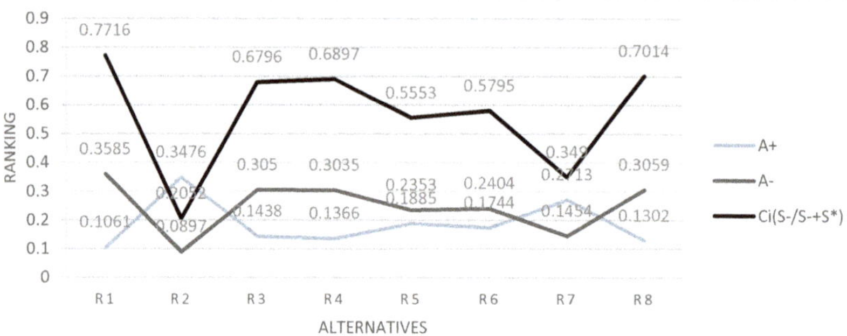

Figure 4. Alternatives' rank.

In the study, the criteria with the highest importance levels are sensitive areas, land structure, population density, ability to expand and develop, and construction cost. As a result, this line, which was selected first, became the foreground in terms of being the longest route and having high population.

5. Results

Route selection and evaluation are very important in terms of the success of metropolitan municipality and urban planning. The objective of this paper is to present an integrated different approach for effective route decisions and determination of the suitable route. Therefore, an integrated approach of ANP- TOPSIS is proposed in order to select the best route and define the relation among the criteria used to select the best route. Urban transportation planning is an important decision for planners and it is the most important activity for managers. Therefore, there are a lot of criteria effecting this selection process. All this criterion should be evaluated with various dimensions. In this study, for 8 route alternatives, four criteria and 15 sub-criteria were evaluated, and the model solution was established with multi-criteria decision-making. Finally, the most suitable routes were ranked. Results show that application of the ANP-TOPSIS methods together provide some important advantages such as the establishing relationship, evaluation of the various factors at the same time and use of tangible and intangible criteria.

In other studies, for the transportation planners and other public institutes in other decision process, the ANP-TOPSIS integrated model can help transport infrastructure project or project selection/ranking. The other MCDM methods such as analytic hierarchy process (AHP), VIKOR (Vise Kriterijumska Optimizacija I Kompromisno Resenje), AHP-TOPSIS, or fuzzy methods can be used, and this model can be also used in the other urban planning processes related to decision-making. In addition, group decision-making approaches can be developed using various MCDM techniques such as AHP or PROMETHEE for selection of transportation projects or route selection in public institutions. Besides, this study can be extended via a mathematical programming-decision model for transport infrastructure projects with resource constraints such as budget.

Public participation in transport planning is important factor for society and it is, therefore, emerging as a basic component in the decision process for rationality [4]. We don't deal with public participation factors in this study. So, in future studies, alternative projects or routes can be evaluated to take account of factors and criteria such as public participation and "community acceptance". The community acceptance can be researched with surveys. Survey-based MCDM techniques can be used to reproduce a participatory process where territorial communities acts as key stakeholders.

Author Contributions: Data curation, M.H.; Methodology, M.H. and T.E.; Writing—review & editing, M.H. and T.E.

Funding: No funding was received.

Acknowledgments: Special thanks to Ankara Metropolitan Municipality and to Faik DİKMEN for allowing us to use their data and to M. Abdullah GENÇER for providing the necessary information regarding his experiments.

Conflicts of Interest: The authors declare that there is no conflict of interest regarding the publication of this paper.

References

1. Cascetta, E.; Carteni, A.; Pagliara, F.; Montanino, M. A new look at planning and designing transportation systems: A decision-making model based on cognitive rationality, stakeholder engagement and quantitative methods. *Transp. Policy* **2015**, *38*, 27–39. [CrossRef]
2. De Dios Ortuzar, J.; Willumsen, L.G. *Modelling Transport*; John Wiley Sons: Hoboken, NJ, USA, 2011.
3. De Luca, S. Public engagement in strategic transportation planning: An analytic hierarchy process based approach. *Transp. Policy* **2014**, *33*, 110–124. [CrossRef]
4. Cascettaa, E.; Pagliaraa, F. Public engagement for planning and designing transportation systems. *Procedia Soc. Behav. Sci.* **2013**, *87*, 103–116. [CrossRef]
5. Shang, J.S.; Tjader, Y.; Ding, Y. A unified framework for multicriteria evaluation of transportation projects. *IEEE Trans. Eng. Manag.* **2004**, *51*, 300–313. [CrossRef]
6. Nassereddine, M.; Eskandari, H. An integrated MCDM approach to evaluate public transportation systems in Tehran. *Transp. Res. Part A Policy Pr.* **2017**, *106*, 427–439. [CrossRef]
7. Shen, J.; Sakata, Y.; Hashimoto, Y. The influence of environmental deterioration and network improvement on transport modal choice. *Environ. Sci. Policy* **2009**, *12*, 338–346. [CrossRef]
8. Delle Site, P.; Filippi, F. Weighting methods in multi-attribute assessment of transport projects. *Eur. Transp. Res. Rev.* **2011**, *1*, 199–206. [CrossRef]
9. Figuera, J.; Greco, S.; Ehrgott, M. (Eds.) *Multiple Criteria Decision Analysis, State of the Art Surveys*; Springer: New York, NY, USA, 2005.

10. Pérez, J.C.; Carrillo, M.H.; Montoya-Torres, J.R. Multi-criteria approaches for urban passenger transport systems: A literature review. *Ann. Oper. Res.* **2015**, *226*, 69–87. [CrossRef]

11. Macharis, C.; Bernardini, A. Reviewing the use of multi-criteria decision analysis for the evaluation of transport projects: Time for a multi-actor approach. *Transp. Policy* **2015**, *37*, 177–186. [CrossRef]

12. De La Vega, D.S.; Vieira, J.G.V.; Toso, E.A.V.; de Faria, R.N. A decision on the truckload and less-than-truckload problem: An approach based on MCDA. *Int. J. Prod. Econ.* **2018**, *195*, 132–145. [CrossRef]

13. Crainic, T.G.; Perboli, G.; Rosano, M. Simulation of intermodal freight transportation systems: A taxonomy. *Eur. J. Oper. Res.* **2018**, *270*, 401–418. [CrossRef]

14. Cavone, G.; Dotoli, M.; Epicoco, N.; Seatzu, C. A decision making procedure for robust train rescheduling based on mixed integer linear programming and Data Envelopment Analysis. *Appl. Math. Model.* **2017**, *52*, 255–273. [CrossRef]

15. Dulebenets, M.A. A comprehensive multi-objective optimization model for the vessel scheduling problem in liner shipping. *Int. J. Prod. Econ.* **2018**, *196*, 293–318. [CrossRef]

16. Ehrgot, M.; Verma, R. A note on solving multicriteria transportation-location problems by fuzzy programming. *Asia Pac. J. Oper. Res.* **2001**, *18*, 149–164.

17. Zhu, Q.; Shah, P.; Sarkis, J. Addition by subtraction: Integrating product deletion with lean and sustainable supply chain management. *Int. J. Prod. Econ.* **2018**, *205*, 201–214. [CrossRef]

18. Acquaye, A.; Ibn-Mohammed, T.; Genovese, A.; Afrifa, G.A.; Yamoah, F.A.; Oppon, E. A quantitative model for environmentally sustainable supply chain performance measurement. *Eur. J. Oper. Res.* **2018**, *269*, 188–205. [CrossRef]

19. Macura, D.; Bošković, B.; Bojović, N.; Milenković, M. A model for prioritization of rail infrastructure projects using ANP. *Int. J. Transp. Econ.* **2011**, *38*, 285–309.

20. Gür, Ş.; Hamurcu, M.; Eren, T. Selecting of Monorail projects with analytic hierarchy process and 0-1 goal programming methods in Ankara. *Pamukkale Univ. J. Eng. Sci.* **2017**, *23*, 437–443. [CrossRef]

21. Hamurcu, M.; Eren, T. Transportation planning with analytic hierarchy process and goal programming. *Int. Adv. Res. Eng. J.* **2018**, *2*, 92–97.

22. Chang, H.W.; Hsieh, H.N. Recreational cycling routes investment selection-Hsinchu Technopolis case by applying ZOGP. *J. East. Asia Soc. Transp. Stud.* **2013**, *10*, 1227–1242.

23. Sadasivuni, R.; O'Hara, C.G.; Nobrega, R.; Dumas, J. A transportation corridor case study for multi-criteria decision analysis. In Proceedings of the American Society of Photogrammetry and remote Sensing Annual Conference, Baltimore, MD, USA, 9–13 March 2009; pp. 11–14.

24. Kuwabara, T.; Hiraishi, M.; Goda, K.; Okamoto, S.; Ito, A.; Sugita, Y. New solution for urban traffic: Small-type monorail system. *Autom. People Mov.* **2001**, *50*, 139.

25. Wang, B. Constructability analysis of monorail project. *Pre-Project Planning & Constructablility Analysis.* Available online: http://www.123seminarsonly.com/Seminar-Reports/2014-03/129660992-Constructability-Analysis-of-Monorail-Project.pdf (accessed on 24 December 2018).

26. Kato, M.; Yamazaki, K.; Amazawa, T.; Tamotsu, T. Straddle-type monorail systems with driverless train operation system. *Hitachi Rev.* **2004**, *53*, 25.

27. Sadatuqu, T.M.N.; Toshio, K. Study on the influences of public transportation on land and building use: The case of Tama Monorail Line. *City Plan. Inst. Jpn.* **2010**, 98–102.

28. Sekitani, T.; Hiraishi, M.; Yamasaki, S.; Tamotsu, T. China's first urban monorail system in Chongqing. *Hitachi Rev.* **2005**, *54*, 193.

29. Kennedy, R.R. *Considering Monorail Rapid Transit for North American Cities*; The Monorail Society: California, CA, USA, 2008.

30. Kimijima, N.; Takahashi, H.; Kawabata, I.; Matsuo, S. New urban transport system for middle east monorail system for Dubai Palm Jumeirah transit system. *Hitachi Rev.* **2010**, *59*, 47.

31. Ghafooripour, A.; Ogwuda, O.; Rezaei, S. An efficient cost analysis of monorail in the Middle East using statistics of existing monorail and metro models. *WIT Trans. Built Environ.* **2012**, *128*, 241–252.

32. Das, A.M.; Yukawa, S.; Ismail, A.; Rahmat, R.A.O.K.; Ladin, M.A. Comparative analysis of monorail system between Kuala Lumpur Malaysia and Kitakyushu Japan. In Proceedings of the Malaysian Universities Transport Research Forum Conference, Bangi, Malaysia, 23–24 December 2013.

33. Marathe, R.; Hajiani, N.D. A review of research on Monorail as an alternative mass rapid transit system. *Int. J. Sci. Res.* **2013**, *4*, 275–277.

34. Parekh, J.A.; Raval, N.G.; Dodiya, D. Overview of monorail rapid transit system. *J. Inf. Knowl. Res. Comput. Eng.* **2013**, *2*, 285–291.

35. Liu, X.; Sun, H.; Liu, F. Study on the application and development of monorail transit system. *Int. J. Eng. Res. Technol.* **2014**, *3*, 213–216.

36. Hussien, M.; Sharawneh, O. Comparison between monorail system and BRT system in Amman City. In Proceedings of the Fourth Jordan International Conference and Exhibition for Roads and Transport, Amman, Jordan, 12–13 March 2014; pp. 1–15.

37. Li, Y.; Xu, Y.; Yan, H.; Wang, K.; Wei, N. Suspended monorail system: A new development of an urban rail transit system with low passenger capacity. In Proceedings of the Fifth International Conference on Transportation Engineering, Dalian, China, 26–27 September 2015; pp. 3180–3186.

38. Timan, P.E. Why monorail systems provide a great solution for metropolitan areas. *Urban Rail Transit* **2015**, *1*, 13–25. [CrossRef]

39. He, X. Application and prospect of straddle monorail transit system in China. *Urban Rail Transit* **2015**, *1*, 26–34. [CrossRef]

40. Hamurcu, M.; Eren, T. Monorail and potential applicability in Turkey. In Proceedings of the International Transportation Technologies Symposium and Fair, Istanbul, Turkey, 17–19 December 2015.

41. Hamurcu, M.; Eren, T. A multicriteria decision-making for monorail route selection in Ankara. *Int. J. Ind. Electron. Electr. Eng.* **2016**, *4*, 121–125.

42. Hamurcu, M.; Alağaş, H.M.; Eren, T. Selection of rail system projects with analytic hierarchy process and goal programming. *J. Eng. Nat. Sci.* **2017**, *8*, 291–302.

43. Hamurcu, M.; Eren, T. Using ANP-TOPSIS methods for route selection of monorail in Ankara. In Proceedings of the 28th European Conference on Operational Research, Poznan, Polland, 3–6 July 2016.

44. Hamurcu, M.; Eren, T. Selection of monorail technology by using multicriteria decision making. *Sigma J. Eng. Nat. Sci.* **2017**, *8*, 303–314.

45. Taş, M.; Özlemiş, Ş.N.; Hamurcu, M.; Eren, T. Selection of Monorail Projects by Using Analytic Hierarchy Process and Goal Programming Combined Model. *Harran Univ. J. Eng.* **2017**, *2*, 24–34.

46. Hamurcu, M.; Eren, T. Prioritization of high-speed rail projects. *Int. Adv. Res. Eng. J.* **2018**, *2*, 98–103.

47. Hamurcu, M.; Eren, T. Decision Making for Rail System Projects with AHP-GP and ANP-GP. *Gazi J. Eng. Sci.* **2017**, *3*, 1–13.

48. Alkubaısı, M.I.T. Predefined evaluating criteria to select the best tramway route. *J. Traffic Logist. Eng.* **2014**, *2*. [CrossRef]

49. Saat, M.R.; Aguilar Serrano, J. Multicriteria high-speed rail route selection: Application to Malaysia's high-speed rail corridor prioritization. *Transp. Plan. Technol.* **2015**, *38*, 200–213. [CrossRef]

50. Kim, H.; Wunneburger, D.; Neuman, M. High-speed rail route and regional mobility with a raster-based decision support system: The Texas Urban Triangle case. *J. Geogr. Inf. Syst.* **2013**, *5*, 559–566.

51. Kosijer, M.; Ivic, M.; Markovic, M.; Belosevic, I. Multicriteria decision-making in railway route planning and design. *Gradevinar* **2012**, *64*, 195–205.

52. Kaysi, I.; Sadek, S.; Al-Naghi, H. A GIS-Based framework for multi-criteria evaluation and ranking of transportation corridor alternatives. In Proceedings of the 11th World Conference on Transport Research, Berkeley, CA, USA, 24–28 June 2007.

53. Banai, R. Public transportation decision-making: A case analysis of the Memphis light rail corridor and route selection with analytic hierarchy process. *J. Public Transp.* **2006**, *9*, 1–24. [CrossRef]

54. Ludin, A.; Latip, S.N.H.M. Using multi-criteria analysis to identify suitable light rail transit route. In *Map Asia Geo ICT for Good Governance*; Geospatial World: Bangkok, Thailand, 2006; Volume 29.

55. Banai, R. Evaluation of land use-transportation systems with the Analytic Network Process. *J. Transp. Land Use* **2010**, *3*, 85–112. [CrossRef]

56. Ahmed, N.G.; Asmael, N.M. A GIS-assisted optimal BAGHDAD metro route selection based on multi criteria decision making. *J. Eng. Sustain. Dev.* **2015**, *19*, 44–58.

57. Dane, G.Z.; Tecím, V. GIS based route determination for light rail systems: A case study in Izmir, Turkey. In Local governance and sustainable development. In Proceedings of the Joint Congress of the European Regional Science Association (47th) and ASRDLF (44th), Paris, France, 29 August–2 September 2007; pp. 1–14.

58. Farkas, A. Route/site selection of urban transportation facilities: An integrated GIS/MCDM approach. In Proceedings of the 7th International Conference on Management, Enterprise and Benchmarking, Budapest, Hungary, 5–6 June 2009; pp. 5–6.

59. Blainey, S.P.; Preston, J.M. A GIS-based appraisal framework for new local railway stations and services. *Transp. Policy* **2013**, *25*, 41–51. [CrossRef]

60. Bueno Cadena, P.C.; Vassallo Magro, J.M. Setting the weights of sustainability criteria for the appraisal of transport projects. *Transport* **2015**, *30*, 298–306. [CrossRef]

61. Kalamaras, G.S.; Brino, L.; Carrieri, G.; Pline, C.; Grasso, P. Application of multicriteria analysis to select the best highway alignment. *Tunn. Undergr. Space Technol.* **2000**, *15*, 415–420. [CrossRef]

62. Piantanakulchai, M.; Saengkhao, N. Evaluation of alternatives in transportation planning using multi-stakeholders' multi-objectives AHP modeling. *Proc. East. Asia Soc. Trans. Stud.* **2000**, *4*, 1613–1628.

63. Ahmed, N.G.; Asmael, N.M. A GIS-Assisted optimal urban route selection based on multi criteria approach. *Iraqi J. Mech. Mater. Eng.* **2009**, *2*, 557–567.

64. Sharifi, M.A.; Boerboom, L.; Shamsudin, K.B.; Veeramuthu, L. Spatial multiple criteria decision analysis in integrated planning for public transport and land use development study in Klang Valley, Malaysia. In Proceedings of the ISPRS Technical Commission II Symposium, Vienna, Austria, 12–16 July 2006; pp. 85–91.

65. Gerçek, H.; Karpak, B.; Kılınçaslan, T. A multiple criteria approach for the evaluation of the rail transit networks in Istanbul. *Transportation* **2004**, *31*, 203–228. [CrossRef]

66. Velasquez, M.; Hester, P.T. An analysis of multi-criteria decision making methods. *Int. J. Oper. Res.* **2013**, *10*, 56–66.

67. Saaty, T.L.; Ozdemir, M.S. Why the magic number seven plus or minus two? *Math. Comput. Model.* **2003**, *38*, 233–244. [CrossRef]

68. Saaty, T.L. Fundamentals of the analytic hierarchy process. In *The Analytic Hierarchy Process In Natural Resource and Environmental Decision Making*; Springer: Dordrecht, The Netherlands, 2001; pp. 15–35.

69. Tsai, W.; Leu, J.; Liu, J.; Lin, S.; Shaw, M. A MCDM approach for sourcing strategy mix decision in IT projects. *Expert Syst. Appl.* **2010**, *37*, 3870–3886. [CrossRef]

70. Lee, J.W.; Kim, S.H. Using analytic network process and goal programming for interdependent information system project selection. *Comput. Oper. Res.* **2000**, *27*, 367–382. [CrossRef]

71. Meade, L.M.; Presley, A.R. R&D project selection using the analytic network process. *IEEE Trans. Eng. Manag.* **2002**, *49*, 59–66.

72. Ravi, V.; Shankar, R.; Tiwari, M.K. Selection of a reverse logistics project for end-of-life computers: ANP and goal programming approach. *Int. J. Prod. Res.* **2008**, *46*, 4849–4870. [CrossRef]

73. Büyüközkan, G.; Öztürkcan, D. An integrated analytic approach for six sigma project selection. *Expert Syst. Appl.* **2010**, *37*, 5835–5847. [CrossRef]

74. Wey, W.M.; Wu, K.Y. Using ANP priorities with goal programming in resource allocation in transportation. *Math. Comput. Model.* **2007**, *46*, 985–1000. [CrossRef]

75. Begičević, N.; Divjak, B.; Hunjak, T. Decision-making on prioritization of projects in higher education institutions using the analytic network process approach. *Central Eur. J. Oper. Res.* **2010**, *18*, 341–364. [CrossRef]

76. El-Abbasy, M.S.; Zayed, T.; Ahmed, M.; Alzraiee, H.; Abouhamad, M. Contractor selection model for highway projects using integrated simulation and analytic network process. *J. Constr. Eng. Manag.* **2013**, *139*, 755–767. [CrossRef]

77. Tuzkaya, U.R.; Yolver, E. R & D project selection by integrated grey analytic network process and grey relational analysis: An implementation for home appliances company. *J. Aeronaut. Space Technol.* **2015**, *8*, 35–41.

78. Saaty, T.L. *The Analytic Hierarquic Process*; RWS Publications: Pittsburg, CA, USA, 1980.

79. Hwang, C.L.; Yoon, K. *Multiple Attribute Decision Making Methods and Applications*; Springer: Berlin/Heidelberg, Germany, 1981.

80. Qin, X.; Huang, G.; Chakma, A.; Nie, X.; Lin, Q. A MCDM-based expert system for climate-change impact assessment and adaptation planning—A case study for the Georgia Basin, Canada. *Expert Syst. Appl.* **2008**, *34*, 2164–2179. [CrossRef]

81. Ic, Y. An experimental design approach using TOPSIS method for the selection of computer-integrated manufacturing technologies. *Robot. Comput. Integr. Manuf.* **2012**, *28*, 245–256. [CrossRef]

82. Tsou, C.H. Multi-Objective inventory planning using MOPSO and TOPSIS. *Expert Syst. Appl.* **2008**, *35*, 136–142. [CrossRef]

83. Önüt, S.; Soner, S. Transshipment site selection using the AHP and TOPSIS approaches under fuzzy environment. *Waste Manag.* **2008**, *28*, 1552–1559. [CrossRef]

84. Abalı, Y.A.; Kutlu, B.S.; Eren, T. Multicriteria decision making methods with selection of scholarship holder: Application in an educational institution. *Atatürk Univ. J. Econ. Adm. Sci.* **2012**, *26*, 259–272.

85. Özbek, A.; Eren, T. Selecting the service provider through multiple criteria decision making techniques. *Acad. Overv.* **2013**, *36*, 1–22.

86. Korkmaz, M. Analysis of economic efficiency at forest enterprises with TOPSIS method. *Turk. J. For.* **2012**, *13*, 14–20.

87. Shih, H.S.; Shyur, H.J.; Lee, E.S. An extension of TOPSIS for group decision making. *Math. Comput. Model.* **2007**, *45*, 801–813. [CrossRef]

88. Kannan, G.; Pokharel, S.; Kumar, P.S. A hybrid approach using ISM and fuzzy TOPSIS for the selection of reverse logistics provider. *Resour. Conserv. Recycl.* **2009**, *54*, 28–36. [CrossRef]

89. Ersöz, F.; Kabak, M.; Yilmaz, Z. A model proposal for course selection for postgraduate students. *J. Fac. Econ. Adm. Sci. Afyon Kocatepe Univ.* **2011**, *13*, 227–249.

90. Shyur, H.J.; Shih, H.S. A hybrid MCDM model for strategic vendor selection. *Math. Comput. Model.* **2006**, *44*, 749–761. [CrossRef]

91. Azimi, R.; Yazdani-Chamzini, A.; Fouladgar, M.M.; Zavadskas, E.K.; Basiri, M.H. Ranking the strategies of mining sector through ANP and TOPSIS in a SWOT framework. *J. Bus. Econ. Manag.* **2011**, *12*, 670–689. [CrossRef]

Article

Staff Task-Based Shift Scheduling Solution with an ANP and Goal Programming Method in a Natural Gas Combined Cycle Power Plant

Emir Hüseyin Özder [1], Evrencan Özcan [2] and Tamer Eren [2,*]

[1] Department of Management Information, Faculty of Commercial Science, Başkent University, 06790 Ankara, Turkey; ehozder@baskent.edu.tr
[2] Department of Industrial Engineering, Faculty of Engineering, Kırıkkale University, 71450 Kırıkkale, Turkey; evrencan.ozcan@kku.edu.tr
* Correspondence: tamereren@gmail.com; Tel.: +90-318-357-3576

Received: 22 November 2018; Accepted: 31 January 2019; Published: 18 February 2019

Abstract: Shift scheduling problems (SSPs) are advanced NP-hard problems which are generally evaluated with integer programming. This study presents an applicable shift schedule of workers in a large-scale natural gas combined cycle power plant (NGCCPP), which realize 35.17% of the total electricity generation in Turkey alone, as at of the end of 2018. This study included 80 workers who worked three shifts in the selected NGCCPP for 30 days. The proposed scheduling model was solved according to the skills of the workers, and there were nine criteria by which the workers were evaluated for their abilities. Analytic network process (ANP) is a method used for obtaining the weights of workers' abilities in a particular skill. These weights are used in the proposed scheduling model as concepts in goal programming (GP). The SSP–ANP–GP model sees employees' everyday preferences as their main feature, bringing high-performance to the highest level, and bringing an objective functionality, and lowering the lowest success of daily choice. At the same time, the model introduced large-scale and soft constraints that reflect the nature of the shift requirements of this program by specifying the most appropriate program. The required data were obtained from the selected NGCCPP and the model solutions were approved by the plant experts. The SSP–ANP–GP model was resolved at a reasonable time. Monthly acquisition time was significantly reduced, and the satisfaction of the employees was significantly increased by using the obtained program. When past studies were examined, it was determined that a shift scheduling problem of this size in the energy sector had not previously been studied.

Keywords: shift scheduling; goal programming; ANP; natural gas combined cycle power plant; energy sector

---

1. Introduction

Personnel efficiency and productivity is one of the basic objectives of many companies and organizations. While trying to find all the existing elements that can affect efficiency and productivity, personnel scheduling is an important problem that must be attentively steered. Personnel scheduling problems can be even more complicated by shifts in plans for industries in which shift personnel work seven days a week, and specific personnel have to work non-routine working hours. Some service providers must continuously operate, such as power plants, hospitals, and security and safety-related units while serving mostly the common good, and scheduling problems such as the above are further complicated.

Electricity generation in power plants and maintenance or repair work require complex structured, specialized, and labor-intensive activities. When considering the high impact and indispensability of electrical energy in daily human life, the process of assigning suitable employees to such specialized work and improving the performance of these employees comes to the fore with the impact on society, in addition to the economic value added. Power plants are continuous generation facilities with the main purpose of achieving uninterrupted, reliable, efficient, economical, and environmentally-friendly energy generation called sustainable energy supply, therefore, it is an indispensable requirement to manage the generation, manpower, material, and maintenance processes within this system by operating the plants in accordance with the operating rules in the direction of this five-legged comprehensive target. In other words, maximizing employee performance related to personnel management through balanced work distribution, appointment of the appropriate employees for the appropriate jobs, etc. has critical importance in terms of sustainable energy supply. With today's constantly developing technology, power plants are operated by shift personnel from a main control center or supervisory control and data acquisition (SCADA) system, which distributes control to the power plant (DCS) remotely. Throughout the world, as in Turkey, although not with the SCADA system currently, personnel-operated power plants are still existing. Compared to advanced technology, the number of shift personnel working in these old plants is naturally higher than that of modern plants, and the probability of operator errors from shift personnel is higher than that of SCADA. Put differently, shift personnel may experience faulty operations due to fatigue, lack of concentration, experience, fair working order, and thus lack of motivation in these power plants. As a result, long-term failures and thus millions of kWh of energy and income loss may occur. In addition, a negative social impact may occur due to the lack of sustainable energy supply. From this point of view, it can be said that personnel scheduling models used for fair and appropriate assignments have critical importance in electricity generation facilities.

In spite of the various models proposed and resolved in SSPs, shift scheduling studies involving personnel in power plants or other related areas are still insufficient. Previous works on SSPs have mainly focused on factory or security personnel, health personnel (i.e., doctors and nurses), teachers, and police officers. For this reason, this study is a new application of models for shift scheduling problems (SSPs) and will be relevant to solving SSPs.

When the distribution of resources used in electricity generation in the world is examined, it is seen that the most widely-used resource for electricity generation is coal, after which natural gas is the second most common resource. When we look at the resources with the largest share of electrical energy generation in the countries with the largest economies, coal is predominantly used in the United States, China, India, and Germany, natural gas in Russia, nuclear energy in France, and renewable energy resources in Canada. Natural gas is the number one resource to generate electricity in Turkey, and this has not changed in the last 10 years. Over the last 10 years in Turkey, power was generated on average of 40% due to natural gas, and the share of this resource in Turkey's energy mix was 35.17% as at of the end of 2018. According to these data, natural gas is the most widely used resource in electricity generation in Turkey [1]. Although there is no resource availability in Turkey for natural gas, NGCCPPs are the most widely used power plants, because these large-scale facilities have important advantages such as having high capacities, low setup times, low installation costs, low environmental impacts, short run times, high yields, ease of operation and maintenance, and long economic lifespans, compared to other fossil-fueled-based power plants. The high share of natural gas in Turkey's total electricity generation with the above-mentioned advantages of the NGCCPPs and the importance of the personnel scheduling in the power plants are taken into consideration together, and the objective of the study is to schedule the shift personnel fairly and according to their capabilities in the large-scale NGCCPP in Turkey.

NGCCPPs consist of generation blocks, and each combined cycle generation block includes two gas turbines, two compressors, two gas turbine generators, two waste heat boilers, two condenser units, one steam turbine and one steam turbine generator. In addition to these, dry-type or sea water-cooled wet type cooling towers, water treatment plant, switchgear, and control and control systems are located in the plant.

Generation blocks operate independently of each other in NGCCPPs. However, blocks use some facilities that are in common. Electricity generation in these plants is carried out in two different stages, as shown in Figure 1. The natural gas is mixed with air and it is been burned in gas turbines. Burned gas turns the gas turbine on the same shaft as a generator and generates electricity in the first stage. Simultaneously, the hot gases generated from this combustion are sent to the waste heat boiler, and steam is generated by this heat. Steam reaching the required pressure and temperature is sent to the steam turbine and the turbine is rotated to generate the second stage electricity by the generator means located on the same shaft as the steam turbine. Steam from the steam turbine is condensed in the condensers with the cooling water from the cooling towers and converted into water. The condensation water, which accumulates in the lower part of the condensers, is sent to the waste heat boilers for reheating. Steam produced in the boiler is sent to the steam turbine and the cycle is completed. There are three different pressure levels (low, medium, high pressure) in the boilers, which is produced by steam force in order to keep the efficiency at maximum level. Thus, the hot gases in the boilers are utilized as much as possible [2].

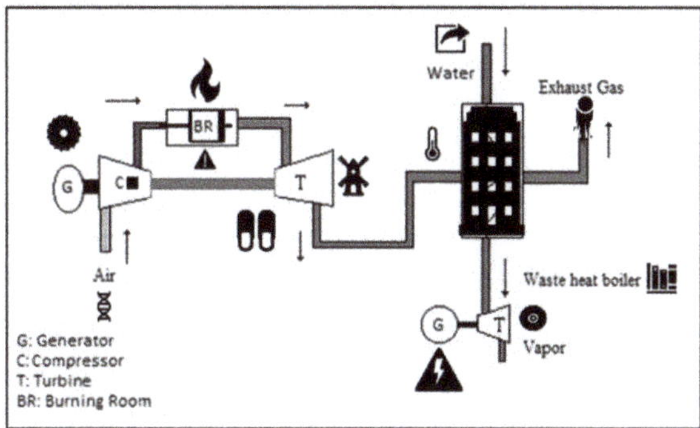

Figure 1. Schematic representation of the working principle of the combined cycle power plant.

Turbines are the most important of the systems mentioned in NGCCPPs. This is, as the basic reasons explain, that electricity is produced in gas-turbine steam turbines. If the turbines are deactivated, the corresponding block disables the unit (each gas turbine is called a unit) and the system stops. On the other hand, some NGCCPPs have by-pass lines, and the hot gas from an accidental gas turbine, which is formed in the waste heat boilers of the steam turbine, can be directly discharged to the atmosphere by means of a by-pass chute and only electricity generation can be continued in gas turbines (single operation). In addition, these systems are structurally more complex than other systems in the plant.

When previous studies were examined, SSP with this size (30 days for 80 personnel) within the energy sector has not been studied before. The problem of the scheduling of worker's shift known as SSP can be complicated, especially in the case of multiple shifts and multiple employees with many employees and skills, due to the sophisticated constraints and solution alternatives. Preparing the schedules of staffs manually is thus infertile and time consuming. Therefore, a methodical approximation for initiating a useful schedule is required within a short period of time. Currently, the scheduling of the operation of the shifts of the NGCCPP is carried out manually. This paper deals with the formulation of a model for the shift scheduling problem for NGCCPP employees working with newly defined constraints that define the structure of the industry, organizational policies, and shift schedules, and the requirements that determine the daily preferences of workers. The model seeks to produce the most appropriate monthly shift plan that maximizes employees' job satisfaction. In this way, stoppages caused by operator error are reduced from 53 hours to 4 hours, and both financial gain is achieved, and the target of sustainable energy supply is reached at this power plant. Furthermore, with this study which provides multi-objective scheduling on the basis of the capabilities of a large group of 80 employees working in NGCCPP for the first time in the literature, a shortcoming in the SSP literature is eliminated.

The paper is shaped as follows: After a brief introduction, the next section gives a literature overview around the problems of shift scheduling. The next section gives information about the perspectives and applications of the study. The last section gives related discussions of the computational results and the conclusion of the study.

2. Shift Scheduling Problem (SSP)

Coordination of workload plans and assignment of personnel and staff planning and personnel scheduling in order to meet the demand for resources that change with time. These problems occur in service industries, such as police officers, call center operators, hospital nurses, transportation personnel (aircraft crews, bus drivers) and so on. It is a very important topic for personnel scheduling. These environments are often prolonged and unsteady, and personnel needs undulate over time. Schedules typically include equipment requirements, trade union rules, etc. It is the subject of various restrictions dictated by inherent features of personnel scheduling problems. The problems that arise tend to be combinatorically difficult. For personnel scheduling problems, Baker proposed one of the first classification methods [3]. In the study of Baker [3] and Pinedo [4] SSP is a brand of the personnel scheduling problem. Personnel scheduling problem's structure can be divided into several categories. General solution method is with integer programming. This method contains a large class of personnel scheduling problem solutions. Besides, there is a specific category of integer programming problems, namely cyclical personnel problems. This problem can be used in terms of class and a combinatorial viewpoint. Apart from these, crew and operator scheduling problems have a different model structure. In the personnel scheduling problem, there are assignment models. In spite of the fact that every assignment mold belongs to a financial reason, it is a reason why the cost is relatively easy. A cycle structure is thinking of a beforehand. With annotative adjustments, the cycle can be a single day, another week or a few weeks. Part of the previous episode is limited to a loop.

In general, the SSP deals with the assignments of employees to each shift determining the number and criteria. Each shift is getting shaped to specific start and finish times. The number and placement of moats time or the long and the short lengths of the shifts are limiting with legal and business rules. The main goal of scheduling shifts is to optimize employee allocations. Besides, it also helps to minimize the total cost. This requires that each shift be precisely assigned to personnel, qualified by several constraints that must be satisfied. SSP is key to ensuring that services continuously operate without any timeframes in specific service sectors, as well as hospitals, police stations, seizure centers, railway stations and airports [5].

One of the main difficulties of shift scheduling concerns with requirements and limitations of existing personnel while a work pattern of each shifts depending on the number of certain regulations. This planning is getting more difficult when multiple shifts and break or idle times and/or lots of skilled and variable shifts are involved, where every shift needs a combination of these to fulfill the request for that shift. When highly skilled employees can accomplish many different activities during the same shift, the problem becomes more difficult. The shift programming problem's mathematical models are formulated to solve, subject to several constraints such as one shift per day, number of permits per day, shift order, gap between shifts, rest day after day and night shifts, consecutive days, and working consecutive night regulations, fixed hours for shifts, restricted working hours, number of workers based on skill or skill, etc. Restrictions taken into account in the SSP may include the ordering or preference of consecutive shifts or preferred shifts to be followed, forbidden shift sequences, constraints on demand, and minimum rest periods between shift changes. For this reason, various SSP variations can be found depending on the restrictions applied. A type of SSP is the Day-off SSP (D-SSP) that focuses on determining the most suitable rest days for an employee's planning horizon. This problem is that the cost of closed molds on different days is different and the goal is to minimize the total labor cost. In this study, model formulation is considered preferences of the workers presented through its objective function and constraints.

In this paper, the SSP is formulated as Analytic Network Process (ANP) and Goal Programming (GP) model for personnel of power plant, referred to as the SSP-ANP-GP model. However, the recommended goal programming model is different from other models for scheduling shift personnel described above because of the different goals. The objective of the SSP-ANP-GP model is to maximize the personnel's needs, which is determined with four goals in the proposed model.

Mathematical programming models for the SSP can be solved using precise algorithms, intuitive approaches, and top-level information. Our model gives an optimal solution, but in the case of large-scale problems, it is hard to apply in terms of operation time and complexity. Because of that reason, metaheuristics and intuitive approaches are more hopeful options for these situations. It is much better in big-size problems when an intuitive, optimization is sufficient as a solution and has optimal solutions in reasonable time. Approximate optimization algorithm frameworks are meta-analyses that can be defined as a primary strategy or a general algorithmic framework that can be applied to a variety of problems with only a few changes. They are commonly used when they are confronted with a complex problem that cannot be resolved by certain algorithms [6]. The SSP-ANP-GP model of our paper is solved using the integer linear programming algorithm of IBM ILOG optimization tool (IBM, Armonk, New York, NY, USA), which employs the heuristics and exact approaches.

In our paper, a different SSP approach is used for the literature review in shift scheduling studies apart from other studies. It shows that our SSP application area is different from other data and application areas. The energy sector is a new sector for the SSP problems. According to the results of the investigated papers, shift scheduling studies are done for the main production or service personnel. In other words, there is a gap in scheduling of people working in auxiliary and complementary processes in the energy sector for personnel scheduling literature.

On the other hand, these processes may directly affect the quality levels of output of production and service systems, customer satisfaction levels, even sectoral positions of enterprises and social welfare level. The most important examples of this are continuous production systems such as electricity generation, petrochemical and cement production, telecommunication, and health services. However, personnel performing auxiliary and complementary processes may be partly or entirely engaged in the same work with the shift team directly involved in the main processes. The most important example in this situation is the power generation plants. Personnel operating and maintaining these facilities, especially those with electrical and electronic origins, can be operated as both shift and maintenance personnel. This provides significant advantages in terms of productivity, quality, uninterrupted production, and costs in terms of the establishment that owns the plant. Because of that reason, shift scheduling studies are very important in the energy sector, and above all, in power generation facilities.

3. Related Works

According to the observations on personnel scheduling, most studies focused on creating appropriate shift schedules or work programs for workers, considering a deterministic workload. The research process of this paper started with an investigation of the review articles [7,8]. After that point, the cited paper was investigated, which were given in the review articles. The whole reference list was checked and completed in related articles. Search phrase combinations were stated: Workforce scheduling, personnel scheduling and staffing.

Personnel scheduling literature showed a variety of investigation methodologies that brought together a specific analytical approach with a solution or evaluation technique. A number of studies, mathematical programming categories such as goal programming, linear programming, dynamic programming, and integer programming or optimization were classified as intuitive. Queuing, constraint programming and simulation were other categories for solution technique.

In general, scheduling has an important role in the production and service sector [9]. Most popular studies are done in the shift scheduling area. Shift scheduling is a widely studied area, which is a process that plays an important role in manufacturing and service industries. SSPs can be very complicated, especially in the case of multiple shifts with a large number of employees due to multiple constraints and skill-based works. Shift scheduling problem is a complex NP-hard integer programming problem. There are some examples about large-scale scheduling problems in the literature. As a literature observation, the SSP study in the energy sector has not been encountered in previous studies before. The SSP can be complex, especially in the case of several shifts, with many employees and skills, due to complex constraints and solution alternatives. Especially in most of the shift scheduling studies, the application results were not given. At this point, our study reveals its difference from other studies. Considering all of these, it is thought that shift scheduling studies on energy facilities are a new area waiting to be explored.

In Figure 2, a general literature overview is shown. Those studies are the most related studies to this paper. As is shown in Figure 2, solution methods and objectives/criteria are investigated. Generally; heuristic methods [10–24], integer programming [25–38] and goal programming [39–49] are the most used techniques for solving SSP. Within the scope of all the studies in Figure 2, some studies which were thought to be most similar to the model structure of our study were selected and explained in detail below.

Aickelin et. al. [10] have studied a heuristic technique which is called Improved Squeaky Wheel Optimization (ISWO) for scheduling problems of the drivers in their work. This method develops the main Squeaky Wheel Optimization (SWO) method.

Aickelin et al. [11] developed a solution to two different staff scheduling problems by discussing and analyzing a heuristic method. The new model improves the effectiveness and execution speed. The results are presented on two different areas of staff scheduling. These are: Bus and train driver shift schedules and nurse shift schedules.

Yunes et. al. [13] have dealt with the issue of shift scheduling for a firm using hybrid algorithms and integer programming methods in their work. In this article, the general team management problem has been discussed, which derives from the daily operation of an urban transit bus company serving the public area. They divided the problem into two distinct subproblems: Shift scheduling and crew rostering. They studied each of these problems using constraint logic programming and mathematical programming approaches. Moreover, hybrid column generation algorithms are developed for solving problems, with combining constraint logic programming methods and mathematical programming.

Akjiratikarl et al. [14], the problem of scheduling home caregivers was solved by the Particle-based Swarm Optimization (PSO), the meta-intuitive technique for population. The proposed methodology is applied, tested and compared with existing solutions on various real problem examples. With these aspects of the study revealed, it showed the difference between the others.

Abbink et. al. [24] have successfully implemented a complex SSP model in their work. The authors described the corresponding analysis methods for scheduling the personnel. The authors have developed an alternative heuristic model that satisfies personnel. New schedules produced according to the alternative production model have reduced personnel costs.

In Alfares [27]'s study, a four day-off scheduling problem for three weeks were resolved. A mathematical model of this problem has been formulated and solved efficiently by setting a limit on the workforce dimension.

Trilling et al. [28] have studied the SSP of the nurses at the hospital in their work. Their study focuses two methods to solve the staff scheduling problem based on constraint and integer programming. Maximization of the fairness of the schedule is the main objective. They tested two techniques in order to be compared.

Authors	Methods	Criteria / Objective
Aickelin et. al. [10]	Heuristic Method	Fair Scheduling
Aickelin et al. [11]	Heuristic Method	Balanced Assignment
Aickelin and Dowsland [12]	Genetic Algorithm	Getting Faster Results
Yunes et. al. [13]	Hybrid Algorithms and Integer Programming	Getting Faster Results
Akjiratikarl et al. [14]	Partial Swarm Optimization	Systematic Scheduling
Ásgeirsson [15]	Hybrid Algorithm	Fast Scheduling
Lei et. al. [16]	Heuristic Method	Personnel Scheduling
Smet et. al. [17]	Hybrid Algorithm	Shift Sch. with Skill-Based
Mısır et. al. [18]	Heuristic Method	Shift Scheduling
Mısıt et. al. [19]	Heuristic Method	Personnel Scheduling
Smet and Vanden Berghe [20]	Metaheuristic Method	Personnel Scheduling
Lee et. al. [21]	Heuristic Method	Personnel Scheduling
Veen et. al. [22]	Heuristic Method	Shift Scheduling
Smet et. al. [23]	Heuristic Method	Shift Scheduling
Abbink et. al. [24]	Heuristic Method	Balanced Assignment
Aickelin and White [25]	Integer Programming	Getting Fair Schedules
Alfares [26]	Integer Programming	Effective Assignment
Alfares [27]	Integer Programming	Workforce Allocation
Trilling et al. [28]	Constraint and Integer Programming	Fair Scheduling
Lezaun et al. [29]	Integer Programming	Workforce Scheduling
Al-Yakoob and Sherali [30]	Mixed Integer Programming	Meeting Workers' Demands
Al-Yakoob and Sherali [31]	Mixed Integer Programming	Cost Minimization
Alfares [32]	Integer Programming	Cost Minimization
Lezaun et al. [33]	Mixed Integer Programming	Production Scheduling
Corominas et al. [34]	Mixed Integer Programming	Fair Scheduling
Bard et al. [35]	Integer Linear Programming	Fair Scheduling
Corominas et al. [36]	Mixed Integer Programming	Balanced Assignment
Henao et. al. [37]	Mixed Integer Programming	Cost Minimization and Fair Sch.
Veldhoven et. al. [38]	Integer Programming	Effective Assignment
Bağ et al. [39]	0-1 Goal Programming	Fair Scheduling
Hung-Tso et al. [40]	Goal Programming	Fair Scheduling
Li et al. [41]	Goal Programming	Fair Scheduling
Kassa and Tizazu [42]	Goal Programming	Fair Scheduling
Louly [43]	Goal Programming	Fair Scheduling
Labadi et al. [44]	Goal Programming	Fair Scheduling
Todovic et al. [45]	Goal Programming	Fair Scheduling
Shuib and Kamarudin [46]	Goal Programming	Shift Scheduling
Özder et al. [47]	Goal Programming	Fair Scheduling
Varlı et al. [48]	Goal Programming	Fair Scheduling
Ernst et al. [50]	Bibliographic Study	Cost Minimization Tech.
Azaies and Al-Sharif [51]	0-1 Goal Programming	Providing Uninterrupted Service
Topaloğlu [52]	Goal Programming	Balanced Assignments
Alfares [53]	Simulation Methods	Cost Minimization
Alfieri et al. [54]	Branch-Price Algorithm	Task Minimization
Chu [55]	Goal Programming	Task Scheduling
Thompson and Pullman [56]	Integer Programming	Fair Workforce Distribution
Sinreich and Jabali [57]	Linear Programming	Idle Time Minimization
Al-Yakoob and Sherali [58]	Mixed Integer Programming	Workforce Assignment
De Matta and Peters [59]	Branch and Price	Shift Scheduling
Tsai and Li [60]	Genetic Algorithm	Workforce Scheduling
Lezaun et al. [61]	Binary Programming	Shift Scheduling
Rönnberg and Larsson [62]	Linear Programming	Shift Scheduling
Zolfaghari et al. [63]	Genetic Algorithm	Effective Scheduling
Alsheddy and Tsang [64]	Linear Programming	Workforce Scheduling
Fırat and Hurkens [65]	Mixed Integer Programming	Fair Scheduling
Asensio-Cuesta et al. [66]	Genetic Algorithm	Job Satisfaction
Veen and Veltman [73]	Branch and Price Method	Shift Scheduling

Figure 2. Literature Overview.

Lezaun et al. [29] studied the shift scheduling of railroad drivers. They presented a practical study on how to provide a more equitable annual work allocation for Metro Bilbao. The proposed model is solved as a series of four types of integer programming problems. The main benefit of this paper is to combine part-time allocation with a workload unevenly distributed on the days of the week and a planning time frame divided into five periods of three different types.

In Alfares [32]'s paper, the Information Technology help desk operators described personnel shift schedules for a large petrochemical company. The aim of the author was to reduce the cost of labor by setting the best staff level and employee weekly tour schedules needed to cover the changing workload during the 24-hour work period. An integer programming model has been formulated and solved to determine tour scheduling assignments. Selected tour scheduling offers better service at a lower cost and for fewer employees.

In another study Lezaun et al. [33], presented a case study commissioned by the Spanish railway carrier for the annual rostering of shift schedules for the station personnel. A mixed rostering process was used in their research. These problems were resolved using the computer software LINGO package. The model of the study was similar to our model structure.

Corominas et al. [34] aimed at eliminating the issue of production planning by making shift schedules of production workers in their work. They proposed a mixed-integer linear program model. The model helped to solve the problem of scheduling the production and the working hours of an expert team which runs in a multi-product process.

Bard et al. [35] investigated the tour scheduling problem in the United States Postal Service for their work. The study suggested a model to tour scheduling problem for decreasing the size of the personnel scheduling, targeted with several scenarios with integer linear programing. The whole mathematical model included not only full-time but also part-time workers, in addition the specific constraints described by the union contract. They solved the problem faster and in a more cost-efficient way.

Corominas et al. [36] have studied problems such as labor balancing and regulation, using a mixed integer programming model in their work. Their study presented the scheduling problem of annual working hours for any employee, where weekly working hours belong to a pre-set finite set and day-off weeks are the same for all employees. The problem is modeled and solved as a mixed integer linear program.

Bağ et al. [39] investigated the SSP of nurses using a 0–1 goal programming and analytic network process in their work. The analytical networking process (ANP) was used to determine the weights of the goal program. The model was implemented in a state hospital.

Hung-Tso et al. [40] investigated the problem of staff scheduling with the goal programming technique in their work. A linear goal programming model is proposed, and three management goals are considered simultaneously for generating a roster.

Li et al. [41], Kassa and Tizazu [42], Louly [43], Labadi et al. [44] and Todovic et al. [45] studied shift scheduling by using goal programming method and intuitive methods together.

Shuib and Kamarudin [46], studied an SSP of the workers at the largest power plant in Malaysia. Their study aimed to define the basic criteria and conditions of the shift planning problem at the power plant in order to formulate the goal programming model for the shift programming problem, which optimized the daytime preferences of the workers and determined the optimal scheduling for the model-based workers. The study focused on three legs which were shift scheduling, day-off scheduling, and scheduling with demands. The study involved scheduling 43 personnel in a specific department of the power plant for 28 days where personnel work in three shifts (morning, evening, and night shifts) and with day-offs.

Ernst et al. [50] conducted a bibliographic study that contributed significantly to the literature on personnel SSPs in their work. They do not review the software packages in their work. Rather, they review scheduling problems in specific areas of application and the models and algorithms reported in the literature for the solutions. They also search for commonly used methods to solve staff problems.

Azaies and Al-Sharif [51] have solved the SSPs of the nurses by using a 0–1 goal programming method in their studies. This study is an example of an NP-hard integer programming problem. The model of the study is similar to our model structure. The constructed model calculates as well as some of the policies proposed in the literature regarding both the hospital targets and the preferences of the nurses. Hospital objectives include providing uninterrupted service with unnecessary nursing skills and staff size and avoiding additional costs for unnecessary overtime.

Topaloğlu [52] studied the problem of shift scheduling health-care personnel with goal programming method. The author suggested a goal programming (GP) model that includes both hard and soft constraints for a monthly planning period. Hard constraints must be strictly observed but may be violated if soft constraints are required. The relative significance values of soft constraints were calculated by the analytic hierarchy process (AHP) used as deviation coefficients from the soft constraints in the objective function. Their models were tested in the emergency room of a large local university hospital. The main results of the study are that realistic problems can be solved quickly, and the schedules created have very high qualifications according to the schedules prepared manually.

In Alfares [53]'s study, a simulation approach was presented for workers' permission schedules when daily labor demands were random variables. The author created a simulation model and stated that the proposed approach is a case study application. The model stated that the employees had a limited number of jobs, included variable workload, and that employees had policy restrictions on the selection of work programs. The shift scheduling model also offered an alternative for leave days, reducing average production times for maintenance work orders by 25% without increasing the number of employees or cost.

Alfieri et al. [54] have defined an intuitive procedure with an intuitive branch-price algorithm based on a dynamic programming algorithm to find a suitable solution for pricing. They tested the results on the timetable of train lines.

Chu [55] solved the SSP of airport staff using the goal programming technique in his work. Chu proposed an integrated crew duty assignment for the luggage department staff at Hong Kong International Airport. The results could be adopted as a good crew schedule. The result showed that it was both feasible and the model satisfied various work conditions and minimized idle shifts.

In Thompson and Pullman's [56] study, the SSP of the staff were solved. The study focused on workforce schedules breaks or reliefs of the workers.

Sinreich and Jabali [57] used a linear programming model and simulation tools to solve the SSP in their work. In their study, they used a linear optimization model (S-model) and a heuristic iterative simulation-based algorithm (SWSSA) for scheduling the resources' work shifts, one resource at a time. Their algorithm was tested using data that was gathered from emergency departments of five general hospital. By using this model, they were able to achieve a reduction in patient's "length of stay" minutes.

The other paper of Al-Yakoob and Sherali [58] concerns the issue of assigning employees to a range of work centers, taking into account the preferences expressed for specific shifts, days and business centers by using a heuristic method. Computational results have shown that the proposed approach can facilitate the creation of good quality scheduling even for large-scale problematic cases at a reasonable time. De Matta and Peters [59] used branch and price; Tsai and Li [60] used genetic algorithm and Lezaun et al. [61] used binary programming for solving scheduling problems.

Rönnberg and Larsson [62] studied the problem of appointing nurses shift scheduling in the Swedish healthcare sector. The authors present a pilot study that aims to determine if it is possible to create an optimization tool that presents a program that is automatically available based on the charts recommended by nurses. The study was conducted in a typical Swedish nursing ward, where we developed a mathematical model and presented timetables. The results of this study are highly encouraging and suggest ongoing studies. Zolfaghari et al. [63] used genetic algorithm and Alsheddy and Tsang [64] used linear programming for solving scheduling problem.

Fırat and Hurkens [65] performed personnel shift schedules with mixed integer programming. The authors assigned technicians to tasks with multi-level skills requirements. The study deals with the scheduling of complex tasks with a non-homogeneous resource set. They created programs by repetitively applying a flexible matching model that selects the tasks to be processed and which creates technician groups assigned to task combinations. The underlying mixed integer programming model is capable of reviewing technician-task distributions and performs very well, especially in the case of rare skills. Asensio-Cuesta et al. [66] used genetic algorithm for solving scheduling problem.

4. Methods

In this study, two different methods were used. One of them was goal programming. The other one was the analytic network process. These two techniques will be briefly summarized below.

4.1. Goal Programming (GP)

Goal programming is a kind of multi-criteria decision-making model. The model is established using both soft constraints and hard constraints. Soft constraints are used to model situations where deviations are acceptable to a desired goal value. Thus, more than one desired situation is provided approximately or fully. Goal programming is a mathematical programming method aiming at minimizing deviations from the goal values determined by turning aims to goals and ranking goals by importance ratings, or weighting each of them. In linear programming, while a single objective function is used, it is aimed to achieve the same goal by targeting multiple goals differently in the goal programming. In 1955, Charnes and his colleagues first worked on goal programming [67]. Later in 1961 and 1977, Charnes and Cooper developed this model [68,69]. Ignizio [70] describes goal programming as follows: Minimize the deviations in the aim thus that each target reaches as far as possible the given goals. The goal programming mathematical representation is is as follows [71]:

$$Minimize\ Z = \sum_{g=1}^{e} (d_g^+ + d_g^-) \tag{1}$$

$$\sum_{s=1}^{y} a_{gs}x_s - d_g^+ + d_g^- = b_g \tag{2}$$

$$d_g^+, d_g^-, x_s \geq 0 \tag{3}$$

$$g = 1 \ldots e \quad s = 1 \ldots y, g: Number\ of\ goals \quad s: Number\ of\ decision\ variables \tag{4}$$

$$x_s: s^{\text{th}}\ decision\ variable,\ s = 1 \ldots y \tag{5}$$

$$a_{gs}: Coefficient\ of\ g^{\text{th}}\ goal\ and\ s^{\text{th}}\ decision\ variable \quad g = 1 \ldots es = 1 \ldots y \tag{6}$$

$$b_g: Desired\ value\ for\ the\ goal\ g \quad g = 1 \ldots e \tag{7}$$

$$d_g^+: Positive\ deviation\ variable\ of\ goal \quad g = 1 \ldots e \tag{8}$$

$$d_g^-: Negative\ deviation\ variable\ of\ goal\ g \quad g = 1 \ldots e \tag{9}$$

4.2. Analytic Network Process (ANP)

Analytical Networking (ANP) is a multi-criteria decision-making technique developed by Thomas L. Saaty as a more general approach than the Analytic Hierarchy Process (AHP) method and works with the dual comparative logic like AHP. The Analytical Networking Process can be used to model decision problems that need to take account of the relationships between factors and to achieve more effective results. In the ANP method, factors affecting a goal and a target are grouped according to their effects on each other and a suitable network is modeled [72]. The ANP differs from the AHP in that it uses a hierarchical structure (network / network form) instead of a hierarchical structure from top to bottom [73]. Furthermore, an important problem encountered in the AHP method is rank reversal. Order change; the alternative priorities determined by a particular set of factors change when a new alternative is added or removed [74]. This problem has been reduced by the ANP method [72]. Steps of the ANP method is like that [75]:

Step 1: Determining the Decision-Making Problem

Step 2: Determining Relationships: Interactions between criteria and sub criteria are identified.

Step 3: Performing Criteria Binary Comparisons

Step 4: Calculating of Consistency: The consistency ratio (CR) of each binary comparison matrix is calculated. For consistency, CR < 0.1.

Step 5: Creating Super Matrices in Order:

- Unweighted Super Matrix: A square matristor consisting of super matrices, vectors that take into account all interactions between the problem criterion, sub criterion, and alternatives. - Weighted Super Matrix: The unweighted super matrix is equal to 1 column sum.
- Limit Super Matrix: The weighted super matrix is formed by taking the strength of the lines until they are not changed.

Step 6: Determination of the Best Alternative: Alternative priorities are obtained by selecting the highest alternative among these values by finding the limit super matrix and criterion weights.

5. Case Study

This study was carried out on a large-scale shift scheduling optimization with the sustainability of the electricity generation of a NGCCPP for 80 workers at their place of destination directly acting considering programming model for 30 days in Turkey. The application area and the magnitude of the problem have been realized by the combination of the methods mentioned for the first time in the literature. The implementation steps of the work are given in Figure 3. Our study was conducted with four levels. The First Level was getting shaped in the name of the data collection and model development. The second stage was generated as ANP calculations and analysis of the result. The third level was generated as a model computation and analysis step. The fourth level was the validation and verification of the model. In the first level, all data were taken from the technical management and human resource department of the power plant.

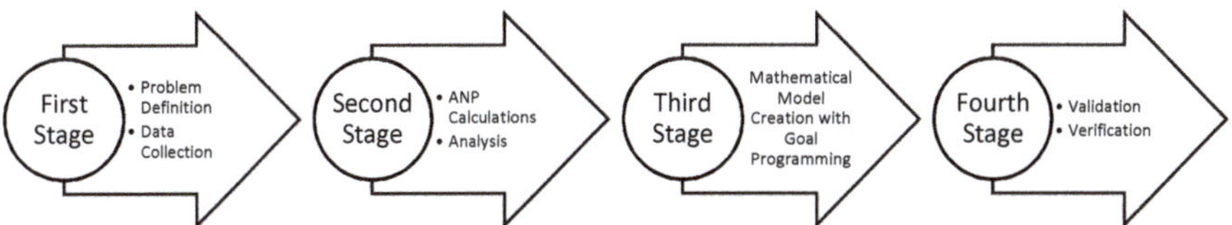

Figure 3. Schematic representation of the study.

In this study, scheduling focuses on three steps. These are demand modeling, scheduling of shifts, and scheduling of the days-off. This paper deals with SSP of the personnel who are in working for generation of the electricity to the power plant. The goal programming method is used for its advantages in many shifts and many employees of many workers.

5.1. Determining the Workers' Skill Weights with Analytic Network Process (ANP)

Analytic Network Process includes decision-maker's judgments value. It can also provide a decision-making process included in the model. Besides significant tangible or intangible factors to be correct, the decision must be examined. ANP gives solution that can accommodate within the complex structure and this method is capable of easy calculation. As mentioned in the first phase of the work, there are a number of situations that stop electricity production if it is checked and not intervened at the required times. The ability of each worker working in the natural gas combined cycle line to intervene in these situations is different. As it seen in Figure 4, ANP steps are applied in order. Schematic representation of the network structure can be seen in Figure 5.

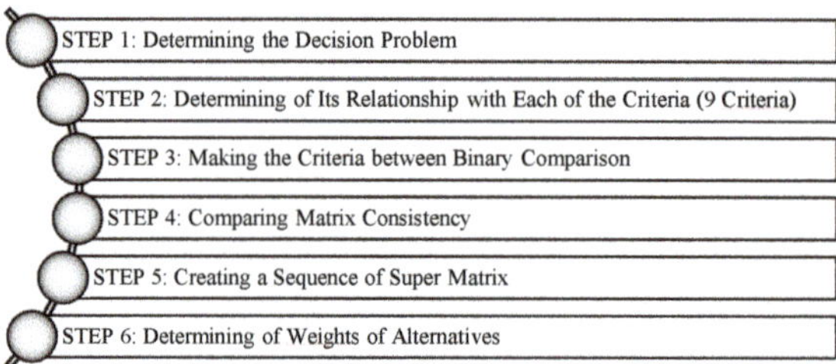

Figure 4. Schematic representation of the analytic network process (ANP) steps.

There are nine criteria of the skills of the workers. Each criterion is determined with the expert of the power plant. The network structure was established based on the internal and external dependencies among the criteria in the hierarchical structure. According to these nine criteria, each personnel's skill weight is determined. These criteria are given in Table 1.

Table 1. Criteria about personnel skills (C1, C2, ... , C9).

C1: Intervention to SCADA system	**C5:** Intervention to hydraulic lubrication system
C2: Subcontracting and removal of execution unit	**C6:** Intervention to central internal demand system
C3: Intervention to fault SCADA faults	**C7:** Intervention to water proofing equipment
C4: Intervention to main power transformer and equipments	**C8:** Interference to generator defects
C9: Intervention to switchgear equipment	

Consistency ratios (CR) of the binary comparison matrix prepared with the central experts were found to be less than 0.10. This means that the matrices were consistent. Three types of super matrix calculations are made in ANP. These are the unweighted supermatrix, weighted supermatrix and limit supermatrix. On a lean quilt; matrices obtained as a result of initial effects. The relative weight vectors obtained from the ANP charts are placed on the columns and rows of this matrix.

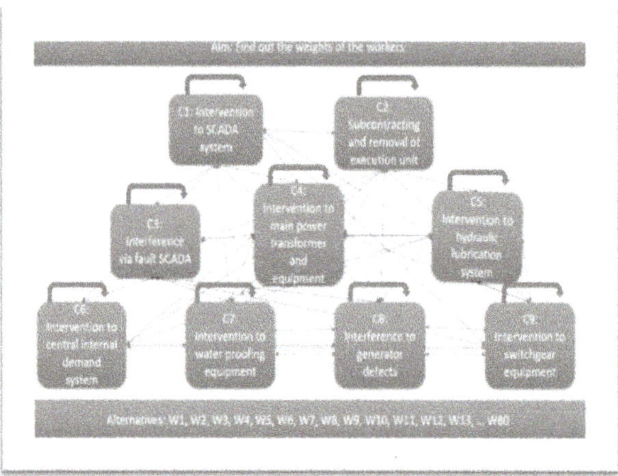

Figure 5. Schematic representation of the network structure.

After transferring the data to the super decisions 3.0 package program, the weighted sequence of the alternatives is reached by obtaining the non-weighted super matrix, the weighted super matrix and the limit super matrix, respectively. Limit super matrix of the ANP calculations is given in Appendix A as Figures A1–A3.

After all calculations, each worker's score is determined with the help of the nine criteria. In Table 2, ANP weights of each worker are given.

Table 2. ANP Weights (t_i) of the workers (P1, P2, P3, ... P80).

P1 to P10	P11 to P20	P21 to P30	P31 to P40	P41 to P50	P51 to P60	P61 to P70	P71 to P80
0.040845070	0.008450700	0.040845070	0.000069897	0.006960000	0.009990010	0.000069897	0.000069897
0.135560010	0.005865000	0.000868010	0.000021550	0.000025010	0.005665500	0.000021550	0.000021550
0.040840010	0.065868010	0.002567600	0.000021554	0.040845070	0.000860010	0.000021554	0.000021554
0.040045070	0.040111070	0.002240010	0.000002150	0.040845070	0.040775070	0.000002150	0.000002150
0.030845070	0.039845060	0.040845070	0.000000454	0.040845070	0.040845070	0.000000454	0.000000454
0.006960000	0.009990010	0.040845070	0.040845070	0.008450700	0.040663070	0.040845007	0.040845070
0.000025010	0.005665500	0.005686800	0.135560010	0.005865001	0.000868010	0.005686800	0.005686800
0.040845070	0.000860010	0.001235600	0.040840010	0.065868010	0.002567600	0.001235600	0.001235600
0.033945070	0.040845070	0.035655000	0.040045070	0.040845070	0.002240010	0.035655000	0.035655000
0.040740070	0.040845070	0.000000450	0.030845070	0.040845070	0.040845070	0.000000450	0.000000450

In this model, a seniority-based calculation was made in order to incorporate the skill basics into the model, and a separate mean value was calculated for each level of seniority level. In calculating these average values, the number of seniorities (4 different levels) and the individual weight values (obtained in ANP calculations) were used. Therefore, the weights are included in the model. According to this, the average values for the shift supervisor, foreman, expert and assistant are given in Table 3 respectively.

Table 3. Seniority-based ANP weights (t_i) of the workers.

Shift Supervisor's Weight	Foremen's Weight	Expert's Weight	Assistant's Weight
0.06432254	0.026957512	0.022802299	0.016094762

The values which are observed in Table 3 will be used in the goal constraints.

5.2. Proposed Goal Programming Model

Satisfying every goal specified by the decision-makers may not always be possible to completely provide. Therefore, goal programming is often referred to as a sequence of procedure in which the multiple goals are satisfied in their priority order [71]. The shift scheduling model will also attempt to satisfy several goals like establishing fairness among personnel. Because of that reason, a unique goal programming model is an excellent approach to solve this problem.

5.2.1. Notations

The workers (P1, P2, P3, ... P80) (*i*) in this department were divided into three shifts (*k*); Morning (S1) shift 8 a.m. to 4 p.m., Evening (S2) shift—4 p.m. to 12 a.m., and Night (S3) shift—12 a.m. to 8 a.m. Scheduling horizon is 30 days (*j*). Workers in the power plant have four different seniority level. These are:

- (1st Seniority-Shift Supervisor, 2nd Seniority-Foreman, 3rd Seniority-Expert, 4th Seniority-Assistant)
- There are 80 People in the department (4 Shift Supervisors, 12 Foremen, 24 Experts, 40 Assistants)
- 3 Shifts (Morning, Evening and Night Shifts)

There are some hard and weak constraints about our mathematical model. There are 10 sets of constraints, six of them correspond to the hard constraints (Constraint 1–Constraint 6 under 5.2.4 Constraints subsection) and the four corresponds to weak constraints (Goal 1–Goal 4 under 5.2.5 Goal Constraints subsection):

Constraint 1: Number of personnel needed for each shift every day. (3 shifts 20 people) Number of personnel assigned to their seniority in each shift.

- At least 1 Shift Supervisor must be present in each shift.   (1st Seniority Level)
- There must be at least 3 in each shift from the Foremen.   (2nd Seniority Level)
- There must be at least 6 in each shift from the Experts.   (3rd Seniority Level)
- There must be at least 10 in each shift from the Assistants. (4th Seniority Level)

Constraint 2: A staff working any day at night shift should not work in the morning and evening shifts the next day.

Constraint 3: A person working on any day of the evening shift should not work the next day in the morning.

Constraint 4: This constraint indicates that every staff member should take one day day-off (at least) in a week. In other words, every staff member should not work more than 6 days in a week:

Constraint 5: Every staff member should not work on his/her the day off.

Constraint 6: In the evening shift, the staff cannot be operated more than 9 days.

Decision variables to be used are X_{ijk} and h_{ij}, here notation of i, j and k are the indices for 80 workers, 30 days and three shifts, respectively. Variable X_{ijk} indicates worker i is assigned to work on day j for shift k and h_{ij}, indicate the assignment of worker i to be in day-off, respectively, on day j.

The complete formulation of the SSP-GP model is as follows:

5.2.2. Parameters: All Parameters Are Given in the Model Below

$$i: \text{Personnel index,} \quad i = 1,2,\dots,l \tag{10}$$

$$j: \text{Day index,} \quad j = 1,2,\dots,m \tag{11}$$

$$k: \text{Shift index,} \quad k = 1,2,\dots,n \tag{12}$$

$$g: \text{Goal index,} \quad g = 1,2,\dots,z \tag{13}$$

$$l: \text{Number of Personnel,} \quad l = 80 \tag{14}$$

$$m: \text{Number of Day,} \quad m = 30 \tag{15}$$

$$n: \text{Number of Shifts,} \quad n = 3 \tag{16}$$

$$t_i: \text{Weights of the skills of each personnel,} \quad i = 1,2,\dots,l \tag{17}$$

$$x_{ijk}: \text{Decision variable for i}^{\text{th}} \text{ personnel, j}^{\text{th}} \text{ day, for shift k } i = 1,2\dots l j = 1\dots m k = 1,2\dots n \tag{18}$$

$$h_{ij}: \text{Decision variable for day-off of i}^{\text{th}} \text{ personnel, j}^{\text{th}} \text{ day} \quad i = 1,2,\dots,l \quad j = 1\dots m \tag{19}$$

$$d^+_{gjk}: \text{Positive deviation variable of g}^{\text{th}} \text{ goal, j}^{\text{th}} \text{ day, for shift k} g = 1,2\dots z$$
$$j = 1,2\dots m k = 1,2\dots n \tag{20}$$

$$d^-_{gjk}: \text{Negative deviation variable of g}^{\text{th}} \text{ goal j}^{\text{th}} \text{ day, for shift k} g = 1,2\dots z$$
$$j = 1,2\dots m k = 1,2\dots n \tag{21}$$

5.2.3. Decision Variables: There Are Two Decision Variables on the Model. Those Are X_{ijk} and h_{ij}

$$X_{ijk} = \begin{cases} 1, & \textit{If personnel i is assigned to day j on shift k} \\ 0, & \textit{otherwise} \end{cases} i = 1,2,\dots,l j = 1,2,\dots,m k = 1,2,\dots,n \tag{22}$$

$$h_{ij} = \begin{cases} 1, & \textit{If the personnel i is on day} - \textit{off in day j} \\ 0, & \textit{otherwise} \end{cases} i = 1,2,\dots,l j = 1,2,\dots,m \tag{23}$$

5.2.4. Constraints

Constraint 1: The constraint that indicating the number of personnel assigned to each grade in accordance with their seniority:

a. Number of Shift Supervisors needed for shift k. $\sum_{i=1}^{4} X_{ijk} \geq 1 j = 1, 2, \ldots, m\, k = 1, 2, \ldots, n$ \hfill (24)

b. Number of Foremen needed for shift k. $\sum_{i=5}^{16} X_{ijk} \geq 3 j = 1, 2, \ldots, m\, k = 1, 2, \ldots, n$ \hfill (25)

c. Number of Expert needed for shift k. $\sum_{i=17}^{40} X_{ijk} \geq 6 j = 1, 2, \ldots, m\, k = 1, 2, \ldots, n$ \hfill (26)

d. Number of Assistant needed for shift k. $\sum_{i=41}^{80} X_{ijk} \geq 10 j = 1, 2, \ldots, m\, k = 1, 2, \ldots, n$ \hfill (27)

Constraint 2: The constraint that indicating if a staff working any day at night shift should not work in the morning and evening shifts the next day:

$$X_{ij3} + X_{i(j+1)1} + X_{i(j+1)2} \leq 1 i = 1, 2, 3, \ldots, l\, j = 1, 2, \ldots, m-1 \tag{28}$$

Constraint 3: The constraint that indicating if a person working on any day of the evening should not work the next day in the morning:

$$X_{ij2} + X_{i(j+1)1} \leq 1 i = 1, 2, 3, \ldots, l\, j = 1, 2, \ldots, m-1 \tag{29}$$

Constraint 4: This constraint indicates that every staff member should take one day day-off (at least) in a week. In other words, every staff member should not work more than 6 days in a week:

$$h_{ij} + h_{i(j+1)} + h_{i(j+2)} + h_{i(j+3)} + h_{i(j+4)} + h_{i(j+5)} + h_{i(j+6)} \geq 1 i = 1, 2, \ldots, l\, j = 1, 2, \ldots, 24 \tag{30}$$

Constraint 5: The constraint that indicating every staff member should not work on his/her the day off:

$$\sum_{k=1}^{3} X_{ijk} + h_{ij} = 1 i = 1, 2, 3, \ldots, l\, j = 1, 2, \ldots, m \tag{31}$$

Constraint 6: The constraint that indicating in the night shift, the staff cannot be operated more than 9 days:

$$X_{ij3} + X_{i(j+1)3} + X_{i(j+2)3} + X_{i(j+3)3} + X_{i(j+4)3} + X_{i(j+5)3} + X_{i(j+6)3} + X_{i(j+7)3} + X_{i(j+8)3} \leq 9$$
$$i = 1, 2, 3, \ldots, l\, j = 1, 2, \ldots, m-8 \tag{32}$$

5.2.4.1. Goal Constraints

The total number of shifts assigned to each staff by their seniority should be as equal as possible.
Goal 1: Constraints for Shift Supervisors

$$\sum_{i=1}^{4} t_i * X_{ijk} - d_{1jk}^{+} + d_{1jk}^{-} = 1 * 0.06432254 j = 1, 2, 3, \ldots, m\, k = 1, 2, \ldots, n \tag{33}$$

Goal 2: Constraints for Foremen

$$\sum_{i=5}^{16} t_i * X_{ijk} - d_{2jk}^{+} + d_{2jk}^{-} = 3 * 0.026957512 j = 1, 2, 3, \ldots, m\, k = 1, 2, \ldots, n \tag{34}$$

Goal 3: Constraints for Experts

$$\sum_{i=17}^{40} t_i * X_{ijk} - d_{3jk}^{+} + d_{3jk}^{-} = 6 * 0.022802299j = 1, 2, 3, \ldots, mk = 1, 2, \ldots, n \tag{35}$$

Goal 4: Constraints for Assistants

$$\sum_{i=41}^{80} t_i * X_{ijk} - d_{4jk}^{+} + d_{4jk}^{-} = 10 * 0.016094762j = 1, 2, 3, \ldots, mk = 1, 2, \ldots, n \tag{36}$$

5.2.4.2. Objective Function

$$min\ Z = \sum_{j=1}^{30} \sum_{k=1}^{3} d_{1jk}^{-} + d_{1jk}^{+} + d_{2jk}^{-} + d_{2jk}^{+} + d_{3jk}^{-} + d_{3jk}^{+} + d_{4jk}^{-} + d_{4jk}^{+} \tag{37}$$

5.2.5. Analysis of The Result

All goals have the same weight. Solving of the model is used with the features computer which processor "Intel (R) Core (TM) i7-2800 CPU@2.00 GH", 16 GB of memory and Windows 10 operating system. The proposed model, ILOG CPLEX Studio IDE is written in the program and is solved with the CPLEX solvent. The proposed schedule is created after running the ILOG CPLEX Studio IDE Solver at a reasonable time.

The complete assignment results can be seen in Figure 6. In order to examine the results in detail, the amount of deviation or not, if any, was calculated. The results show that a 1.56% positive deviation was observed from the second goal. A 0.48% negative deviation was observed to the fourth goal. There was no deviation from the first and third goals. As can be seen here, the deviation rates are very low, and the model has yielded positive and efficient results.

The model of this study increases the balanced assignments of the workers and that the program needs to adapt these workers on shift scheduling. Thus, it will increase the satisfaction of workers in terms of considering their shift schedule. As it seen from the Table 4, workloads of the personnel are unstable and irregular when the schedules are done in the manual. It has been determined that staff assignments are made by ignoring the hard and soft constraints.

In the Table 5, the workloads of each personnel can be seen after scheduling is done with a mathematical model. All hard constraints are satisfied and many of the soft constraints are satisfied. Worker's preferences are not ignored.

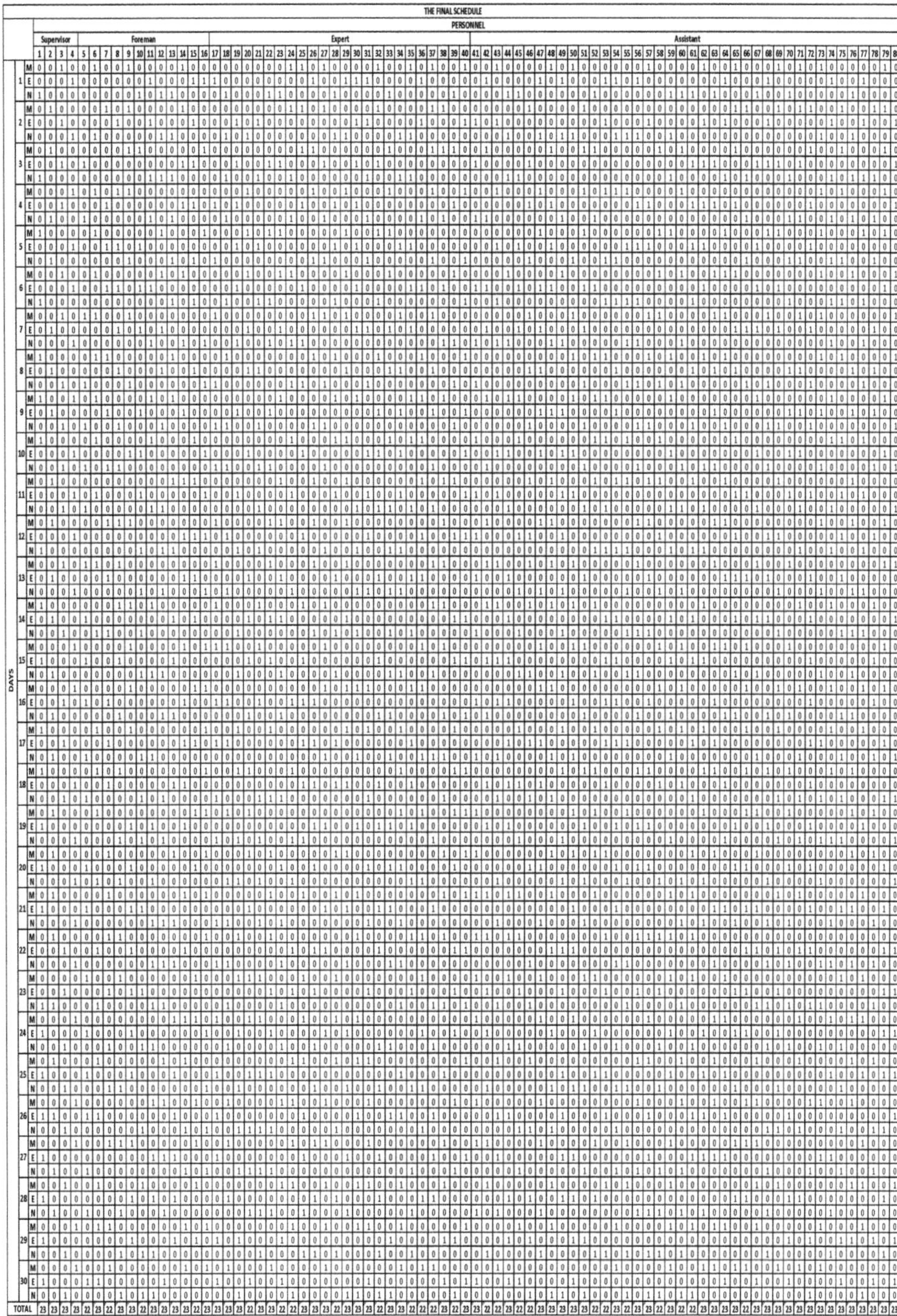

Figure 6. The final schedule.

Table 4. Workloads of each personnel with manual scheduling.

P1	27	P9	19	P17	26	P25	25	P33	20	P41	22	P49	20	P57	19	P65	20	P73	22
P2	11	P10	23	P18	19	P26	20	P34	14	P42	22	P50	26	P58	27	P66	20	P74	30
P3	17	P11	26	P19	24	P27	27	P35	16	P43	21	P51	24	P59	27	P67	17	P75	30
P4	20	P12	25	P20	25	P28	26	P36	21	P44	27	P52	17	P60	21	P68	15	P76	19
P5	25	P13	26	P21	27	P29	28	P37	23	P45	11	P53	30	P61	16	P69	19	P77	22
P6	21	P14	20	P22	20	P30	17	P38	22	P46	25	P54	12	P62	22	P70	21	P78	16
P7	26	P15	19	P23	26	P31	25	P39	26	P47	21	P55	30	P63	23	P71	27	P79	23
P8	22	P16	24	P24	23	P32	10	P40	24	P48	22	P56	24	P64	26	P72	22	P80	22

Table 5. Workloads of each personnel after scheduling is done with mathematical model.

P1	23	P9	23	P17	23	P25	23	P33	23	P41	23	P49	23	P57	20	P65	23	P73	23
P2	23	P10	22	P18	23	P26	23	P34	23	P42	22	P50	23	P58	23	P66	22	P74	23
P3	23	P11	23	P19	23	P27	23	P35	22	P43	22	P51	23	P59	23	P67	23	P75	23
P4	23	P12	23	P20	22	P28	22	P36	22	P44	23	P52	22	P60	22	P68	23	P76	23
P5	22	P13	23	P21	23	P29	23	P37	22	P45	22	P53	22	P61	22	P69	23	P77	23
P6	23	P14	23	P22	23	P30	23	P38	23	P46	22	P54	22	P62	23	P70	23	P78	23
P7	22	P15	22	P23	22	P31	23	P39	22	P47	23	P55	23	P63	23	P71	23	P79	23
P8	23	P16	23	P24	22	P32	22	P40	23	P48	23	P56	22	P64	23	P72	23	P80	23

Computational results are given in Table 6. When the results are examined, the ratio of the demands that are met before the work is done and the rates after the work are done are different from each other. According to gathering results, fair scheduling requests were met, and fairness was provided from shift schedules. The preferences mentioned in the study constitute the entire hard and soft constraints. In the previous manual scheduling, 4 of 10 constraints, some hard, some soft, were not available. After the solution of the proposed mathematical model, all hard and weak constraints were met. 24 of the total 80 employees work 22 days and 56 of them work 23 days in a month for proposed scheduling mathematical model.

Table 6. Computational result.

Criteria	Existing Schedule	Proposed Schedule
Unsatisfied preference	4	0
Satisfied preference	6	10
Total preference	10	10
Percentage of satisfaction	**%60.00**	**%100.00**

This scheduling study, together with shift scheduling issues, has made very high contributions to reducing costs. In order to express the contributions of this work in decreasing costs, it is more appropriate to first explain the calculation over the capacity of the plant. We can say that if the power plant runs at 100% capacity, installed capacity of the power plant is 1000 MW. According to this, the monthly capacity can be calculated as 720 hours (24 hours × 30 days). 1 MW electricity is selling by 17.1 Kuruş (One Turkish Lira (₺) is equal to 100 Turkish Kuruş) from the Energy Market Inspection Agency (EPDK, Ankara, Turkey). This value is the wholesale electricity sales price for 2018 in Turkey. We can find the monthly 100% total generation capacity of the NGCCPP as 720.000 MW (1000 MW × 24 Hours × 30 Days).

There are some shutdown problems of the power plant also included, which arise from doing shift scheduling manually. When the shift scheduling is done manually, on average 53 hours were lost because of SSP. Due to the faults from shift scheduling, 53 hours shutdown caused 90.630.000 ₺ expenses to the NGCCPP (53 Hours × 1000 MW × 17,1 Krş × 100). After the proposed schedule was applied in six months in the power plant, some results were gathered and calculated from the engineering department. According to this, 53 hours loss has been decreased to 4 hours loss after the proposed model of shift scheduling was done. After our implementation of the new schedule, only 4 hours shutdown were seen in the NGCCPP because of the faults from shift scheduling errors. 4 hours shutdown cause 6.840.000 ₺ expenses to the NGCCPP (4 Hours × 1000 MW × 17,1 Krş × 100). According to this calculations, 83.790.000 ₺ total profit (90.630.000 ₺ − 6.840.000 ₺) is gathering from proposed model of shift scheduling. For this reason, this article differs from other studies in the literature in the context of giving the results of application.

6. Conclusions

NGCCPPs have significant advantages such as having high capacities, low setup times, low installation costs, low environmental impacts, short run times, high yields, ease of operation and maintenance, and a long economic lifespan compared to the other types of power plants, which are especially fossil fueled and providing base load requirements. Despite the lack of resource availability, because of these advantages, these power plants are located in first place with a third of Turkey's energy mix as at of the end of 2018.

The main purpose of NGCCPPs, as in all others, is to provide a sustainable energy supply and one of the most important and problematic pillars of this comprehensive goal is the uninterruptedness. Because, many factors such as operating the power plants without considering the operation and maintenance directives, fuel quality, grid failure, atmospheric conditions, and lack of water cause the NGCCPs' shutdowns. In addition to this, in NGCCPPs without a SCADA system, the power plant is operated from a main control room or in place by intervening on the equipment spread over the power plant. In this context, the continuity of generation is directly related to the operators, or in other words, to their attention. Because fatigue, unwillingness, and a lack of motivation may cause the operators not to perform the necessary interventions on time. This may lead to longer shutdowns in the power plant. An important way to increase operator motivation in power plants is to establish a fair work distribution and to make competency-based assignments. Therefore, in this study, talent based SSP is handled in one of the large-scale NGCCPP in Turkey and a monthly shift schedule is obtained by solving the proposed multi-objective goal programming model supported with ANP, which is used for calculating the operators' competencies for decreasing the generation shutdowns due to operator errors.

The power plant has previously been operated by schedules which were unplanned, arbitrary, and without considering the operator qualifications. Therefore, the motivation and work requests of the operators had been lost, and their attention levels decreased considerably. Thus, the power plant had to interrupt generation for 53 hours due to operator error. This loss means millions of kWh of energy; equaling to a significant monetary magnitude. The shift schedule produced by the proposed model within this study has maximized the operator motivation on the basis of fair work distribution and capabilities. Thus, a 92.5% improvement was achieved in operator-centered generation downtimes and millions of TL losses were prevented.

Two of the most important reasons for operators' motivation loss are that operators are assigned to the shifts without a fair and balanced distribution and without regard to their level of expertise, regardless of their seniority. This fact is agreed upon by all managers and operators at the power plant where this study was carried out. The impact of their level of expertise on motivation is related to each operator's feeling safe. As mentioned above, the fact that the generation carried out in the power plant is removed from sustainability can cause significant financial losses and social problems. In addition, each shift is a team of four seniority levels, and team members think that a balanced distribution of expertise levels will ensure that sustainable energy supply will not be interrupted. This is consistent with the 92.5% improvement in the proposed model's generation downtimes. From this point of view, it can be interpreted that operator motivation is increased by taking the working competencies of the proposed model into consideration in the shift schedules produced in NGCCPPs.

The model was solved by using IBM ILOG Optimization Tool in a reasonable time by considering the complexity of the model and the use of precise and intuitive algorithms rather than metaheuristic approaches. However, the SSP-ANP-GP model significantly reduces the time it takes to obtain a monthly schedule based on the current manual scheduling. On the basis of the 30-day program obtained with the SSP-ANP-GP model, the satisfaction of the workers for their day-off preferences increased significantly compared to the current schedule. In addition, according to the current schedule and practice, each worker has to work with a fixed group of people throughout the year. By contrast, in addition to having preferred leisure time choices, the workers have the opportunity to work with a variety of people for their shifts. It has been shown that optimal results that maximizes the day-off preferences can be obtained using the formulated SSP-ANP-GP model.

The proposed SSP-ANP-GP model has the potential to be adapted to the other power plants by changing some possible constraints such as holidays, operator groups, shift number etc. In other words, all types of power plants have specific conditions, requirements, or constraints based on their technology, and the proposed model can be directly applied to the NGCCPPs without a DCS system. However, computational results consistent with real life power plant management of the proposed model show that this study comes to the fore from other studies in the literature.

The examination of the results of the study and the possibility of long-term implementation are important for the contribution to the literature. In practice, SSPs tend to be intertwined with other factory programming problems. For example, the reservation date etc. (it is not possible to change). In the literature, these more general problems (the integration of machine planning and personnel planning) have not been addressed yet. However, there are a number of programming systems on the market at this time. Regarding the solution method, we can see that the literature has multiplied the mathematical programming approaches and metaheuristic approaches.

In spite of the significant increase in productivity of complete algorithms in order to solve integer programming problems in recent years, they are often not applicable to solve practical problems in the medium and large size due to excessive working times and memory requirements. For this reason, for the future application of the SSP-ANP-GP model in the electricity generation of workers' shift schedules, heuristics can be considered to obtain a complete or approximate solution for shorter calculation times. Other proposals for future work to expand or investigate the SSP for future applications may include strategically reducing the number of decision variables and constraints to make it more practical in the field of application for overtime. In a different point of view, researchers may create a user-friendly interface for using the mathematical model and knowledge based on the solutions obtained by the users to solve SSPs quickly.

Author Contributions: The analyses were carried out by E.H.Ö. This paper was written by E.H.Ö., E.Ö. and T.E. supervised the conducted the research, reviewed and revised the paper. T.E. and E.Ö. reviewed, revised, and improved the paper.

Funding: This research received no external funding.

Conflicts of Interest: "The authors declare no conflict of interest." "The funders had no role in the design of the study; in the collection, analyses, or interpretation of data; in the writing of the manuscript, or in the decision to publish the results".

Appendix A

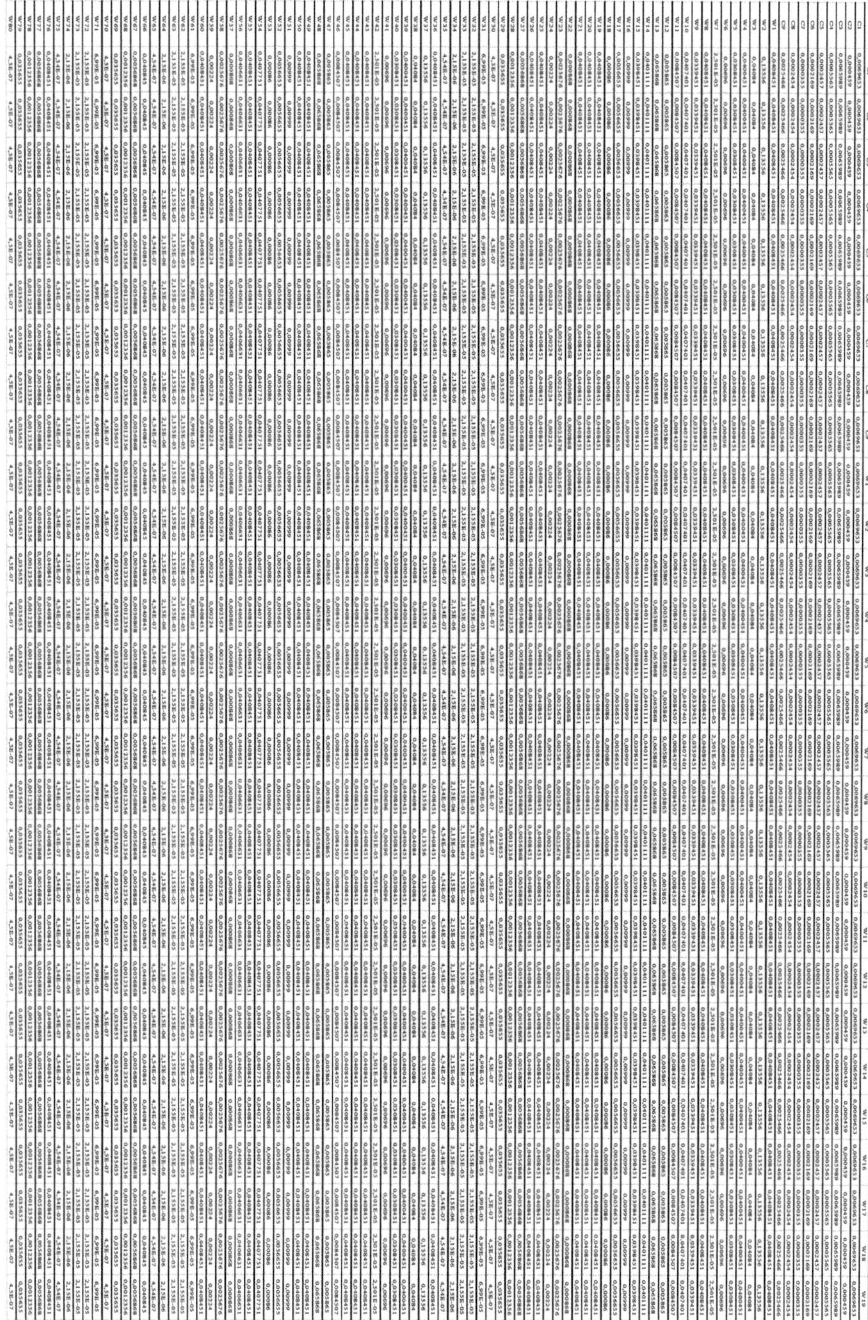

Figure A1. Limit super matrix of the ANP calculations.

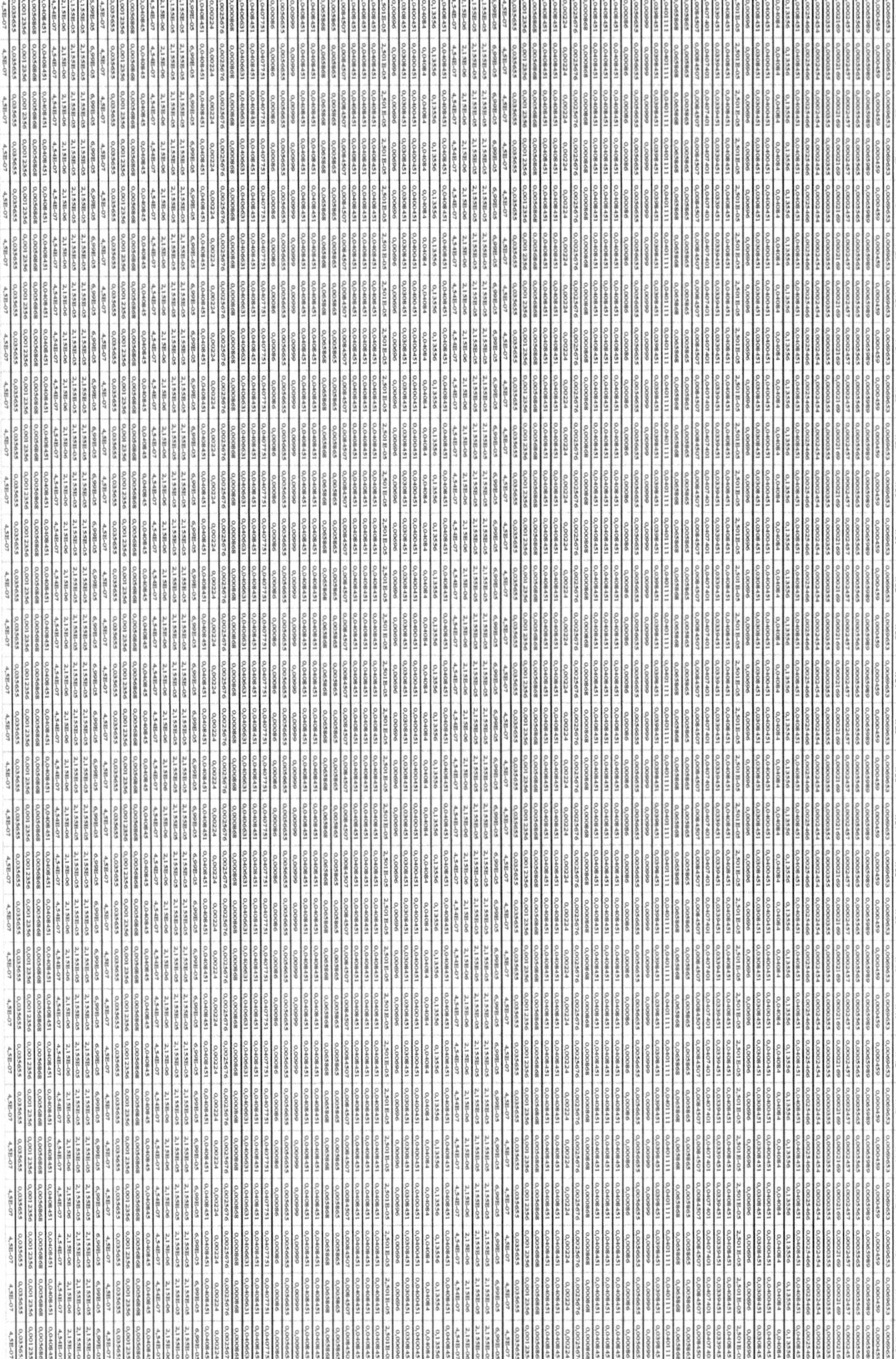

Figure A2. Limit super matrix of the ANP calculations (cont.).

W49	W50	W51	W52	W53	W54	W55	W56	W57	W58	W59	W60	W61	W62	W63	W64	W65	W66	W67	W68	W69	W70	W71	W72	W73	W74	W75	W76	W77	W78	W79	W80
0,0069653	0,0069653	0,0069653	0,0069653	0,0069653	0,0069653	0,0069653	0,0069653	0,0069653	0,0069653	0,0069653	0,0069653	0,0069653	0,0069653	0,0069653	0,0069653	0,0069653	0,0069653	0,0069653	0,0069653	0,0069653	0,0069653	0,0069653	0,0069653	0,0069653	0,0069653	0,0069653	0,0069653	0,0069653	0,0069653	0,0069653	0,0069653
0,000459	0,000459	0,000459	0,000459	0,000459	0,000459	0,000459	0,000459	0,000459	0,000459	0,000459	0,000459	0,000459	0,000459	0,000459	0,000459	0,000459	0,000459	0,000459	0,000459	0,000459	0,000459	0,000459	0,000459	0,000459	0,000459	0,000459	0,000459	0,000459	0,000459	0,000459	0,000459
0,0065989	0,0065989	0,0065989	0,0065989	0,0065989	0,0065989	0,0065989	0,0065989	0,0065989	0,0065989	0,0065989	0,0065989	0,0065989	0,0065989	0,0065989	0,0065989	0,0065989	0,0065989	0,0065989	0,0065989	0,0065989	0,0065989	0,0065989	0,0065989	0,0065989	0,0065989	0,0065989	0,0065989	0,0065989	0,0065989	0,0065989	0,0065989
0,0005565	0,0005565	0,0005565	0,0005565	0,0005565	0,0005565	0,0005565	0,0005565	0,0005565	0,0005565	0,0005565	0,0005565	0,0005565	0,0005565	0,0005565	0,0005565	0,0005565	0,0005565	0,0005565	0,0005565	0,0005565	0,0005565	0,0005565	0,0005565	0,0005565	0,0005565	0,0005565	0,0005565	0,0005565	0,0005565	0,0005565	0,0005565
0,0002457	0,0002457	0,0002457	0,0002457	0,0002457	0,0002457	0,0002457	0,0002457	0,0002457	0,0002457	0,0002457	0,0002457	0,0002457	0,0002457	0,0002457	0,0002457	0,0002457	0,0002457	0,0002457	0,0002457	0,0002457	0,0002457	0,0002457	0,0002457	0,0002457	0,0002457	0,0002457	0,0002457	0,0002457	0,0002457	0,0002457	0,0002457
0,0002169	0,0002169	0,0002169	0,0002169	0,0002169	0,0002169	0,0002169	0,0002169	0,0002169	0,0002169	0,0002169	0,0002169	0,0002169	0,0002169	0,0002169	0,0002169	0,0002169	0,0002169	0,0002169	0,0002169	0,0002169	0,0002169	0,0002169	0,0002169	0,0002169	0,0002169	0,0002169	0,0002169	0,0002169	0,0002169	0,0002169	0,0002169
0,0000355	0,0000355	0,0000355	0,0000355	0,0000355	0,0000355	0,0000355	0,0000355	0,0000355	0,0000355	0,0000355	0,0000355	0,0000355	0,0000355	0,0000355	0,0000355	0,0000355	0,0000355	0,0000355	0,0000355	0,0000355	0,0000355	0,0000355	0,0000355	0,0000355	0,0000355	0,0000355	0,0000355	0,0000355	0,0000355	0,0000355	0,0000355
0,0002454	0,0002454	0,0002454	0,0002454	0,0002454	0,0002454	0,0002454	0,0002454	0,0002454	0,0002454	0,0002454	0,0002454	0,0002454	0,0002454	0,0002454	0,0002454	0,0002454	0,0002454	0,0002454	0,0002454	0,0002454	0,0002454	0,0002454	0,0002454	0,0002454	0,0002454	0,0002454	0,0002454	0,0002454	0,0002454	0,0002454	0,0002454
0,0025466	0,0025466	0,0025466	0,0025466	0,0025466	0,0025466	0,0025466	0,0025466	0,0025466	0,0025466	0,0025466	0,0025466	0,0025466	0,0025466	0,0025466	0,0025466	0,0025466	0,0025466	0,0025466	0,0025466	0,0025466	0,0025466	0,0025466	0,0025466	0,0025466	0,0025466	0,0025466	0,0025466	0,0025466	0,0025466	0,0025466	0,0025466
0,0408451	0,0408451	0,0408451	0,0408451	0,0408451	0,0408451	0,0408451	0,0408451	0,0408451	0,0408451	0,0408451	0,0408451	0,0408451	0,0408451	0,0408451	0,0408451	0,0408451	0,0408451	0,0408451	0,0408451	0,0408451	0,0408451	0,0408451	0,0408451	0,0408451	0,0408451	0,0408451	0,0408451	0,0408451	0,0408451	0,0408451	0,0408451
0,13556	0,13556	0,13556	0,13556	0,13556	0,13556	0,13556	0,13556	0,13556	0,13556	0,13556	0,13556	0,13556	0,13556	0,13556	0,13556	0,13556	0,13556	0,13556	0,13556	0,13556	0,13556	0,13556	0,13556	0,13556	0,13556	0,13556	0,13556	0,13556	0,13556	0,13556	0,13556
0,04084	0,04084	0,04084	0,04084	0,04084	0,04084	0,04084	0,04084	0,04084	0,04084	0,04084	0,04084	0,04084	0,04084	0,04084	0,04084	0,04084	0,04084	0,04084	0,04084	0,04084	0,04084	0,04084	0,04084	0,04084	0,04084	0,04084	0,04084	0,04084	0,04084	0,04084	0,04084
0,0400451	0,0400451	0,0400451	0,0400451	0,0400451	0,0400451	0,0400451	0,0400451	0,0400451	0,0400451	0,0400451	0,0400451	0,0400451	0,0400451	0,0400451	0,0400451	0,0400451	0,0400451	0,0400451	0,0400451	0,0400451	0,0400451	0,0400451	0,0400451	0,0400451	0,0400451	0,0400451	0,0400451	0,0400451	0,0400451	0,0400451	0,0400451
0,0308451	0,0308451	0,0308451	0,0308451	0,0308451	0,0308451	0,0308451	0,0308451	0,0308451	0,0308451	0,0308451	0,0308451	0,0308451	0,0308451	0,0308451	0,0308451	0,0308451	0,0308451	0,0308451	0,0308451	0,0308451	0,0308451	0,0308451	0,0308451	0,0308451	0,0308451	0,0308451	0,0308451	0,0308451	0,0308451	0,0308451	0,0308451
0,00696	0,00696	0,00696	0,00696	0,00696	0,00696	0,00696	0,00696	0,00696	0,00696	0,00696	0,00696	0,00696	0,00696	0,00696	0,00696	0,00696	0,00696	0,00696	0,00696	0,00696	0,00696	0,00696	0,00696	0,00696	0,00696	0,00696	0,00696	0,00696	0,00696	0,00696	0,00696
2,501E-05	2,501E-05	2,501E-05	2,501E-05	2,501E-05	2,501E-05	2,501E-05	2,501E-05	2,501E-05	2,501E-05	2,501E-05	2,501E-05	2,501E-05	2,501E-05	2,501E-05	2,501E-05	2,501E-05	2,501E-05	2,501E-05	2,501E-05	2,501E-05	2,501E-05	2,501E-05	2,501E-05	2,501E-05	2,501E-05	2,501E-05	2,501E-05	2,501E-05	2,501E-05	2,501E-05	2,501E-05
0,0408451	0,0408451	0,0408451	0,0408451	0,0408451	0,0408451	0,0408451	0,0408451	0,0408451	0,0408451	0,0408451	0,0408451	0,0408451	0,0408451	0,0408451	0,0408451	0,0408451	0,0408451	0,0408451	0,0408451	0,0408451	0,0408451	0,0408451	0,0408451	0,0408451	0,0408451	0,0408451	0,0408451	0,0408451	0,0408451	0,0408451	0,0408451
0,0339451	0,0339451	0,0339451	0,0339451	0,0339451	0,0339451	0,0339451	0,0339451	0,0339451	0,0339451	0,0339451	0,0339451	0,0339451	0,0339451	0,0339451	0,0339451	0,0339451	0,0339451	0,0339451	0,0339451	0,0339451	0,0339451	0,0339451	0,0339451	0,0339451	0,0339451	0,0339451	0,0339451	0,0339451	0,0339451	0,0339451	0,0339451
0,0407401	0,0407401	0,0407401	0,0407401	0,0407401	0,0407401	0,0407401	0,0407401	0,0407401	0,0407401	0,0407401	0,0407401	0,0407401	0,0407401	0,0407401	0,0407401	0,0407401	0,0407401	0,0407401	0,0407401	0,0407401	0,0407401	0,0407401	0,0407401	0,0407401	0,0407401	0,0407401	0,0407401	0,0407401	0,0407401	0,0407401	0,0407401
0,0084507	0,0084507	0,0084507	0,0084507	0,0084507	0,0084507	0,0084507	0,0084507	0,0084507	0,0084507	0,0084507	0,0084507	0,0084507	0,0084507	0,0084507	0,0084507	0,0084507	0,0084507	0,0084507	0,0084507	0,0084507	0,0084507	0,0084507	0,0084507	0,0084507	0,0084507	0,0084507	0,0084507	0,0084507	0,0084507	0,0084507	0,0084507
0,005865	0,005865	0,005865	0,005865	0,005865	0,005865	0,005865	0,005865	0,005865	0,005865	0,005865	0,005865	0,005865	0,005865	0,005865	0,005865	0,005865	0,005865	0,005865	0,005865	0,005865	0,005865	0,005865	0,005865	0,005865	0,005865	0,005865	0,005865	0,005865	0,005865	0,005865	0,005865
0,065868	0,065868	0,065868	0,065868	0,065868	0,065868	0,065868	0,065868	0,065868	0,065868	0,065868	0,065868	0,065868	0,065868	0,065868	0,065868	0,065868	0,065868	0,065868	0,065868	0,065868	0,065868	0,065868	0,065868	0,065868	0,065868	0,065868	0,065868	0,065868	0,065868	0,065868	0,065868
0,0401111	0,0401111	0,0401111	0,0401111	0,0401111	0,0401111	0,0401111	0,0401111	0,0401111	0,0401111	0,0401111	0,0401111	0,0401111	0,0401111	0,0401111	0,0401111	0,0401111	0,0401111	0,0401111	0,0401111	0,0401111	0,0401111	0,0401111	0,0401111	0,0401111	0,0401111	0,0401111	0,0401111	0,0401111	0,0401111	0,0401111	0,0401111
0,0398451	0,0398451	0,0398451	0,0398451	0,0398451	0,0398451	0,0398451	0,0398451	0,0398451	0,0398451	0,0398451	0,0398451	0,0398451	0,0398451	0,0398451	0,0398451	0,0398451	0,0398451	0,0398451	0,0398451	0,0398451	0,0398451	0,0398451	0,0398451	0,0398451	0,0398451	0,0398451	0,0398451	0,0398451	0,0398451	0,0398451	0,0398451
0,00999	0,00999	0,00999	0,00999	0,00999	0,00999	0,00999	0,00999	0,00999	0,00999	0,00999	0,00999	0,00999	0,00999	0,00999	0,00999	0,00999	0,00999	0,00999	0,00999	0,00999	0,00999	0,00999	0,00999	0,00999	0,00999	0,00999	0,00999	0,00999	0,00999	0,00999	0,00999
0,0056655	0,0056655	0,0056655	0,0056655	0,0056655	0,0056655	0,0056655	0,0056655	0,0056655	0,0056655	0,0056655	0,0056655	0,0056655	0,0056655	0,0056655	0,0056655	0,0056655	0,0056655	0,0056655	0,0056655	0,0056655	0,0056655	0,0056655	0,0056655	0,0056655	0,0056655	0,0056655	0,0056655	0,0056655	0,0056655	0,0056655	0,0056655
0,00086	0,00086	0,00086	0,00086	0,00086	0,00086	0,00086	0,00086	0,00086	0,00086	0,00086	0,00086	0,00086	0,00086	0,00086	0,00086	0,00086	0,00086	0,00086	0,00086	0,00086	0,00086	0,00086	0,00086	0,00086	0,00086	0,00086	0,00086	0,00086	0,00086	0,00086	0,00086
0,0408451	0,0408451	0,0408451	0,0408451	0,0408451	0,0408451	0,0408451	0,0408451	0,0408451	0,0408451	0,0408451	0,0408451	0,0408451	0,0408451	0,0408451	0,0408451	0,0408451	0,0408451	0,0408451	0,0408451	0,0408451	0,0408451	0,0408451	0,0408451	0,0408451	0,0408451	0,0408451	0,0408451	0,0408451	0,0408451	0,0408451	0,0408451
0,0408451	0,0408451	0,0408451	0,0408451	0,0408451	0,0408451	0,0408451	0,0408451	0,0408451	0,0408451	0,0408451	0,0408451	0,0408451	0,0408451	0,0408451	0,0408451	0,0408451	0,0408451	0,0408451	0,0408451	0,0408451	0,0408451	0,0408451	0,0408451	0,0408451	0,0408451	0,0408451	0,0408451	0,0408451	0,0408451	0,0408451	0,0408451
0,0408451	0,0408451	0,0408451	0,0408451	0,0408451	0,0408451	0,0408451	0,0408451	0,0408451	0,0408451	0,0408451	0,0408451	0,0408451	0,0408451	0,0408451	0,0408451	0,0408451	0,0408451	0,0408451	0,0408451	0,0408451	0,0408451	0,0408451	0,0408451	0,0408451	0,0408451	0,0408451	0,0408451	0,0408451	0,0408451	0,0408451	0,0408451
0,000868	0,000868	0,000868	0,000868	0,000868	0,000868	0,000868	0,000868	0,000868	0,000868	0,000868	0,000868	0,000868	0,000868	0,000868	0,000868	0,000868	0,000868	0,000868	0,000868	0,000868	0,000868	0,000868	0,000868	0,000868	0,000868	0,000868	0,000868	0,000868	0,000868	0,000868	0,000868
0,0025676	0,0025676	0,0025676	0,0025676	0,0025676	0,0025676	0,0025676	0,0025676	0,0025676	0,0025676	0,0025676	0,0025676	0,0025676	0,0025676	0,0025676	0,0025676	0,0025676	0,0025676	0,0025676	0,0025676	0,0025676	0,0025676	0,0025676	0,0025676	0,0025676	0,0025676	0,0025676	0,0025676	0,0025676	0,0025676	0,0025676	0,0025676
0,00224	0,00224	0,00224	0,00224	0,00224	0,00224	0,00224	0,00224	0,00224	0,00224	0,00224	0,00224	0,00224	0,00224	0,00224	0,00224	0,00224	0,00224	0,00224	0,00224	0,00224	0,00224	0,00224	0,00224	0,00224	0,00224	0,00224	0,00224	0,00224	0,00224	0,00224	0,00224
0,0408451	0,0408451	0,0408451	0,0408451	0,0408451	0,0408451	0,0408451	0,0408451	0,0408451	0,0408451	0,0408451	0,0408451	0,0408451	0,0408451	0,0408451	0,0408451	0,0408451	0,0408451	0,0408451	0,0408451	0,0408451	0,0408451	0,0408451	0,0408451	0,0408451	0,0408451	0,0408451	0,0408451	0,0408451	0,0408451	0,0408451	0,0408451
0,0408451	0,0408451	0,0408451	0,0408451	0,0408451	0,0408451	0,0408451	0,0408451	0,0408451	0,0408451	0,0408451	0,0408451	0,0408451	0,0408451	0,0408451	0,0408451	0,0408451	0,0408451	0,0408451	0,0408451	0,0408451	0,0408451	0,0408451	0,0408451	0,0408451	0,0408451	0,0408451	0,0408451	0,0408451	0,0408451	0,0408451	0,0408451
0,0056868	0,0056868	0,0056868	0,0056868	0,0056868	0,0056868	0,0056868	0,0056868	0,0056868	0,0056868	0,0056868	0,0056868	0,0056868	0,0056868	0,0056868	0,0056868	0,0056868	0,0056868	0,0056868	0,0056868	0,0056868	0,0056868	0,0056868	0,0056868	0,0056868	0,0056868	0,0056868	0,0056868	0,0056868	0,0056868	0,0056868	0,0056868
0,0012356	0,0012356	0,0012356	0,0012356	0,0012356	0,0012356	0,0012356	0,0012356	0,0012356	0,0012356	0,0012356	0,0012356	0,0012356	0,0012356	0,0012356	0,0012356	0,0012356	0,0012356	0,0012356	0,0012356	0,0012356	0,0012356	0,0012356	0,0012356	0,0012356	0,0012356	0,0012356	0,0012356	0,0012356	0,0012356	0,0012356	0,0012356
0,035655	0,035655	0,035655	0,035655	0,035655	0,035655	0,035655	0,035655	0,035655	0,035655	0,035655	0,035655	0,035655	0,035655	0,035655	0,035655	0,035655	0,035655	0,035655	0,035655	0,035655	0,035655	0,035655	0,035655	0,035655	0,035655	0,035655	0,035655	0,035655	0,035655	0,035655	0,035655
4,5E-07	4,5E-07	4,5E-07	4,5E-07	4,5E-07	4,5E-07	4,5E-07	4,5E-07	4,5E-07	4,5E-07	4,5E-07	4,5E-07	4,5E-07	4,5E-07	4,5E-07	4,5E-07	4,5E-07	4,5E-07	4,5E-07	4,5E-07	4,5E-07	4,5E-07	4,5E-07	4,5E-07	4,5E-07	4,5E-07	4,5E-07	4,5E-07	4,5E-07	4,5E-07	4,5E-07	4,5E-07
6,99E-05	6,99E-05	6,99E-05	6,99E-05	6,99E-05	6,99E-05	6,99E-05	6,99E-05	6,99E-05	6,99E-05	6,99E-05	6,99E-05	6,99E-05	6,99E-05	6,99E-05	6,99E-05	6,99E-05	6,99E-05	6,99E-05	6,99E-05	6,99E-05	6,99E-05	6,99E-05	6,99E-05	6,99E-05	6,99E-05	6,99E-05	6,99E-05	6,99E-05	6,99E-05	6,99E-05	6,99E-05
2,155E-05	2,155E-05	2,155E-05	2,155E-05	2,155E-05	2,155E-05	2,155E-05	2,155E-05	2,155E-05	2,155E-05	2,155E-05	2,155E-05	2,155E-05	2,155E-05	2,155E-05	2,155E-05	2,155E-05	2,155E-05	2,155E-05	2,155E-05	2,155E-05	2,155E-05	2,155E-05	2,155E-05	2,155E-05	2,155E-05	2,155E-05	2,155E-05	2,155E-05	2,155E-05	2,155E-05	2,155E-05
2,155E-05	2,155E-05	2,155E-05	2,155E-05	2,155E-05	2,155E-05	2,155E-05	2,155E-05	2,155E-05	2,155E-05	2,155E-05	2,155E-05	2,155E-05	2,155E-05	2,155E-05	2,155E-05	2,155E-05	2,155E-05	2,155E-05	2,155E-05	2,155E-05	2,155E-05	2,155E-05	2,155E-05	2,155E-05	2,155E-05	2,155E-05	2,155E-05	2,155E-05	2,155E-05	2,155E-05	2,155E-05
2,15E-06	2,15E-06	2,15E-06	2,15E-06	2,15E-06	2,15E-06	2,15E-06	2,15E-06	2,15E-06	2,15E-06	2,15E-06	2,15E-06	2,15E-06	2,15E-06	2,15E-06	2,15E-06	2,15E-06	2,15E-06	2,15E-06	2,15E-06	2,15E-06	2,15E-06	2,15E-06	2,15E-06	2,15E-06	2,15E-06	2,15E-06	2,15E-06	2,15E-06	2,15E-06	2,15E-06	2,15E-06
4,54E-07	4,54E-07	4,54E-07	4,54E-07	4,54E-07	4,54E-07	4,54E-07	4,54E-07	4,54E-07	4,54E-07	4,54E-07	4,54E-07	4,54E-07	4,54E-07	4,54E-07	4,54E-07	4,54E-07	4,54E-07	4,54E-07	4,54E-07	4,54E-07	4,54E-07	4,54E-07	4,54E-07	4,54E-07	4,54E-07	4,54E-07	4,54E-07	4,54E-07	4,54E-07	4,54E-07	4,54E-07
0,0408451	0,0408451	0,0408451	0,0408451	0,0408451	0,0408451	0,0408451	0,0408451	0,0408451	0,0408451	0,0408451	0,0408451	0,0408451	0,0408451	0,0408451	0,0408451	0,0408451	0,0408451	0,0408451	0,0408451	0,0408451	0,0408451	0,0408451	0,0408451	0,0408451	0,0408451	0,0408451	0,0408451	0,0408451	0,0408451	0,0408451	0,0408451
0,13556	0,13556	0,13556	0,13556	0,13556	0,13556	0,13556	0,13556	0,13556	0,13556	0,13556	0,13556	0,13556	0,13556	0,13556	0,13556	0,13556	0,13556	0,13556	0,13556	0,13556	0,13556	0,13556	0,13556	0,13556	0,13556	0,13556	0,13556	0,13556	0,13556	0,13556	0,13556
0,04084	0,04084	0,04084	0,04084	0,04084	0,04084	0,04084	0,04084	0,04084	0,04084	0,04084	0,04084	0,04084	0,04084	0,04084	0,04084	0,04084	0,04084	0,04084	0,04084	0,04084	0,04084	0,04084	0,04084	0,04084	0,04084	0,04084	0,04084	0,04084	0,04084	0,04084	0,04084
0,0400451	0,0400451	0,0400451	0,0400451	0,0400451	0,0400451	0,0400451	0,0400451	0,0400451	0,0400451	0,0400451	0,0400451	0,0400451	0,0400451	0,0400451	0,0400451	0,0400451	0,0400451	0,0400451	0,0400451	0,0400451	0,0400451	0,0400451	0,0400451	0,0400451	0,0400451	0,0400451	0,0400451	0,0400451	0,0400451	0,0400451	0,0400451
0,0308451	0,0308451	0,0308451	0,0308451	0,0308451	0,0308451	0,0308451	0,0308451	0,0308451	0,0308451	0,0308451	0,0308451	0,0308451	0,0308451	0,0308451	0,0308451	0,0308451	0,0308451	0,0308451	0,0308451	0,0308451	0,0308451	0,0308451	0,0308451	0,0308451	0,0308451	0,0308451	0,0308451	0,0308451	0,0308451	0,0308451	0,0308451
0,00696	0,00696	0,00696	0,00696	0,00696	0,00696	0,00696	0,00696	0,00696	0,00696	0,00696	0,00696	0,00696	0,00696	0,00696	0,00696	0,00696	0,00696	0,00696	0,00696	0,00696	0,00696	0,00696	0,00696	0,00696	0,00696	0,00696	0,00696	0,00696	0,00696	0,00696	0,00696
2,501E-05	2,501E-05	2,501E-05	2,501E-05	2,501E-05	2,501E-05	2,501E-05	2,501E-05	2,501E-05	2,501E-05	2,501E-05	2,501E-05	2,501E-05	2,501E-05	2,501E-05	2,501E-05	2,501E-05	2,501E-05	2,501E-05	2,501E-05	2,501E-05	2,501E-05	2,501E-05	2,501E-05	2,501E-05	2,501E-05	2,501E-05	2,501E-05	2,501E-05	2,501E-05	2,501E-05	2,501E-05
0,0408451	0,0408451	0,0408451	0,0408451	0,0408451	0,0408451	0,0408451	0,0408451	0,0408451	0,0408451	0,0408451	0,0408451	0,0408451	0,0408451	0,0408451	0,0408451	0,0408451	0,0408451	0,0408451	0,0408451	0,0408451	0,0408451	0,0408451	0,0408451	0,0408451	0,0408451	0,0408451	0,0408451	0,0408451	0,0408451	0,0408451	0,0408451
0,0408451	0,0408451	0,0408451	0,0408451	0,0408451	0,0408451	0,0408451	0,0408451	0,0408451	0,0408451	0,0408451	0,0408451	0,0408451	0,0408451	0,0408451	0,0408451	0,0408451	0,0408451	0,0408451	0,0408451	0,0408451	0,0408451	0,0408451	0,0408451	0,0408451	0,0408451	0,0408451	0,0408451	0,0408451	0,0408451	0,0408451	0,0408451
0,0408451	0,0408451	0,0408451	0,0408451	0,0408451	0,0408451	0,0408451	0,0408451	0,0408451	0,0408451	0,0408451	0,0408451	0,0408451	0,0408451	0,0408451	0,0408451	0,0408451	0,0408451	0,0408451	0,0408451	0,0408451	0,0408451	0,0408451	0,0408451	0,0408451	0,0408451	0,0408451	0,0408451	0,0408451	0,0408451	0,0408451	0,0408451
0,0084507	0,0084507	0,0084507	0,0084507	0,0084507	0,0084507	0,0084507	0,0084507	0,0084507	0,0084507	0,0084507	0,0084507	0,0084507	0,0084507	0,0084507	0,0084507	0,0084507	0,0084507	0,0084507	0,0084507	0,0084507	0,0084507	0,0084507	0,0084507	0,0084507	0,0084507	0,0084507	0,0084507	0,0084507	0,0084507	0,0084507	0,0084507
0,005865	0,005865	0,005865	0,005865	0,005865	0,005865	0,005865	0,005865	0,005865	0,005865	0,005865	0,005865	0,005865	0,005865	0,005865	0,005865	0,005865	0,005865	0,005865	0,005865	0,005865	0,005865	0,005865	0,005865	0,005865	0,005865	0,005865	0,005865	0,005865	0,005865	0,005865	0,005865
0,065868	0,065868	0,065868	0,065868	0,065868	0,065868	0,065868	0,065868	0,065868	0,065868	0,065868	0,065868	0,065868	0,065868	0,065868	0,065868	0,065868	0,065868	0,065868	0,065868	0,065868	0,065868	0,065868	0,065868	0,065868	0,065868	0,065868	0,065868	0,065868	0,065868	0,065868	0,065868
0,0408451	0,0408451	0,0408451	0,0408451	0,0408451	0,0408451	0,0408451	0,0408451	0,0408451	0,0408451	0,0408451	0,0408451	0,0408451	0,0408451	0,0408451	0,0408451	0,0408451	0,0408451	0,0408451	0,0408451	0,0408451	0,0408451	0,0408451	0,0408451	0,0408451	0,0408451	0,0408451	0,0408451	0,0408451	0,0408451	0,0408451	0,0408451
0,0408451	0,0408451	0,0408451	0,0408451	0,0408451	0,0408451	0,0408451	0,0408451	0,0408451	0,0408451	0,0408451	0,0408451	0,0408451	0,0408451	0,0408451	0,0408451	0,0408451	0,0408451	0,0408451	0,0408451	0,0408451	0,0408451	0,0408451	0,0408451	0,0408451	0,0408451	0,0408451	0,0408451	0,0408451	0,0408451	0,0408451	0,0408451
0,00999	0,00999	0,00999	0,00999	0,00999	0,00999	0,00999	0,00999	0,00999	0,00999	0,00999	0,00999	0,00999	0,00999	0,00999	0,00999	0,00999	0,00999	0,00999	0,00999	0,00999	0,00999	0,00999	0,00999	0,00999	0,00999	0,00999	0,00999	0,00999	0,00999	0,00999	0,00999
0,0056655	0,0056655	0,0056655	0,0056655	0,0056655	0,0056655	0,0056655	0,0056655	0,0056655	0,0056655	0,0056655	0,0056655	0,0056655	0,0056655	0,0056655	0,0056655	0,0056655	0,0056655	0,0056655	0,0056655	0,0056655	0,0056655	0,0056655	0,0056655	0,0056655	0,0056655	0,0056655	0,0056655	0,0056655	0,0056655	0,0056655	0,0056655
0,00086	0,00086	0,00086	0,00086	0,00086	0,00086	0,00086	0,00086	0,00086	0,00086	0,00086	0,00086	0,00086	0,00086	0,00086	0,00086	0,00086	0,00086	0,00086	0,00086	0,00086	0,00086	0,00086	0,00086	0,00086	0,00086	0,00086	0,00086	0,00086	0,00086	0,00086	0,00086
0,0407751	0,0407751	0,0407751	0,0407751	0,0407751	0,0407751	0,0407751	0,0407751	0,0407751	0,0407751	0,0407751	0,0407751	0,0407751	0,0407751	0,0407751	0,0407751	0,0407751	0,0407751	0,0407751	0,0407751	0,0407751	0,0407751	0,0407751	0,0407751	0,0407751	0,0407751	0,0407751	0,0407751	0,0407751	0,0407751	0,0407751	0,0407751
0,0408451	0,0408451	0,0408451	0,0408451	0,0408451	0,0408451	0,0408451	0,0408451	0,0408451	0,0408451	0,0408451	0,0408451	0,0408451	0,0408451	0,0408451	0,0408451	0,0408451	0,0408451	0,0408451	0,0408451	0,0408451	0,0408451	0,0408451	0,0408451	0,0408451	0,0408451	0,0408451	0,0408451	0,0408451	0,0408451	0,0408451	0,0408451
0,0406631	0,0406631	0,0406631	0,0406631	0,0406631	0,0406631	0,0406631	0,0406631	0,0406631	0,0406631	0,0406631	0,0406631	0,0406631	0,0406631	0,0406631	0,0406631	0,0406631	0,0406631	0,0406631	0,0406631	0,0406631	0,0406631	0,0406631	0,0406631	0,0406631	0,0406631	0,0406631	0,0406631	0,0406631	0,0406631	0,0406631	0,0406631
0,000868	0,000868	0,000868	0,000868	0,000868	0,000868	0,000868	0,000868	0,000868	0,000868	0,000868	0,000868	0,000868	0,000868	0,000868	0,000868	0,000868	0,000868	0,000868	0,000868	0,000868	0,000868	0,000868	0,000868	0,000868	0,000868	0,000868	0,000868	0,000868	0,000868	0,000868	0,000868
0,0025676	0,0025676	0,0025676	0,0025676	0,0025676	0,0025676	0,0025676	0,0025676	0,0025676	0,0025676	0,0025676	0,0025676	0,0025676	0,0025676	0,0025676	0,0025676	0,0025676	0,0025676	0,0025676	0,0025676	0,0025676	0,0025676	0,0025676	0,0025676	0,0025676	0,0025676	0,0025676	0,0025676	0,0025676	0,0025676	0,0025676	0,0025676
0,00224	0,00224	0,00224	0,00224	0,00224	0,00224	0,00224	0,00224	0,00224	0,00224	0,00224	0,00224	0,00224	0,00224	0,00224	0,00224	0,00224	0,00224	0,00224	0,00224	0,00224	0,00224	0,00224	0,00224	0,00224	0,00224	0,00224	0,00224	0,00224	0,00224	0,00224	0,00224
0,0408451	0,0408451	0,0408451	0,0408451	0,0408451	0,0408451	0,0408451	0,0408451	0,0408451	0,0408451	0,0408451	0,0408451	0,0408451	0,0408451	0,0408451	0,0408451	0,0408451	0,0408451	0,0408451	0,0408451	0,0408451	0,0408451	0,0408451	0,0408451	0,0408451	0,0408451	0,0408451	0,0408451	0,0408451	0,0408451	0,0408451	0,0408451
6,99E-05	6,99E-05	6,99E-05	6,99E-05	6,99E-05	6,99E-05	6,99E-05	6,99E-05	6,99E-05	6,99E-05	6,99E-05	6,99E-05	6,99E-05	6,99E-05	6,99E-05	6,99E-05	6,99E-05	6,99E-05	6,99E-05	6,99E-05	6,99E-05	6,99E-05	6,99E-05	6,99E-05	6,99E-05	6,99E-05	6,99E-05	6,99E-05	6,99E-05	6,99E-05	6,99E-05	6,99E-05
2,155E-05	2,155E-05	2,155E-05	2,155E-05	2,155E-05	2,155E-05	2,155E-05	2,155E-05	2,155E-05	2,155E-05	2,155E-05	2,155E-05	2,155E-05	2,155E-05	2,155E-05	2,155E-05	2,155E-05	2,155E-05	2,155E-05	2,155E-05	2,155E-05	2,155E-05	2,155E-05	2,155E-05	2,155E-05	2,155E-05	2,155E-05	2,155E-05	2,155E-05	2,155E-05	2,155E-05	2,155E-05
2,155E-05	2,155E-05	2,155E-05	2,155E-05	2,155E-05	2,155E-05	2,155E-05	2,155E-05	2,155E-05	2,155E-05	2,155E-05	2,155E-05	2,155E-05	2,155E-05	2,155E-05	2,155E-05	2,155E-05	2,155E-05	2,155E-05	2,155E-05	2,155E-05	2,155E-05	2,155E-05	2,155E-05	2,155E-05	2,155E-05	2,155E-05	2,155E-05	2,155E-05	2,155E-05	2,155E-05	2,155E-05
2,15E-06	2,15E-06	2,15E-06	2,15E-06	2,15E-06	2,15E-06	2,15E-06	2,15E-06	2,15E-06	2,15E-06	2,15E-06	2,15E-06	2,15E-06	2,15E-06	2,15E-06	2,15E-06	2,15E-06	2,15E-06	2,15E-06	2,15E-06	2,15E-06	2,15E-06	2,15E-06	2,15E-06	2,15E-06	2,15E-06	2,15E-06	2,15E-06	2,15E-06	2,15E-06	2,15E-06	2,15E-06
4,54E-07	4,54E-07	4,54E-07	4,54E-07	4,54E-07	4,54E-07	4,54E-07	4,54E-07	4,54E-07	4,54E-07	4,54E-07	4,54E-07	4,54E-07	4,54E-07	4,54E-07	4,54E-07	4,54E-07	4,54E-07	4,54E-07	4,54E-07	4,54E-07	4,54E-07	4,54E-07	4,54E-07	4,54E-07	4,54E-07	4,54E-07	4,54E-07	4,54E-07	4,54E-07	4,54E-07	4,54E-07
0,040845	0,040845	0,040845	0,040845	0,040845	0,040845	0,040845	0,040845	0,040845	0,040845	0,040845	0,040845	0,040845	0,040845	0,040845	0,040845	0,040845	0,040845	0,040845	0,040845	0,040845	0,040845	0,040845	0,040845	0,040845	0,040845	0,040845	0,040845	0,040845	0,040845	0,040845	0,040845
0,0056868	0,0056868	0,0056868	0,0056868	0,0056868	0,0056868	0,0056868	0,0056868	0,0056868	0,0056868	0,0056868	0,0056868	0,0056868	0,0056868	0,0056868	0,0056868	0,0056868	0,0056868	0,0056868	0,0056868	0,0056868	0,0056868	0,0056868	0,0056868	0,0056868	0,0056868	0,0056868	0,0056868	0,0056868	0,0056868	0,0056868	0,0056868
0,0012356	0,0012356	0,0012356	0,0012356	0,0012356	0,0012356	0,0012356	0,0012356	0,0012356	0,0012356	0,0012356	0,0012356	0,0012356	0,0012356	0,0012356	0,0012356	0,0012356	0,0012356	0,0012356	0,0012356	0,0012356	0,0012356	0,0012356	0,0012356	0,0012356	0,0012356	0,0012356	0,0012356	0,0012356	0,0012356	0,0012356	0,0012356
0,035655	0,035655	0,035655	0,035655	0,035655	0,035655	0,035655	0,035655	0,035655	0,035655	0,035655	0,035655	0,035655	0,035655	0,035655	0,035655	0,035655	0,035655	0,035655	0,035655	0,035655	0,035655	0,035655	0,035655	0,035655	0,035655	0,035655	0,035655	0,035655	0,035655	0,035655	0,035655
4,5E-07	4,5E-07	4,5E-07	4,5E-07	4,5E-07	4,5E-07	4,5E-07	4,5E-07	4,5E-07	4,5E-07	4,5E-07	4,5E-07	4,5E-07	4,5E-07	4,5E-07	4,5E-07	4,5E-07	4,5E-07	4,5E-07	4,5E-07	4,5E-07	4,5E-07	4,5E-07	4,5E-07	4,5E-07	4,5E-07	4,5E-07	4,5E-07	4,5E-07	4,5E-07	4,5E-07	4,5E-07
6,99E-05	6,99E-05	6,99E-05	6,99E-05	6,99E-05	6,99E-05	6,99E-05	6,99E-05	6,99E-05	6,99E-05	6,99E-05	6,99E-05	6,99E-05	6,99E-05	6,99E-05	6,99E-05	6,99E-05	6,99E-05	6,99E-05	6,99E-05	6,99E-05	6,99E-05	6,99E-05	6,99E-05	6,99E-05	6,99E-05	6,99E-05	6,99E-05	6,99E-05	6,99E-05	6,99E-05	6,99E-05
2,155E-05	2,155E-05	2,155E-05	2,155E-05	2,155E-05	2,155E-05	2,155E-05	2,155E-05	2,155E-05	2,155E-05	2,155E-05	2,155E-05	2,155E-05	2,155E-05	2,155E-05	2,155E-05	2,155E-05	2,155E-05	2,155E-05	2,155E-05	2,155E-05	2,155E-05	2,155E-05	2,155E-05	2,155E-05	2,155E-05	2,155E-05	2,155E-05	2,155E-05	2,155E-05	2,155E-05	2,155E-05
2,155E-05	2,155E-05	2,155E-05	2,155E-05	2,155E-05	2,155E-05	2,155E-05	2,155E-05	2,155E-05	2,155E-05	2,155E-05	2,155E-05	2,155E-05	2,155E-05	2,155E-05	2,155E-05	2,155E-05	2,155E-05	2,155E-05	2,155E-05	2,155E-05	2,155E-05	2,155E-05	2,155E-05	2,155E-05	2,155E-05	2,155E-05	2,155E-05	2,155E-05	2,155E-05	2,155E-05	2,155E-05
2,15E-06	2,15E-06	2,15E-06	2,15E-06	2,15E-06	2,15E-06	2,15E-06	2,15E-06	2,15E-06	2,15E-06	2,15E-06	2,15E-06	2,15E-06	2,15E-06	2,15E-06	2,15E-06	2,15E-06	2,15E-06	2,15E-06	2,15E-06	2,15E-06	2,15E-06	2,15E-06	2,15E-06	2,15E-06	2,15E-06	2,15E-06	2,15E-06	2,15E-06	2,15E-06	2,15E-06	2,15E-06
4,54E-07	4,54E-07	4,54E-07	4,54E-07	4,54E-07	4,54E-07	4,54E-07	4,54E-07	4,54E-07	4,54E-07	4,54E-07	4,54E-07	4,54E-07	4,54E-07	4,54E-07	4,54E-07	4,54E-07	4,54E-07	4,54E-07	4,54E-07	4,54E-07	4,54E-07	4,54E-07	4,54E-07	4,54E-07	4,54E-07	4,54E-07	4,54E-07	4,54E-07	4,54E-07	4,54E-07	4,54E-07
0,0408451	0,0408451	0,0408451	0,0408451	0,0408451	0,0408451	0,0408451	0,0408451	0,0408451	0,0408451	0,0408451	0,0408451	0,0408451	0,0408451	0,0408451	0,0408451	0,0408451	0,0408451	0,0408451	0,0408451	0,0408451	0,0408451	0,0408451	0,0408451	0,0408451	0,0408451	0,0408451	0,0408451	0,0408451	0,0408451	0,0408451	0,0408451
0,0056868	0,0056868	0,0056868	0,0056868	0,0056868	0,0056868	0,0056868	0,0056868	0,0056868	0,0056868	0,0056868	0,0056868	0,0056868	0,0056868	0,0056868	0,0056868	0,0056868	0,0056868	0,0056868	0,0056868	0,0056868	0,0056868	0,0056868	0,0056868	0,0056868	0,0056868	0,0056868	0,0056868	0,0056868	0,0056868	0,0056868	0,0056868
0,0012356	0,0012356	0,0012356	0,0012356	0,0012356	0,0012356	0,0012356	0,0012356	0,0012356	0,0012356	0,0012356	0,0012356	0,0012356	0,0012356	0,0012356	0,0012356	0,0012356	0,0012356	0,0012356	0,0012356	0,0012356	0,0012356	0,0012356	0,0012356	0,0012356	0,0012356	0,0012356	0,0012356	0,0012356	0,0012356	0,0012356	0,0012356
0,035655	0,035655	0,035655	0,035655	0,035655	0,035655	0,035655	0,035655	0,035655	0,035655	0,035655	0,035655	0,035655	0,035655	0,035655	0,035655	0,035655	0,035655	0,035655	0,035655	0,035655	0,035655	0,035655	0,035655	0,035655	0,035655	0,035655	0,035655	0,035655	0,035655	0,035655	0,035655
4,5E-07	4,5E-07	4,5E-07	4,5E-07	4,5E-07	4,5E-07	4,5E-07	4,5E-07	4,5E-07	4,5E-07	4,5E-07	4,5E-07	4,5E-07	4,5E-07	4,5E-07	4,5E-07	4,5E-07	4,5E-07	4,5E-07	4,5E-07	4,5E-07	4,5E-07	4,5E-07	4,5E-07	4,5E-07	4,5E-07	4,5E-07	4,5E-07	4,5E-07	4,5E-07	4,5E-07	4,5E-07

Figure A3. Limit super matrix of the ANP calculations (cont.).

References

1. *Electricity Generation Company of Turkey*; Yearly Activity Report; EUAS: Ankara, Turkey, 2018.
2. Oocities Internet Source: Gas Turbines with Steam Injection. Available online: http://www.oocities.org/siliconvalley/platform/6647/kojenen2.htm (accessed on 11 April 2018).
3. Baker, K.R. Workforce allocation in cyclical scheduling problems: A survey. *Oper. Res. Q.* **1976**, *27*, 155–167. [CrossRef]
4. Pinedo, M. *Planning and Scheduling in Manufacturing and Services*; Springer: New York, NY, USA, 2005.
5. Talbi, E.G. *Metaheuristics: From Design to Implementation*; Wiley Series on Parallel and Distributed Computing; John Wiley & Sons Inc.: Hoboken, NJ, USA, 2009.
6. Morton, T.E.; Pentico, D.W. *Heuristic Scheduling Systems: With Applications to Production Systems and Project Management*; Wiley: New York, NY, USA, 1993.
7. Brucker, P.; Qu, R.; Burke, E. Personnel scheduling: Models and complexity. *Eur. J. Oper. Res.* **2011**, *210*, 467–473. [CrossRef]
8. Bergh, J.V.; Beliën, J.; Bruecker, P.; Demeulemeester, E.; Boeck, L. Personnel scheduling: A literature review. *Eur. J. Oper. Res.* **2013**, *226*, 367–385. [CrossRef]
9. Gür, Ş.; Eren, T. Scheduling and Planning in Service Systems with Goal Programming: Literature Review. *Mathematics* **2018**, *6*, 265. [CrossRef]
10. Aickelin, U.; Burke, E.K.; Li, J.P. Improved squeaky wheel optimisation for driver scheduling, Parallel Problem Solving from Nature—Ppsn Ix. *Lect. Notes Comput. Sci.* **2006**, *4193*, 182–191. [CrossRef]
11. Aickelin, U.; Burke, E.K.; Li, J.P. An evolutionary squeaky wheel optimization approach to personnel scheduling. *IEEE Trans. Evol. Comput.* **2009**, *13*, 433–443. [CrossRef]
12. Aickelin, U.; Dowsland, K.A. An indirect genetic algorithm for a nurse scheduling problem. *Comput. Oper. Res.* **2004**, *31*, 761–778. [CrossRef]
13. Yunes, T.H.; Moura, A.V.; De Souza, C.C. Hybrid column generation approaches for urban transit crew management problems. *Transp. Sci.* **2005**, *39*, 273–288. [CrossRef]
14. Akjiratikarl, C.; Yenradee, P.; Drake, P.R. PSO-based algorithm for home care worker scheduling in the UK. *Comput. Ind. Eng.* **2007**, *53*, 55–583. [CrossRef]
15. Ásgeirsson, E. Bridging the gap between self-schedules and feasible schedules in staff scheduling. *Ann. Oper. Res.* **2014**. [CrossRef]
16. Lei, L.; Pinedo, M.; Qi, L.; Wang, S.; Yang, J. Personnel scheduling and supplies provisioning in emergency relief operations. *Ann. Oper. Res.* **2015**, *235*, 487–515. [CrossRef]
17. Smet, P.; Wauters, T.; Mihaylov, M.; Vanden Berghe, G. The shift minimisation personnel task scheduling problem: A new hybrid approach and computational insights. *Omega* **2014**, *46*, 64–73. [CrossRef]
18. Mısır, M.; Smet, P.; Vanden Berghe, G. An analysis of generalised heuristics for vehicle routing and personnel rostering problems. *J. Oper. Res. Soc.* **2015**, *66*, 858–870. [CrossRef]
19. Mısır, M.; Smet, P.; Verbeeck, K.; Vanden Berghe, G. Security personnel routing and rostering: A hyper-heuristic approach. In Proceedings of the 3rd International Conference on Applied Operational Research, İstanbul, Turkey, 3 August 2011.
20. Smet, P.; Vanden Berghe, G. A matheuristic approach to the shift minimisation personnel task scheduling problem. In Proceedings of the 9th International Conference on the Practice and Theory of Automated Timetabling, Son, Norway, 28–31 August 2012; pp. 145–160.
21. Lee, C.Y.; Lei, L.; Pinedo, M. Current trends in deterministic scheduling. *Ann. Oper. Res.* **1997**, *70*, 1–41. [CrossRef]
22. Veen, E.; Hans, E.W.; Post, G.F.; Veltman, B.; Dumitrescu, S. Shift rostering using decomposition: Assign weekend shifts first. *J. Sched.* **2015**, *18*, 29–43. [CrossRef]
23. Smet, P.; Bilgin, B.; De Causmaecker, P.; Vanden Berghe, G. Modelling and evaluation issues in nurse rostering. *Ann. Oper. Res.* **2014**, *218*, 303–326. [CrossRef]
24. Abbink, E.; Fischetti, M.; Kroon, L.; Timmer, G.; Vromans, M. Reinventing crew scheduling at Netherlands railways. *Interfaces* **2005**, *35*, 393–401. [CrossRef]
25. Aickelin, U.; White, P. Building better nurse scheduling algorithms. *Ann. Oper. Res.* **2004**, *128*, 159–177. [CrossRef]

26. Alfares, H.K. Survey, categorization, and comparison of recent tour scheduling literature. *Ann. Oper. Res.* **2004**, *127*, 145–175. [CrossRef]

27. Alfares, H.K. Compressed workweek scheduling with days-off consecutivity weekend-off frequency and work stretch constraints. *Informatics* **2006**, *44*, 175–189. [CrossRef]

28. Trilling, L.; Guinet, A.; Le Magny, D. Nurse scheduling using integer linear programming and constraint programming. In Proceedings of the 12th IFAC International Symposium, Saint-Etienne, France, 17–19 May 2006; Elsevier: Amsterdam, The Netherlands, 2006; Volume 3, pp. 651–656. [CrossRef]

29. Lezaun, M.; Perez, G.; De La Maza, E.S. Crew rostering problem in a public transport company. *J. Oper. Res. Soc.* **2006**, *57*, 1173–1179. [CrossRef]

30. Al-Yakoob, S.M.; Sherali, H.D. Mixed-integer programming models for an employee scheduling problem with multiple shifts and work locations. *Ann. Oper. Res.* **2007**, *155*, 119–142. [CrossRef]

31. Al-Yakoob, S.M.; Sherali, H.D. Multiple shift scheduling of hierarchical workforce with multiple work centers. *Informatics* **2007**, *18*, 325–342.

32. Alfares, H.K. Operator staffing and scheduling for an IT-help call centre. *Eur. J. Ind. Eng.* **2007**, *1*, 414–430. [CrossRef]

33. Lezaun, M.; Perez, G.; De La Maza, E.S. Staff rostering for the station personnel of a railway company. *J. Sched.* **2007**, *10*, 245–254. [CrossRef]

34. Corominas, A.; Lusa, A.; Pastor, R. Using a MILP model to establish a framework for an annualised hours agreement. *Eur. J. Oper. Res.* **2007**, *177*, 1495–1506. [CrossRef]

35. Bard, J.F.; Binici, C.; De Silva, A.H. Staff scheduling at the United States Postal Service. *Comput. Oper. Res.* **2003**, *30*, 745–771. [CrossRef]

36. Corominas, A.; Lusa, A.; Pastor, R. Planning annualised hours with a finite set of weekly working hours and joint holidays. *Ann. Oper. Res.* **2004**, *128*, 217–233. [CrossRef]

37. Henao, C.A.; Muñoz, J.C.; Ferrer, J.C. The impact of multi-skilling on personnel scheduling in the service sector: A retail industry case. *J. Oper. Res. Soc.* **2015**, *66*, 1949–1959. [CrossRef]

38. Veldhoven, S.; Post, G.; van der Veen, E.; Curtois, T. An assessment of a days off decomposition approach to personnel shift scheduling. *Ann. Oper. Res.* **2016**, *239*, 207–223. [CrossRef]

39. Bağ, N.; Özdemir, M.; Eren, T. 0-1 Hedef Programlama ve ANP Yöntemi ile Hemşire Çizelgeleme Problemi Çözümü. *Int. J. Eng. Res. Dev.* **2012**, *1*, 2–6.

40. Hung-Tso, L.; Yen-Ting, C.; Tsung-Yu, C.; Yi-Chun, L. Crew rostering with multiple goals: An empirical study. *Comput. Ind. Eng.* **2012**, *63*, 483–493. [CrossRef]

41. Li, J.; Burke, E.K.; Curtois, T.; Petrovic, S.; Rong, Q. The falling tide algorithm: A new multi-objective approach for complex workforce scheduling. *Omega* **2012**, *40*, 283–293. [CrossRef]

42. Kassa, B.A.; Tizazu, A.E. Personnel scheduling using an integer programming model- an application at Avanti Blue-Nile Hotels. *SpringerPlus* **2013**, *2*, 1–7. [CrossRef] [PubMed]

43. Louly, M.A. A goal programming model for staff scheduling at a telecommunications center. *J. Math. Model. Algorithms Oper. Res.* **2013**, *12*, 167–178. [CrossRef]

44. Labidi, M.; Mrad, M.; Gharbi, A.; Louly, M.A. Scheduling IT Staff at a Bank: A Mathematical Programming Approach. *Sci. World J.* **2014**, *1*, 1–10. [CrossRef] [PubMed]

45. Todovic, D.; Makajic-Nikolic, D.; Kostic-Stankovic, M.; Martic, M. Police officer scheduling using goal programming. *Polic. Int. J. Police Strateg. Manag.* **2015**, *38*, 295–313. [CrossRef]

46. Shuib, A.; Kamarudin, F.I. Solving shift scheduling problem with days-off preference for power station workers using binary integer goal programming model. *Ann. Oper. Res.* **2019**, *272*, 355–372. [CrossRef]

47. Özder, E.H.; Varlı, E.; Eren, T. A Model Suggestion for Cleaning Staff Scheduling Problem with Goal Programming Approach. *Black Sea J. Sci.* **2017**, *7*, 114–127. [CrossRef]

48. Varlı, E.; Alağaş, H.; Eren, T.; Özder, E.H. Goal Programming Solution of the Examiner Assignment Problem. *Bilge Int. J. Sci. Technol. Res.* **2017**, *1*, 105–118.

49. Akbari, M.; Zandieh, M.; Dorri, B. Scheduling part-time and mixed-skilled workers to maximize employee satisfaction. *Int. J. Adv. Manuf. Technol.* **2013**, *64*, 1017–1027. [CrossRef]

50. Ernst, A.T.; Jiang, H.; Krishnamoorthy, M.; Sier, D. Staff scheduling and rostering: A review of applications, methods and models. *Eur. J. Oper. Res.* **2004**, *153*, 3–27. [CrossRef]

51. Azaiez, M.N.; Al-Sharif, S.S. A 0–1 goal programming model for nurse scheduling. *Comput. Oper. Res.* **2005**, *32*, 491–507. [CrossRef]

52. Topaloğlu, S. A. multi-objective programming model for scheduling emergency medicine residents. *Comput. Ind. Eng.* **2006**, *51*, 375–388. [CrossRef]

53. Alfares, H.K. A simulation approach for stochastic employee days-off scheduling. *Int. J. Model. Simul.* **2007**, *27*, 9–15. [CrossRef]

54. Alfieri, A.; Kroon, L.; Van de Velde, S. Personnel scheduling in a complex logistic system: A railway application case. *J. Intell. Manuf.* **2007**, *18*, 223–232. [CrossRef]

55. Chu, S.C.K. Generating, scheduling and rostering of shift crew-duties: Applications at the Hong Kong International Airport. *Eur. J. Oper. Res.* **2007**, *177*, 1764–1778. [CrossRef]

56. Thompson, G.M.; Pullman, M.E. Scheduling workforce relief breaks in advance versus in real-time. *Eur. J. Oper. Res.* **2007**, *181*, 139–155. [CrossRef]

57. Sinreich, D.; Jabali, O. Staggered workshifts: A way to downsize and restructure an emergency department workforce yet maintain current operational performance. *Health Care Manag. Sci.* **2007**, *10*, 293–308. [CrossRef]

58. Al-Yakoob, S.M.; Sherali, H.D. A column generation approach for an employee scheduling problem with multiple shifts and work locations. *J. Oper. Res. Soc.* **2008**, *59*, 34–43. [CrossRef]

59. De Matta, R.; Peters, E. Developing work schedules for an inter-city transit system with multiple driver types and fleet types. *Eur. J. Oper. Res.* **2009**, *192*, 852–865. [CrossRef]

60. Tsai, C.C.; Li, S.H.A. A two-stage modeling with genetic algorithms for the nurse scheduling problem. *Expert Syst. Appl.* **2009**, *36*, 9506–9512. [CrossRef]

61. Lezaun, M.; Perez, G.; De La Maza, E.S. Rostering in a rail passenger carrier. *J. Sched.* **2007**, *10*, 245–254. [CrossRef]

62. Rönnberg, E.; Larsson, T. Automating the self-scheduling process of nurses in Swedish healthcare: A pilot study. *Health Care Manag. Sci.* **2010**, *13*, 35–53. [CrossRef] [PubMed]

63. Zolfaghari, S.; Quan, V.; El-Bouri, A.; Khashayardoust, M. Application of a genetic algorithm to staff scheduling in retail sector. *Int. J. Ind. Syst. Eng.* **2010**, *5*, 20–47. [CrossRef]

64. Alsheddy, A.; Tsang, E.P.K. Empowerment scheduling for a field workforce. *J. Sched.* **2011**, *14*, 639–654. [CrossRef]

65. Fırat, M.; Hurkens, C.A.J. An improved MIP-based approach for a multi-skill workforce scheduling problem. *J. Sched.* **2011**, *15*, 363–380. [CrossRef]

66. Asensio-Cuesta, S.; Diego-Mas, J.A.; Canos-Daros, L.; Andres-Romano, C. A genetic algorithm for the design of job rotation schedules considering ergonomic and competence criteria. *Int. J. Adv. Manuf. Technol.* **2012**, *60*, 1161–1174. [CrossRef]

67. Charnes, A.; Cooper, W.W.; Ferguson, R. Optimal Estimation of Executive Compensation by Linear Programming. *Manag. Sci.* **1955**, *1*, 138–151. [CrossRef]

68. Charnes, A.; Cooper, W.W. *Management Models and Industrial Applications of Linear Programming*; Wiley: New York, NY, USA, 1961.

69. Charnes, A.; Cooper, W.W. Goal programming and multiple objective optimizations. *Eur. J. Oper. Res.* **1977**, *1*, 39–54. [CrossRef]

70. Ignizio, J. *Introduction to Goal Programming*; Sage Publications Inc.: Beverley Hills, CA, USA, 1985.

71. Lee, S.M. *Goal Programming for Decision Analysis of Multiple Objectives*; Auerback: Philadelphia, PA, USA, 1972.

72. Saaty, T.L. *The Analytic Hierarchy Process*; McGraw-Hill: New York, NY, USA, 1980.

73. Veen, E.; Veltman, B. Rostering from Staffing Levels: A Branch-And-Price Approach. *Int. J. Health Manag. Inf.* **2011**, *2*, 41–52.

74. Hamurcu, M.; Eren, T. Raylı Sistem Projeleri Kararında Ahs-Hp ve Aas-Hp Kombinasyonu. *Gazi Mühendislik Bilimleri Derg.* **2017**, *3*, 1–13.

75. Hamurcu, M.; Gür, Ş.; Özder, E.H.; Eren, T. A Multicriteria Decision Making for Monorail Projects with Analytic Network Process and 0–1 Goal Programming. *Int. J. Adv. Electron. Comput. Sci.* **2016**, *3*, 8–12.

 *mathematics*

Article

AHP-Group Decision Making Based on Consistency

Juan Aguarón, María Teresa Escobar, José María Moreno-Jiménez * and Alberto Turón

Grupo Decisión Multicriterio Zaragoza (GDMZ), Facultad de Economía y Empresa, Universidad de Zaragoza, Gran Vía 2, 50005 Zaragoza, Spain; aguaron@unizar.es (J.A.); mescobar@unizar.es (M.T.E.); turon@unizar.es (A.T.)

* Correspondence: moreno@unizar.es

Received: 7 February 2019; Accepted: 25 February 2019; Published: 7 March 2019

Abstract: The Precise consistency consensus matrix (PCCM) is a consensus matrix for AHP-group decision making in which the value of each entry belongs, simultaneously, to all the individual consistency stability intervals. This new consensus matrix has shown significantly better behaviour with regards to consistency than other group consensus matrices, but it is slightly worse in terms of compatibility, understood as the discrepancy between the individual positions and the collective position that synthesises them. This paper includes an iterative algorithm for improving the compatibility of the PCCM. The sequence followed to modify the judgments of the PCCM is given by the entries that most contribute to the overall compatibility of the group. The procedure is illustrated by means of its application to a real-life situation (a local context) with three decision makers and four alternatives. The paper also offers, for the first time in the scientific literature, a detailed explanation of the process followed to solve the optimisation problem proposed for the consideration of different weights for the decision makers in the calculation of the PCCM.

Keywords: Analytic Hierarchy Process (AHP); group decision making; consistency; compatibility

---

1. Introduction

One of the multicriteria decision making techniques that best responds to the challenges and needs of the Knowledge Society [1], especially the consideration of intangible aspects and decision-making with multiple actors, is the Analytical Hierarchy Process (AHP). AHP was proposed by Thomas L. Saaty in the early 1970s (20th century) [2]. This multicriteria technique incorporates the intangible aspects associated with the human factor through the use of pairwise comparisons. In group decision-making, where all the actors work as a single unit, AHP usually follows one of the two most traditional approaches [3–5]: the Aggregation of Individual Judgements (AIJ) and the Aggregation of Individual Priorities (AIP).

Both methods present two important limitations that have been addressed in some of the most recent proposals: the certainty of the data and the use of the geometric mean as the synthesising procedure of the considered values (judgments in AIJ and priorities in AIP). Escobar and Moreno-Jiménez [6] consider the principle of certainty and incorporate the context effect through the procedure called the Aggregation of Individual Preference Structures (AIPS). Altuzarra et al. [7] advance a Bayesian approach as a prioritisation procedure and a group decision-making aggregation procedure.

The concept of consistency [2] is one of the characteristics that distinguishes AHP from the other multicriteria techniques and gives coherence to the method; Moreno-Jiménez et al. [8–10] used this to design a new procedure for group decision making: the Consistency Consensus Matrix (CCM). Under certain conditions, the CCM automatically provides an interval judgement matrix where each entry reflects the range of values in which all decision makers would simultaneously be consistent in their initial matrices.

One limitation of this new decision-making tool is that the CCM is sometimes incomplete. The Precise Consistency Consensus Matrix (PCCM) has been proposed [11,12] to respond to this limitation by including more judgments in the group consensus matrix and allowing decision makers to have different weights assigned in the resolution of the problem. This new consensus matrix has, by construction, demonstrated good behaviour with respect to consistency, but it can be improved with respect to compatibility, understood as the discrepancy between the individual positions and the collective position that synthesises them.

This work presents a procedure to improve the compatibility of the PCCM guaranteeing that the consistency does not exceed a predetermined level. Compatibility is improved by modifying those judgments of the PCCM that most contribute to the global compatibility, with the idea of reducing this contribution. The combination of what happens to the consistency and the compatibility will allow selecting, as the preferred option, the one that most improves the cumulative relative changes of the two criteria (consistency and compatibility).

The paper is structured as follows: Section 2 gives the background to the developments; Section 3 describes the PCCM and the algorithm that solves the optimisation problem that aims to find the precise value that maximises the slack of consistency that remains free for the following steps when the actors have different weights; Section 4 explains the proposal for improving the compatibility of the PCCM and applies it to a case study; Section 5 highlights the most important conclusions of the study.

2. Background

2.1. Multiactor Decision Making (MADM)

As previously mentioned, consistency (coherence of decision makers in eliciting their judgments) and a good behaviour in the decision-making with multiple actors are two of the most important properties for multicriteria decision making techniques. [6,13] distinguish three areas in multi-actor decision-making: (i) Group Decision Making (GDM); (ii) Negotiated Decision Making (NDM); and (iii) Systemic Decision Making (SDM).

In GDM, individuals work together in pursuit of a common goal under the *principle of consensus*. Consensus refers to the approach, model, tools, and procedures for deriving the collective position or final group priority vector.

NDM is based on the *principle of agreement* and the assumption that all the actors follow the same scientific approach. The individuals resolve the problem separately, the zones of agreement and disagreement between the actors are identified and agreement paths (sometimes known as consensus paths) are constructed by changing, in a personal, semiautomatic or automatic way, one or several judgements.

SDM follows the *principle of tolerance*: the individual acts independently and the individual preferences, expressed as probability distributions, are aggregated to form a collective one—the tolerance distribution. This new approach integrates all the preferences, even if they are provided from different 'individual theoretical models'; the only requirement is that they must be expressed as some kind of probability distribution.

The systemic situation for dealing with multiactor decision making allows capturing the holistic vision of reality and the subjacent ideas of lateral thinking [14]. The information provided by the tolerance distribution can be used to construct tolerance paths to produce a more democratic and representative final decision. In other words, a decision will be accepted by a greater number of actors or by a number of actors with greater weighting in the decisional process [15,16].

2.2. Analytic Hierarchy Process

The Analytic Hierarchy Process is one of the most widely utilised multicriteria decision making techniques. Its methodology consists of three phases [2]: (a) modelling, (b) valuation, and, (c) prioritisation and synthesis.

(a) *Modelling* refers to the construction of a hierarchy of different levels that represent the relevant aspects of the problem (scenarios, actors, criteria, alternatives). The mission or goal hangs on the highest level. The subsequent levels contain the criteria, the first order subcriteria, the second order, etc. This continues to the last order subcriteria or attributes (characteristics of the reality that are susceptible to be measured for the alternatives); the alternatives hang from the lowest subcriteria level (attributes).

(b) *Valuation* involves the incorporation of the preferences of the decision makers via pairwise comparisons of the elements that hang from the nodes of the hierarchy in relation to the common node. The judgements follow Saaty's fundamental scale [2] and reflect the relative importance of one element with respect to another with regards to the criterion that is considered. They are expressed in reciprocal pairwise comparison matrices.

(c) *Prioritisation and synthesis* determines the local, global and total priorities. Local priorities (priorities of the elements of the hierarchy with regards to the node from which they hang) are obtained from the pairwise comparison matrices using any of the existing prioritisation procedures. The Eigenvector (EGV) and the Row Geometric Mean (RGM) are the two most commonly employed. Global priorities (the priorities of the elements of the hierarchy with regards to the mission) are obtained through the principle of hierarchical composition, whilst Total priorities (the priorities of the alternatives with regards to the mission) are obtained by a multiadditive aggregation of the global priorities of each alternative.

In the AHP-group decision making context, the two techniques traditionally used are: (i) the Aggregation of Individual Judgements (AIJ) and (ii) the Aggregation of Individual Priorities (AIP); firstly, it is necessary to specify the notation that will be utilised. Given a local context (one criteria in the hierarchy) with n alternatives $(A_1,...,A_n)$ and r decision makers $(D1,...,Dr)$, let $A^{(k)} = (a_{ij}^{(k)})$ be the pairwise comparison matrix of decision-maker Dk $(k = 1,...,r; i, j = 1,...,n)$ and π_k be the relative importance in the group $(\pi_k \geq 0, \sum_{k=1}^{r} \pi_k = 1)$.

The priorities following the two approaches AIJ and AIP are obtained as follows:

- Aggregation of Individual Judgements: The individual pairwise comparison matrices $A^{(k)} k = 1,...,r$, are first aggregated to obtain a new judgement matrix for the group $A^{(G)} = (a_{ij}^{(G)})$. Then, the priority vector $w^{(G/J)} = (w_i^{(G/J)})$ is derived from this new matrix using one of the existing prioritisation methods.

- Aggregation of Individual Priorities: The priority vectors are first obtained for each individual, $w^{(k)} = (w_i^{(k)})$ and $k = 1,...,r$, using one of the existing prioritisation methods and then aggregated to obtain the priorities of the group $w^{(G/P)} = (w_i^{(G/P)})$.

Using the Weighted Geometric Mean Method (WGMM) as the aggregation procedure, the group judgement matrix and the group priority vector are given by:

- $A^{(G)} = (a_{ij}^{(G)})$ with $a_{ij}^{(G)} = \prod_{k=1}^{r} \left(a_{ij}^{(k)}\right)^{\pi_k}$, $i,j = 1,\ldots,n$
- $w^{(G/P)} = (w_i^{(G/P)})$ with $w_i^{(G/P)} = \prod_{k=1}^{r} \left(w_i^{(k)}\right)^{\pi_k}$, $i = 1,\ldots,n$

When the WGMM aggregation procedure is employed and the priorities are obtained using the RGM, the two approaches, AIJ and AIP, provide the same solution [17,18].

2.3. Consistency and Compatibility in AHP

AHP allows for the evaluation of the consistency of the decision-maker when the judgements are introduced into the pairwise comparison matrices. Saaty [2] defined consistency in AHP as the cardinal transitivity of the judgements included in the pairwise comparison matrices, that is to say, the reciprocal pairwise comparison matrix $A_{nxn} = (a_{ij})$ is *consistent* if $\forall i,j,k = 1,...,n$ satisfies $a_{ij} \cdot a_{jk} = a_{ik}$.

Consistency is associated with the (internal) coherence of the decision makers when their judgements are considered in the pairwise comparison matrices. Consistency is usually evaluated —depending on the prioritisation procedure that is used— as the 'representativeness' of the local priorities vector derived from the pairwise comparison matrices (a_{ij} is an estimation of w_i/w_j).

In the case of the EGV and RGM, the inconsistency indicators are given, respectively, by the Consistency Index (CI) and the Geometric Consistency Index (GCI) [19]:

$$CI = \frac{1}{n(n-1)} \sum_{i,j=1}^{n} (e_{ij} - 1) \tag{1a}$$

$$GCI = \frac{2}{(n-1)(n-2)} \sum_{i<j} log^2 e_{ij} \tag{1b}$$

where $e_{ij} = a_{ij}(w_j/w_i)$. Obviously, if the matrix is consistent, both indicators of inconsistency are null, thus errors $e_{ij} = 1$ ($a_{ij} = w_i/w_j$).

The *Consistency Interval Judgement matrix for the group* (GCIJA) is an interval matrix GCIJA = $([\underline{a}_{ij}, \overline{a}_{ij}])$ where the entries correspond to the range of values for which all the decision makers will not exceed the maximum inconsistency allowed and will belong to the Saaty's fundamental scale range of values [1/9, 9].

The values that determine the limits of each entry of the GCIJA are given by $\underline{a}_{ij} = \underset{k}{Max}\left\{ \underline{a}_{ij}^{(k)}; 1/9 \right\}$ and $\overline{a}_{ij} = \underset{k}{Min}\left\{ \overline{a}_{ij}^{(k)}; 9 \right\}$, where $\underline{a}_{ij}^{(k)}$ and $\overline{a}_{ij}^{(k)}$ are the limits of the individual consistency stability interval for $a_{ij}^{(k)}$ ($[\underline{a}_{ij}^{(k)}, \overline{a}_{ij}^{(k)}]$) with $\Delta^{(k)} = GCI^* - GCI^{(k)}$, GCI^* being the maximum inconsistency allowed for the problem and $GCI^{(k)}$ the Geometric Consistency Index for the individual matrix $A^{(k)}$ [20].

Compatibility refers to the (internal) coherence of the group when selecting its priority vector ($w^{(G)} = (w_1^{(G)}, \dots, w_n^{(G)})$), that is to say, its representativeness in relation to the individual positions ($w^{(k)} = (w_1^{(k)}, \dots, w_n^{(k)})$). To evaluate the compatibility of an individual k ($w^{(k)}$), $k = 1, \dots, r$, with the collective position or group priority vector($w^{(G)}$), it is sufficient to adapt the previous expression of the GCI, taking $eij = a_{ij}^{(k)}(w_j^{(G)}/w_i^{(G)})$ in local context or $eij = (w_i^{(k)}/w_j^{(k)})(w_j^{(G)}/w_i^{(G)})$ in a global one. The concept of compatibility reflects the distance between the individual and collective positions and is calculated automatically, without the express intervention of the individual with the exception of the emission of the initial judgements of the pairwise comparison matrices. [21] advanced the *Geometric Compatibility Index* (GCOMPI) in order to evaluate the compatibility of the individual positions with respect of the collective position provided by any of the existing procedures. The expression of the GCOMPI for a decision maker k in a local context (one criterion) is given by:

$$GCOMPI^{(k,\,G)} = \frac{2}{(n-1)(n-2)} \sum_{i=1}^{n-1} \sum_{j=i+1}^{n} log^2\left(a_{ij}^{(k)} \frac{w_j^{(G)}}{w_i^{(G)}}\right) \tag{2}$$

and in a global context (hierarchy) by:

$$GCOMPI^{(k,\,G)} = \frac{2}{(n-1)(n-2)} \sum_{i=1}^{n-1} \sum_{j=i+1}^{n} log^2\left(\frac{w_i^{(k)}}{w_j^{(k)}} \frac{w_j^{(G)}}{w_i^{(G)}}\right) \tag{3}$$

The *GCOMPI* for the group is given by:

$$GCOMPI^{(G)} = \sum_{k=1,\dots r} \pi_k GCOMPI^{(k,G)} = \frac{2}{(n-1)(n-2)} \sum_{i=1}^{n-1} \sum_{j=i+1}^{n} \sum_{k=1}^{r} \pi_k log^2\left(a_{ij}^{(k)} \frac{w_j^{(G)}}{w_i^{(G)}}\right) \tag{4}$$

In addition to the use of the GCI and the GCOMPI, two more indicators are used in the literature to compare the behaviour of the different procedures with respect to consistency and compatibility [11,12]: the Number of Violations in Consistency (CVN) for the consistency; and the Number of Violations in Priorities (PVN) for the compatibility.

The CVN considers the mean number of entries of the group pairwise comparison matrix that do not belong to the corresponding consistency stability interval judgement of each individual, calculated for the inconsistency threshold considered in the problem. The Consistency Violation Number (CVN) for the group is given by $CVN^{(G)} = \Sigma_k \pi_k CVN^{(k,G)}$, where

$$CVN^{(k,G)} = \frac{2}{n(n-1)} \sum_{\substack{i<j}}^{n} I_{ij}(CIJA^{(k)}/A^{(G)}) \tag{5}$$

and

$$I_{ij}(CIJA^{(k)}/A^{(G)}) = \begin{cases} 1 & if\ a_{ij}^{(G)} \notin \left[\underline{a}_{ij}^{(k)}, \overline{a}_{ij}^{(k)}\right] \\ 0 & otherwise \end{cases} \tag{6}$$

The PVN measures the ordinal compatibility of each AHP-GDM procedure by means of the minimum number of violations [22].

The Priority Violation Number (PVN) for the group is given by $PVN^{(G)} = \Sigma_k \pi_k PVN^{(k,G)}$, where

$$PVN^{(k,G)} = PVN(A^{(k)}/A^{(G)}) = \frac{2}{(n-1)(n-2)} \sum_{\substack{i<j}}^{n} I_{ij}(A^{(k)}/A^{(G)}) \tag{7}$$

and

$$I_{ij}(A^{(k)}/A^{(G)}) = \begin{cases} 1 & if\ a_{ij}^{(k)} > 1\ and\ w_i^{(G)} < w_j^{(G)} \\ 0.5 & if\ a_{ij}^{(k)} = 1\ and\ w_i^{(G)} \neq w_j^{(G)} \\ 0.5 & if\ a_{ij}^{(k)} \neq 1\ and\ w_i^{(G)} = w_j^{(G)} \\ 0 & otherwise \end{cases} \tag{8}$$

3. The Precise Consensus Consistency Matrix (PCCM)

Moreno-Jiménez et al. [9,10] proposed a decisional tool, the Consistency Consensus Matrix (CCM), which identifies the core of consistency of the group decision using an interval matrix that may not be complete or connected. In [12], the same authors refined this tool and introduced the PCCM, which selects a precise value for each interval judgement in such a way that the quantity of slack that remains free for successive algorithm iterations is the maximum possible.

Escobar et al. [11] extended the PCCM to allow the assignment of different weights to the decision makers and to guarantee that the group consensus values were acceptable to the individuals in terms of inconsistency. In the same work, these authors put forward a number of methods for completing the PCCM matrix if it were incomplete.

The improved version of the algorithm for constructing the PCCM proposed in [11] starts by calculating the variance of the logarithms of the corresponding judgements, taking into account the fact that decision makers may have different weights. It also provides (Step 1) the initial Consistency Stability Intervals [20] for the individuals and for the group (GCIJA). The judgement with least variance (Step 2) that has a non-null intersection for the initial individual consistency stability intervals is selected. The consistency stability intervals for each decision maker are calculated for this judgement (Step 3) and the intersection of all these intervals is obtained (Step 4). In this common interval, it is guaranteed that the individual judgements can oscillate without the GCI exceeding a previously fixed level of inconsistency. The intersection of the previous interval with the range of values [1/9,9] and the initial consistency stability intervals is then calculated (Step 5). This avoids taking a value distanced from the initial judgements of all the decision makers more than the amount allowed for the fixed inconsistency level. The algorithm determines a precise value that belongs to the common interval

(Step 6). Any judgment in this interval will have an acceptable inconsistence. Some of the matrices will be more inconsistent than others and they will therefore admit less slack for the following iterations. In order to address this point, the algorithm selects the value that provides the greatest slack for the most inconsistent matrix (the value that minimises the GCI of the most inconsistent matrix). Finally, the value obtained is included as an entry of the PCCM and serves to update the initial individual judgment matrices (Step 7). The detailed version of the algorithm can be seen in [11].

The consideration of different weights for the decision makers has notably increased the difficulty of the optimisation model (9) solved in Step 6. This non-trivial optimisation problem is solved using an iterative procedure which searches for the intersection points of the parabolas (the second order equations associated with the GCI($A^{(k)}$) functions).

$$Min_{\alpha_{rs}} Max_k \pi_k \left(GCI^{(k)} + \frac{2}{n(n-1)}\left[\left(\alpha_{rs} - \alpha_{rs}^{(k)}\right)^2 + \frac{2n}{n-2}\left(\alpha_{rs} - \alpha_{rs}^{(k)}\right)\varepsilon_{rs}^{(k)}\right]\right) \tag{9}$$

with $\alpha_{rs} \in \left[\log \underline{a}_{rs}^t, \log \overline{a}_{rs}^t\right]$, where $\alpha_{rs} = \log a_{rs}$, $\alpha_{rs}^{(k)} = \log a_{rs}^{(k)}$ and $\varepsilon_{rs}^{(k)} = \log e_{rs}^{(k)}$

When all decision makers have the same weight (initial version of the algorithm [12]), all the parabolas have the same 'width' (the same coefficient of the quadratic term). In that situation, the parabolas may intersect in one point or none. But when the decision makers have different weights [11], the parabolas may have different coefficients for their respective quadratic terms. Each pair of parabolas may intersect in one or two points, or none. Moreover, in this case, some parabolas can be tangential. The resolution of the optimisation model (9) should consider all these possibilities and carefully analyse each intersection point. A more detailed explanation of the procedure followed to solve this optimisation model (9) can be seen in Appendix A.

4. Improving the PCCM's Compatibility

4.1. Iterative Procedure

The PCCM decisional tool has been applied to decisional problems [11,12] and the values of consistency are significantly better than those obtained with other GDM approaches (AIJ, Dong procedure [23]), but they are slightly worse in terms of compatibility.

This paper suggests an iterative procedure to improve compatibility without significantly worsening consistency (keeping it below a preset threshold). If the PCCM is constructed by sequentially considering the judgements from the least to the greatest variance, the proposed improvement of compatibility will sequentially consider the judgments with the greatest contribution (participation) to the global compatibility measure employed (4). This value corresponds to the entry p_{rs} of the PCCM for which:

$$Max_{ij} \sum_{k=1}^{r} \pi_k log^2 \left(a_{ij}^{(k)} \frac{v_j^{(G)}}{v_i^{(G)}}\right) \tag{10}$$

where $v^{(G)}$ is the priority vector derived from the PCCM using the RGM method.

The procedure will modify the selected judgement p_{rs} approaching it to the ratio $\frac{w_r^{(G)}}{w_s^{(G)}}$ of the priorities derived for the AIJ matrix; following a similar idea of that employed in the Dong procedure [23].

$$p'_{rs} = (p_{rs})^\theta \cdot \left(\frac{w_r^{(G)}}{w_s^{(G)}}\right)^{1-\theta} \quad , \theta \in [0,1] \tag{11}$$

In any case, the modified value would never exceed the limits of the consistency stability intervals for this judgment, guaranteeing that the level of inconsistency for each decision maker is acceptable.

In what follows, the new iterative procedure for improving compatibility is explained in detail. It is described for any judgement matrix P; it will be applied to P = PCCM, as following Algorithm 1.

Algorithm 1

Let $A^{(k)} = (a_{ij}^{(k)})$ be the pairwise comparison matrix of decision maker Dk ($k = 1,...,r; i, j = 1,...,n$) and π_k its relative importance in the group ($\pi_k \geq 0$, $\sum_{k=1}^{r} \pi_k = 1$); $\underline{a}_{ij}, \overline{a}_{ij}$ ($i, j = 1, \dots, n$) the limits of the intervals of the Consistency Interval Judgement matrix for the group; $\theta \in [0,1]$; $w^{(G)}$ the priority vector obtained when applying the RGM to the AIJ matrix; P a judgement matrix; and v the priority vector derived from P using the RGM method.

Step 0: Initialisation

Let $t = 0$, $P^{(0)} = P$, $J = \{(i,j),$ with $i < j\}$ and calculate for all $(i,j) \in J$:

$$d_{ij} = \sum_k \pi_k log^2 \frac{a_{ij}^{(k)} v_j}{v_i}$$

Step 1: Selection of the judgement

Let (r, s) be the entry for which $d_{rs} = \max_{(i,j) \in J} d_{ij}$

$$J = J - \{(r,s)\}$$

Step 2: Obtaining a PCCM entry

$$P^{(t+1)} = P^{(t)}$$

Let $z = \left(p_{rs}^{(t)}\right)^{\theta} \left(\frac{w_r^{(G)}}{w_s^{(G)}}\right)^{1-\theta}$

$$p_{rs}^{(t+1)} = \begin{cases} \underline{a}_{rs} & \text{if} & z < \underline{a}_{rs} \\ z & \text{if} & \underline{a}_{rs} \leq z \leq \overline{a}_{rs} \\ \overline{a}_{rs} & \text{if} & z > \overline{a}_{rs} \end{cases}$$

Step 3: Finalisation

$J = \varnothing$, then Stop
Else let $t = t + 1$ and go to Step 1.

4.2. Case Study

The previous procedure was applied to a case study which has been widely employed in the literature [11,12,24,25]: three decision makers (DM1, DM2 and DM3) must compare 5 alternatives (A1, ... , A5). The individual pairwise comparison matrices are given in Table 1. The decision makers were given different weights ($\pi_1 = 5$; $\pi_2 = 4$; and $\pi_3 = 2$).

Table 1. Pairwise comparison matrices for the three decision makers.

DM1	A1	A2	A3	A4	A5	DM2	A1	A2	A3	A4	A5	DM3	A1	A2	A3	A4	A5
A1	1	3	5	8	6	A1	1	3	7	9	5	A1	1	5	7	7	5
A2	-	1	3	5	4	A2	-	1	3	7	1	A2	-	1	1	5	1
A3	-	-	1	3	2	A3	-	-	1	5	1/5	A3	-	-	1	5	1/3
A4	-	-	-	1	1/3	A4	-	-	-	1	1/5	A4	-	-	-	1	1/5
A5	-	-	-	-	1	A5	-	-	-	-	1	A5	-	-	-	-	1

Table 2 gives the resulting priorities using the RGM for each of the three individual matrices and their corresponding rankings.

Table 2. Individual priority vectors, rankings and GCIs.

Priorities	DM1	DM2	DM3
A1	0.513	0.520	0.560
A2	0.251	0.195	0.135
A3	0.115	0.072	0.101
A4	0.042	0.030	0.035
A5	0.079	0.182	0.168
Rankings	1-2-3-5-4	1-2-5-3-4	1-5-2-3-4
GCIs	0.143	0.303	0.298

The PCCM matrix that is obtained by applying the procedure explained in [11,12] is shown in Table 3.

Table 3. Precise Consistency Consensus Matrix (PCCM).

PCCM	A1	A2	A3	A4	A5
A1	1	2.05	5.51	9	3.17
A2	0.49	1	3	6.08	1.74
A3	0.18	0.33	1	2.71	0.68
A4	0.11	0.16	0.37	1	0.35
A5	0.32	0.58	1.47	2.4	1

Two other AHP-GDM procedures have been applied: the AIJ that was explained in Section 2, and the Dong procedure [23]. Table 4 shows the priority vectors obtained with the three AHP-GDM procedures. It can be observed that the ranking of the alternatives is the same for the three procedures.

Table 4. Priority vectors and rankings for the AHP-GDM procedures.

Priorities	PCCM	AIJ	Dong
A1	0.467	0.533	0.531
A2	0.255	0.208	0.216
A3	0.095	0.096	0.099
A4	0.044	0.037	0.038
A5	0.139	0.125	0.116
Rankings	1-2-5-3-4	1-2-5-3-4	1-2-5-3-4

Table 5 shows the consistency and compatibility indicator values for the three AHP-GDM procedures.

Table 5. Consistency and compatibility indicator values (the best value of the methods is in bold, for each indicator).

	PCCM	AIJ	Dong
GCI	**0.023**	0.122	0.069
CVN	**0**	0.018	0.018
GCOMPI	0.529	**0.464**	0.472
PVN	**0.136**	**0.136**	**0.136**

With respect to the indicators that measure consistency (GCI and CVN), the values obtained with the PCCM are considerably better than those obtained with the other two approaches. The values of the GCI for the AIJ procedure (0.122) and for the Dong procedure (0.069) are, respectively, more than five times (535.7%) and three times (304.2%) greater than that of the PCCM (0.023). The behaviour of the CVN is also better for the PCCM (CVN(PCCM) = 0 while CVN(AIJ) = CVN(Dong) = 0.018).

With respect to the compatibility, the value of the GCOMPI for the AIJ procedure (0.464) is better than those of the PCCM (the value 0.529 is 14% greater than the AIJ) and the Dong procedure (the value 0.472 is 1.7% greater). Finally, in the analysis of the number of violations (ordinal compatibility), the three methods gave the same result (0.136).

Having observed that the PCCM is the procedure (among the three being compared) that achieves the highest value for the GCOMPI indicator, the iterative procedure proposed at Section 4.1 is applied with the aim of detecting an improvement in the compatibility of the PCCM.

The iterative procedure was applied with different values of θ ($\theta = 0.75$; $\theta = 0.5$; $\theta = 0.25$; and $\theta = 0$); the PCCM corresponds to $\theta = 1$. In order to compare the results obtained for the combinations considered, the focus is on the two cardinal indicators—the GCI for consistency and the GCOMPI for compatibility.

Tables 6–9 show the sequence of iterations followed when applying the procedure (each column) and the values obtained for the two indicators for each iteration. The second row specifies the judgement that is modified in the corresponding iteration. The values for the original PCCM are shown in the first column as it corresponds to the starting point of the iterative procedure ($t = 0$). The modified values for each entry can be seen in Table 10. The values of the GCI and GCOMPI for the judgment (1,4), $t = 8$, are empty because modifying this judgement will lead to a figure out of the matrix GCIJA. The initial value is maintained and the procedure continues, selecting the following judgement.

Table 6. Results for the iterative procedure with $\theta = 0.75$ (the best value of the methods is in bold, for each indicator).

Iteration°	$t = 0$	$t = 1$	$t = 2$	$t = 3$	$t = 4$	$t = 5$	$t = 6$	$t = 7$	$t = 8$	$t = 9$	$t = 10$
Modif. Judg.	-	(3,5)	(2,5)	(3,4)	(1,2)	(1,5)	(2,3)	(4,5)	(1,4)	(1,3)	(2,4)
GCI	0.023	0.023	0.023	0.022	0.025	0.023	0.021	0.019	°	0.020	**0.019**
GCOMPI	0.529	0.528	0.527	0.528	0.522	0.517	0.512	0.511	°	0.511	**0.511**

Table 7. Results for the iterative procedure with $\theta = 0.5$ (the best value of the methods is in bold, for each indicator).

Iteration°	$t = 0$	$t = 1$	$t = 2$	$t = 3$	$t = 4$	$t = 5$	$t = 6$	$t = 7$	$t = 8$	$t = 9$	$t = 10$
Modif. Judg.	-	(3,5)	(2,5)	(3,4)	(1,2)	(1,5)	(2,3)	(4,5)	(1,4)	(1,3)	(2,4)
GCI	0.023	0.023	0.024	0.022	0.028	0.025	0.020	0.019	°	0.019	**0.018**
GCOMPI	0.529	0.527	0.526	0.527	0.516	0.507	0.499	0.496	°	**0.496**	0.496

Table 8. Results for the iterative procedure with $\theta = 0.25$ (the best value of the methods is in bold, for each indicator).

Iteration°	$t = 0$	$t = 1$	$t = 2$	$t = 3$	$t = 4$	$t = 5$	$t = 6$	$t = 7$	$t = 8$	$t = 9$	$t = 10$
Modif. Judg.	-	(3,5)	(2,5)	(3,4)	(1,2)	(1,5)	(2,3)	(4,5)	(1,4)	(1,3)	(2,4)
GCI	0.023	0.024	0.024	0.023	0.031	0.029	0.021	0.021	°	0.021	**0.019**
GCOMPI	0.529	0.526	0.525	0.526	0.510	0.499	0.489	0.485	°	**0.485**	0.486

Table 9. Results for the iterative procedure with $\theta = 0$ (the best value of the methods is in bold, for each indicator).

Iteration°	$t = 0$	$t = 1$	$t = 2$	$t = 3$	$t = 4$	$t = 5$	$t = 6$	$t = 7$	$t = 8$	$t = 9$	$t = 10$
Modif. Judg.	-	(3,5)	(2,5)	(3,4)	(1,2)	(1,5)	(2,3)	(4,5)	(1,4)	(1,3)	(2,4)
GCI	0.023	0.024	0.025	0.023	0.036	0.033	0.024	0.025	°	0.025	**0.023**
GCOMPI	0.529	0.525	0.523	0.525	0.505	0.493	0.484	0.477	°	**0.477**	0.479

From Tables 6–9, it is possible to make the following observations:

- The order of the entrance of the judgements is the same for all the values considered for θ. This does not always have to happen.
- According to the proposal followed in expression (11), the values of the compatibility indicator improve when the value of the parameter θ decreases.
- In addition to improving compatibility, the final result also improves consistency.
- The value of the compatibility indicator almost always decreases with the iterations. In just a few cases, for judgements (3,4) and (2,4), compatibility is slightly worse. The lowest value of the GCOMPI (0.477) is obtained with θ = 0 and applying the iteration procedure until the penultimate iteration. This value is only 2.8% greater than the one obtained with the AIJ.
- The consistency indicator oscillates a little until achieving the highest value at iteration t = 4 (modifying judgement (1,2)). The next iteration (modifying judgement (1,5)) is a turning point and from there the GCI reduces its value significantly (this can also be seen in Figures 1 and 2). It can also be observed that the lower values of the GCI tend to be those obtained with high and intermediate values of θ. The lowest value of the GCI (0.018) is obtained with θ =0.5 and applying the iteration procedure until the last iteration. This value is 15.8% lower than that obtained with the PCCM procedure.
- The fact that in the iteration t = 8 the judgement (1,4) is not modified means that the value of the GCI is not null at the last iteration of the procedure for θ = 0.

Figures 1 and 2 give the information provided in Tables 6–9 in the form of graphs; they illustrate the relationship between the two indicator values, GCI and GCOMPI. Figure 1 shows the paths that these values follow in the iterative procedure (as new judgements are modified) for each value of θ separately, while Figure 2 shows all the paths in the same graphic presentation. The graphic visualisations help us to understand the relationship between the two indicators and to identify the steps of the algorithm that provide the biggest changes.

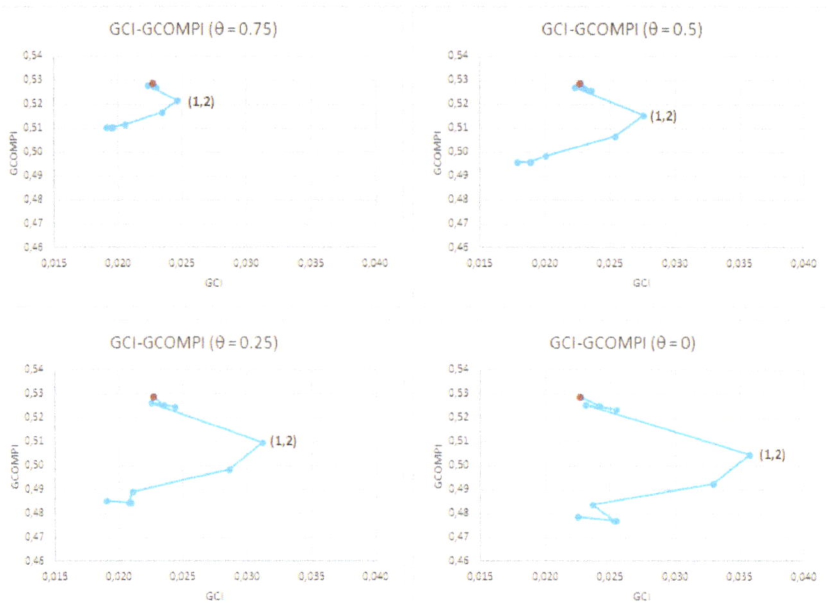

Figure 1. GCI and GCOMPI values for the iterative procedure for different values of θ. Red points correspond to the initial values of the iterative process.

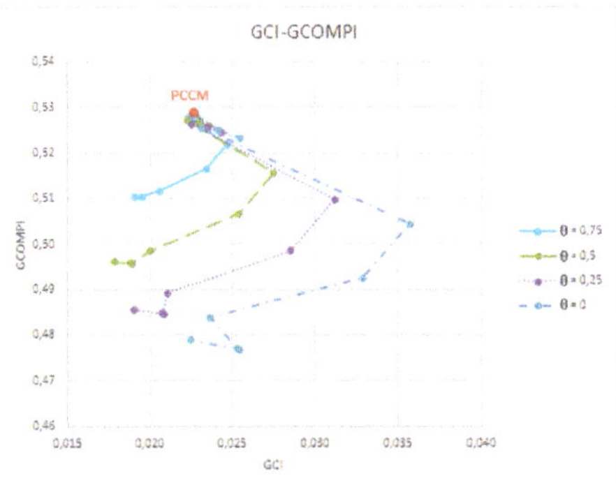

Figure 2. GCI and GCOMPI values for the iterative procedure (all the values of θ).

It can be observed that for all the values of θ, there is a turning point where the value of the GCI begins to decrease significantly. It corresponds to iteration t = 5 when judgment (1,5) is modified. Moreover, when θ decreases, the value of the GCOMPI also decreases and the variability of the GCI increases (Figure 2).

Table 10 includes the modified PCCMs corresponding to the last iteration for each value of θ. Table 11 gives the priorities associated to these matrices; all the priority vectors have the same ranking and their range increases when the value of θ decreases.

Table 10. Modified PCCMs for different values of θ.

Modified PCCM (θ = 0.75)	A1	A2	A3	A4	A5	Modified PCCM (θ = 0.5)	A1	A2	A3	A4	A5
A1	1	2.17	5.52	9	3.41	A1	1	2.29	5.52	9	3.67
A2	0.46	1	2.76	5.97	1.72	A2	0.44	1	2.55	5.86	1.70
A3	0.18	0.36	1	2.68	0.70	A3	0.18	0.39	1	2.66	0.72
A4	0.11	0.17	0.37	1	0.34	A4	0.11	0.17	0.38	1	0.32
A5	0.29	0.58	1.42	2.97	1	A5	0.27	0.59	1.38	3.11	1
Modified PCCM (θ = 0.25)	A1	A2	A3	A4	A5	Modified PCCM (θ = 0)	A1	A2	A3	A4	A5
A1	1	2.42	5.53	9	3.96	A1	1	2.56	5.54	9	4.26
A2	0.41	1	2.35	5.75	1.68	A2	0.39	1	2.16	5.64	1.66
A3	0.18	0.43	1	2.64	0.75	A3	0.18	0.46	1	2.61	0.77
A4	0.11	0.17	0.38	1	0.31	A4	0.11	0.18	0.38	1	0.29
A5	0.25	0.59	1.34	3.25	1	A5	0.23	0.60	1.30	3.39	1

Table 11. Priority vectors and rankings for the modified PCCMs with different values of θ.

Priorities	PCCM (θ = 1)	θ = 0.75	θ = 0.5	θ = 0.25	θ = 0
A1	0.467	0.477	0.488	0.498	0.508
A2	0.255	0.245	0.236	0.227	0.218
A3	0.095	0.096	0.098	0.099	0.101
A4	0.044	0.044	0.043	0.043	0.042
A5	0.139	0.137	0.135	0.133	0.131
Rankings	1-2-5-3-4	1-2-5-3-4	1-2-5-3-4	1-2-5-3-4	1-2-5-3-4

5. Conclusions

This paper has addressed issues related to the calculation and exploitation of the PCCM decisional tool employed in group decision making with AHP.

There are two particular contributions to the literature:

(i) For the first time, the procedure followed to solve the optimisation problem that arises in each iteration of the calculation algorithm of the PCCM has been explained in detail. The consideration of different weights for decision makers greatly increases the difficulty of the optimisation problem, and it has been necessary to study all of the possible situations that could occur.

(ii) The work presents a proposal to improve the compatibility of PCCM matrices. As previously mentioned, whilst the PCCM gives much better values than other procedures with regards to consistency, its behavior in terms of compatibility is worse. Following a sequential procedure in line with their contribution to the GCOMPI, the judgments of the PCCM are modified using a combination of the initial value of the PCCM and the ratio of the priorities obtained with the AIJ procedure.

The case study proved that compatibility substantially improves, reaching values close to those of the AIJ procedure. Consistency also improved, guaranteeing that the judgments of the consensus matrix belong to the consistency stability intervals of all decision makers.

Although the proposal made in this paper has been focused on improving the compatibility of the PCCM, the procedure can be adapted and applied to any consensus matrix.

Future research will seek to establish other criteria that determine the sequence in which the judgments of the group consensus matrix are selected for modification. At the same time, future extensions of this research will include a comparison of the proposal set out in this paper with the recently published improvements made by the authors of [23] to their methodology.

Author Contributions: The paper has been elaborated jointly by the four authors.

Funding: This research was funded by the Spanish Ministry of Economy and Competitiveness and FEDER funding, project ECO2015-66673-R.

Acknowledgments: The authors would like to acknowledge the work of English translation professional David Jones in preparing the final text.

Conflicts of Interest: The authors declare no conflict of interest.

Appendix A

The following is a description of the procedure followed to solve the optimisation problem posed in Step 6 and given by expression (9):

$$\min_{\alpha_{rs}} \max_{k} \pi_k \left(GCI^{(k)} + \frac{2}{n(n-1)} \left[\left(\alpha_{rs} - \alpha_{rs}^{(k)} \right)^2 + \frac{2n}{n-2} \left(\alpha_{rs} - \alpha_{rs}^{(k)} \right) \varepsilon_{rs}^{(k)} \right] \right)$$

with $\alpha_{rs} \in \left[\log \underline{a}_{rs}^t, \log \overline{a}_{rs}^t \right]$, where $\alpha_{rs} = \log a_{rs}$, $\alpha_{rs}^{(k)} = \log a_{rs}^{(k)}$ and $\varepsilon_{rs}^{(k)} = \log e_{rs}^{(k)}$

This problem can be written as:

$$\min_{\alpha_{rs} \in [l,\, u]} \max_{k} p_k(\alpha_{rs})$$

where $p_k(\alpha_{rs}) = \pi_k \left(GCI^{(k)} + \frac{2}{n(n-1)} \left[\left(\alpha_{rs} - \alpha_{rs}^{(k)} \right)^2 + \frac{2n}{n-2} \left(\alpha_{rs} - \alpha_{rs}^{(k)} \right) \varepsilon_{rs}^{(k)} \right] \right)$ are second degree polynomials with α_{rs} the variable and the coefficient for the quadratic term positive.

Therefore, the problem can be rewritten as:

$$\min_{x \in [l,\, u]} \max_{k} p_k(x) \tag{A1}$$

where $p_k(x) = a_k x^2 + b_k x + c_k$ with $a_k > 0$.

There are three cases:

a. The polynomial that is dominant in $x = l$ is an increasing function at this point. In this case, the solution to the optimisation problem (9) is $x^* = l$ (Figure A1a).

b. The polynomial that is dominant in $x = u$ is a decreasing function at this point. In this case, the solution to the optimisation problem (9) is $x^* = u$ (Figure A1b).

c. The polynomial that is dominant in $x = l$ is a decreasing function at this point. In this case, we start from this initial point and move forward until we find a section/segment in which the dominant polynomial is an increasing function (Figure A1c).

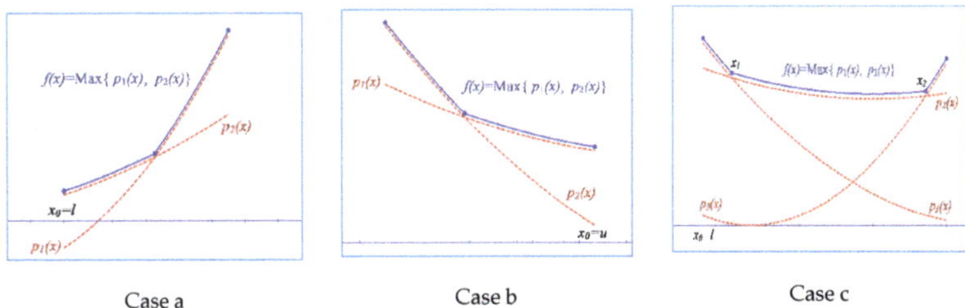

Case a   Case b   Case c

Figure A1. Graphical representation of the three cases.

Case a. *When we say that a polynomial p_r is dominant in $x = l$, we refer to the situation in which the polynomial p_r provides the maximum value in a neighbourhood $[l, l + \varepsilon)$:*

$$p_r(x) = \max_k p_k(x) \, with \, x \in [l, \, l + \varepsilon)$$

In order to obtain this dominant polynomial, we start by analysing the values $[l, l + \varepsilon)$. The polynomial $p_r(x)$ that provides the maximum value at this point is the polynomial that we were looking for. Nevertheless, it is possible that some ties exist. In this case, we should look for the polynomial that is dominant in the neighbourhood $[l, l + \varepsilon)$.

If $p_i(l) = p_j(l)$, in order to determine which of the two polynomials is dominant in $[l, l + \varepsilon)$, we examine the first derivative: if $p_i'(l) > p_j'(l)$, the polynomial p_i is dominant.

If $p_i(l) = p_j(l)$ and $p_i'(l) = p_j'(l)$, the dominant polynomial will be that one with the maximum value in the second derivative.

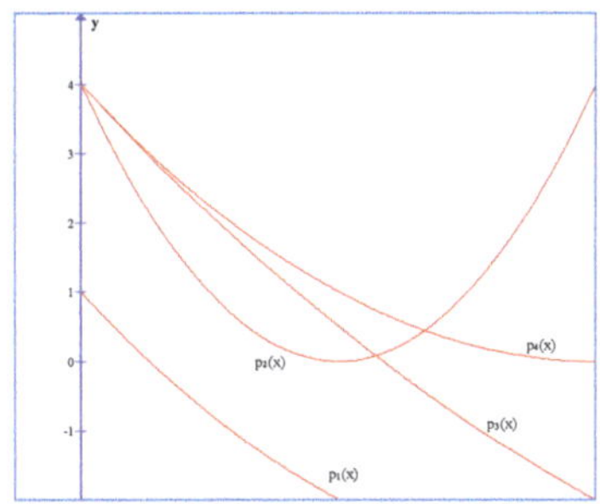

Figure A2. Several domination relationships.

In the Figure A2 it can be appreciated that $p_1(l) = 1$ and $p_2(l) = p_3(l) = p_4(l) = 4$. It can also be observed that $p_2'(l) < p_3'(l)$ and $p_2'(l) < p_4'(l)$, thus the polynomial $p_2(x)$ is not dominant. Polynomials $p_3(x)$ and $p_4(x)$ have the same value as their derivative in $x = l$, but the polynomial $p_4(x)$ has a greater value in the second derivative (greater coefficient of x^2), it is therefore the dominant polynomial.

In short, if p_r is the polynomial that is dominant in $x = l$, the following condition should be fulfilled:

$$(p_r(l), p\prime_r(l), p_r''(l)) = lex \max_k (p_k(l), p\prime_k(l), p_k''(l))$$

Case b. *The polynomial that is dominant in $x = u$ can be determined in an analogous way:*

$$(p_s(u), -p\prime_s(u), p\prime\prime_s(u)) = lex \max_k (p_k(u), -p\prime_k(u), p\prime\prime_k(u))$$

By determining the polynomials which are dominant in l and u, and calculating their respective derivatives we will be able to identify cases a and b.

Case c. *Starting from point $x0 = l$ and from the corresponding polynomial that is dominant at this point, we determine point $x1$ in which this polynomial becomes dominated. If the following polynomial that is dominant is a decreasing function at this point, we continue the process, updating $x0 = x1$. At the moment in which we go out from the interval under study, or when the new polynomial that is dominant is an increasing function, we have finished the iterative stage and we only have to calculate the minimum of the present dominant polynomial in the interval under study.*

In Figure A1c we can see that p1 is the polynomial that is dominant in $x = 1$. It continues to be the dominant polynomial until x1 where the new polynomial that is dominant is p2. Again, this polynomial is dominant until point x2. But the polynomial that is dominant from this point is an increasing function at x2, and therefore it is not necessary to continue. The solution to the optimisation problem is given by the minimum of the polynomial p2 in the interval $[x1, x2]$.

Finding the points where a polynomial ceases to be dominant is determined by exploring the possible cut points with the rest of the polynomials in the problem (those that are within the interval under study) as following Algorithm A1.

Algorithm A1

$$\min_{x\in[l,\,u]} \max_k p_k(x) \text{ where } p_k(x) = a_k x^2 + b_k x + c_k \text{ with } a_k > 0$$

Step 1: Find r and with

$$(p_r(l), p\prime_r(l), p\prime\prime_r(l)) = lex \max_k (p_k(l), p\prime_k(l), p\prime\prime_k(l))$$

$$(p_s(u), -p\prime_s(u), p\prime\prime_s(u)) = lex \max_k (p_k(u), -p\prime_k(u), p\prime\prime_k(u))$$

If $p\prime_r(l) \geq 0$ then $x^* = l$. STOP

If $p\prime_s(u) \leq 0$ then $x^* = u$. STOP

Step 2: Start from point $x_0 = l$, where the polynomial that is dominant, $p_r(x)$, is a decreasing function at this point.

Step 3: Calculate $I = \{i$ such that $\exists\, t_i \in (x_0, u]$ with $p_i(t_i) = p_r(t_i)$ and $p_i\prime(t_i) > p_r\prime(t_i)\}$

Step 4: If $I = \varnothing$ the optimal solution is given by:

$$\min_{x\in[x_0,\,u]} p_r(x)$$

Step 5: Let j be the value such that

$$(-t_j, p\prime_j(t_j), p\prime\prime_j(t_j)) = lex \max_{i\in I} (-t_i, p\prime_i(t_i), p\prime\prime_i(t_i))$$

If $p\prime_j(t_j) > 0$ the optimal solution is given by:

$$\min_{x\in[x_0,\,t_j]} p_r(x)$$

Otherwise, update and $r = j$ and go to Step 3.

References

1. Moreno-Jiménez, J.M.; Vargas, L.G. Cognitive Multicriteria Decision Making and the Legacy of AHP. *Estudios de Economía Aplicada* **2018**, *36*, 67–80.

2. Saaty, T.L. *The Analytic Hierarchy Process*; 2nd impression 1990, RSW Pub. Pittsburgh; Mc Graw-Hill: New York, NY, USA, 1980.

3. Saaty, T.L. Group decision making and the AHP. In *The Analytic Hierarchy Process*; Golden, B.L., Wasil, E.A., Harker, P.T., Eds.; Springer: Berlin/Heidelberg, Germany, 1989; pp. 59–67.

4. Ramanathan, R.; Ganesh, L.S. Group preference aggregation methods employed in AHP: An evaluation and intrinsic process for deriving members' weightages. *Eur. J. Oper. Res.* **1994**, *79*, 249–265. [CrossRef]

5. Forman, E.; Peniwati, K. Aggregating individual judgements and priorities with the analytic hierarchy process. *Eur. J. Oper. Res.* **1998**, *108*, 165–169. [CrossRef]

6. Escobar, M.T.; Moreno-Jiménez, J.M. Aggregation of Individual Preference Structures. *Group Decis. Negotiat.* **2007**, *16*, 287–301. [CrossRef]

7. Altuzarra, A.; Moreno-Jiménez, J.M.; Salvador, M. A Bayesian priorization procedure for AHP-group decision making. *Eur. J. Oper. Res.* **2007**, *182*, 367–382. [CrossRef]

8. Moreno-Jiménez, J.M.; Aguarón, J.; Escobar, M.T. Decisional Tools for Consensus Building in AHP-Group Decision Making. In Proceedings of the 12th Mini Euro Conference, Brussels, Belgium, 2–5 April 2002.

9. Moreno-Jiménez, J.M.; Aguarón, J.; Raluy, A.; Turón, A. A spreadsheet module for consistent AHP-consensus building. *Group Decis. Negotiat.* **2005**, *14*, 89–108.

10. Moreno-Jiménez, J.M.; Aguarón, J.; Escobar, M.T. The core of consistency in AHP-group decision making. *Group Decis. Negotiat.* **2008**, *17*, 249–265.

11. Escobar, M.T.; Aguarón, J.; Moreno-Jiménez, J.M. Some extensions of the Precise Consistency Consensus Matrix. *Decis. Support Syst.* **2015**, *74*, 67–77. [CrossRef]

12. Aguarón, J.; Escobar, M.T.; Moreno-Jiménez, J.M. The precise consistency consensus matrix in a local AHP-group decision making context. *Ann. Oper. Res.* **2016**, *245*, 245–259. [CrossRef]

13. Moreno-Jiménez, J.M. Los Métodos Estadísticos en el Nuevo Método Científico. In *Información Económica y Técnicas de Análisis en el Siglo XXI*; Casas, J.M., Pulido, A., Eds.; INE: Madrid, Spain, 2003; pp. 331–348.

14. De Bono, E. *Lateral Thinking: Creativity Step by Step*; Harper & Row: New York, NY, USA, 1970.

15. Salvador, M.; Altuzarra, A.; Gargallo, M.P.; Moreno-Jiménez, J.M. A Bayesian Approach for maximising inner compatibility in AHP-Systemic Decision Making. *Group Decis. Negotiat.* **2015**, *24*, 655–673. [CrossRef]

16. Moreno-Jiménez, J.M.; Gargallo, M.P.; Salvador, M.; Altuzarra, A. Systemic Decision Making: A Bayesian Approach in AHP. *Ann. Oper. Res.* **2016**, *245*, 261–284. [CrossRef]

17. Barzilai, J.; Golany, B. AHP rank reversal, normalization and aggregation rules. *INFOR* **1994**, *32*, 57–63. [CrossRef]

18. Escobar, M.T.; Aguarón, J.; Moreno-Jiménez, J.M. A Note on AHP Group Consistency for the Row Geometric Mean Priorization Procedure. *Eur. J. Oper. Res.* **2004**, *153*, 318–322. [CrossRef]

19. Aguarón, J.; Moreno-Jiménez, J.M. The geometric consistency index: Approximate thresholds. *Eur. J. Oper. Res.* **2003**, *147*, 137–145. [CrossRef]

20. Aguarón, J.; Escobar, M.T.; Moreno-Jiménez, J.M. Consistency stability intervals for a judgement in AHP decision support systems. *Eur. J. Oper. Res.* **2003**, *145*, 382–393. [CrossRef]

21. Moreno-Jiménez, J.M. An AHP/ANP Multicriteria Methodology to Estimate the Value and Transfers Fees of Professional Football Players. In Proceedings of the ISAHP 2011, Sorrento, Italy, 14–18 June 2011.

22. Golany, B.; Kress, M. A multicriteria evaluation of methods for obtaining weights from ratio-scale matrices. *Eur. J. Oper. Res.* **1993**, *69*, 210–220. [CrossRef]

23. Dong, Y.; Zhang, G.; Hong, W.C.; Xu, Y. Consensus models for AHP group decision making under row geometric mean prioritization method. *Decis. Support Syst.* **2010**, *49*, 281–289. [CrossRef]

24. Moreno-Jiménez, J.M.; Gómez-Bahillo, C.; Sanaú, J. Viabilidad Integral de Proyectos de Inversión Pública. Valoración económica de los aspectos sociales. In *Anales de Economía Aplicada XXIII*; Pires, J.R., Monteiro, J.D., Eds.; Delta: Madrid, Spain, 2009; pp. 2551–2562.

25. Turón, A.; Aguarón, J.; Escobar, M.T.; Moreno-Jiménez, J.M. A Decision Support System and Visualisation Tools for AHP-GDM. *Int. J. Decis. Supports Syst. Technol.* **2019**, *11*, 1–19. [CrossRef]

Article

Evaluation Based on Distance from Average Solution Method for Multiple Criteria Group Decision Making under Picture 2-Tuple Linguistic Environment

Siqi Zhang [1], Hui Gao [1], Guiwu Wei [1], Yu Wei [2,*] and Cun Wei [1,3]

[1] School of Business, Sichuan Normal University, Chengdu 610101, China; zsq1277@163.com (S.Z.); gaohuisxy@sicnu.edu.cn (H.G.); weiguiwu1973@sicnu.edu.cn (G.W.); weicun1990@163.com (C.W.)
[2] School of Finance, Yunnan University of Finance and Economics, Kunming 650221, China
[3] School of Science, Southwest Petroleum University, Chengdu 610500, China
* Correspondence: weiyusy@126.com or ywei@home.swjtu.edu.cn

Received: 31 January 2019; Accepted: 5 March 2019; Published: 8 March 2019

Abstract: In this paper, we design the EDAS (evaluation based on distance from average solution) model with picture 2-tuple linguistic numbers (P2TLNs). First, we briefly reviewed the definition of P2TLSs and introduced the score function, accuracy function, and operational laws of P2TLNs. Then, we combined the traditional EDAS model for multiple criteria group decision making (MCGDM) with P2TLNs. Our presented model was more accurate and effective for considering the conflicting attributes. Finally, a numerical case for green supplier selection was given to illustrate this new model, and some comparisons were also conducted between the picture 2-tuple linguistic weighted averaging (P2TLWA), picture 2-tuple linguistic weighted geometric (P2TLWG) aggregation operators and EDAS model with P2TLNs, to further illustrate the advantages of the new method.

Keywords: multiple criteria group decision making (MCGDM) problems; picture fuzzy sets (PFSs); picture 2-tuple linguistic numbers (P2TLNs); picture 2-tuple linguistic sets (P2TLSs); EDAS model; green supplier selection

---

1. Introduction

The traditional EDAS (evaluation based on distance from average solution) method [1], which can consider the conflicting attributes, has been studied in many multi-attribute decision making (MADM) problems. By computing the average solution (AV), this model can describe the difference between all the alternatives and the AV based on two distance measures which are namely PDA (Positive Distance from Average) and NDA (Negative Distance from Average), the alternative with higher values of PDA and lower values of NDA is the best choice. Until now, lots of MADM methods such as the VIKOR (VIseKriterijumska Optimizacija I KOmpromisno Resenje) method [2,3], the ELECTRE (ELimination and Choice Expressing the Reality) method [4], the TOPSIS (Technique for Order Preference by Similarity to Ideal Solution) method [5], the PROMETHEE (Preference Ranking Organisation Method for Enrichment Evaluations) method [6,7], the GRA (Grey relational Analysis) method [8], the MULTIMOORA method [9] and the TODIM (an acronym in Portuguese for Interactive Multi-Criteria Decision Making) method [10–12] were broadly investigated by a large amount of scholars. Compared to the existing work, the EDAS model owns the merit of only taking AVs into account with respect to the intangibility of decision maker (DM) and the uncertainty of the decision making environment to obtain more accurate and effective aggregation results.

Atanassov [13] introduced the concept of intuitionistic fuzzy sets (IFSs), which is a generalization of the concept of fuzzy sets [14]. Atanassov and Gargov [15], and Atanassov [16] proposed the concept of interval-valued intuitionistic fuzzy sets (IVIFSs), which are characterized by a membership function, a non-membership function, and a hesitancy function whose values are intervals. Recently, Cuong and Kreinovich [17] proposed picture fuzzy sets (PFSs) and investigated some basic operations and properties of PFSs. The PFS is characterized by three functions expressing the degree of membership, the degree of neutral membership, and the degree of non-membership. The only constraint is that the sum of the three degrees must not exceed 1. Singh [18] presented the geometrical interpretation of PFSs and proposed correlation coefficients for PFSs. Son [19] presented a novel distributed picture fuzzy clustering (FC) method on PFSs. Thong and Son [20] proposed the model between picture FC and intuitionistic fuzzy recommender systems for medical diagnosis. Thong and Son [21] proposed automatic picture fuzzy clustering (AFC-PFS) for determining the most suitable number of clusters for AFC-PFS. Wei [22] proposed the MADM method based on the proposed picture fuzzy cross entropy. Son [23] defined the generalized picture distance measures and picture association measures. Son and Thong [24] developed some novel hybrid forecast models with picture FC for weather nowcasting from satellite image sequences. Wei [25] gave some cosine similarity measures of PFSs for strategic decision making on the basis of traditional similarity measures [26,27]. Wei [28] proposed some aggregation operators for MADM based on the PFSs based on traditional aggregation operators [29–35]. Wei [36] defined some similarity measures for PFSs. Wei [12] proposed the TODIM method for picture fuzzy MADM. Wei and Gao [37] developed the generalized dice similarity measures for PFSs. Wei [38] proposed some picture fuzzy Hamacher aggregation operators in MADM with traditional Hamacher operations [39–42]. Wei et al. [43] designed the projection models for MADM with picture fuzzy information. Wei et al. [44] proposed some picture 2-tuple linguistic operators in MADM. Wei [45] proposed some Bonferroni mean (BM) operators with P2TLNs in MADM. Wei [46] defined some picture uncertain linguistic BM operators for MADM.

Turskis et al. [1] originally defined the EDAS method for multi-criteria inventory classification. Keshavarz Ghorabaee et al. [47] proposed the extended EDAS method for supplier selection. Kahraman et al. [48] established the EDAS model under intuitionistic fuzzy information for solid waste disposal site selection. Keshavarz Ghorabaee et al. [49] extended the EDAS method with interval type-2 fuzzy sets. Keshavarz Ghorabaee et al. [50] defined the multi-criteria EDAS model with interval type-2 fuzzy sets. Keshavarz Ghorabaee et al. [51] proposed the stochastic EDAS method for MADM with normally distributed data. Peng and Liu [52] resolved the neutrosophic soft decision making method based on EDAS. Ecer [53] gave third-party logistics provider selection with the fuzzy AHP and the fuzzy EDAS integrated method. Feng et al. [54] developed the EDAS method for hesitant fuzzy linguistic MADM. Ilieva [55] assigned the group decision models with EDAS for interval fuzzy sets. Karasan and Kahraman [56] defined the interval-valued neutrosophic EDAS method. Keshavarz-Ghorabaee et al. [57] developed the dynamic fuzzy EDAS method for multi-criteria subcontractor evaluation. Stevic et al. [58] gave the selection of carpenter manufacturer using the fuzzy EDAS method. Keshavarz-Ghorabaee et al. [59] gave a comparative analysis of the rank reversal phenomenon with the EDAS and TOPSIS methods.

Wei et al. [44] introduced the concept of P2TLSs based on PFSs [17] and the 2-tuple linguistic information processing model [60], and developed some BM and geometric BM operators with P2TLNs. However, no studies using the EDAS model with P2TLNs were found in the literature. Hence, it was necessary to take the picture 2-tuple linguistic EDAS model into account. The purpose of our work is to establish an extended EDAS model according to the traditional EDAS method and P2TLNs to study multiple criteria group decision making (MCGDM) problems more effectively. Thus, the main contributions of this paper are (1) to extend EDAS models to picture 2-tuple linguistic sets; (2) to combine the traditional EDAS model for MCGDM with P2TLNs; (3) to provide a numerical case for green supplier selection to illustrate this new model and conduct some comparisons between the EDAS model with P2TLNs, and P2TLWA and P2TLWG aggregation operators to further illustrate advantages of the new method.

The structure of our paper is organized as follows: definition, score function, accuracy function, and operational formulas of P2TLNs are briefly introduced in Section 2. We introduce some aggregation operators of P2TLNs in Section 3. We combine the traditional EDAS model for MCGDM with P2TLNs, and the computing steps are simply depicted in Section 4. In Section 5, a numerical example for green supplier selection has been given to illustrate this new model, and some comparisons between the use of P2TLWA and P2TLWG operators in the EDAS model with P2TLNs were also conducted to further illustrate the advantages of the new method. Section 6 describes some conclusions of our work.

2. Preliminaries

In the following, we introduced some basic concepts related to 2-tuple linguistic term sets and PFSs.

2.1. 2-Tuple Linguistic Term Sets

Let $S = \{s_i \,|i = 1, 2, \cdots, t\}$ be a linguistic term set with odd cardinality. s_i represents a possible value for a linguistic variable, and it should satisfy the following characteristics [61]:

(1) The set is ordered: $s_i > s_j$, if $i > j$; (2) Max operator: $\max(s_i, s_j) = s_i$, if $s_i \geq s_j$; (3) Min operator: $\min(s_i, s_j) = s_i$, if $s_i \leq s_j$. For example, S can be defined as

$$S = \{s_1 = \text{extremely poor}, s_2 = \text{very poor}, s_3 = \text{poor}, s_4 = \text{medium},$$
$$s_5 = \text{good}, s_6 = \text{very good}, s_7 = \text{extremely good}\}.$$

Herrera and Martinez [60] defined the 2-tuple fuzzy linguistic representation model based on the concept of symbolic translation. It is utilized for depicting the linguistic information with a 2-tuple (s_i, α_i), where s_i is a linguistic label from predefined linguistic term set S, and α_i is the value of symbolic translation, and $\alpha_i \in [-0.5, 0.5)$.

2.2. Picture Fuzzy Sets (PFSs)

Definition 1 ([17]). *A PFS on the universe. X is an object of the form*

$$A = \{\langle x, \mu_A(x), \eta_A(x), \nu_A(x)\rangle | x \in X\}, \tag{1}$$

where $\mu_A(x) \in [0, 1]$ is called the "degree of positive membership of A", $\eta_A(x) \in [0, 1]$ is defined as the "degree of neutral membership of A", and $\nu_A(x) \in [0, 1]$ is defined as the "degree of negative membership of A", and $\mu_A(x), \eta_A(x), \nu_A(x)$ satisfy the following condition: $0 \leq \mu_A(x) + \eta_A(x) + \nu_A(x) \leq 1, \forall\, x \in X$. Then, for $x \in X$, $\pi_A(x) = 1 - (\mu_A(x) + \eta_A(x) + \nu_A(x))$ could be defined as the degree of refusal membership of x in A.

Definition 2 ([17]). *Let $\alpha = (\mu_\alpha, \eta_\alpha, \nu_\alpha)$ and $\beta = (\mu_\beta, \eta_\beta, \nu_\beta)$ be two PFNs, the operation formula of them can be given:*

(1) $\quad \alpha \oplus \beta = (\mu_\alpha + \mu_\beta - \mu_\alpha\mu_\beta, \eta_\alpha\eta_\beta, \nu_\alpha\nu_\beta);$

(2) $\quad \alpha \otimes \beta = (\mu_\alpha\mu_\beta, \eta_\alpha + \eta_\beta - \eta_\alpha\eta_\beta, \nu_\alpha + \nu_\beta - \nu_\alpha\nu_\beta);$

(3) $\quad \lambda\alpha = \left(1 - (1 - \mu_\alpha)^\lambda, \eta_\alpha^\lambda, \nu_\alpha^\lambda\right), \lambda > 0;$

(4) $\quad \alpha^\lambda = \left(\mu_\alpha^\lambda, 1 - (1 - \eta_\alpha)^\lambda, 1 - (1 - \nu_\alpha)^\lambda\right), \lambda > 0.$

According to Definition 2, the operation laws have the following properties [17].

$$\alpha \oplus \beta = \beta \oplus \alpha, \alpha \otimes \beta = \beta \otimes \alpha, \left((\alpha)^{\lambda_1}\right)^{\lambda_2} = (\alpha)^{\lambda_1\lambda_2}; \tag{2}$$

$$\lambda(\alpha \oplus \beta) = \lambda\alpha \oplus \lambda\beta, (\alpha \otimes \beta)^\lambda = (\alpha)^\lambda \otimes (\beta)^\lambda; \tag{3}$$

$$\lambda_1 \alpha \oplus \lambda_2 \alpha = (\lambda_1 + \lambda_2)\alpha, \ (\alpha)^{\lambda_1} \otimes (\alpha)^{\lambda_2} = (\alpha)^{(\lambda_1 + \lambda_2)}. \tag{4}$$

2.3. Picture 2-Tuple Linguistic Sets (P2TLSs)

In the following, we introduce the concepts and basic operations of the P2TLSs based on the PFSs [17] and 2-tuple linguistic information model [60].

Definition 3 ([44,45]). *A P2TLS A in X is given*

$$A = \left\{ \left(s_{\theta(x)}, \rho \right), (\mu_A(x), \eta_A(x), \nu_A(x)), x \in X \right\}, \tag{5}$$

where $\left(s_{\theta(x)}, \rho \right) \in S$, $\rho \in [-0.5, 0.5)$, $u_A(x) \in [0,1]$, $\eta_A(x) \in [0,1]$, and $v_A(x) \in [0,1]$, with the condition $0 \leq u_A(x) + \eta_A(x) + v_A(x) \leq 1$, $\forall x \in X$, $s_{\theta(a)} \in S$, and $\rho \in [-0.5, 0.5)$. The numbers $\mu_A(x), \eta_A(x), \nu_A(x)$ represent, respectively, the degree of positive membership, degree of negative membership, and degree of negative membership of the element x to 2-tuple linguistic variable $\left(s_{\theta(x)}, \rho \right)$.

For convenience, we call $\tilde{\alpha} =< \left(s_{\theta(a)}, \rho \right), (u(a), \eta(a), v(a)) >$ a P2TLN, where $\mu_\alpha \in [0,1]$, $\eta_\alpha \in [0,1]$, $v_\alpha \in [0,1]$, $\mu_\alpha + \eta_\alpha + v_\alpha \leq 1$, $s_{\theta(a)} \in S$ and $\rho \in [-0.5, 0.5)$.

Definition 4 ([44]). *Let $\tilde{a} =< \left(s_{\theta(a)}, \rho \right), (u(a), \eta(a), v(a)) >$ be a P2TLN, and a score function $\tilde{a}$ can be defined as follows:*

$$S(\tilde{a}) = \Delta \left(\Delta^{-1} \left(s_{\theta(a)}, \rho \right) \cdot \frac{1 + \mu_\alpha - v_\alpha}{2} \right), \ \Delta^{-1}(S(\tilde{a})) \in [1, t]. \tag{6}$$

Definition 5 ([44]). *Let $\tilde{a} =< \left(s_{\theta(a)}, \rho \right), (u(a), \eta(a), v(a)) >$ be a P2TLN, and the accuracy function can be defined as follows:*

$$H(\tilde{a}) = \Delta \left(\Delta^{-1} \left(s_{\theta(a)}, \rho \right) \cdot \frac{\mu_\alpha + \eta_\alpha + v_\alpha}{2} \right), \ \Delta^{-1}(H(\tilde{a})) \in [1, t]. \tag{7}$$

Definition 6 ([44]). *Let $\tilde{a}_1 =< \left(s_{\theta(a_1)}, \rho_1 \right), (u(a_1), \eta(a_1), v(a_1)) >$ and $\tilde{a}_2 =< \left(s_{\theta(a_2)}, \rho_2 \right), (u(a_2), \eta(a_2), v(a_2)) >$ be two P2TLNs, $S(\tilde{a}_1) = \Delta \left(\Delta^{-1} \left(s_{\theta(a_1)}, \rho_1 \right) \cdot \frac{1 + \mu_{\alpha_1} - v_{\alpha_1}}{2} \right)$ and $S(\tilde{a}_2) = \Delta \left(\Delta^{-1} \left(s_{\theta(a_2)}, \rho_2 \right) \cdot \frac{1 + \mu_{\alpha_2} - v_{\alpha_2}}{2} \right)$ be the scores of $\tilde{a}_1$ and $\tilde{a}_2$, respectively, and let $H(\tilde{a}_1) = \Delta \left(\Delta^{-1} \left(s_{\theta(a_1)}, \rho_1 \right) \cdot \frac{\mu_{\alpha_1} + \eta_{\alpha_1} + v_{\alpha_1}}{2} \right)$ and $H(\tilde{a}_2) = \Delta \left(\Delta^{-1} \left(s_{\theta(a_2)}, \rho_2 \right) \cdot \frac{\mu_{\alpha_2} + \eta_{\alpha_2} + v_{\alpha_2}}{2} \right)$ be the accuracy degrees of $\tilde{a}_1$ and $\tilde{a}_2$, respectively, then if $S(\tilde{a}_1) < S(\tilde{a}_2)$, $\tilde{a}_1 < \tilde{a}_2$; if $S(\tilde{a}_1) = S(\tilde{a}_2)$, then (1) if $H(\tilde{a}_1) = H(\tilde{a}_2)$, then $\tilde{a}_1 = \tilde{a}_2$; (2) if $H(\tilde{a}_1) < H(\tilde{a}_2)$, then, $\tilde{a}_1 < \tilde{a}_2$.*

Some operational laws of P2TLNs are defined as follows:

Definition 7 ([44]). *Let* $\tilde{a}_1 = \left\langle \left(s_{\theta(a_1)}, \rho_1\right), (u(a_1), \eta(a_1), v(a_1)) \right\rangle$ *and* $\tilde{a}_2 = \left\langle \left(s_{\theta(a_2)}, \rho_2\right), (u(a_2), \eta(a_2), v(a_2)) \right\rangle$ *be two P2TLNs, then*

$$
\begin{aligned}
\tilde{a}_1 \oplus \tilde{a}_2 &= \left\langle \Delta\left(\Delta^{-1}\left(s_{\theta(a_1)}, \rho_1\right) + \Delta^{-1}\left(s_{\theta(a_2)}, \rho_2\right)\right), \right.\\
&\quad \left. (u(a_1) + u(a_2) - u(a_1)u(a_2), \eta(a_1)\eta(a_2), v(a_1)v(a_2)) \right\rangle; \\
\tilde{a}_1 \otimes \tilde{a}_2 &= \left\langle \Delta\left(\Delta^{-1}\left(s_{\theta(a_1)}, \rho_1\right) \cdot \Delta^{-1}\left(s_{\theta(a_2)}, \rho_2\right)\right), \right.\\
&\quad \left. (u(a_1)u(a_2), \eta(a_1) + \eta(a_2) - \eta(a_1)\eta(a_2), v(a_1) + v(a_2) - v(a_1)v(a_2)) \right\rangle; \\
\lambda\tilde{a}_1 &= \left\langle \Delta\left(\lambda\Delta^{-1}\left(s_{\theta(a_1)}, \rho_1\right)\right), \left(1 - (1 - u(a_1))^{\lambda}, \eta(a_1)^{\lambda}, v(a_1)^{\lambda}\right) \right\rangle; \\
(\tilde{a}_1)^{\lambda} &= \left\langle \Delta\left(\left(\Delta^{-1}\left(s_{\theta(a_1)}, \rho_1\right)\right)^{\lambda}\right), \left(u(a_1)^{\lambda}, 1 - (1 - \eta(a_1))^{\lambda}, 1 - (1 - v(a_1))^{\lambda}\right) \right\rangle.
\end{aligned}
$$

3. Picture 2-Tuple Linguistic Aggregation Operators

In this section, we propose some aggregation operators with P2TLNs, such as the P2TLWA operator and the P2TLWG operator.

Definition 8. *Let* $\tilde{\alpha}_j = \left\langle (s_j, \rho_j), (\mu_j, \eta_j, v_j) \right\rangle (j = 1, 2, \cdots, n)$ *be a collection of P2TLNs, and the P2TLWA operator can be represented as*

$$
\text{P2TLWA}_\omega(\tilde{\alpha}_1, \tilde{\alpha}_2, \cdots, \tilde{\alpha}_n) = \bigoplus_{j=1}^{n} (\omega_j\tilde{\alpha}_j), \tag{8}
$$

where $\omega = (\omega_1, \omega_2, \ldots, \omega_n)^T$ *is the weight vector of* $\tilde{\alpha}_j (j = 1, 2, \ldots, n)$ *and* $\omega_j > 0, \sum_{j=1}^{n} \omega_j = 1$.

Based on the Definition 8, we can get the following result:

Theorem 1. *The aggregated value by using the P2TLWA operator is also a P2TLN, where*

$$
\begin{aligned}
\text{P2TLWA}_\omega(\tilde{\alpha}_1, \tilde{\alpha}_2, \cdots, \tilde{\alpha}_n) &= \bigoplus_{j=1}^{n} (\omega_j\tilde{\alpha}_j) \\
&= \left\langle \Delta\left(\sum_{j=1}^{n} \omega_j\Delta^{-1}(s_j, \rho_j)\right), \left(1 - \prod_{j=1}^{n} (1 - \mu_j)^{\omega_j}, \prod_{j=1}^{n} (\eta_j)^{\omega_j}, \prod_{j=1}^{n} (v_j + \eta_j)^{\omega_j} - \prod_{j=1}^{n} (\eta_j)^{\omega_j}\right) \right\rangle
\end{aligned} \tag{9}
$$

where $\omega = (\omega_1, \omega_2, \ldots, \omega_n)^T$ *is the weight vector of* $\tilde{\alpha}_j (j = 1, 2, \ldots, n)$ *and* $\omega_j > 0, \sum_{j=1}^{n} \omega_j = 1$.

Definition 9. *Let* $\tilde{\alpha}_j = \left\langle (s_j, \rho_j), (\mu_j, \eta_j, v_j) \right\rangle (j = 1, 2, \cdots, n)$ *be a collection of P2TLNs, the P2TLWG operator can be represented as*

$$
\text{P2TLWG}_\omega(\tilde{\alpha}_1, \tilde{\alpha}_2, \cdots, \tilde{\alpha}_n) = \bigotimes_{j=1}^{n} (\tilde{\alpha}_j)^{\omega_j} \tag{10}
$$

where $\omega = (\omega_1, \omega_2, \ldots, \omega_n)^T$ *is the weight vector of* $\tilde{\alpha}_j (j = 1, 2, \ldots, n)$ *and* $\omega_j > 0, \sum_{j=1}^{n} \omega_j = 1$.

Based on Definition 9, we can get the following result:

Theorem 2. *The aggregated value by using the P2TLWG operator is also a P2TLN, where*

$$\text{P2TLWG}_\omega(\tilde{\alpha}_1, \tilde{\alpha}_2, \cdots, \tilde{\alpha}_n) = \overset{n}{\underset{j=1}{\otimes}} (\tilde{\alpha}_j)^{\omega_j}$$

$$= \left\langle \Delta\left(\prod_{j=1}^{n}\left(\Delta^{-1}(s_j, \rho_j)^{\omega_j}\right)\right), \left(\prod_{j=1}^{n}\left(\mu_{\alpha_j} + \eta_{\alpha_j}\right)^{\omega_j} - \prod_{j=1}^{n}\left(\eta_{\alpha_j}\right)^{\omega_j}, \prod_{j=1}^{n}\left(\eta_{\alpha_j}\right)^{\omega_j}, 1 - \prod_{j=1}^{n}\left(1 - v_{\alpha_j}\right)^{\omega_j}\right)\right\rangle \tag{11}$$

where $\omega = (\omega_1, \omega_2, \ldots, \omega_n)^T$ is the weight vector of $\alpha_j(j = 1, 2, \ldots, n)$ and $\omega_j > 0$, $\sum_{j=1}^{n} \omega_j = 1$.

4. The EDAS Model with P2TLNs

The traditional EDAS method [1], which can consider the conflicting attributes, has been studied in many MCDM problems. By computing the average solution (AV), this model can describe the difference between all the alternatives and the AV based on two distance measures which are namely PDA (positive distance from average) and NDA (negative distance from average); the alternative with higher values of PDA and lower values of PDA is the best choice. To combine the EDAS model with P2TLNs, we construct the EDAS model so the evaluation values are presented by P2TLNs. The computing steps of our proposed model can be established as follows.

Suppose there are m alternatives $\{\delta_1, \delta_2, \ldots \delta_m\}$, n attributes $\{G_1, G_2, \ldots G_n\}$, and r experts $\{a_1, a_2, \ldots a_r\}$, let $\{\omega_1, \omega_2, \ldots \omega_n\}$ and $\{\theta_1, \theta_2, \ldots \theta_r\}$ be the attribute's weighting vector and expert's weighting vector which satisfy $\omega_i \in [0, 1], \theta_i \in [0, 1]$ and $\sum_{i=1}^{n} \omega_i = 1, \sum_{i=1}^{t} \theta_i = 1$. Then:

Step 1. Construct the picture 2-tuple linguistic decision matrix $\tilde{R} = \left(\tilde{r}_{ij}\right)_{m\times n} = \left\langle(s_{ij}, \rho_{ij}), (\mu_{ij}, \eta_{ij}, v_{ij})\right\rangle_{m\times n}, i = 1, 2, \ldots, m, j = 1, 2, \ldots, n$, which can be depicted as follows.

$$\tilde{R} = \left(\tilde{r}_{ij}\right)_{m\times n} = \begin{bmatrix} \tilde{r}_{11} & \tilde{r}_{12} & \cdots & \tilde{r}_{1n} \\ \tilde{r}_{21} & \tilde{r}_{22} & \cdots & \tilde{r}_{2n} \\ \vdots & \vdots & \vdots & \vdots \\ \tilde{r}_{m1} & \tilde{r}_{m2} & \cdots & \tilde{r}_{mn} \end{bmatrix}, \tag{12}$$

where $\tilde{r}_{ij}$ denotes the P2TLNs of alternative ϑ_i on attribute U_j by expert q^r.

Step 2. Normalize the evaluation matrix $\tilde{R} = \left(\tilde{r}_{ij}\right)_{m\times n}$ to $\tilde{R}' = \left(\tilde{r}'_{ij}\right)_{m\times n}$.

For benefit attributes:

$$\tilde{r}'_{ij} = \tilde{r}_{ij} = \left\langle(s_{ij}, \rho_{ij}), (\mu_{ij}, \eta_{ij}, v_{ij})\right\rangle, i = 1, 2, \ldots, m, j = 1, 2, \ldots, n \tag{13}$$

For cost attributes:

$$\tilde{r}'_{ij} = \left(\tilde{r}_{ij}\right)^c = \left\langle\Delta\left(T - \Delta^{-1}(s_{ij}, \rho_{ij})\right), (v_{ij}, \eta_{ij}, \mu_{ij})\right\rangle, i = 1, 2, \ldots, m, j = 1, 2, \ldots, n. \tag{14}$$

Step 3. According to the decision making matrix $\tilde{R}' = \left(\tilde{r}'_{ij}\right)_{m\times n}$ and expert's weighting vector $\{\delta_1, \delta_2, \ldots \delta_r\}$, we can utilize overall $\tilde{r}'_{ij}$ to r'_{ij} by using P2TLWA or P2TLWG aggregation operators, and the computing results can be presented as follows.

$$R = \left[r'_{ij}\right]_{m\times n} = \begin{bmatrix} r'_{11} & r'_{12} & \cdots & r'_{1n} \\ r'_{21} & r'_{22} & \cdots & r'_{2n} \\ \vdots & \vdots & \vdots & \vdots \\ r'_{m1} & r'_{m2} & \cdots & r'_{mn} \end{bmatrix} \tag{15}$$

Step 4. Compute the value of AV based on all proposed attributes;

$$\text{AV} = \left[\text{AV}_j\right]_{1\times n} = \left[\frac{\sum_{i=1}^{m} r'_{ij}}{m}\right]_{1\times n}. \tag{16}$$

Based on Definition 8,

$$\sum_{i=1}^{m} r'_{ij} = \left\langle \Delta\left(\sum_{i=1}^{m} \Delta^{-1}(s_i, \rho_i)\right), \ \left(1 - \prod_{i=1}^{m}\left(1 - \mu'_{ij}\right), \prod_{i=1}^{m} \eta'_{ij}, \prod_{i=1}^{m}\left(\nu'_{ij} + \eta'_{ij}\right) - \prod_{i=1}^{m} \eta'_{ij}\right)\right\rangle \tag{17}$$

$$\begin{aligned}
\text{AV} = \left[\text{AV}_j\right]_{1\times n} &= \left[\frac{\sum_{i=1}^{m} r'_{ij}}{m}\right]_{1\times n} \\
&= \left\langle \Delta\left(\sum_{i=1}^{m} \frac{1}{m}\Delta^{-1}(s_i, \rho_i)\right), \ \left(1 - \prod_{i=1}^{m}\left(1 - \mu'_{ij}\right)^{\frac{1}{m}}, \prod_{i=1}^{m}\left(\eta'_{ij}\right)^{\frac{1}{m}}, \prod_{i=1}^{m}\left(\nu'_{ij} + \eta'_{ij}\right)^{\frac{1}{m}} - \prod_{i=1}^{m}\left(\eta'_{ij}\right)^{\frac{1}{m}}\right)\right\rangle
\end{aligned} \tag{18}$$

Step 5. According to the results of AV, we can compute the PDA and NDA by using the following formula:

$$\text{PDA}_{ij} = \left[\text{PDA}_{ij}\right]_{m\times n} = \frac{\max\left(0, \left(r'_{ij} - \text{AV}_j\right)\right)}{\text{AV}_j}, \tag{19}$$

$$\text{NDA}_{ij} = \left[\text{NDA}_{ij}\right]_{m\times n} = \frac{\max\left(0, \left(\text{AV}_j - r'_{ij}\right)\right)}{\text{AV}_j}. \tag{20}$$

For convenience, we can use the score function of P2TLNs presented in Definition 4 to determine the results of PDA and NDA as follows.

$$\text{PDA}_{ij} = \left[\text{PDA}_{ij}\right]_{m\times n} = \frac{\max\left(0, \left(s\left(r'_{ij}\right) - s\left(\text{AV}_j\right)\right)\right)}{s\left(\text{AV}_j\right)} \tag{21}$$

$$\text{NDA}_{ij} = \left[\text{NDA}_{ij}\right]_{m\times n} = \frac{\max\left(0, \left(s\left(\text{AV}_j\right) - s\left(r'_{ij}\right)\right)\right)}{s\left(\text{AV}_j\right)} \tag{22}$$

Step 6. Calculate the values of SP_i and SN_i which denotes the weighted sum of PDA and NDA, the computing formula are provided as follows.

$$\text{SP}_i = \sum_{j=1}^{n} w_j PDA_{ij}, \ \text{SN}_i = \sum_{j=1}^{n} w_j NDA_{ij} \tag{23}$$

Step 7. The results of Equation (23) can be normalized as

$$\text{NSP}_i = \frac{\text{SP}_i}{\max_{i}(\text{SP}_i)}, \ \text{NSN}_i = 1 - \frac{\text{SN}_i}{\max_{i}(\text{SN}_i)}. \tag{24}$$

Step 8. Compute the values of appraisal score (AS) based on each alternative's NSP_i and NSN_i.

$$\text{AS}_i = \frac{1}{2}(\text{NSP}_i + \text{NSN}_i) \tag{25}$$

Step 9. According to the calculating results of the AS, we can rank all the alternatives; the bigger the value of AS is, the better the selected alternative will be.

5. The Numerical Example

5.1. Numerical for MCGDM Problems with PFNs

In this section, we provide a numerical example for green supplier selection by using EDAS models with P2TLNs. Assuming that five possible green suppliers $\vartheta_i(i = 1, 2, 3, 4, 5)$ are to be selected and there are four criteria to assess these green suppliers: ① U_1 is the price factor; ② U_2 is the delivery factor; ③ U_3 is the environmental factors; ④ U_4 is the product quality factor. The five possible green suppliers $\vartheta_i (i = 1, 2, 3, 4, 5)$ are to be evaluated with P2TLNs with the four criteria by three experts, a^r (attributes weight $\omega = (0.22, 0.36, 0.28, 0.14)$, expert's weight $\delta = (0.24, 0.45, 0.31)$.).

Step 1. Construct the evaluation matrix $\widetilde{R} = \left(\widetilde{r}_{ij}\right)_{m \times n}, i = 1, 2, \ldots, m, j = 1, 2, \ldots, n$ for each of the three experts, which are listed in Tables 1–3.

Table 1. Picture 2-tuple linguistic evaluation information by q^1.

	U_1	U_2
ϑ_1	$\langle (S_3, 0), (0.41, 0.26, 0.33) \rangle$	$\langle (S_5, 0), (0.54, 0.36, 0.10) \rangle$
ϑ_2	$\langle (S_6, 0), (0.72, 0.11, 0.17) \rangle$	$\langle (S_3, 0), (0.25, 0.17, 0.58) \rangle$
ϑ_3	$\langle (S_1, 0), (0.35, 0.26, 0.39) \rangle$	$\langle (S_2, 0), (0.28, 0.16, 0.56) \rangle$
ϑ_4	$\langle (S_3, 0), (0.47, 0.22, 0.31) \rangle$	$\langle (S_1, 0), (0.16, 0.38, 0.46) \rangle$
ϑ_5	$\langle (S_5, 0), (0.58, 0.17, 0.25) \rangle$	$\langle (S_3, 0), (0.39, 0.21, 0.40) \rangle$
	U_3	U_4
ϑ_1	$\langle (S_1, 0), (0.33, 0.35, 0.32) \rangle$	$\langle (S_2, 0), (0.59, 0.16, 0.25) \rangle$
ϑ_2	$\langle (S_4, 0), (0.59, 0.15, 0.26) \rangle$	$\langle (S_5, 0), (0.68, 0.21, 0.11) \rangle$
ϑ_3	$\langle (S_7, 0), (0.13, 0.24, 0.63) \rangle$	$\langle (S_3, 0), (0.27, 0.31, 0.42) \rangle$
ϑ_4	$\langle (S_3, 0), (0.56, 0.19, 0.25) \rangle$	$\langle (S_4, 0), (0.41, 0.29, 0.30) \rangle$
ϑ_5	$\langle (S_1, 0), (0.28, 0.39, 0.33) \rangle$	$\langle (S_2, 0), (0.75, 0.17, 0.08) \rangle$

Table 2. Picture 2-tuple linguistic evaluation information by q^2.

	U_1	U_2
ϑ_1	$\langle (S_2, 0), (0.27, 0.28, 0.45) \rangle$	$\langle (S_1, 0), (0.50, 0.24, 0.26) \rangle$
ϑ_2	$\langle (S_7, 0), (0.59, 0.17, 0.24) \rangle$	$\langle (S_4, 0), (0.66, 0.21, 0.13) \rangle$
ϑ_3	$\langle (S_2, 0), (0.46, 0.25, 0.29) \rangle$	$\langle (S_3, 0), (0.22, 0.13, 0.65) \rangle$
ϑ_4	$\langle (S_1, 0), (0.34, 0.10, 0.56) \rangle$	$\langle (S_5, 0), (0.34, 0.42, 0.24) \rangle$
ϑ_5	$\langle (S_5, 0), (0.34, 0.10, 0.56) \rangle$	$\langle (S_4, 0), (0.18, 0.25, 0.57) \rangle$
	U_3	U_4
ϑ_1	$\langle (S_4, 0), (0.39, 0.38, 0.23) \rangle$	$\langle (S_3, 0), (0.42, 0.18, 0.40) \rangle$
ϑ_2	$\langle (S_6, 0), (0.60, 0.16, 0.24) \rangle$	$\langle (S_5, 0), (0.75, 0.10, 0.15) \rangle$
ϑ_3	$\langle (S_3, 0), (0.38, 0.11, 0.51) \rangle$	$\langle (S_4, 0), (0.48, 0.29, 0.23) \rangle$
ϑ_4	$\langle (S_5, 0), (0.29, 0.31, 0.40) \rangle$	$\langle (S_3, 0), (0.57, 0.25, 0.18) \rangle$
ϑ_5	$\langle (S_2, 0), (0.57, 0.26, 0.17) \rangle$	$\langle (S_1, 0), (0.63, 0.21, 0.16) \rangle$

Table 3. Picture 2-tuple linguistic evaluation information by q^3.

	U_1	U_2
ϑ_1	$\langle (S_4, 0), (0.19, 0.33, 0.48) \rangle$	$\langle (S_3, 0), (0.32, 0.29, 0.39) \rangle$
ϑ_2	$\langle (S_5, 0), (0.51, 0.37, 0.12) \rangle$	$\langle (S_5, 0), (0.77, 0.11, 0.12) \rangle$
ϑ_3	$\langle (S_3, 0), (0.59, 0.25, 0.16) \rangle$	$\langle (S_4, 0), (0.35, 0.25, 0.40) \rangle$
ϑ_4	$\langle (S_7, 0), (0.57, 0.19, 0.24) \rangle$	$\langle (S_3, 0), (0.27, 0.24, 0.49) \rangle$
ϑ_5	$\langle (S_1, 0), (0.22, 0.21, 0.57) \rangle$	$\langle (S_2, 0), (0.41, 0.36, 0.23) \rangle$
	U_3	U_4
ϑ_1	$\langle (S_2, 0), (0.59, 0.24, 0.17) \rangle$	$\langle (S_5, 0), (0.74, 0.16, 0.10) \rangle$
ϑ_2	$\langle (S_4, 0), (0.64, 0.13, 0.23) \rangle$	$\langle (S_7, 0), (0.78, 0.15, 0.07) \rangle$
ϑ_3	$\langle (S_2, 0), (0.49, 0.17, 0.34) \rangle$	$\langle (S_1, 0), (0.53, 0.28, 0.19) \rangle$
ϑ_4	$\langle (S_1, 0), (0.34, 0.31, 0.35) \rangle$	$\langle (S_3, 0), (0.59, 0.21, 0.20) \rangle$
ϑ_5	$\langle (S_4, 0), (0.71, 0.19, 0.10) \rangle$	$\langle (S_2, 0), (0.34, 0.42, 0.24) \rangle$

Step 2. Normalize the evaluation matrix $\widetilde{R} = \left[\widetilde{r}_{ij}\right]_{m \times n}$ to $\widetilde{R}' = \left[\widetilde{r}'_{ij}\right]_{m \times n}$; if all the attributes are benefitted, then normalization is not needed.

Step 3. According to the decision making matrix $\widetilde{R}' = \left[\widetilde{r}'_{ij}\right]_{m \times n}$ and expert's weighting vector $\{\delta_1, \delta_2, \ldots \delta_r\}$, utilize overall $\widetilde{r}'_{ij}$ to r'_{ij} by using the P2TLWA aggregation operator, and the computing results can be presented as follows in Table 4.

Table 4. The fused values by using picture 2-tuple linguistic weighted averaging (P2TLWA) operator.

	U_1	U_2
ϑ_1	$\langle (S_3, -0.1), (0.2836, 0.2894, 0.4269) \rangle$	$\langle (S_3, -0.4), (0.4609, 0.2805, 0.2586) \rangle$
ϑ_2	$\langle (S_6, 0.1), (0.6046, 0.1949, 0.2005) \rangle$	$\langle (S_4, 0.1), (0.6358, 0.1634, 0.2008) \rangle$
ϑ_3	$\langle (S_2, 0.1), (0.4816, 0.2524, 0.2660) \rangle$	$\langle (S_3, 0.1), (0.2769, 0.1673, 0.5558) \rangle$
ϑ_4	$\langle (S_3, 0.3), (0.4517, 0.1474, 0.4008) \rangle$	$\langle (S_3, 0.4), (0.2785, 0.3447, 0.3768) \rangle$
ϑ_5	$\langle (S_4, -0.2), (0.3893, 0.2479, 0.3628) \rangle$	$\langle (S_3, 0.1), (0.3103, 0.2684, 0.4213) \rangle$
	U_3	U_4
ϑ_1	$\langle (S_3, -0.3), (0.4484, 0.3231, 0.2285) \rangle$	$\langle (S_3, 0.4), (0.5838, 0.1687, 0.2474) \rangle$
ϑ_2	$\langle (S_5, -0.1), (0.6106, 0.1477, 0.2417) \rangle$	$\langle (S_6, -0.4), (0.7450, 0.1355, 0.1195) \rangle$
ϑ_3	$\langle (S_4, -0.3), (0.3670, 0.1518, 0.4812) \rangle$	$\langle (S_3, -0.2), (0.4533, 0.2915, 0.2552) \rangle$
ϑ_4	$\langle (S_3, 0.3), (0.3812, 0.2756, 0.3432) \rangle$	$\langle (S_3, 0.2), (0.5429, 0.2454, 0.2117) \rangle$
ϑ_5	$\langle (S_2, 0.4), (0.5693, 0.2600, 0.1707) \rangle$	$\langle (S_2, -0.4), (0.5970, 0.2475, 0.1555) \rangle$

Step 4. According to Table 4, we can obtain the value of the AV based on all proposed attributes by Formula (16), which is listed in Table 5.

Table 5. The value of the average solution (AV).

	Average Solution
U_1	$\langle (S_4, -0.4), (0.4526, 0.0119, 0.5355) \rangle$
U_2	$\langle (S_3, 0.3), (0.4114, 0.2345, 0.3541) \rangle$
U_3	$\langle (S_3, 0.4), (0.4850, 0.2203, 0.2947) \rangle$
U_4	$\langle (S_3, 0.3), (0.5966, 0.2096, 0.1938) \rangle$

Step 5. According to the results of the AV, we can compute the PDA and the NDA by using the Formulas (19) and (20), which are listed in Tables 6–8.

Table 6. The score values of ϑ'_{ij} and AV_j.

	U_1	U_2	U_3	U_4
ϑ_1	$(S_1, 0.2251)$	$(S_2, -0.4490)$	$(S_2, -0.3775)$	$(S_2, 0.2585)$
ϑ_2	$(S_4, 0.3106)$	$(S_3, -0.0798)$	$(S_3, 0.3536)$	$(S_5, -0.4321)$
ϑ_3	$(S_1, 0.2582)$	$(S_1, 0.1069)$	$(S_2, -0.3834)$	$(S_2, -0.3047)$
ϑ_4	$(S_2, -0.2450)$	$(S_2, -0.4581)$	$(S_2, -0.2976)$	$(S_2, 0.1565)$
ϑ_5	$(S_2, -0.0701)$	$(S_1, 0.3958)$	$(S_2, -0.3356)$	$(S_1, 0.1172)$
AV	$(S_2, -0.3337)$	$(S_2, -0.2787)$	$(S_2, 0.0080)$	$(S_2, 0.3314)$

Table 7. The results of PDA_{ij}.

	U_1	U_2	U_3	U_4
ϑ_1	0.0000	0.0000	0.0000	0.0000
ϑ_2	1.5869	0.6965	0.6702	0.9593
ϑ_3	0.0000	0.0000	0.0000	0.0000
ϑ_4	0.0532	0.0000	0.0000	0.0000
ϑ_5	0.1582	0.0000	0.0000	0.0000

Table 8. The results of NDA_{ij}.

	U_1	U_2	U_3	U_4
ϑ_1	0.2648	0.0990	0.1920	0.0313
ϑ_2	0.0000	0.0000	0.0000	0.0000
ϑ_3	0.2449	0.3569	0.1949	0.2729
ϑ_4	0.0000	0.1043	0.1522	0.0750
ϑ_5	0.0000	0.1891	0.1711	0.5208

Step 6. By calculating the values of SP_i and SN_i by Equation (23) and the attributes weighting vector $\omega = (0.22,\ 0.36,\ 0.28,\ 0.14)$, we can obtain the results as

$$SP_1 = 0.0000,\ SP_2 = 0.9218,\ SP_3 = 0.0000,\ SP_4 = 0.0117,\ SP_5 = 0.0348$$
$$SN_1 = 0.1520, SN_2 = 0.0000, SN_3 = 0.2752, SN_4 = 0.0907, SN_5 = 0.1889$$

Step 7. The results of Step 6 can be normalized by Formula (24) and are listed as

$$NSP_1 = 0.0000,\ NSP_2 = 1.0000,\ NSP_3 = 0.0000,\ NSP_4 = 0.0127,\ NSP_5 = 0.0378$$
$$NSN_1 = 0.4475, NSN_2 = 1.0000, NSN_3 = 0.0000, NSN_4 = 0.6705, NSN_5 = 0.3135$$

Step 8. Based on each alternative's NSP_i and NSN_i, compute the values of AS;

$$AS_1 = 0.2238, AS_2 = 1.0000, AS_3 = 0.0000, AS_4 = 0.3416, AS_5 = 0.1756.$$

Step 9. According to the calculated results of AS, we can rank all the alternatives; the bigger the value of AS is, the better the selected alternative will be. Clearly, the rank of all alternatives is $\vartheta_2 > \vartheta_4 > \vartheta_1 > \vartheta_5 > \vartheta_3$, and ϑ_2 is the best green supplier.

5.2. Compare P2TLNs EDAS Method with Some Aggregation Operators with P2TLNs

In this section, we compare our proposed picture 2-tuple linguistic EDAS method when using either the P2TLWA operator or the P2TLWG operator. According to the results of Table 4 and attributes weighting vector $\omega = (0.22, 0.36, 0.28, 0.14)$, we can utilize overall r'_{ij} to r'_i by using the P2TLWA and P2TLWG operators, which is listed in Table 9.

Table 9. The fused values by using some picture 2-tuple linguistic number (P2TLN) aggregation operators.

	P2TLWA	P2TLWG
ϑ_1	$\langle (S_3, -0.2), (0.4430, 0.2737, 0.2834) \rangle$	$\langle (S_3, -0.2), (0.4362, 0.2737, 0.2901) \rangle$
ϑ_2	$\langle (S_5, 0), (0.6405, 0.1608, 0.1986) \rangle$	$\langle (S_5, -0.1), (0.6375, 0.1608, 0.2017) \rangle$
ϑ_3	$\langle (S_3, 0), (0.3774, 0.1926, 0.4300) \rangle$	$\langle (S_2, -0.1), (0.3644, 0.1926, 0.4430) \rangle$
ϑ_4	$\langle (S_3, 0.3), (0.3896, 0.2561, 0.3542) \rangle$	$\langle (S_3, 0.3), (0.3918, 0.2561, 0.3520) \rangle$
ϑ_5	$\langle (S_3, -0.2), (0.4541, 0.2585, 0.2874) \rangle$	$\langle (S_3, -0.3)(0.4308, 0.2585, 0.3107) \rangle$

According to the score function of P2TLNs, we can obtain the alternative score results which are shown in Table 10.

The ranking of alternatives by some P2TLN aggregation operators are listed in Table 11.

Comparing the results of the picture 2-tuple linguistic EDAS model using either P2TLWA or P2TLWG operators, the aggregation results are slightly different in the ranking of alternatives, and the best alternatives are the same. However, the picture 2-tuple linguistic EDAS model has the valuable characteristic of considering the conflicting attributes, and can be more accurate and effective in the application of MCGDM problems.

Table 10. Score results of alternatives ϑ_i.

	P2TLWA	**P2TLWG**
$s(\vartheta_1)$	$(S_2, -0.3905)$	$(S_2, -0.4158)$
$s(\vartheta_2)$	$(S_4, -0.4134)$	$(S_4, -0.4751)$
$s(\vartheta_3)$	$(S_1, 0.4111)$	$(S_1, 0.3459)$
$s(\vartheta_4)$	$(S_2, -0.2719)$	$(S_2, -0.2649)$
$s(\vartheta_5)$	$(S_2, -0.3427)$	$(S_2, -0.4662)$

Table 11. Rank of alternatives by some P2TLN aggregation operators.

	Order
P2TLWA operator	$\vartheta_2 > \vartheta_4 > \vartheta_5 > \vartheta_1 > \vartheta_3$
P2TLWG operator	$\vartheta_2 > \vartheta_4 > \vartheta_1 > \vartheta_5 > \vartheta_3$
P2TLNs EDAS model	$\vartheta_2 > \vartheta_4 > \vartheta_1 > \vartheta_5 > \vartheta_3$

6. Conclusions

In this paper, we present the picture fuzzy EDAS model for MCGDM based on the traditional EDAS model and some fundamental theories of P2TLNs. First, we briefly reviewed the definition of P2TLNs and introduced the score function, accuracy function, and operational laws of P2TLNs. Next, to fuse the P2TLNs, we introduced some aggregation operators of P2TLNs. Furthermore, we combined the traditional EDAS model with P2TLNs, the picture fuzzy EDAS model for MCGDM was established, and the computing steps were simply depicted. Our presented model was more accurate and effective for considering the conflicting attributes. Finally, a numerical example for green supplier selection was given to illustrate this new model and some comparisons between P2TLWA and P2TLWG operators using the P2TLN EDAS model were also conducted to further illustrate advantages of the new method. In the future, the picture fuzzy EDAS model can be applied to risk analysis, MADM problems [62–65], and many other uncertain and fuzzy environments [44,66–71].

Author Contributions: S.Z., H.G., G.W., Y.W. and C.W. conceived and worked together to achieve this work, S.Z. compiled the computing work by Excel and analyzed the data, S.Z. and G.W. wrote the paper. Finally, all the authors have read and approved the final manuscript.

Funding: The work was supported by the National Natural Science Foundation of China under Grant No. 71571128 and the Humanities and Social Sciences Foundation of Ministry of Education of the People's Republic of China (16XJA630005) and the Construction Plan of Scientific Research Innovation Team for Colleges and Universities in Sichuan Province (15TD0004).

Conflicts of Interest: The authors declare no conflict of interest.

References

1. Keshavarz Ghorabaee, M.; Zavadskas, E.K.; Olfat, L.; Turskis, Z. Multi-criteria inventory classification using a new method of evaluation based on distance from average solution (EDAS). *Informatica* **2015**, *26*, 435–451. [CrossRef]

2. Wang, J.; Wei, G.W.; Lu, M. An extended VIKOR method for multiple criteria group decision making with triangular fuzzy neutrosophic numbers. *Symmetry* **2018**, *10*, 497. [CrossRef]

3. Wei, G.; Zhang, N.A. A multiple criteria hesitant fuzzy decision making with Shapley value-based VIKOR method. *J. Intell. Fuzzy Syst.* **2014**, *26*, 1065–1075.

4. Chen, N.; Xu, Z.S. Hesitant fuzzy ELECTRE II approach: A new way to handle multi-criteria decision making problems. *Inf. Sci.* **2015**, *292*, 175–197. [CrossRef]

5. Wei, G.W. Extension of TOPSIS method for 2-tuple linguistic multiple attribute group decision making with incomplete weight information. *Knowl. Inf. Syst.* **2010**, *25*, 623–634. [CrossRef]

6. Hajlaoui, S.; Halouani, N. Hesitant-fuzzy-Promethee method. In Proceedings of the 2013 5th International Conference on Modeling, Simulation and Applied Optimization, Hammamet, Tunisia, 28–30 April 2013.

7. Liao, H.C.; Xu, Z.S. Multi-criteria decision making with intuitionistic fuzzy PROMETHEE. *J. Intell. Fuzzy Syst.* **2014**, *27*, 1703–1717.
8. Li, X.Y.; Wei, G.W. GRA method for multiple criteria group decision making with incomplete weight information under hesitant fuzzy setting. *J. Intell. Fuzzy Syst.* **2014**, *27*, 1095–1105.
9. Liu, H.C.; You, J.X.; Lu, C.; Shan, M.M. Application of interval 2-tuple linguistic MULTIMOORA method for health-care waste treatment technology evaluation and selection. *Waste Manag.* **2014**, *34*, 2355–2364. [CrossRef]
10. Huang, Y.H.; Wei, G.W. TODIM method for Pythagorean 2-tuple linguistic multiple attribute decision making. *J. Intell. Fuzzy Syst.* **2018**, *35*, 901–915. [CrossRef]
11. Wang, J.; Wei, G.W.; Lu, M. TODIM method for multiple attribute group decision making under 2-tuple linguistic neutrosophic environment. *Symmetry* **2018**, *10*, 486. [CrossRef]
12. Wei, G.W. TODIM method for picture fuzzy multiple attribute decision making. *Informatica* **2018**, *29*, 555–566. [CrossRef]
13. Atanassov, K.T. Intuitionistic fuzzy sets. *Fuzzy Sets Syst.* **1986**, *20*, 87–96. [CrossRef]
14. Zadeh, L.A. Fuzzy sets. *Inf. Control* **1965**, *8*, 338–353. [CrossRef]
15. Atanassov, K.; Gargov, G. Interval valued intuitionistic fuzzy-sets. *Fuzzy Sets Syst.* **1989**, *31*, 343–349. [CrossRef]
16. Atanassov, K.T. Operators over interval-valued intuitionistic fuzzy sets. *Fuzzy Sets Syst.* **1994**, *64*, 159–174. [CrossRef]
17. Cuong, B.C.; Kreinovich, V. Picture fuzzy sets—A new concept for computational intelligence problems. In Proceedings of the 2013 Third World Congress on Information and Communication Technologies (WICT 2013), Hanoi, Vietnam, 15–18 December 2013; pp. 1–6.
18. Singh, P. Correlation coefficients for picture fuzzy sets. *J. Intell. Fuzzy Syst.* **2015**, *28*, 591–604.
19. Son, L.H. DPFCM: A novel distributed picture fuzzy clustering method on picture fuzzy sets. *Expert Syst. Appl.* **2015**, *42*, 51–66. [CrossRef]
20. Thong, N.T.; Son, L.H. HIFCF: An effective hybrid model between picture fuzzy clustering and intuitionistic fuzzy recommender systems for medical diagnosis. *Expert Syst. Appl.* **2015**, *42*, 3682–3701. [CrossRef]
21. Thong, P.H.; Son, L.H. A novel automatic picture fuzzy clustering method based on particle swarm optimization and picture composite cardinality. *Knowl. Based Syst.* **2016**, *109*, 48–60. [CrossRef]
22. Wei, G.W. Picture fuzzy cross-entropy for multiple attribute decision making problems. *J. Bus. Econ. Manag.* **2016**, *17*, 491–502. [CrossRef]
23. Son, L.H. Measuring analogousness in picture fuzzy sets: From picture distance measures to picture association measures. *Fuzzy Optim. Decis. Mak.* **2017**, *16*, 359–378. [CrossRef]
24. Son, L.H.; Thong, P.H. Some novel hybrid forecast methods based on picture fuzzy clustering for weather nowcasting from satellite image sequences. *Appl. Intell.* **2017**, *46*, 1–15. [CrossRef]
25. Wei, G.W. Some cosine similarity measures for picture fuzzy sets and their applications to strategic decision Making. *Informatica* **2017**, *28*, 547–564. [CrossRef]
26. Wei, G.W.; Lin, R.; Wang, H.J. Distance and similarity measures for hesitant interval-valued fuzzy sets. *J. Intell. Fuzzy Syst.* **2014**, *27*, 19–36.
27. Szmidt, E.; Kacprzyk, J. A similarity measure for intuitionistic fuzzy sets and its application in supporting medical diagnostic reasoning. In Proceedings of the Artificial Intelligence and Soft Computing—ICAISC 2004, Zakopane, Poland, 7–11 June 2004; pp. 388–393.
28. Wei, G.W. Picture fuzzy aggregation operators and their application to multiple attribute decision making. *J. Intell. Fuzzy Syst.* **2017**, *33*, 713–724. [CrossRef]
29. Deng, X.M.; Wei, G.W.; Gao, H.; Wang, J. Models for safety assessment of construction project with some 2-tuple linguistic Pythagorean fuzzy Bonferroni mean operators. *IEEE Access* **2018**, *6*, 52105–52137. [CrossRef]
30. Gao, H.; Lu, M.; Wei, G.W.; Wei, Y. Some novel Pythagorean fuzzy interaction aggregation operators in multiple attribute decision making. *Fundamenta Informaticae* **2018**, *159*, 385–428. [CrossRef]
31. Li, Z.X.; Wei, G.W.; Lu, M. Pythagorean fuzzy hamy mean operators in multiple attribute group decision making and their application to supplier selection. *Symmetry* **2018**, *10*, 505. [CrossRef]
32. Wang, J.; Wei, G.W.; Wei, Y. Models for green supplier selection with some 2-tuple linguistic neutrosophic number Bonferroni mean operators. *Symmetry* **2018**, *10*, 131. [CrossRef]

33. Wei, G.W.; Alsaadi, F.E.; Hayat, T.; Alsaedi, A. Bipolar fuzzy Hamacher aggregation operators in multiple attribute decision making. *Int. J. Fuzzy Syst.* **2018**, *20*, 1–12. [CrossRef]
34. Wei, G.W.; Gao, H.; Wang, J.; Huang, Y.H. Research on risk evaluation of enterprise human capital investment with interval-valued bipolar 2-tuple linguistic information. *IEEE Access* **2018**, *6*, 35697–35712. [CrossRef]
35. Wu, S.J.; Wang, J.; Wei, G.W.; Wei, Y. Research on construction engineering project risk assessment with some 2-tuple linguistic neutrosophic hamy mean operators. *Sustainability* **2018**, *10*, 1536. [CrossRef]
36. Wei, G.W. Some similarity measures for picture fuzzy sets and their applications. *Iran. J. Fuzzy Syst.* **2018**, *15*, 77–89.
37. Wei, G.W.; Gao, H. The generalized dice similarity measures for picture fuzzy sets and their applications. *Informatica* **2018**, *29*, 107–124. [CrossRef]
38. Wei, G.W. Picture fuzzy Hamacher aggregation operators and their application to multiple attribute decision making. *Fundamenta Informaticae* **2018**, *157*, 271–320. [CrossRef]
39. Lu, M.; Wei, G.W.; Alsaadi, F.E.; Hayat, T.; Alsaedi, A. Hesitant Pythagorean fuzzy Hamacher aggregation operators and their application to multiple attribute decision making. *J. Intell. Fuzzy Syst.* **2017**, *33*, 1105–1117. [CrossRef]
40. Wei, G.W.; Lu, M. Dual hesitant Pythagorean fuzzy Hamacher aggregation operators in multiple attribute decision making. *Arch. Control Sci.* **2017**, *27*, 365–395. [CrossRef]
41. Wu, S.J.; Wei, G.W. Pythagorean fuzzy Hamacher aggregation operators and their application to multiple attribute decision making. *Int. J. Knowl. Based Intell. Eng. Syst.* **2017**, *21*, 189–201. [CrossRef]
42. Gao, H.; Wei, G.W.; Huang, Y.H. Dual hesitant bipolar fuzzy Hamacher prioritized aggregation operators in multiple attribute decision making. *IEEE Access* **2018**, *6*, 11508–11522. [CrossRef]
43. Wei, G.W.; Alsaadi, F.E.; Hayat, T.; Alsaedi, A. Projection models for multiple attribute decision making with picture fuzzy information. *Int. J. Mach. Learn. Cybern.* **2018**, *9*, 713–719. [CrossRef]
44. Wei, G.W.; Alsaadi, F.E.; Hayat, T.; Alsaedi, A. Picture 2-tuple linguistic aggregation operators in multiple attribute decision making. *Soft Comput.* **2018**, *22*, 989–1002. [CrossRef]
45. Wei, G.W. Picture 2-tuple linguistic Bonferroni mean operators and their application to multiple attribute decision making. *Int. J. Fuzzy Syst.* **2017**, *19*, 997–1010. [CrossRef]
46. Wei, G.W. Picture uncertain linguistic Bonferroni mean operators and their application to multiple attribute decision making. *Kybernetes* **2017**, *46*, 1777–1800. [CrossRef]
47. Keshavarz Ghorabaee, M.; Zavadskas, E.K.; Amiri, M.; Turskis, Z. Extended EDAS method for fuzzy multi-criteria decision-making: An application to supplier selection. *Int. J. Comput. Commun. Control* **2016**, *11*, 358–371. [CrossRef]
48. Kahraman, C.; Keshavarz Ghorabaee, M.; Zavadskas, E.K.; Onar, S.C.; Yazdani, M.; Oztaysi, B. Intuitionistic fuzzy EDAS method: An application to solid waste disposal site selection. *J. Environ. Eng. Landsc. Manag.* **2017**, *25*, 1–12. [CrossRef]
49. Keshavarz Ghorabaee, M.; Amiri, M.; Zavadskas, E.K.; Turskis, Z. Multi-criteria group decision-making using an extended EDAS method with interval type-2 fuzzy sets. *E & M Ekonomie a Management* **2017**, *20*, 48–68.
50. Keshavarz Ghorabaee, M.; Amiri, M.; Zavadskas, E.K.; Turskis, Z.; Antucheviciene, J. A new multi-criteria model based on interval type-2 fuzzy sets and EDAS method for supplier evaluation and order allocation with environmental considerations. *Comput. Ind. Eng.* **2017**, *112*, 156–174. [CrossRef]
51. Keshavarz Ghorabaee, M.; Amiri, M.; Zavadskas, E.K.; Turskis, Z.; Antucheviciene, J. Stochastic EDAS method for multi-criteria decision-making with normally distributed data. *J. Intell. Fuzzy Syst.* **2017**, *33*, 1627–1638. [CrossRef]
52. Peng, X.D.; Liu, C. Algorithms for neutrosophic soft decision making based on EDAS, new similarity measure and level soft set. *J. Intell. Fuzzy Syst.* **2017**, *32*, 955–968. [CrossRef]
53. Ecer, F. Third-party logistics (3PLS) provider selection via fuzzy ahp and EDAS integrated model. *Technol. Econ. Dev. Econ.* **2018**, *24*, 615–634. [CrossRef]
54. Feng, X.Q.; Wei, C.P.; Liu, Q. EDAS method for extended hesitant fuzzy linguistic multi-criteria decision making. *Int. J. Fuzzy Syst.* **2018**, *20*, 2470–2483. [CrossRef]
55. Ilieva, G. Group decision analysis algorithms with EDAS for interval fuzzy sets. *Cybern. Inf. Technol.* **2018**, *18*, 51–64. [CrossRef]

56. Karasan, A.; Kahraman, C. A novel interval-valued neutrosophic EDAS method: Prioritization of the United Nations national sustainable development goals. *Soft Comput.* **2018**, *22*, 4891–4906. [CrossRef]

57. Keshavarz-Ghorabaee, M.; Amiri, M.; Zavadskas, E.K.; Turskis, Z.; Antucheviciene, J. A dynamic fuzzy approach based on the EDAS method for multi-criteria subcontractor evaluation. *Information* **2018**, *9*, 68. [CrossRef]

58. Stevic, Z.; Vasiljevic, M.; Zavadskas, E.K.; Sremac, S.; Turskis, Z. Selection of carpenter manufacturer using fuzzy EDAS method. *Eng. Econ.* **2018**, *29*, 281–290. [CrossRef]

59. Keshavarz-Ghorabaee, M.; Amiri, M.; Zavadskas, E.K.; Turskis, Z.; Antucheviciene, J. A comparative analysis of the rank reversal phenomenon in the EDAS and TOPSIS methods. *Econ. Comput. Econ. Cybern. Stud. Res.* **2018**, *52*, 121–134.

60. Herrera, F.; Martinez, L. The 2-tuple linguistic computational model. Advantages of its linguistic description, accuracy and consistency. *Int. J. Uncertain. Fuzziness Knowl. Based Syst.* **2001**, *9*, 33–48. [CrossRef]

61. Herrera, F.; Martinez, L. A 2-tuple fuzzy linguistic representation model for computing with words. *IEEE Trans. Fuzzy Syst.* **2000**, *8*, 746–752.

62. Wei, G.W.; Lu, M. Pythagorean fuzzy power aggregation operators in multiple attribute decision making. *Int. J. Intell. Syst.* **2018**, *33*, 169–186. [CrossRef]

63. Wei, G.W.; Wei, Y. Similarity measures of Pythagorean fuzzy sets based on the cosine function and their applications. *Int. J. Intell. Syst.* **2018**, *33*, 634–652. [CrossRef]

64. Wei, G.W.; Zhang, Z.P. Some single-valued neutrosophic Bonferroni power aggregation operators in multiple attribute decision making. *J. Ambient Intell. Humaniz. Comput.* **2019**, *10*, 863–882. [CrossRef]

65. Wei, G.W. Pythagorean fuzzy Hamacher power aggregation operators in multiple attribute decision making. *Fundamenta Informaticae* **2019**, *166*, 57–85.

66. Wang, J.; Wei, G.W.; Gao, H. Approaches to multiple attribute decision making with interval-valued 2-tuple linguistic Pythagorean fuzzy information. *Mathematics* **2018**, *6*, 201. [CrossRef]

67. Wei, G.; Wei, Y. Some single-valued neutrosophic dombi prioritized weighted aggregation operators in multiple attribute decision making. *J. Intell. Fuzzy Syst.* **2018**, *35*, 2001–2013. [CrossRef]

68. Hashemkhani Zolfani, S.; Saparauskas, J. New application of SWARA method in prioritizing sustainability assessment indicators of energy system. *Eng. Econ.* **2013**, *24*, 408–414.

69. Wang, J.; Gao, H.; Wei, G.W.; Wei, Y. Methods for multiple-attribute group decision making with q-rung interval-valued orthopair fuzzy information and their applications to the selection of green suppliers. *Symmetry* **2019**, *11*, 56. [CrossRef]

70. Wang, R.; Wang, J.; Gao, H.; Wei, G.W. Methods for MADM with picture fuzzy muirhead mean operators and their application for evaluating the financial investment risk. *Symmetry* **2019**, *11*, 6. [CrossRef]

71. Sarkar, B.; Mahapatra, A.S. Periodic review fuzzy inventory models with variable lead time and fuzzy demand. *Int. Trans. Op. Res.* **2017**, *24*, 1197–1227. [CrossRef]

 *mathematics*

Article

Comparative Evaluation and Ranking of the European Countries Based on the Interdependence between Human Development and Internal Security Indicators

Aleksandras Krylovas [1], Rūta Dadelienė [2], Natalja Kosareva [1] and Stanislav Dadelo [3,*]

[1] Department of Mathematical Modelling, Vilnius Gediminas Technical University, Saulėtekio al. 11, LT-10223 Vilnius, Lithuania; aleksandras.krylovas@vgtu.lt (A.K.); natalja.kosareva@vgtu.lt (N.K.)

[2] Institute of Health Science, Department of Rehabilitation, Physical and Sports Medicine, Vilnius University, Universiteto g. 3, LT-01513 Vilnius, Lithuania; ruta.dadeliene@gmail.com

[3] Vilnius Gediminas Technical University, Saulėtekio al. 11, LT-10223 Vilnius, Lithuania

* Correspondence: stanislav.dadelo@vgtu.lt

Received: 23 January 2019; Accepted: 11 March 2019; Published: 21 March 2019

Abstract: New solutions and techniques for developing country policies are used under real conditions. The present study aims to propose a new approach for evaluating and ranking the European countries by using the interrelation between two groups of criteria, associated with the Human Development Index (HDI) and the World Internal Security and Police Index (WISPI). HDI and its components rank countries by value and detail the values of the components of longevity, education and income per capita. WISPI focuses on the effective rendering of security services and the outcome of rendered services. The priority of criteria is determined in the descending order of their correlation values with other group criteria. The criteria weights are set simultaneously for both groups by applying the weight balancing method WEBIRA. The methodology based on minimising sum of squared differences of the weighted sums within groups is used. Finally, the generalised criteria measuring the level of the country are calculated using the SAW method. Cluster analysis of the countries was carried out and compared with MCDM results. The study revealed that WEBIRA ranking of countries is basically consistent with the results of cluster analysis. The proposed methodology can be applied to develop the management policy of the countries, as well as to their evaluation and ranking by using various indices, criteria and procedures. The results of this research can also be used to reveal national policy choices, to point out government policy priorities.

Keywords: human development; internal security; MCDM; weight balancing; WEBIRA; cluster analysis

1. Introduction

The European countries are exposed to various hybrid dangers (e.g., political differences, military aggression, financial and economic crises, natural and technogenic catastrophes, social upheavals, criminal offences, etc.). This problem is closely associated with the internal security of the states. The institutions, ensuring the internal security of the states, also guarantee their economic stability. The European Union (EU) is becoming a centre of sustainability of the European values and stability. The systematic and consistent attitudes towards the topical issues of internal security are formed by the EU member states through their joint regulations. However, the particular EU states develop their security systems depending on numerous internal and external factors. Therefore, their internal security systems have some specific features. The states not belonging to the EU also demonstrate their distinctive properties, though in the world involved in the global processes similar tendencies of changes in the internal safety systems can be observed [1]. This field of research has been barely explored. The World Internal Security and Police Index (WISPI) [2] focuses on describing both the effective rendering of security services and the outcome of the rendered services. WISPI is considered the first international index for measuring the indicators of the internal security worldwide, as well as ranking the states according to their ability to provide security services and boost security performance.

The economic and human development of a particular country relate to the state of security in this country [3]. Higher development levels, in terms of GDP per capita, are capable of providing social and individual prosperity or human development. It is not clear whether other interrelations between prosperity indicators exist on different levels of economic development. Social and human development and security status indicators improve with economic development. Public well-being increases with income rise at all levels of economic development [4]. Focusing on people instead of economic outcomes provides a wider range of options for policy-makers [5]. Countries' development policies should strive to remove any obstacles that impede people's freedoms: political freedom provides individuals to enjoy the freedom of political expression; economic facilities allow the use of economic resources for the purpose of consumption, production or exchange; social opportunities are made possible access to education and health; transparency guarantees relate to openness and the prevention of any type of corruption; and protective security allows a social safety network that protects individuals from misery [6]. Human development paves the way for economic development and security. State policy not only secures educational programs but also promotes development through innovation and expansion of new programs [7]. Meanwhile relationship exists between pro-government militias and various types of human rights violations [5]. A relationship between a given government regime's security repertoire and the likelihood of control and violence against civilians exists [8]. Uncontrollable human rights violations have a harmful effect on the positive country's image [9]. Since the government is a source of legitimate authority, laws and regulations also provide important cues about which course is supported and protected by the government. A legal country system that protects certain interests with certain methods sends a signal to world societal participants that these interests and these methods should be determined as a dominant image of the country [10].

Human development is a process, which seeks to expand the possibilities to create an environment where people can live long, healthy and creative lives. Human Development Index (HDI) [11] is one of the most widely used composite indicators of socioeconomic development of a country. People who have achieved high or very high human development level represent 51 percent of the global population. Researchers are extremely interested in factors influencing HDI [12]. Overwhelming evidence of the direct positive effects of economic freedom on human development is provided by a large number of the cross-country studies [13]. However, the 'original sin' of HDI involves neglecting the environmental and social sustainability and personal security issues [14].

A wide variety of approaches and evaluation techniques are used in the field of security research; however, there are some gaps, particularly if researchers aim to study the internal security of the whole country. The aim of the present study is to propose a new approach to identify a method of ranking the countries for evaluating the internal security of the European countries, using indicators such as the Human Development Index [11] and the Internal Security and Police Index [2]. Thus, the combining of HDI indicators with the World Internal Security and Police Index can provide an integrated evaluation approach for filling this gap. However, the conventional security system's modelling tools and models, such as expert-based or other approaches, do not propose any integral internal security metric, covering all types of threats, to which the countries and citizens are exposed.

There is not much research in the literature dedicated to studying HDI and particularly WISPI by means of mathematical modelling. Most research is related to the separate dimensions of HDI—public health, economic development and quality of life. The most commonly used methods are various tools of mathematical statistics, i.e., correlation, regression analysis and some econometric models.

The study by Zaborskis et al. [15] introduces several methods for measuring family affluence inequality in adolescent life satisfaction (LS) and assesses its relationship with macrolevel indices

(Gross National Income, Human Development Index and the mean Overall Life Satisfaction score). Poisson regression estimations and correlation analysis were used in this research. Murray et al. [16] investigated how preterm delivery rates differ in a country with a very high human development index and explored rural vs. urban environmental and socioeconomic factors that may be responsible for this variation. A multiple linear regression was used for this purpose. The study by Liu et al. [17] employs a panel smooth transition vector error correction model (PST-VECM) to explore the education-health causality. The paper by Sayed et al. [18] discusses the rank reversal issue in multicriteria decision-making (MCDM) techniques. The proposed methodology of the Goal Programming Benefit-of-the-Doubt (GP-BOD) aims to overcome this problem and obtain consistent and stable rankings for the human development index (HDI) framework. The paper by Carvalhal Monteiro et al. [19] proposes a new Human Development Index (HDI) classification method using the combination of the ELECTRE TRI method with statistical tools to define classes and class profiles for the HDI.

We could not find any quantitative investigation of WISPI in the literature. The synergy of HDI and WISPI as a research object is unprecedented in the scientific literature. However, the task of ranking the countries according to HDI and WISPI interrelation is an interesting and relevant issue.

MCDM methods usually rank countries by set of homogeneous (having the same nature) indicators. If several criteria groups having different nature exist, for example, subjective and objective, external and internal evaluations of alternatives, other methodologies should be proposed. KEMIRA [20] is the MCDM method implemented by maximising compatibility of two or more subsets of criteria, thus it is naturally appropriate for solving our task. In this research, a modification of KEMIRA called WEBIRA [21] has been applied to the case of two groups of evaluation criteria. The advantage of WEBIRA is that its efficiency does not decrease with increasing number of alternatives as other MCDM methods [22]. It also remains stable with increasing number of criteria [21].

Prioritisation of criteria is a separate issue of the WEBIRA method that needs to be addressed before solving the optimisation task. In this sense, WEBIRA is not a fully objective method for determining criteria weights. The problem of criteria prioritisation can be solved by applying wide range of objective or subjective (expert-based) methods. Examples of expert-based methods are Analytic Hierarchy Process (AHP) [23], Kemeny median method [24], Stepwise Weight Assessment Ratio Analysis (SWARA) [25], a fuzzy inference system (FIS) approach [26], etc. However, when dealing with country rating task, we need to look for alternative methods for prioritising criteria, because we do not have information about criteria assessments by experts. Objective methods for criteria weighting are based on initial data values and their structure (entropy-based methods [27], mathematical programming models [28], IDOCRIW [29], etc.).

There are three main steps of WEBIRA: (1) criteria priority setting separately in every subset; (2) criteria weight determining by solving optimisation problem; and (3) ranking of alternatives by applying one of MCDM methods. One of the novelty elements of this article is to use correlation analysis to set criteria priority. Statistical methods are traditionally used in weighting attributes. Thus, CRICTIC (Criteria Importance Through Intercriteria Correlation), developed by Diakoulaki et al. [30], aims to determine objective weights of relative importance in MCDM problems by considering correlation coefficient values between criteria and standard deviations of each criterion for alternatives. High correlation is considered as some kind of double counting, so assigned weights are inversely proportional to the correlation coefficient value. Our methodological assumption is based on the maximisation of compatibility between two different groups of indicators. A suitable way to measure compatibility is to apply intergroup correlation coefficients. Unlike correlations within groups, where attributes with strong correlation are undesirable, high correlation of the attribute with the attributes of other groups indicates that the interdependence between the two group's indicators became higher; such indicator is more preferable in the decision-making process.

This idea arose from the ultimate goal of this research—to evaluate countries by combining several dimensions: economic prosperity, comprehensive education, healthy lifestyle, safe environment and human security. Thus, two groups of criteria—X and Y—were distinguished and the optimisation task has been solved according to weight balancing procedure. This procedure ensures that the criteria for the two groups in the final order of alternatives are maximally aligned with each other.

2. Materials and Methods

2.1. Criteria and Their Definitions

The Human Development Index is a summary measure of average achievement in key dimensions of human development: (1) a long and healthy life; (2) knowledge; and (3) a decent standard of living. The knowledge dimension consists of two subdimensions: (1) mean of years of schooling for adults aged 25 years and more and (2) expected years of schooling for children of school entering age. The HDI is the geometric mean of normalised indices for each of the three dimensions [11]. In the present work, four components of HDI are used: the ability to lead a long and healthy life, measured by life expectancy at birth (years) (y_1); the ability to acquire knowledge, measured by the mean number of years of schooling (y_2); the expected years of schooling (y_3); and the ability to achieve a decent standard of living, measured by the gross national income (GNI) per capita (PPP $) ($y_4$) [11]. HDI makes an assessment of diverse countries with very different price levels. To compare economic statistics across countries, the data must first be converted into a common currency. For this reason GNI per capita is measured in purchasing power parity (PPP) international dollars (PPP $). One PPP dollar (or international dollar) has the same purchasing power in the domestic economy of any country as US$1 has in the US economy.

World Internal Security and Police Index (WISPI) measures the capacity and efficiency of police and security service providers to address the internal security issues worldwide through the four domains, i.e., capacity, process, legitimacy and outcomes (Table 1) [2]. Domain content can be explained by answering these questions:

Capacity: Do security providers have the resources needed to address security violation?
Process: Are the resources directed towards violence prevention used effectively?
Legitimacy: Are security providers trusted by the people? Do they abuse their position?
Outcomes: Do people feel safe in their neighbourhoods? Are crime rates low?

Each WISPI domain acquires values from 0 to 1. The higher the numerical value of the country's respective domain, the higher the position of that country in the corresponding rating. WISPI measures the ability of police and internal security service to protect society as well as provides broader measure of human security.

Table 1. World Internal Security and Police Index, Domains and Indicators [2].

Domain	Indicator	Definition
Capacity	Police	Number of Police and Internal Security Officers per 100,000 people
	Armed Forces	Number of Armed Service Personnel per 100,000 people
	Private Security	Number of Private Security Contractors per 100,000 people
	Prison Capacity	Ratio of Prisoners to Official Prison Capacity
Process	Corruption	Control of Corruption
	Effectiveness	Criminal Justice effectiveness, impartial, respects rights
	Bribe Payments to Police	% of Respondents who paid a bribe to a police officer in the past year
	Underreporting	Ratio of police reported thefts to survey reported thefts
Legitimacy	Due Process	Due process of law and rights of the accused
	Confidence in Police	% of Respondents who have confidence in their local police
	Public Use, Private Gain	Government officials in the police and the military do not use public office for private gain
	Political Terror	Use of Force by Government Against Its Own Citizens
Outcomes	Homicide	Number of Intentional Homicides per 100,000 people
	Violent Crime	% Assaulted or mugged in the last year
	Terrorism	Composite measure of deaths, injuries and incidents of terrorism
	Public Safety Perceptions	Perceptions of safety walking alone at night

The initial data matrix, maximum and minimum values of indicators are presented in Table 2.

Table 2. The initial values of World Internal Security and Police Index (WISPI) indices [2] and Human Development Index (HDI) factors [11].

Country	Mean Years of Schooling (years)	Expected Years of Schooling (years)	Life Expectancy at Birth (years)	GNI per Capita (PPP $)	Outcomes	Capacity	Legitimacy	Process
Albania	10.00	14.80	78.50	11.89	0.72	0.647	0.562	0.297
Armenia	11.70	13.00	74.80	9.14	0.893	0.921	0.516	0.479
Austria	12.10	16.10	81.80	45.42	0.894	0.77	0.899	0.817
Azerbaijan	10.70	12.70	72.10	15.60	0.871	0.723	0.487	0.295
Belarus	12.30	15.50	73.10	16.32	0.686	0.975	0.486	0.472
Belgium	11.80	19.80	81.30	42.16	0.807	0.71	0.847	0.79
Bosnia and Herzegovina	9.70	14.20	77.10	11.72	0.824	0.916	0.642	0.465
Bulgaria	11.80	14.80	74.90	18.74	0.753	0.985	0.556	0.494
Cyprus	12.10	14.60	80.70	31.57	0.77	0.736	0.794	0.634
Croatia	11.30	15.00	77.80	22.16	0.854	0.939	0.695	0.605
Czech Republic	12.70	16.90	78.90	30.59	0.827	0.875	0.772	0.638
Denmark	12.60	19.10	80.90	47.92	0.885	0.648	0.904	0.948
Estonia	12.70	16.10	77.70	28.99	0.734	0.967	0.804	0.754
Finland	12.40	17.60	81.50	41.00	0.893	0.674	0.919	0.922
France	11.50	16.40	82.70	39.25	0.783	0.773	0.817	0.734
Georgia	12.80	15.00	73.40	9.19	0.766	0.823	0.752	0.593
Germany	14.10	17.00	81.20	46.14	0.852	0.778	0.867	0.876
Greece	10.80	17.30	81.40	24.65	0.704	0.783	0.691	0.583
Hungary	11.90	15.10	76.10	25.39	0.793	0.541	0.647	0.632
Iceland	12.40	19.30	82.90	45.81	0.906	0.635	0.893	0.81
Ireland	12.50	19.60	81.60	53.75	0.805	0.841	0.852	0.78
Italy	10.20	16.30	83.20	35.30	0.761	0.724	0.725	0.681
Latvia	12.80	15.80	74.70	25.00	0.695	0.934	0.691	0.558
Lithuania	13.00	16.10	74.80	28.31	0.68	0.903	0.733	0.605
Montenegro	11.30	14.90	77.30	16.78	0.833	0.914	0.681	0.481
Netherlands	12.20	18.00	82.00	47.90	0.866	0.707	0.858	0.898
Norway	12.60	17.90	82.30	68.01	0.801	0.658	0.916	0.908
Poland	12.30	16.40	77.80	26.15	0.858	0.848	0.738	0.676
Portugal	9.20	16.30	81.40	27.32	0.834	0.909	0.732	0.679
Romania	11.00	14.30	75.60	22.65	0.805	0.835	0.616	0.535
Russian Federation	12.00	15.50	71.20	24.23	0.449	0.984	0.33	0.415

Table 2. *Cont.*

Country	Mean Years of Schooling (years)	Expected Years of Schooling (years)	Life Expectancy at Birth (years)	GNI per Capita (PPP $)	Outcomes	Capacity	Legitimacy	Process
Serbia	11.10	14.60	75.30	13.02	0.851	0.886	0.587	0.462
Slovakia	12.50	15.00	77.00	29.47	0.825	0.945	0.773	0.564
Slovenia	12.20	17.20	81.10	30.59	0.903	0.91	0.758	0.703
Spain	9.80	17.90	83.30	34.26	0.849	0.854	0.837	0.627
Sweden	12.40	17.60	82.60	47.77	0.848	0.611	0.886	0.92
Switzerland	13.40	16.20	83.50	57.63	0.864	0.674	0.9	0.824
United Kingdom	12.90	17.40	81.70	39.12	0.771	0.654	0.84	0.828
Maximum	14.10	19.80	83.50	68.01	0.91	0.99	0.92	0.95
Minimum	9.20	12.70	71.20	9.14	0.45	0.54	0.33	0.30

2.2. General Description of WEBIRA Method

Let the initial data be the results of the performed measurements, expert evaluations, etc., presented in $m \times n$-dimension matrix $X = \left(x_{ij}\right)_{m \times n}$. The element x_{ij} of the decision-making matrix is the estimate of the alternative i $(i = 1, 2, \ldots, m)$ based on using the criteria j $(j = 1, 2, \ldots, n)$. A data normalisation procedure is required because there are different criteria measurement units. There is a variety of data normalisation formulas, in this case a min–max normalisation was used:

$$\begin{cases} \tilde{x}_{ij} = \dfrac{x_{ij} - \min\limits_{1 \leq i \leq m} x_{ij}}{\max\limits_{1 \leq i \leq m} x_{ij} - \min\limits_{1 \leq i \leq m} x_{ij}}, & \text{for the direct normalisation,} \\[2ex] \tilde{x}_{ij} = \dfrac{\max\limits_{1 \leq i \leq m} x_{ij} - x_{ij}}{\max\limits_{1 \leq i \leq m} x_{ij} - \min\limits_{1 \leq i \leq m} x_{ij}}, & \text{for inverse normalisation.} \end{cases}$$

The choice of min–max normalisation is based on the results of previous studies [22,31], which revealed that min–max normalisation ensures the best stability of the SAW method compared to other well-known normalisation procedures such as max, sum, vector, logarithmic, etc. Stability of the min–max method was the highest for cases of both more and less separable alternatives. All the variables after the min–max normalisation gain their values between 0 and 1.

If new countries with values not in the range analysed initially would be introduced, normalisation would be performed again and all values after normalisation would also range between 0 and 1. However, after introducing new countries (cases) and recalculation of correlation coefficients the priority of criteria, data structure and, subsequently, the overall rating of the countries, could be changed. For this reason, only European countries were involved in the investigation.

The normalised decision-making matrix has the following form, $\tilde{X} = \left(\tilde{x}_{ij}\right)_{m \times n}, 0 \leq \tilde{x}_{ij} \leq 1$. Let $w_j, j = 1, 2, \ldots, n$ be criteria weights, satisfying the conditions as follows

$$\sum_{j=1}^{n} w_j = 1, \ 0 \leq w_j \leq 1. \tag{1}$$

The Simple Additive Weighting (SAW) method [32] is a well-known and widely used MCDM tool. SAW with the weights $w_j, j = 1, 2, \ldots, n$ can be applied to solve MCDM problem. The aggregated value based on using the SAW criteria was calculated for each alternative as follows

$$S_i = \sum_{j=1}^{n} w_j \tilde{x}_{ij}, \ i = 1, 2, \ldots, m. \tag{2}$$

The values $0 \leq \tilde{x}_{ij} \leq 1$ in Equation (2) were normalised so that the higher $\tilde{x}_{ij}$ value would correspond to the better evaluation of the i-th alternative S_i.

The weighted coefficients (1) are usually determined by using various methods that could be based on expert judgement (subjective methods) or the objective weight assessing methods [33]. WEBIRA is objective weight assessing method which is appropriate for solving our problem for two reasons. The first is the absence of highly qualified expert judgements. This prevents the use of subjective methods such as Analytic Hierarchy Process (AHP), Delphi, Stepwise Weight Assessment Ratio Analysis (SWARA), etc. The second reason is the structure of the data. The set of criteria (indicators) naturally and logically could be divided to two groups of criteria. The idea of WEBIRA method is weights determining procedure when the rankings of alternatives in the few groups of criteria maximally match each other. This goal is achieved by performing a so-called weight balancing procedure aimed at minimising a certain objective function.

Suppose that n criteria are being divided to r groups. The coefficient calculation scheme is introduced when there are r normalised data matrices X^k:

$$X^k = \|x_{ij}^k\|_{m \times n^k}, 0 \leq x_{ij}^k \leq 1, \ k = 1, 2, \ldots, r, \ \sum_{k=1}^{r} n^k = n. \tag{3}$$

The aggregated values for each matrix $k = 1, 2, \ldots, r$ obtained by using SAW criteria were as follows

$$S_i^k = \sum_{j=1}^{n^k} w_j^k \tilde{x}_{ij}^k, \ i = 1, 2, \ldots, m, \ k = 1, 2, \ldots, r. \tag{4}$$

The coefficients w_j^k in Formula (4) satisfy the inequalities

$$1 \geq w_1^k \geq w_2^k \geq \ldots \geq w_{n^k}^k \geq 0, \ k = 1, 2, \ldots, r. \tag{5}$$

The optimisation problem is formulated where the minimum value of the function has to be found:

$$F\left(W^1, W^2, \ldots, W^r\right) = \sum_{k=1}^{r-1} \sum_{l=k+1}^{r} \sum_{i=1}^{m} \left| S_i^k - S_i^l \right|^{\delta}$$

by checking the value of the function above with each vector $W^k = \left(w_1^k, w_2^k, \ldots, w_{n^k}^k\right)$, $k = 1, 2, \ldots, r$ satisfying the inequalities (5) and the relationships (1). The parameter's δ value is $\delta = 2$ throughout the paper. The inequalities (5) can be determined by using various methods of processing the expert assessments; in Krylovas et al. [24], it has been proposed to apply Kemeny median [34] for this purpose. This method for prioritising criteria and determining weights which satisfy Formulas (1) and (5) is named the KEmeny Median Indicator Ranks Accordance (KEMIRA) method. The order of preference of the weighted coefficients can be determined by using other methods. In this paper correlation analysis is applied to the solution of this problem. Therefore, a group of the methods given in Krylovas et al. [21] is referred to as WEBIRA (WEight Balancing Indicator Ranks Accordance).

Suppose that $A = \{1, 2, \ldots, m\}$ is a set of the available alternatives, while the subsets of the set A are denoted as follows

$$A_{\alpha}^{+} = \left\{ i \in A : S_i^1 > \alpha, S_i^2 > \alpha, \ldots, S_i^r > \alpha \right\},$$
$$A_{\alpha}^{-} = \left\{ i \in A : S_i^1 \leq \alpha, S_i^2 \leq \alpha, \ldots, S_i^r \leq \alpha \right\},$$
$$A_{\alpha}^{\pm} = A \backslash (A_{\alpha}^{+} \cup A_{\alpha}^{-}),$$

A_{α}^{+} denotes the sets of the undoubtedly superior alternatives, A_{α}^{-} are the sets of undoubtedly inferior alternatives and $A_{\alpha}^{\pm}$ denotes the sets of alternatives whose assessment is doubtful. Note that when $0 \leq S_i^k \leq 1$, $A_0^{+} = A_1^{-} = A$, $A_1^{+} = A_0^{-} = \varnothing$. The functions $F^{+}(\alpha), F^{-}(\alpha), \ F^{\pm}(\alpha)$ are determined as the number of elements of the respective sets $A_{\alpha}^{+}, A_{\alpha}^{-}, A_{\alpha}^{\pm}$. It is obvious that $F^{+}(\alpha) + F^{-}(\alpha) + F^{\pm}(\alpha) = m$. $F^{+}(\alpha), F^{-}(\alpha), \ F^{\pm}(\alpha)$ are stepwise functions, having the first type points of discontinuity. The values of the functions can help assess the quality of weight balancing. In the ideal case, $F^{\pm}(\alpha) \equiv 0$. In this research, the authors deal with $A_0^{+} = A_1^{-} = A$.

3. Results

A problem of determining the ranks of the European countries based on two groups of criteria ($r = 2$), including internal security and human development, was solved. In the first step, correlation analysis was applied to the data in Table 2 for establishing the priority of the criteria, such as internal security, $X = (x_1, x_2, x_3, x_4)$ and human development, $Y = (y_1, y_2, y_3, y_4)$, in each group. The larger the absolute value of the correlation coefficient of the respective criterion with the criteria of the other group, the higher its priority order. The values of the Pearson correlation coefficients are presented in Table 3.

Process criteria has the highest priority value in the Int_Sec_Group, due to higher correlation with Y group criteria (0.868), followed by Legitimacy (0.823), Capacity (−0.537) and, finally, Outcomes criterion (0.467). In the human development group, GNI per Capita (0.868) has the highest priority value, Life Expectancy at birth (0.823) is second, the Expected Years of Schooling (0.772) is third and the Mean Years of Schooling (0.506) is last. Therefore, the priority order of the considered criteria is as follows

$$x_4 \succ x_1 \succ x_3 \succ x_2, \; y_4 \succ y_1 \succ y_3 \succ y_2$$

and the respective weight priority (5) is

$$1 \geq w_4^X \geq w_1^X \geq w_3^X \geq w_2^X \geq 0, \; 1 \geq w_4^Y \geq w_1^Y \geq w_3^Y \geq w_2^Y \geq 0. \tag{6}$$

Table 3. Values of Pearson correlation coefficients of the criteria (first row) and *p*-values (second row).

Factors (Criteria)	Life Expectancy at Birth y_1	Mean Years of Schooling y_2	Expected Years of Schooling y_3	GNI per Capita y_4	Legitimacy x_1	Outcomes x_2	Capacity x_3
GNI per capita y_4	0.757 ** 0.000	0.451 ** 0.005	0.762 ** 0.000	1			
Legitimacy x_1	0.823 ** 0.000	0.390 * 0.016	0.711 ** 0.000	0.794 ** 0.000	1		
Outcomes x_2	0.467 ** 0.003	−0.003 0.987	0.159 0.339	0.268 0.103	0.539 ** 0.000	1	
Capacity x_3	−0.537 ** 0.001	−0.116 0.487	−0.384 * 0.017	−0.531 ** 0.001	−0.492 ** 0.002	−0.319 0.051	1
Process x_4	0.754 ** 0.000	0.506 ** 0.001	0.772 ** 0.000	0.868 ** 0.000	0.890 ** 0.000	0.404 * 0.012	−0.498 ** 0.001

** Correlation is significant at the 0.01 level (2-tailed). * Correlation is significant at the 0.05 level (2-tailed).

The second step in solving the MCDM problem is normalising the decision-making matrix elements x_{ij} and y_{ij}. Min–max normalisation equations were used in both cases. This method demonstrated the highest accuracy and was most stable compared to other normalisation techniques, when applied with SAW [30]. In the case of direct normalisation, the equations were as follows

$$\widetilde{x}_{ij} = \frac{x_{ij} - \min\limits_{1 \leq i \leq m} x_{ij}}{\max\limits_{1 \leq i \leq m} x_{ij} - \min\limits_{1 \leq i \leq m} x_{ij}}, \; \widetilde{y}_{ij} = \frac{y_{ij} - \min\limits_{1 \leq i \leq m} y_{ij}}{\max\limits_{1 \leq i \leq m} y_{ij} - \min\limits_{1 \leq i \leq m} y_{ij}},$$

while the inverse normalisation equation was applied only to the Capacity criterion, having an opposite direction with respect to the goal:

$$\widetilde{x}_{ij} = \frac{\max\limits_{1 \leq i \leq m} x_{ij} - x_{ij}}{\max\limits_{1 \leq i \leq m} x_{ij} - \min\limits_{1 \leq i \leq m} x_{ij}}.$$

In Table 4 the normalised values of the criteria $\widetilde{x}_{ij}, \widetilde{y}_{ij}$ are given.

Then, the procedure of weight balancing was carried out. A possible set of weights, satisfying the conditions (1) and (6), is presented in Table 5. The elements of this set were reselected and the weighted sums $S_i^X = \sum_{j=1}^4 w_j^x \widetilde{x}_{ij}, S_i^Y = \sum_{j=1}^4 w_j^y \widetilde{y}_{ij}$ were calculated for each alternative. The minimum value of the target function was obtained for the optimal weight values $W^{X*} = \left(w_1^{x*}, w_2^{x*}, w_3^{x*}, w_4^{x*} \right)$ and $W^{Y*} = \left(w_1^{y*}, w_2^{y*}, w_3^{y*}, w_4^{y*} \right)$:

$$F\left(W^{X*}, W^{Y*} \right) = \min_{W^X, W^Y} \sum_{i=1}^m \left(S_i^X - S_i^Y \right)^2 = \min_{W^X, W^Y} \sum_{i=1}^m \left(\sum_{j=1}^4 w_j^x \widetilde{x}_{ij} - \sum_{j=1}^4 w_j^y \widetilde{y}_{ij} \right)^2. \tag{7}$$

$F\left(W^{X*}, W^{Y*} \right)$ is the minimum value of a disagreement measure between two alternative rankings (according to the criteria values X and Y). It could be interpreted as the function of assessing the weight balancing quality.

Table 4. Normalised decision-making matrix.

Country (*Alternative*)	Outcomes	Capacity	Legitimacy	Process	Mean Years of Schooling	Expected Years of Schooling	Life Expectancy at Birth	GNI per Capita
Factors	x_2	x_3	x_1	x_4	y_2	y_3	y_1	y_4
Albania	0.5929	0.7612	0.3938	0.0030	0.1632	0.2957	0.5934	0.0465
Armenia	0.9715	0.1441	0.3157	0.2817	0.5102	0.0422	0.2926	0
Austria	0.9737	0.4842	0.9660	0.7993	0.5918	0.4788	0.8617	0.6161
Azerbaijan	0.9234	0.5900	0.2665	0	0.3061	0	0.0731	0.1096
Belarus	0.5185	0.0225	0.2648	0.2710	0.6326	0.3943	0.1544	0.1219
Belgium	0.7833	0.6193	0.8777	0.7580	0.5306	1	0.8211	0.5607
Bosnia and Herzegovina	0.8205	0.1554	0.5297	0.2603	0.1020	0.2112	0.4796	0.0436
Bulgaria	0.6652	0	0.3837	0.3047	0.5306	0.2957	0.3008	0.1630
Cyprus	0.7024	0.5608	0.7877	0.5191	0.5918	0.2676	0.7723	0.3809
Croatia	0.8862	0.1036	0.6196	0.4747	0.4285	0.3239	0.5365	0.2211
Czech Republic	0.8271	0.2477	0.7504	0.5252	0.7142	0.5915	0.6260	0.3642
Denmark	0.9540	0.7590	0.9745	1	0.6938	0.9014	0.7886	0.6586
Estonia	0.6236	0.0405	0.8047	0.7029	0.7142	0.4788	0.5284	0.3371
Finland	0.9715	0.7004	1	0.9601	0.6530	0.6901	0.8373	0.5411
France	0.7308	0.4774	0.8268	0.6722	0.4693	0.5211	0.9349	0.5114
Georgia	0.6936	0.3648	0.7164	0.4563	0.7346	0.3239	0.1788	0.0007
Germany	0.8818	0.4662	0.9117	0.8897	1	0.6056	0.8130	0.6283
Greece	0.5579	0.4549	0.6129	0.4410	0.3265	0.6478	0.8292	0.2633
Hungary	0.7527	1	0.5382	0.5160	0.5510	0.3380	0.3983	0.2760
Iceland	1	0.7882	0.9558	0.7886	0.6530	0.9295	0.9512	0.6228
Ireland	0.7789	0.3243	0.8862	0.7427	0.6734	0.9718	0.8455	0.7577
Italy	0.6827	0.5878	0.6706	0.5911	0.2040	0.5070	0.9756	0.4442
Latvia	0.5382	0.1148	0.6129	0.4027	0.7346	0.4366	0.2845	0.2693
Lithuania	0.5054	0.1846	0.6842	0.4747	0.7755	0.4788	0.2926	0.3256
Montenegro	0.8402	0.1599	0.5959	0.2848	0.4285	0.3098	0.4959	0.1296
Netherlands	0.9124	0.6261	0.8964	0.9234	0.6122	0.7464	0.8780	0.6583
Norway	0.7702	0.7364	0.9949	0.9387	0.6938	0.7323	0.9024	1
Poland	0.8949	0.3085	0.6926	0.5834	0.6326	0.5211	0.5365	0.2888
Portugal	0.8424	0.1711	0.6825	0.5880	0	0.5070	0.8292	0.3086
Romania	0.7789	0.3378	0.4855	0.3675	0.3673	0.2253	0.3577	0.2293
Russian Federation	0	0.0022	0	0.1837	0.5714	0.3943	0	0.2563
Serbia	0.8796	0.2229	0.4363	0.2557	0.3877	0.2676	0.3333	0.0658
Slovakia	0.8227	0.0900	0.7521	0.4119	0.6734	0.3239	0.4715	0.3452
Slovenia	0.9934	0.1689	0.7266	0.6248	0.6122	0.6338	0.8048	0.3643
Spain	0.8752	0.2950	0.8607	0.5084	0.1224	0.7323	0.9837	0.4266
Sweden	0.8730	0.8423	0.9439	0.9571	0.6530	0.6901	0.9268	0.6560
Switzerland	0.9080	0.7004	0.9677	0.8101	0.8571	0.4929	1	0.8235
United Kingdom	0.7045	0.7454	0.8658	0.8162	0.7551	0.6619	0.8536	0.5091

Table 5. The set of possible weight values $1 \geq w_4^{x,y} \geq w_1^{x,y} \geq w_3^{x,y} \geq w_2^{x,y} \geq 0$ for the criteria X and Y.

No	$w_2^{x,y}$	$w_3^{x,y}$	$w_1^{x,y}$	$w_4^{x,y}$	No	$w_2^{x,y}$	$w_3^{x,y}$	$w_1^{x,y}$	$w_4^{x,y}$
1	0	0	0	1	13	0	0.1	0.4	0.5
2	0	0	0.1	0.9	14	0	0.2	0.3	0.5
3	0	0	0.2	0.8	15	0.1	0.1	0.3	0.5
4	0	0.1	0.1	0.8	16	0.1	0.2	0.2	0.5
5	0	0	0.3	0.7	17	0	0.2	0.4	0.4
6	0	0.1	0.2	0.7	18	0.1	0.1	0.4	0.4
7	0.1	0.1	0.1	0.7	19	0	0.3	0.3	0.4
8	0	0	0.4	0.6	20	0.1	0.2	0.3	0.4
9	0	0.1	0.3	0.6	21	0.2	0.2	0.2	0.4
10	0	0.2	0.2	0.6	22	0.1	0.3	0.3	0.3
11	0.1	0.1	0.2	0.6	23	0.2	0.2	0.3	0.3
12	0	0	0.5	0.5					

The number of possible weight combinations and, accordingly, the values of the target function (7) is $23 \times 23 = 529$. The function $F(W^X, W^Y)$ gained its minimum value 0.397 for the respective weight values:

$$w_4^{X*} = 0.6, \; w_1^{X*} = 0.2, w_3^{X*} = 0.2, w_2^{X*} = 0; \; w_4^{Y*} = 0.3, \; w_1^{Y*} = 0.3, w_3^{Y*} = 0.2, w_2^{Y*} = 0.2. \tag{8}$$

Next, the step length 0.05 (twice as small as in Table 5) was chosen and the optimisation procedure was repeated. However, the authors failed to get a better result. The minimum value of the target function (7) remained the same with the same weights (8).

At the last step of WEBIRA, the weighted sum values of the criteria X and Y were calculated for each alternative as follows

$$Q_i\left(W^{X*}, W^{Y*}\right) = S_i^{X*} + S_i^{Y*}, \; i = 1, 2, \ldots, m$$

and the ranking of the alternatives based on these values was performed. The final results are presented in Table 6.

When assessing the criteria of ranking the countries, it is important to take into consideration the mutual distribution of HDI and WISPI components (the difference between the ranks of HDI and WISPI of the countries) (Figure 1). The difference between the ranks of HDI and WISPI reflects the development priorities of the countries (Table 6). Appraisal of changes in the difference between these indicators allows forecasting the trend of the country's development (e.g., development of economic potential and increasing the welfare of the population, development associated with strengthening the policy and recognising the security priorities, or harmonious development). The minimal difference between HDI and WISPI shows a balanced internal policy pursued by the countries, implying that the countries allocate their resources to the internal security and public welfare in a balanced way.

Table 6. European countries ranking results based on the alternative methods of WEBIRA, HDI, WISPI and cluster analysis results.

Country (*Alternative*)	$Q_i=S_i^{X^*}+S_i^{Y^*}$ WEBIRA	$S_i^{Y^*}$ (HDI)	$S_i^{X^*}$ (WISPI)	WEBIRA Rank (HDI + WISPI)	HDI Rank	WISPI Rank	HDI Minus WISPI	k-Means
Norway	1.7655	0.8560	0.9095	1	1	4	−3	4
Denmark	1.6999	0.7532	0.9467	2	6	1	5	4
Sweden	1.675	0.7435	0.9315	3	7	2	5	4
Switzerland	1.6368	0.8171	0.8197	4	2	7	−5	4
Iceland	1.6107	0.7887	0.8220	5	4	6	−2	4
Finland	1.5984	0.6822	0.9162	6	11	3	8	4
Netherlands	1.5913	0.7327	0.8586	7	8	5	3	4
Germany	1.5629	0.7535	0.8094	8	5	9	−4	4
United Kingdom	1.5043	0.6923	0.8120	9	10	8	2	4
Ireland	1.4978	0.8101	0.6877	10	3	12	−9	4
Belgium	1.4749	0.7207	0.7542	11	9	11	−2	4
Austria	1.4272	0.6575	0.7697	12	12	10	2	4
France	1.2962	0.6320	0.6642	13	13	13	0	3
Italy	1.1746	0.5682	0.6064	14	16	15	1	3
Slovenia	1.154	0.6000	0.5540	15	14	18	−4	3
Spain	1.1303	0.5941	0.5362	16	15	20	−5	3
Cyprus	1.0991	0.5179	0.5812	17	19	17	2	3
Estonia	1.0891	0.4983	0.5908	18	20	16	4	3
Czech Republic	1.0731	0.5583	0.5148	19	17	22	−5	3
Poland	1.0287	0.4784	0.5503	20	21	19	2	3
Greece	1.0009	0.5227	0.4782	21	18	24	−6	3
Hungary	0.9974	0.3801	0.6173	22	26	14	12	1
Portugal	0.9664	0.4428	0.5236	23	23	21	2	3
Lithuania	0.895	0.4364	0.4586	24	24	25	−1	2
Slovakia	0.8601	0.4445	0.4156	25	22	27	−5	3
Croatia	0.8073	0.3778	0.4295	26	27	26	1	3
Latvia	0.7876	0.4004	0.3872	27	25	28	−3	2
Georgia	0.7557	0.2656	0.4901	28	34	23	11	1
Romania	0.6799	0.2947	0.3852	29	30	29	1	1
Montenegro	0.6575	0.3354	0.3221	30	28	30	−2	1
Bulgaria	0.564	0.3044	0.2596	31	29	34	−5	2
Serbia	0.5361	0.2508	0.2853	32	35	32	3	1
Albania	0.5167	0.2838	0.2329	33	32	35	−3	1
Bosnia and Herzegovina	0.5129	0.2197	0.2932	34	36	31	5	1
Belarus	0.5084	0.2883	0.2201	35	31	36	−5	2
Armenia	0.4594	0.1983	0.2611	36	37	33	4	1
Russian Federation	0.3808	0.2701	0.1107	37	33	38	−5	2
Azerbaijan	0.2874	0.1161	0.1713	38	38	37	1	1

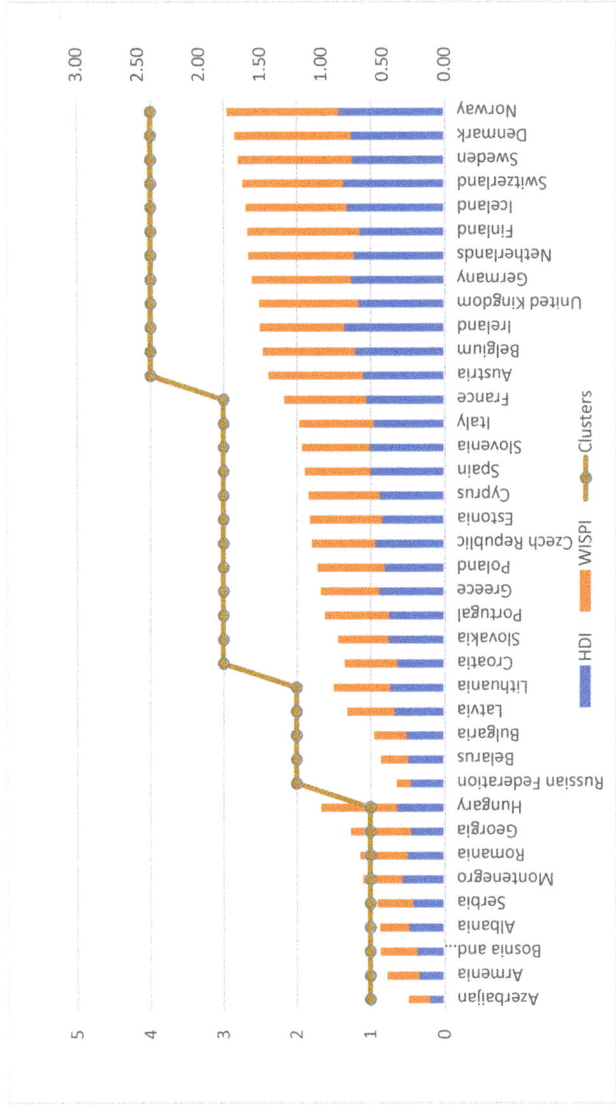

Figure 1. Countries' ranking according to WEBIRA and its components HDI and WISPI vs. cluster analysis results.

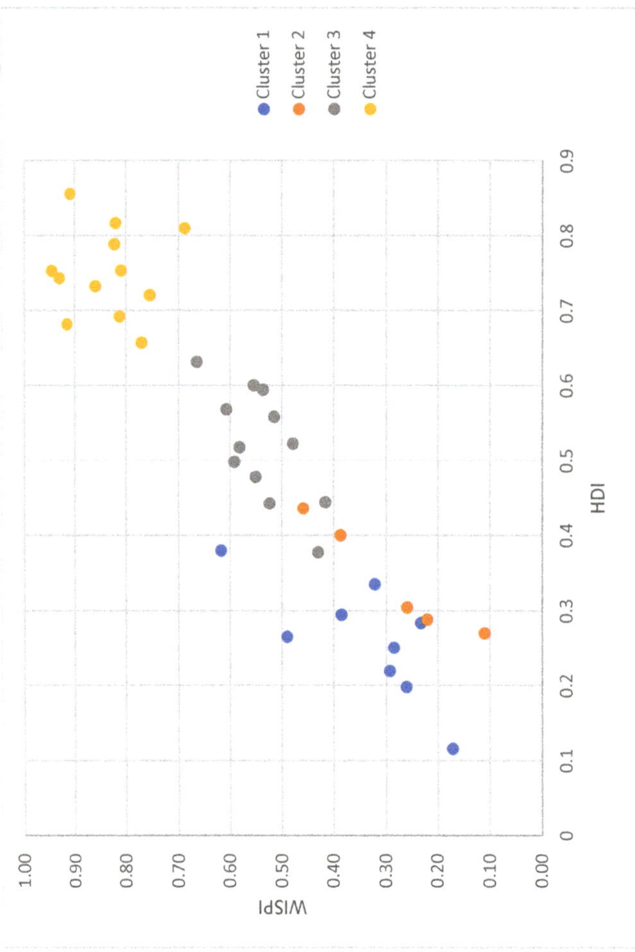

Figure 2. Cluster analysis results on HDI and WISPI axes.

In the following, the performed comparative analysis of WEBIRA method results with the results of cluster analysis is discussed. The clustering procedure was executed with standardised data. Based on the silhouette method, the European countries could be divided into four clusters. The partition into clusters was done by several hierarchical clustering methods and also techniques like k-means cluster analysis. The challenge when applying the hierarchical clustering methods is determining the proper distance measure. In our work, we have tested different distance measures, but there are alternative methods like A-BIRCH [35]. All of the used methods gave very similar results, so we chose the k-means method. Therefore, neither clustering approach can ultimately judge the actual quality of clustering; this needs human evaluation [36], which is highly subjective [37]. Because of these shortcomings, cluster analysis can only be used as rough initial test before applying more accurate methods. Table 6 shows the ranking of the European countries by using methods WEBIRA, HDI, WISPI and cluster analysis results. Table 6 and Figure 1 show that the WEBIRA ranking of countries is basically consistent with the results of cluster analysis. Spearman correlation coefficient of WEBIRA ranks and k-means cluster analysis results show very high correlation between them $\rho_S = 0.912$. All countries with the highest HDI and WISPI indicators have entered cluster 4. Cluster 3 consists of slightly lower HDI and WISPI countries with one exception—Hungary—which is the member of cluster 1. The countries with lowest HDI and WISPI are assigned to cluster 1. The most diverse is cluster 2, which consists of countries which at first glance do not have much in common, i.e., Lithuania, Latvia, Belarus, Bulgaria and Russian Federation. Figure 2 represents cluster analysis results on HDI and WISPI axes. It also shows that WISPI indicators in all cluster 2 countries except Lithuania are lower than in other countries adjacent to these in WEBIRA ranking.

4. Discussion

It is important to note that motivation to use correlations for the criteria prioritisation is based on the reasoning that the criteria of the two groups x_i and y_j describe the same phenomenon—the well-being of the population in the broad sense. Criteria prioritisation procedure consider correlations of x_i with y_j (not correlations in internal groups for x_i and y_j). Ideally, values of x_i must not be correlated with each other as well as values of y_j. However, the correlation between the criteria of different groups may be. If this assumption were completely wrong, WEBIRA would probably fail to balance the weight of the criteria and the final rating would be not logical. However, the final rating of alternatives is consistent with the results of cluster analysis. So, there is no reason to assert that priorities have been wrongly identified.

The question is: Can intergroup correlations be spurious? A spurious correlation can often be created by an antecedent which impacts both variables. In the current situation we do not have such causal relationships; our belief is that this negates the hypothesis of false correlations. Furthermore, security indicators correlation with national health indicators is ascertained in the literature [38].

The resulting weighted sum values S^X and S^Y are strongly correlated ($r = 0.861$) and the alternative for the Formula (7) may be the maximum of the correlation coefficient. There may be other alternatives to the Formula (7). Benchmarking of methods for setting priorities (6) is an interesting task looking forward to further research. To approve the use of correlations for setting the relative importance of the evaluation criteria other well-known objective methods, for example, the entropy method, would be applied. A sensitivity analysis of weights would be performed in order to demonstrate the stability of the results. This is also planned by the authors in their further research.

Security is not only a mighty driver of economic activity worldwide but also has a strong influence on public social welfare. Therefore, it is one of the most significant topics of discussion in the global society today [1]. On the other hand, the relevant problem is to assess the feasibility of identifying country threats in the economic, social and other spheres of society, based on the correlation and consistency of definitions of security and socioeconomic indicators according to their content and logical relationship [39]. Human development indicators are integrated part of economic, social and other spheres of public life and are related to the level of internal security of the countries. These are key elements in assessing the economic, social and internal security aspects of countries. Investigation shows that indicators of WISPI and HDI are closely related and their correlation is reasonable.

This should be considered in performing socioeconomic reforms in the EU. However, when the countries were categorised into more and less developed ones, based on the Human Development Index, they have different effects on their police systems [40]. Ranking the countries based on both indices has revealed the differences between the countries in this respect. It allows the authors to conclude that political strategies in the EU countries differ considerably. Political strategies in the EU member states can be focused on the most significant (weighty) indicators of HDI and WISPI, described in this study. It shows that there is no balance in the development strategies of these countries. According to our insight, this is a preliminary distribution that helps to understand the prevailing trends in the countries (HDI dominates in one group of countries, WISPI dominates in another group of countries and WISPI and HDI harmonise in a third group of countries). This allows us to distinguish countries by their distance from harmonious development according to HDI and WISPI. This question requires further and deeper research and validation.

There are various multiple criteria decision-making techniques available for the analysis of the alternatives based on a set of criteria. They often yield different ranking results of the alternatives. The question arises, which approach is most suitable? It is clear that it depends on the investigated problem and the goals to be achieved. In this research, the problem of ranking the countries by using not only Human Development, but also the criteria describing Internal Security and Police has been solved applying WEBIRA method. The MCDM method WEBIRA meets the objective pursued because it allows the researchers to carry out the weight balancing procedure by solving the optimisation problem and simultaneously determining the weights of the criteria of both groups. Then, the ranking procedure has been performed by applying the SAW method. Brute force (i.e., the total reselection) algorithm implementation was chosen for this particular task. However, the optimisation task could be solved by using other heuristic techniques.

5. Conclusions

In recent years, the events taking place in Europe have come into the focus of attention. Now, people, nations and economies, as well as the global development issues we are facing, have become more closely connected than ever. A completely new modelling algorithm has been proposed, which has not been implemented yet in ranking counties according to internal security criteria of the countries. This improves the understanding of how the methodology should be applied. Thus, the considered methodology is an advancement compared to the methods used in previous studies and provides a comprehensive approach to the analysis of the internal security of the states. When estimating the results of ranking the countries obtained by using various criteria and techniques (WEBIRA, HDI and WISPI), reliable correlations can be observed. It should be noted that the integral WEBIRA ranking method allows for objective determination of the considered states' distribution, as well as assigning weights to the criteria. Generalised criteria measuring the level of the country is being calculated by using the SAW method. Cluster analysis of the countries was carried out and compared with MCDM results. Cluster analysis approved the results of WEBIRA ranking. Thus, the clustering results of the countries correspond to their positions in the WEBIRA ranking. Moreover, their Spearman rank correlation coefficient value is very high (0.912).

This enables the objective evaluation of the security systems of the countries. It should also be mentioned that though HDI does not assess the internal security indicators of the countries, and while WISPI does not determine the level of their human development, they are similar to a great extent. The established strong correlation between HDI and WISPI rankings allows the authors to argue that the internal security of a state mainly depends on the trends of its human development and vice versa. The question arises, which factor, HDI or WISPI, prevails? The answer to this question is given by the correlations between these indicators and the integral combining index. The determination of the correlations between HDI and WISPI and WEBIRA (HDI+WISPI) rank allowed for establishing a stronger correlation between WEBIRA rank and HDI. This, in turn, allows the authors to conclude that the internal security of a state largely depends on the well-being of its citizens. Thus, to increase the internal security of a state, it is necessary not only to strengthen the police and security forces, but also to pay more attention to the well-being of its citizens.

Author Contributions: Conceptualization, S.D.; Data curation, A.K., S.D. and N.K.; Formal analysis, R.D.; Methodology, A.K.; Supervision, S.D.; Visualization, N.K.; Writing—original draft, R.D.; Writing—review & editing, N.K. and S.D.

Funding: No external funding.

Conflicts of Interest: No any arrangement that would compromise the perception of our impartiality.

References

1. Hollis, S. The Global construction of EU. *Dev. Pol. J. Eur. Integr.* **2014**, *36*, 567–583. [CrossRef]
2. Abdelmottlep, M.A. World Internal Security and Police Index 2016. International Police Science Association (IPSA), 2016. Available online: http://insyde.org.mx/wp-content/uploads/WISPI-Report_EN_WEB_0.pdf (accessed on 21 January 2019).
3. Abbott, P.; Teti, A. A Generation in waiting for jobs and justice: Young people not in education employment or training in North Africa. *Arab Transform. Work. Pap.* **2017**. [CrossRef]
4. Fritza, M.; Koch, M. Economic development and prosperity patterns around the world: Structural challenges for a global steady-state economy. *Glob. Environ. Chang.* **2016**, *38*, 41–48. [CrossRef]
5. Rivera, M.A. The synergies between human development, economic growth, and tourism within a developing country: An empirical model for Ecuador. *J. Dest. Mark. Manag.* **2017**, *6*, 221–232. [CrossRef]
6. Sen, A. *Development as Freedom*; Oxford University Press: Oxford, UK, 1999.
7. Asongu, S.A.; Nwachukwu, J.C.; Pyke, C. The right to life: Global evidence on the role of security officers and the police in modulating the effect of insecurity on homicide. *Soc. Indic. Res.* **2018**, *140*, 1–14. [CrossRef]
8. Mitchell, N.J.; Carey, S.C.; Butler, C.K. The impact of pro-government militias on human rights violations. *Int. Interact.* **2014**, *40*, 812–826. [CrossRef]
9. Asongu, S.A.; Nwachukwu, J.C. Mitigating externalities of terrorism on tourism: Global evidence from police, security officers and armed service personnel. *Curr. Issues Tour.* **2018**. [CrossRef]
10. Abdelzaher, D.; Fernandez, W.D.; Schneper, W.D. Legal rights, national culture and social networks: Exploring the uneven adoption of United Nations Global Compact. *Int. Bus. Rev.* **2019**, *28*, 12–24. [CrossRef]
11. Human Development Indices and Indicators 2018 Statistical Update. HDRO (Human Development Report Office) United Nations Development Programme. Retrieved 14 September 2018. Available online: http://hdr.undp.org/sites/default/files/2018_human_development_statistical_update.pdf (accessed on 21 January 2019).
12. Lestari, W.W.; Sanar, V.E. Analysis indicator of factors affecting human development index (Ipm). *Geosfera Indonesia* **2018**, *2*, 11–18. [CrossRef]
13. Naanwaab, C. does economic freedom promote human development? New evidence from a cross-national study. *J. Dev. Areas* **2018**, *52*, 183–198. [CrossRef]
14. Hirai, T. *The Creation of the Human Development Approach*; Palgrave Macmillan: London, UK, 2017.
15. Zaborskis, A.; Grincaite, M.; Lenzi, M.; Tesler, R.; Moreno-Maldonado, C.; Mazur, J. Social inequality in adolescent life satisfaction: Comparison of measure approaches and correlation with macro-level indices in 41 countries. *Soc. Indic. Res.* **2019**, *141*, 1055–1079. [CrossRef]
16. Murray, S.R.; Juodakis, J.; Bacelis, J.; Sand, A.; Norman, J.E.; Sengpiel, V.; Jacobsson, B. Geographical differences in preterm delivery rates in Sweden: A population-based cohort study. *Acta Obstet. Gynecol. Scand.* **2019**, *98*, 106–116. [CrossRef]
17. Liu, S.Y.; Wu, P.C.; Huang, T.Y. Nonlinear Causality between Education and Health: The Role of Human Development Index. *Appl. Res. Qual. Life* **2018**, *13*, 761–777. [CrossRef]
18. Sayed, H.; Hamed, R.; Hosny, S.H.; Abdelhamid, A.H. Avoiding ranking contradictions in human development index using goal programming. *Soc. Indic. Res.* **2018**, *138*, 405–442. [CrossRef]
19. Carvalhal Monteiro, R.L.; Pereira, V.; Costa, H.G. A multicriteria approach to the human development index classification. *Soc. Indic. Res.* **2018**, *136*, 417–438. [CrossRef]

20. Krylovas, A.; Dadelo, S.; Kosareva, N.; Zavadskas, E.K. Entropy-KEMIRA approach for MCDM problem solution in human resources selection task. *Int. J. Inf. Technol. Decis. Mak.* **2017**, *16*, 1151–1154. [CrossRef]

21. Krylovas, A.; Kosareva, N.; Zavadskas, E.K. WEBIRA—Comparative analysis of weight balancing method. *Int. J. Comput. Commun. Control* **2017**, *12*, 238–253. [CrossRef]

22. Kosareva, N.; Krylovas, A.; Zavadskas, E.K. Statistical analysis of MCDM data normalization methods using Monte Carlo approach. The case of ternary estimates matrix. *Econ. Comput. Econ. Cybern. Stud. Res.* **2018**, *52*, 159–175. [CrossRef]

23. Saaty, T.L. *The Analytic Hierarchy Process*; McGraw-Hill: New York, NY, USA, 1980.

24. Krylovas, A.; Zavadskas, E.K.; Kosareva, N.; Dadelo, S. New KEMIRA method for determining criteria priority and weights in solving MCDM problem. *Int. J. Inf. Technol. Decis. Mak.* **2014**, *13*, 1119–1134. [CrossRef]

25. Hashemkhani Zolfani, S.; Saparauskas, J. New Application of SWARA Method in Prioritizing Sustainability Assessment Indicators of Energy System. *Inzinerine Ekonomika—Eng. Econ.* **2013**, *24*, 408–414. [CrossRef]

26. Garcia, N.; Puente, J.; Fernandez, I.; Priore, P. Suitability of a consensual Fuzzy inference system to evaluate suppliers of strategic products. *Symmetry* **2018**, *10*, 22. [CrossRef]

27. Shannon, C.E. A mathematical theory of communication. *Bell Syst. Technol. J.* **1948**, *27*, 379–423. [CrossRef]

28. Pekelman, D.; Sen, S.K. Mathematical programming models for the determination of attribute weights. *Manag. Sci.* **1974**, *20*, 1217–1229. [CrossRef]

29. Zavadskas, E.K.; Podvezko, V. Integrated determination of objective criteria weights in MCDM. *Int. J. Inf. Technol. Decis. Mak.* **2016**, *15*, 267–283. [CrossRef]

30. Diakoulaki, D.; Mavrotas, G.; Papayannakis, L. Determining objective weights in multiple criteria problems: The CRITIC method. *Comput. Oper. Res.* **1995**, *22*, 763–770. [CrossRef]

31. Krylovas, A.; Kosareva, N.; Zavadskas, E.K. Scheme for statistical analysis of some parametric normalization classes. *Int. J. Comput. Commun. Control* **2018**, *13*, 972–987. [CrossRef]

32. MacCrimmon, K.R. Decision Making among Multiple-Attribute Alternatives: A Survey and Consolidated Approach, No. RM-4823-ARPA, Santa Monica: RAND Corporation, 1968. Available online: http://citeseerx.ist.psu.edu/viewdoc/download?doi=10.1.1.924.1201&rep=rep1&type=pdf (accessed on 21 January 2019).

33. Vinogradova, I.; Podvezko, V.; Zavadskas, E.K. The recalculation of the weights of criteria in MCDM methods using the Bayes approach. *Symmetry* **2018**, *10*, 205. [CrossRef]

34. Kemeny, J.G.; Snell, J.L. *Mathematical Models in the Social Sciences*; MIT Press Classic: New York, NY, USA, 1963.

35. Lorbeer, B.; Kosareva, A.; Deva, B.; Softić, D.; Ruppel, P.; Küpper, A. Variations on the Clustering Algorithm BIRCH. *Big Data Res.* **2017**, *11*, 44–53. [CrossRef]

36. Feldman, R.; Sanger, J. *The Text Mining Handbook: Advanced Approaches in Analyzing Unstructured Data*; Cambridge University Press: Cambridge, UK, 2007; Available online: https://wtlab.um.ac.ir/images/e-library/text_mining/The%20Text%20Mining%20HandBook.pdf (accessed on 21 January 2019).

37. Weiss, S.M.; Indurkhya, N.; Zhang, T.; Damerau, F.J. *Text Mining: Predictive Methods for Analyzing Unstructured Information*; Springer: Berlin/Heidelberg, Germany, 2005.

38. Boulton, F.; Louise, N. Can the health of a nation be correlated to its state of internal peace? *Med. Confl. Surviv.* **2016**, *32*, 70–79. [CrossRef] [PubMed]

39. Menshikov, V.; Volkova, O.; Stukalo, N.; Simakhova, A. Social economy as a tool to ensure national security. *J. Secur. Sustain. Issue* **2017**, *2*, 11–231. [CrossRef]

40. Lowatcharina, G.; Stallmann, J.I. The differential effects of decentralization on policeintensity: A cross-national comparison. *Soc. Sci. J.* **2018**. [CrossRef]

Article

Homogeneous Groups of Actors in an AHP-Local Decision Making Context: A Bayesian Analysis

Alfredo Altuzarra, Pilar Gargallo, José María Moreno-Jiménez * and Manuel Salvador

Grupo Decisión Multicriterio Zaragoza (GDMZ), Facultad de Economía y Empresa, Universidad de Zaragoza, Gran Vía 2, 50005 Zaragoza, Spain; altuzarr@unizar.es (A.A.); pigarga@unizar.es (P.G.); salvador@unizar.es (M.S.)
* Correspondence: moreno@unizar.es

Received: 15 February 2019; Accepted: 16 March 2019; Published: 21 March 2019

Abstract: The two procedures traditionally followed for group decision making with the Analytical Hierarchical Process (AHP) are the Aggregation of Individual Judgments (AIJ) and the Aggregation of Individual Priorities (AIP). In both cases, the geometric mean is used to synthesise judgments and individual priorities into a collective position. Unfortunately, positional measures (means) are only representative if dispersion is reduced. It is therefore necessary to develop decision tools that allow: (i) the identification of groups of actors that present homogeneous and differentiated behaviours; and, (ii) the aggregation of the priorities of the near groups to reach collective positions with the greatest possible consensus. Following a Bayesian approach to AHP in a local context (a single criterion), this work proposes a methodology to solve these problems when the number of actors is not high. The method is based on Bayesian comparison and selection of model tools which identify the number and composition of the groups as well as their priorities. This information can be very useful to identify agreement paths among the decision makers that can culminate in a more representative decision-making process. The proposal is illustrated by a real-life case study.

Keywords: Analytic Hierarchy Process (AHP); group decision making; homogeneous groups; Bayesian analysis

---

1. Introduction

Two of the most outstanding characteristics of the Knowledge Society (KS), understood as a space for the talent, imagination and creativity of human beings, are [1]: (i) the collaborative predisposition of citizens in the resolution of complex problems, motivated by the existence of a more educated and participative society that wants to be involved in decision-making processes; and, (ii) the relevance of the human factor and the need for the formal models to incorporate the subjective, intangible and emotional aspects inherent to the human being, alongside the tangible and rational objectives of the traditional scientific method.

In order to take advantage of the characteristics of the KS and to provide an adequate response to new challenges and needs, it is necessary to develop appropriate analytical and computing decision-making tools to solve the complex problems characterised by the existence of multiple scenarios, actors and both tangible and intangible criteria. One of the most utilised multicriteria techniques that best incorporates intangible aspects and multiple actors is the Analytic Hierarchy Process (AHP) proposed by Thomas L. Saaty in the mid-1970s [2]. The methodology consists of three stages: (i) Hierarchical modelling; (ii) Valuation; (iii) Prioritisation and Synthesis.

AHP incorporates the intangible through the judgments issued when assessing the matrices of paired comparisons considered in the problem. The two most commonly used methods in AHP for the calculus of collective priorities in multi-actor decision making are [3–5]: the Aggregation of Individual Judgments (AIJ) and the Aggregation of Individual Priorities (AIP). They obtain, as averages (geometric mean), the collective judgments or collective priorities. The average is not a representative indicator of the collective when it is not homogeneous. It is therefore necessary to evaluate the compatibility (an objective measure obtained automatically) or the agreement (a subjective measure that requires personal intervention) between the individual positions of the actors and the position of the group.

If not all actors are compatible with the collective priority vector, the different positions and the actors that support them must be identified in order to initiate posterior negotiation processes to achieve final decisions that are as representative as possible. This paper presents a Bayesian procedure to solve this problem in a local context (a single criterion). The procedure adopts the Bayesian hierarchical approach (Stochastic AHP) proposed in reference [6]. It allows the estimation of the priorities of the group by incorporating restrictions on the maximum level of inconsistency of the actors involved in the problem. The procedure quantifies the homogeneity of the groups by means of the marginal density of the judgments elicited by their respective actors. The density evaluates the goodness of fit of the model and allows for comparisons between different partitions of the set of actors. The work further puts forward an algorithm for the exhaustive search of homogeneous groups with respect to their priorities based on the Bayesian comparison and selection of models. The methodology is illustrated by a practical example.

The paper is structured as follows: Section 2 presents the model used to determine the priorities of a group of homogeneous decision makers; Section 3 describes the algorithm that identifies decision groups with homogeneous preferences; Section 4 applies the methodology to a case study; and, Section 5 concludes by highlighting the most relevant aspects of the work and possible extensions.

2. Bayesian Local Priorities in an AHP-Multi-Actor Decision Making Context

This section deals with the problem of determining the total priorities of a group of actors with homogeneous opinions regarding a decision criterion. A Bayesian statistical approach based on the use of a log-linear model similar to that used in reference [6] is employed to describe the process of the issuing judgments by the decision makers of a group. The posterior distribution of the group's priorities is calculated by means of Bayes' Theorem; this is followed by a description of how to make inferences about the most preferred alternative (alpha problem—P.α) and the priority ranking (gamma problem—P.γ).

2.1. Problem Formulation

First, the log-linear model that is used to determine the priorities of the groups of decision makers is explained: in what follows, N (μ, σ) denotes the univariate normal distribution of mean μ and standard deviation σ; N_p (μ, Σ) denotes the p-variant normal distribution of mean vector μ and the matrix of variances and covariances Σ; T_p (μ, Σ, υ) denotes the p-variant Student t distribution with mean vector μ, scale matrix Σ and degrees of freedom υ; Gamma(p, a) denotes the gamma distribution with shape parameter p and scale parameter $1/a$; χ^2_υ denotes the chi-squared distribution with υ degrees of freedom; I_A denotes the indicator function of set A; $\propto$ indicates 'proportional to' and [Y | X] denotes the density function of the conditional distribution of Y given X.

Let G = $\{D^{[1]}, \ldots, D^{[K]}\}$ be a group of K homogeneous decision makers (k = 1, $\ldots$, K), A = $\{A_1,$ $\ldots$, $A_n\}$ be a set of n alternatives and $\mathbf{R}^{(k)} = \left(r^{(k)}_{ij}\right)$; k = 1, $\ldots$, K be the nxn paired comparison matrices issued by each decision maker.

We assume, without loss of generality, that the matrices of judgments are complete—all paired comparisons have been made. If some of the r_{ij} comparisons are missing, the proposed methodology could be analogously adapted, as shown in reference [6].

We further assume that the decision makers of **G** have homogeneous opinions regarding the priorities of each of the alternatives of **A** so that:

$$y^{(k)}_{ij} = \mu^{(G)}_i - \mu^{(G)}_j + \varepsilon^{(k)}_{ij}; k = 1,\ldots,K; 1 \leq i < j \leq n; \tag{1}$$

with $y_{ij}^{(k)} = \log\left(r_{ij}^{(k)}\right)$, where:

(a) $\mu_i^{(G)} = \log\left(v_i^{(G)}\right)$; $i = 1, \ldots, n$ being $v_i^{(G)}$ the priority (without normalising) given to the alternative A_i by the members of the group $\mathbf{G}$

(b) $v_n^{(G)} = 1$ (that is to say, $\mu_n^{(G)} = 0$) to avoid identifiability problems

(c) $\varepsilon_{ij}^{(k)} \sim N\left(0, \sigma^{(G)}\right)$; $k = 1, \ldots, K$; $1 \leq i < j \leq n$ independents

The normalised priorities of the group will be given by the vector $\mathbf{w}^{(G)} = \left(w_i^{(G)}; i = 1, \ldots, n\right)'$ where $w_i^{(G)} = \dfrac{v_i^{(G)}}{\sum\limits_{i=1}^{n} v_i^{(G)}}$; $i = 1, \ldots, n$. Likewise, the level of homogeneity will be determined by the standard deviation of the errors $\sigma^{(G)}$ that quantifies the inconsistency level of all decision makers in the group with the priority vector $\mathbf{w}^{(G)}$.

2.2. Estimation of the Local Priorities for the Group

The estimation of the vector of group priorities $\mathbf{w}^{(G)}$ uses a Bayesian approach that allows exact inferences about their values. We adopt, as a prior, a normal-gamma distribution given by:

$$\boldsymbol{\mu}^{(G)} = \left(\mu_1^{(G)}, \ldots, \mu_{n-1}^{(G)}\right)' \big| \tau^{(G)} \sim N_{n-1}\left(0, \frac{1}{c_0 \tau^{(G)}} I_{n-1}\right) \text{ with } c_0 > 0 \tag{2}$$

$$\tau^{(G)} = \frac{1}{\sigma^{(G)2}} \sim \text{Gamma}\left(\frac{n_0}{2}, \frac{n_0 s_0^2}{2}\right) \tag{3}$$

that is the standard conjugate distribution used in Bayesian literature [7]. The constants c_0, n_0 and s_0^2 determine the degree of strength of the prior distribution. In the illustrative example we have taken $c_0 = 0.1$ so that the influence of the prior distribution of $\boldsymbol{\mu}^{(G)}$ is not significant. The hyper-parameters n_0 and s_0^2 are determined from the maximum levels of inconsistency σ_{max}^2 allowed for each decision maker so that

$$P\left[0 \leq \sigma^{(G)2} \leq \sigma_{max}^2\right] = 1 - \alpha$$

being, $1 - \alpha$ ($0 < \alpha < 1$) the level of credibility that we want to achieve. The value of σ_{max}^2 has been set using the consistency thresholds of the geometric consistency index (GCI) proposed by [8]. In our illustrative example, and given that $n = 4$, we take $\sigma_{max}^2 = 0.35$ and $\alpha = 0.05$, which resulted in $n_0 = 0.1$ and $s_0^2 = 0.0014$.

Using Bayes' theorem, and taking into account (1)–(3), we calculate the posterior distribution of $(\boldsymbol{\mu}^{(G)}, \tau^{(G)})$ whose density is given by:

$$\left[\boldsymbol{\mu}^{(G)} \Big| \tau^{(G)}, \left\{y^{(k)}; k \in \{1, \ldots, K\}\right\}\right] \propto \prod_{1 \leq i < j \leq n} \left[y_{ij}^{(k)} \Big| \boldsymbol{\mu}^{(G)}, \tau^{(G)}\right]\left[\boldsymbol{\mu}^{(G)} \Big| \tau^{(G)}\right]\left[\tau^{(G)}\right] \propto$$

$$\propto \prod_{k=1}^{K} \prod_{1 \leq i < j \leq n} \left(\tau^{(G)}\right)^{\frac{1}{2}} \exp\left[-\frac{\tau^{(G)}}{2}\left(y_{ij}^{(k)} - \mu_i^{(G)} + \mu_j^{(G)}\right)^2\right]\left(\tau^{(G)}\right)^{\frac{n-1}{2}} \exp\left[-\frac{c_0 \tau^{(G)}}{2}\left(\sum_{i=1}^{n-1}\left(\mu_i^{(G)}\right)^2\right)\right]$$

$$\times \left(\tau^{(G)}\right)^{\frac{n_0}{2} - 1} \exp\left[-\frac{\tau^{(G)}}{2} n_0 s_0^2\right] I_{(0,\infty)}\left(\tau^{(G)}\right) = \tag{4}$$

$$= \left(\tau^{(G)}\right)^{\frac{JK+n-1+n_0}{2} - 1} \exp\left[-\frac{\tau^{(G)}}{2}\left[n_0 s_0^2 + \sum_{k=1}^{K} \sum_{1 \leq i < j \leq n}\left(y_{ij}^{(k)} - \mu_i^{(G)} + \mu_j^{(G)}\right)^2 + c_0 \left(\sum_{i=1}^{n-1}\left(\mu_i^{(G)}\right)^2\right)\right]\right] I_{(0,\infty)}\left(\tau^{(G)}\right)$$

$$= \left(\tau^{(G)}\right)^{\frac{JK+n-1+n_0}{2} - 1} \exp\left[-\frac{\tau^{(G)}}{2}\left[n_0 s_0^2 + \sum_{k=1}^{K}\left(y^{(k)} - X\boldsymbol{\mu}^{(G)}\right)'\left(y^{(k)} - X\boldsymbol{\mu}^{(G)}\right) + c_0 \left(\boldsymbol{\mu}^{(G)'}\boldsymbol{\mu}^{(G)}\right)\right]\right] I_{(0,\infty)}\left(\tau^{(G)}\right)$$

where $\mathbf{y}^{(k)} = \left(y_{ij}^{(k)}; 1 \leq i < j \leq n\right)'$ for $k = 1, \ldots, K$ and $X = (x_{ij})$ ($J \times (n - 1)$) with $J = \frac{n(n-1)}{2}$ is the regression matrix of model (1) so that:

- $x_{ij} = 1$ if the ith judgement is y_{jk} with $k \neq j$;

- $x_{ij} = -1$ if the ith judgement is y_{kj} with $k \neq j$;
- $x_{ij} = 0$ in any other case.

It follows that:

$$\left[\mu^{(G)}\Big|\tau^{(G)}, \left\{\mathbf{y}^{(k)}; k \in \{1,\ldots,K\}\right\}\right] \propto$$

$$\propto \left(\tau^{(G)}\right)^{\frac{JK+n_0}{2}-1} \exp\left[-\frac{\tau^{(G)}}{2}\left\{n_0 s_0^2 + \sum_{k=1}^{K} \mathbf{y}^{(k)\prime}\mathbf{y}^{(k)} - \mathbf{m}^{(g)\prime}(K(\mathbf{X}'\mathbf{X}) + c_0 I_{n-1})\mathbf{m}^{(g)}\right\}\right] \times \tag{5}$$

$$\times \left(\tau^{(G)}\right)^{\frac{n-1}{2}-1} \exp\left[-\frac{\tau^{(G)}}{2}\left\{\left(\mu^{(G)} - \mathbf{m}^{(G)}\right)'(K(\mathbf{X}'\mathbf{X}) + c_0 I_{n-1})\left(\mu^{(G)} - \mathbf{m}^{(G)}\right)\right\}\right] I_{(0,\infty)}\left(\tau^{(G)}\right)$$

where $\mathbf{m}^{(G)} = (K(\mathbf{X}'\mathbf{X}) + c_0 I_{n-1})^{-1}\left(\mathbf{X}'\left(\sum_{k=1}^{K}\mathbf{y}^{(k)}\right)\right)$.

Therefore, it follows from (5) that:

$$\mu^{(G)}\big|\tau^{(G)}, \left\{\mathbf{y}^{(k)}; k \in \{1,\ldots,K\}\right\} \sim N_{n-1}\left(\mathbf{m}^{(G)}, \frac{1}{\tau^{(G)}}(K(\mathbf{X}'\mathbf{X}) + c_0 I_{n-1})^{-1}\right) \tag{6}$$

$$\tau^{(G)}\big|\left\{\mathbf{y}^{(k)}; k \in \{1,\ldots,K\}\right\} \sim \text{Gamma}\left(\frac{n_0 + JK}{2}, \left(\frac{n_0 + JK}{2}\right)s^{(G)2}\right) \tag{7}$$

where $s^{2(G)} = \dfrac{n_0 s_0^2 + \sum_{k=1}^{K} y^{(k)\prime}y^{(k)} - \mathbf{m}^{(G)\prime}(K(\mathbf{X}'\mathbf{X}) + c_0 I_{n-1})\mathbf{m}^{(G)}}{n_0 + JK}$.

Integrating with respect to $\tau^{(G)}$ in (5), it follows that:

$$\mu^{(G)}\big|\left\{\mathbf{y}^{(k)}; k \in \{1,\ldots,K\}\right\} \sim T_{n-1}\left(\mathbf{m}^{(G)}, s^{(G)2}(K(\mathbf{X}'\mathbf{X}) + c_0 I_{n-1})^{-1}, n_0 + JK\right) \tag{8}$$

From the posterior distributions (7) and (8), point estimates and credibility intervals of $\mathbf{w}^{(G)}$ and $\sigma^{(G)}$ can be obtained using the posterior median and the corresponding quantiles.

In the case of $\sigma^{2(G)}$, and taking into account that from (7) $\tau^{(G)} \sim \dfrac{\chi^2_{n_0+JK}}{(n_0+JK)s^{(G)2}}$, a $100(1-\alpha)\%$, the Bayesian credibility interval for $\sigma^{2(G)}$ is given by $\left[\dfrac{s^{2(G)}(n_0+JK)}{\chi^2_{n_0+JK,1-\frac{\alpha}{2}}}, \dfrac{s^{2(G)}(n_0+JK)}{\chi^2_{n_0+JK,\frac{\alpha}{2}}}\right]$ where $\chi^2_{\nu,\alpha}$ denotes the $(1-\alpha)$th quantile of the distribution χ^2_ν.

To calculate a credibility interval for $w_i^{(G)}$ ($1 \leq i \leq n$) the Monte Carlo method is applied by extracting a sample $\left\{\mu^{(G,s)} = \left(\mu_1^{(G,s)},\ldots,\mu_{n-1}^{(G,s)}\right)'; s = 1,\ldots,S\right\}$ from (8) and calculating a sample of the posterior distribution of $\mathbf{w}^{(G)}$, $\mathbf{W}^{(G)} = \left\{\mathbf{w}^{(G,s)} = \left(w_1^{(G,s)},\ldots,w_n^{(G,s)}\right)'; s = 1,\ldots,S\right\}$ with $w_i^{(G,s)} = \dfrac{\exp\left[\mu_i^{(G,s)}\right]}{\sum_{j=1}^{n}\exp\left[\mu_j^{(G,s)}\right]}$ where $\mu_n^{(G,s)} = 0$. From this sample a credibility interval for $w_i^{(G)}$ is given by $\left[w_i^{(G)}\left(\frac{\alpha}{2}\right), w_i^{(G)}\left(1-\frac{\alpha}{2}\right)\right]$ where $w_i^{(G)}(\alpha)$ is the αth quantile of the sample $\mathbf{W}^{(G)}$.

Alpha distributions could also be calculated $\mathbf{P}_\alpha^{\mathbf{G}_g} = \left(P_{\alpha,1}^{\mathbf{G}_g},\ldots,P_{\alpha,n}^{\mathbf{G}_g}\right)$ with:

$$P_{\alpha,i}^{\mathbf{G}} = P\left[w_i^{(G)} = \max_{1 \leq j \leq n} w_j^{(G)}\Big|\left\{\mathbf{y}^{(k)}; k = 1,\ldots,K\right\}\right]; i = 1,\ldots,n \tag{9}$$

and gamma distributions with:

$$P_{\gamma,\gamma_h}^{\mathbf{G}} = P\left[w_{\gamma_{h,1}}^{(G)} \leq \cdots \leq w_{\gamma_{h,n}}^{(G)}\Big|\left\{\mathbf{y}^{(k)}; k = 1,\ldots,K\right\}\right]; h = 1,\ldots,n! \tag{10}$$

where $\gamma_h = (\gamma_{h,1}, \ldots, \gamma_{h,n})$ is the hth permutation of the elements of A sorted according to the lexicographical order. The approximate calculation of these probabilities is from the sample $\mathbf{W}^{(G)}$ by means of the expressions:

$$\hat{P}^G_{\alpha,i} = \frac{1}{S}\sum_{s=1}^{S} I_{w_i^{(G,s)}=\max_{1\le j\le n} w_j^{(G,s)}}(s) \tag{11}$$

$$\hat{P}^G_{\gamma,\gamma_h} = \frac{1}{S}\sum_{s=1}^{S} I_{w_{\gamma_{h,1}}^{(G,s)}\le\cdots\le w_{\gamma_{h,n}}^{(G,s)}}(s) \tag{12}$$

These distributions report on preferences as well as the most preferred alternative and ranking for the group.

3. Identification of Homogeneous Groups of Actors

The procedure for estimating the priorities of a group, as detailed in the previous section, is based on the hypothesis of the similarity of the opinions of the decision makers. However, it is quite possible that there will be different opinions. In this case, and in order to facilitate a subsequent negotiation process, it is useful to identify the different opinions within the group and the actors that support them; this section presents a systematic procedure for doing this. It utilises a Bayesian oriented tool selection model based on the use of the Bayes factor as a selection element.

3.1. Problem Formulation

Let $\mathbf{D} = \{D^{[1]}, \ldots, D^{[K]}\}$ be a group of K decision makers, $\mathcal{G} = \{G_1, \ldots, G_m\}$ be a partition of $\mathbf{D}$, with $G_g = \left\{D^{[i_{g,1}]}, \ldots, D^{[i_{g,n_g}]}\right\} \subseteq \mathbf{D}$; $g = 1, \ldots, m$, $G_g \cap G_{g'} = \varnothing$ if $g \ne g'$, $\mathbf{D} = \overset{m}{\underset{g=1}{\cup}} G_g$. To avoid identifiability problems we impose that $i_{g,1} < \ldots < i_{g,n_g}$ and $i_{g,1} < i_{g',1}$ if $g < g'$.

The problem is to select the $\mathcal{G}$ partitions that best describe the opinions expressed by the decision makers about the alternative to be chosen from the judgments issued $Y = \left\{y^{(k)}; k = 1, \ldots, K\right\}$. To this end we extend the approach made in the previous section to the case of several groups, assuming that the decision makers of each group $\{G_g; g = 1, \ldots, m\}$ of the partition $\mathcal{G}$ have homogeneous opinions regarding the priorities of each alternative of set $\mathbf{A}$ so that:

$$y_{ij}^{(k)} = \mu_i^{(g(k))} - \mu_j^{(g(k))} + \varepsilon_{ij}^{(k)} \text{ with } \varepsilon_{ij}^{(k)} \sim N\left(0, \sigma^{g(k)}\right); k = 1, \ldots, K; 1 \le i < j \le n \tag{13}$$

(i) $D^{[k]} \in G_{g(k)}$ with $g(k) \in \{1, \ldots, m\}$ the group to which the decision maker $D^{[k]}$ belongs

(ii) $\mu_i^{(g)} = \log\left(v_i^{(g)}\right); i = 1, \ldots, n$ being $v_i^{(g)}$ the priority (without normalising) given to the alternative A_i by the members of the group G_g

(iii) $v_n^{(g)} = 1$ (that is to say, $\mu_n^{(g)} = 0$) to avoid identifiability problems

(iv) $\varepsilon_{ij}^{(k)} \sim N\left(0, \sigma^{g(k)}\right); k = 1, \ldots, K; 1 \le i < j \le n$ and independent

Finally, we take the following prior distributions for the parameters of the normal-gamma model:

$$\mu^{(g)} = \left(\mu_1^{(g)}, \ldots, \mu_{n-1}^{(g)}\right)' | \tau^{(g)} \sim N_{n-1}\left(0, \frac{1}{c_0\tau^{(g)}}I_{n-1}\right) \text{ with } c_0 > 0 \tag{14}$$

$$\tau^{(g)} = \frac{1}{\sigma^{2(g)}} \sim \text{Gamma}\left(\frac{n_0}{2}, \frac{n_0 s_0^2}{2}\right) \tag{15}$$

independents for $g = 1, \ldots, m$.

3.2. Goodness of Fit Evaluation of $\mathcal{G}$

The selection of the best $\mathcal{G}$ partitions is made by an evaluation of their adjustment to the $\mathbf{Y}$ issued judgments. We use $[\mathbf{Y} \mid \mathcal{G}]$, the prior marginal density of the model (13)–(15), which is one of the tools utilised in the Bayesian inference to quantify it, so that, the higher the value, the greater the degree of fit of $\mathcal{G}$ as a description of the existing opinions in $\mathbf{D}$ and the greater is the explanatory power of the $\mathbf{Y}$ issued judgments.

This density is given by:

$$[\mathbf{Y}|\mathcal{G}] = \prod_{g=1}^{m}\left[\left\{\mathbf{y}^{(k)};k:g(k)=g\right\}|G^g\right]$$

$$= \prod_{g=1}^{m}\int_0^\infty \int_{\mathbb{R}^{n-1}}\left[\left\{\mathbf{y}^{(k)};k:g(k)=g\right\}|\mu^{(g)},\tau^{(g)}\right]\left[\mu^{(g)}|\tau^{(g)}\right]\left[\tau^{(g)}\right]d\mu^{(g)}d\tau^{(g)} =$$

$$= \prod_{g=1}^{m}\int_0^\infty\left[\left\{\mathbf{y}^{(k)};k:g(k)=g\right\}|\tau^{(g)}\right]\left[\tau^{(g)}\right]d\tau^{(g)}$$

(16)

Taking into account that:

$$\left[\left\{\mathbf{y}^{(k)};k:g(k)=g\right\}|\tau^{(g)}\right] = \frac{\left[\left\{\mathbf{y}^{(k)};k:g(k)=g\right\},\mu^{(g)},\tau^{(g)}\right]}{\left[\mu^{(g)}\left|\left\{\mathbf{y}^{(k)};k:g(k)=g\right\},\tau^{(g)}\right.\right]}$$

from (6), (13), (14) and (15) it follows that

$$\left[\left\{\mathbf{y}^{(k)};k:g(k)=g\right\}\left|\tau^{(g)}\right.\right] =$$

$$\frac{\left(\tau^{(g)}\right)^{\frac{Jn_g+n-1+n_0}{2}-1}\exp\left[-\frac{\tau^{(g)}}{2}\left[n_0s_0^2+\sum_{k:g(k)=g}\left(\mathbf{y}^{(k)}-\mathbf{X}\mu^{(g)}\right)'\left(\mathbf{y}^{(k)}-\mathbf{X}\mu^{(g)}\right)+c_0\left(\mu^{(g)\prime}\mu^{(g)}\right)\right]\right]I_{\mathbb{R}^{n-1}}\left(\mu^{(g)}\right)I_{(0,\infty)}\left(\tau^{(g)}\right)}{\left(\tau^{(g)}\right)^{\frac{n-1}{2}}\left|\left(n_g\left(\mathbf{X}'\mathbf{X}\right)+c_0I_{n-1}\right)\right|^{\frac{1}{2}}\exp\left[-\frac{\tau^{(g)}}{2}\left(\mu^{(g)}-\mathbf{m}^{(g)}\right)'\left(n_g\left(\mathbf{X}'\mathbf{X}\right)+c_0I_{n-1}\right)\left(\mu^{(g)}-\mathbf{m}^{(g)}\right)\right]I_{\mathbb{R}^{n-1}}\left(\mu^{(g)}\right)}$$

$$\times\frac{(2\pi)^{-\frac{Jn_g+n-1}{2}}\left(\frac{n_0s_0^2}{2}\right)^{\frac{n_0}{2}}c_0^{\frac{n-1}{2}}}{(2\pi)^{-\frac{n-1}{2}}\Gamma\left(\frac{n_0}{2}\right)}$$

(17)

where $\mathbf{m}^{(G)} = \left(n_g\left(\mathbf{X}'\mathbf{X}\right)+c_0I_{n-1}\right)^{-1}\left(\mathbf{X}'\left(\sum_{k:g(k)=g}\mathbf{y}^{(k)}\right)\right)$. It follows that

$$\left[\left\{\mathbf{y}^{(k)};k:g(k)=g\right\}\left|\tau^{(g)}\right.\right] =$$

$$\frac{\left(\tau^{(g)}\right)^{\frac{Jn_g+n_0}{2}-1}\exp\left[-\frac{\tau^{(g)}}{2}\left[n_0s_0^2+\left(\mu^{(g)}-\mathbf{m}^{(g)}\right)'\left(n_g\left(\mathbf{X}'\mathbf{X}\right)+c_0I_{n-1}\right)\left(\mu^{(g)}-\mathbf{m}^{(g)}\right)\right]\right]}{\left|\left(n_g\left(\mathbf{X}'\mathbf{X}\right)+c_0I_{n-1}\right)\right|^{\frac{1}{2}}\exp\left[-\frac{\tau^{(g)}}{2}\left(\mu^{(g)}-\mathbf{m}^{(g)}\right)'\left(n_g\left(\mathbf{X}'\mathbf{X}\right)+c_0I_{n-1}\right)\left(\mu^{(g)}-\mathbf{m}^{(g)}\right)\right]}\times$$

$$\times\exp\left[-\frac{\tau^{(g)}}{2}\left\{\sum_{k:g(k)=g}\mathbf{y}^{(k)\prime}\mathbf{y}^{(k)}-\mathbf{m}^{(g)\prime}\left(n_g\left(\mathbf{X}'\mathbf{X}\right)+c_0I_{n-1}\right)\mathbf{m}^{(g)}\right\}\right]\frac{(2\pi)^{-\frac{Jn_g}{2}}\left(\frac{n_0s_0^2}{2}\right)^{\frac{n_0}{2}}c_0^{\frac{n-1}{2}}}{\Gamma\left(\frac{n_0}{2}\right)}I_{(0,\infty)}\left(\tau^{(g)}\right) =$$

(18)

$$\frac{\left(\frac{n_0s_0^2}{2}\right)^{\frac{n_0}{2}}\left(\tau^{(g)}\right)^{\frac{n_0+Jn_g}{2}-1}}{(2\pi)^{\frac{Jn_g}{2}}\Gamma\left(\frac{n_0}{2}\right)\left|\left(\frac{n_g}{c_0}\left(\mathbf{X}'\mathbf{X}\right)+I_{n-1}\right)\right|^{\frac{1}{2}}}\exp\left[-\frac{\tau^{(g)}}{2}Q^{(g)}\right]I_{(0,\infty)}\left(\tau^{(g)}\right)$$

where $Q^{(g)} = n_0s_0^2 + \sum_{k:g(k)=g}\mathbf{y}^{(k)\prime}\mathbf{y}^{(k)} - \mathbf{m}^{(g)\prime}\left(n_g\left(\mathbf{X}'\mathbf{X}\right)+c_0I_{n-1}\right)\mathbf{m}^{(g)}$.

Substituting in (16) it follows that:

$$
\begin{aligned}
[\mathbf{Y}|\mathcal{G}] &= \frac{\left(\frac{n_0 s_0^2}{2}\right)^{\frac{n_0}{2}}}{(2\pi)^{\frac{Jng}{2}}\Gamma\left(\frac{n_0}{2}\right)\left|\left(\frac{n_g}{c_0}(\mathbf{X}'\mathbf{X})+I_{n-1}\right)\right|^{\frac{1}{2}}}\prod_{g=1}^{m}\int_0^\infty \left(\tau^{(g)}\right)^{\frac{n_0+Jng}{2}-1}\exp\left[-\frac{\tau^{(g)}}{2}Q^{(g)}\right]d\tau^{(g)} \\[2mm]
&= \frac{\left(\frac{n_0 s_0^2}{2}\right)^{m\frac{n_0}{2}}}{\left(\Gamma\left(\frac{n_0}{2}\right)\right)^m}\prod_{g=1}^{m}\Gamma\left(\frac{n_0+Jng}{2}\right)\left|\left(\frac{n_g}{c_0}(\mathbf{X}'\mathbf{X})+I_{n-1}\right)\right|^{-\frac{1}{2}}\left(\frac{2}{Q^{(g)}}\right)^{\frac{n_0+Jng}{2}} \propto \\[2mm]
&\propto \frac{\left(\frac{n_0 s_0^2}{2}\right)^{m\frac{n_0}{2}}}{\left(\Gamma\left(\frac{n_0}{2}\right)\right)^m}\prod_{g=1}^{m}\frac{\Gamma\left(\frac{n_0+Jng}{2}\right)\left|\left(\frac{n_g}{c_0}(\mathbf{X}'\mathbf{X})+I_{n-1}\right)\right|^{-\frac{1}{2}}}{\left(Q^{(g)}\right)^{\frac{n_0+Jng}{2}}}
\end{aligned}
\tag{19}
$$

3.3. Location of Opinion Groups

Now that the evaluation of the adjustment of a $\mathcal{G}$ partition to the issued judgements is completed, in this section we describe the process followed to determine the most representative partitions. We use Bayesian selection models and the Bayes factor, $\frac{[\mathbf{Y}|\mathcal{G}']}{[\mathbf{Y}|\mathcal{G}]}$ as a tool of comparison of two elements $\mathcal{G}$ and $\mathcal{G}'$ of $\wp(\mathbf{D})$, the set of possible partitions of $\mathbf{D}$.

We set a threshold β $(0 < \beta < 1)$ to discriminate if there are significant differences in the data adjustment of the partitions $\mathcal{G}$ and $\mathcal{G}'$ so that if $\frac{[\mathbf{Y}|\mathcal{G}']}{[\mathbf{Y}|\mathcal{G}]} < \beta$ then the degree of fit of $\mathcal{G}'$ is significantly worse than that of $\mathcal{G}$ and, therefore, $\mathcal{G}$ is more representative than $\mathcal{G}'$. In this case, and in line with [9], we take $\beta = 0.05$.

The problem is to determine $\mathcal{G} \in \wp(\mathbf{D})$ so that:

$$
\frac{[\mathbf{Y}|\mathcal{G}]}{[\mathbf{Y}|\mathcal{G}_{\max}]} \geq \beta
\tag{20}
$$

where $[\mathbf{Y}|\mathcal{G}_{\max}] = \max_{\mathcal{G}\in\wp(\mathbf{D})}[\mathbf{Y}|\mathcal{G}]$ and gives us the 'Occam's window' of our problem [10]. The partitions could be taken as starting points for subsequent negotiation processes in order to reach an agreement among the decision makers that is as representative as possible. In our case, we look for the partitions of the window that have the least number of groups, since it can be foreseen that the fewer groups there will be, the easier it will be to reach more representative agreements because there are fewer disparate opinions. In order to do this, we use an exhaustive search algorithm that calculates the values of $[\mathbf{Y}\,|\,\mathcal{G}]$ for all the elements of $\wp(\mathbf{D})$ using expression (19). Then $\mathcal{G}_{\max}$ is determined and, from this, the partitions of Occam's window that verify (20) are identified. Other methods for consensus searching in group decision making can be seen in references [11–16].

Figure 1 shows the main steps for determining the groups with homogeneous opinions.

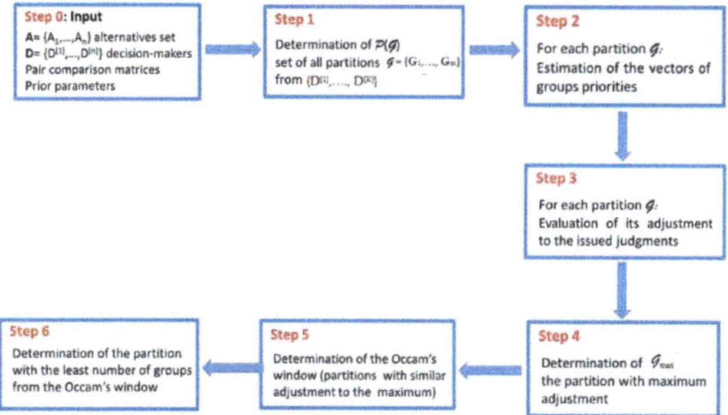

Figure 1. Steps of the proposed methodology.

4. Case Study

In September 2006 the Government of Aragón and the Zaragoza City Council advanced the 'Zaragoza Plan for Sustainable Mobility', which led to the construction of the city's Tram Line 1 that was completed in 2013. The construction was controversial and in the following municipal elections, the political parties presented their proposals for improving transport in the city. Representatives of the main political parties (PSOE, PP and CHA) attended the university to explain their preferred alternative. The case study concerns a citizen participation project in a local community which contemplated alternatives put forward by the political parties during the municipal elections for the extension of the city tram network.

Eleven students ($K = 11$) from the 'Electronic-Government and Public Decisions' course of the Faculty of Economics and Business at the University of Zaragoza (Spain) were involved in its implementation.

There were 4 alternatives:

A1 Build a new tram line
A2 Use a tram and bus combination called Tranbus
A3 Use a tram combination with commuter lines
A4 Do nothing

As the construction of Line 1 had a high economic cost, and the selection of any of first three alternatives assumes a significant investment, a fourth alternative (no cost) was included. The selection problem was solved using the Analytic Hierarchy Process [2]. The hierarchical model comprised four levels (the goal, 3 criteria, 9 attributes and 4 alternatives).

The results of the Investment Cost attribute have been used to illustrate the proposed methodology. The prior distribution parameters were $c_0 = 0.1$, $n_0 = 1$ and $s_0^2 = 0.0014$ (corresponding to take $\sigma_{max}^2 = 0.35$ and $\alpha = 0.05$) and $\beta = 0.05$. The number of partitions was equal to the 678,570 that were processed in 98.84 s of CPU by a Toshiba Ultrabook KIRA with Intel (R) Core™ i7-4510U CPU @ 2.00GHz 2.60 GHz (64 bits) and 8 Gb of RAM.

The resulting number of groups was equal to five and the most probable composition was: $G_1 = \{D_1, D_2, D_3, D_7\}$, $G_2 = \{D_4, D_{10}\}$, $G_3 = \{D_5, D_6\}$, $G_4 = \{D_9\}$, $G_5 = \{D_{11}\}$. The results obtained are shown in Tables 1–3. More specifically, Table 1 contains the posterior medians of the priorities of each alternative for each decision maker and each group. Table 2 shows the posterior probabilities that each alternative would be the most preferred, corresponding to the $P\alpha$ distributions. Table 3 gives the probabilities for each ranking corresponding to the $P\gamma$ distributions. The values were calculated from (7)–(8) and (11)–(12), as described in Section 2.2, using $S = 10,000$ simulations.

So, for example, group G1 made up of decision makers D_1, D_2, D_3 and D_7, gives the highest priority (0.4502) to alternative A_4, followed by the alternatives A_2 (priority 0.3076), A_3 (0.1518) and A_1 (0.0879) (see Table 1). This ranking is also reflected by its $P\alpha$ and $P\gamma$ distributions, which give the maximum posterior probabilities to the A_4 alternative (97.31%, see Table 2) and the ranking 4231 (96.94%, see Table 3). Even though the individual opinion of D_3 is different to the rest of the members of the group (their preferred alternative is A_2 and the ranking is 2431), the consistency of the group G_1 (0.2790) is good, being lower than the maximum level of inconsistency 0.35. This is due to the high priority of D_3 for alternative A_4 (0.2705) and the similarity of their priorities to alternatives A_1 and A_3 which means that G1 can be considered as a homogeneous group.

From the tables, it can be observed that the compositions of the groups are very much determined by their similarity to the most preferred alternative. The decision makers from groups G_1 and G_3 mostly prefer the alternative A_4, those from group G_2 prefer alternative A_1, those from G_3 prefer alternative A_3 and those from G_4 prefer alternative A_2 (see Tables 1 and 2). However, groups G_1 and G_3 differ in the rankings (see Table 3). The decision makers from the group G_1 tend to prefer the 4231 ranking, while those from the group G_3 prefer 4123. Nevertheless, preferences within each group are not completely homogeneous; in group G_1, decision-maker D_3 shows a greater preference for alternative A_2. This opinion is shared with decision maker D_{11} of group $\mathbf{G_4}$, although it assigns a priority 0.2705 to alternative A_4 that justifies its inclusion in group $\mathbf{G_1}$. Something similar happens in group $\mathbf{G_3}$, in which decision maker D_5 shows a greater preference for alternative A_1, although it assigns a non-negligible priority (0.3613) to alternative A_4 and to the ranking 4123, hence its inclusion in group $\mathbf{G_3}$.

The consistency levels of the actors and most of the groups are acceptable (<0.35). The only exception is G_3 whose consistency is estimated as 0.4071, but with a 95% credibility interval [0.20, 1.43] that does not reject the consistency hypothesis $\sigma^{2(G_3)} \leq 0.35$.

The ambiguities of the opinions are revealed if we analyse the partitions selected by Occam's window with the smaller number of groups, included in Table 4 and Figure 2. Figure 2 incorporates, in each node, the groups of decision makers that are classified together in all the selected partitions and includes a link between two nodes if their components are classified together in some of the partitions. Most of the decision makers are linked because of their preferences for alternatives A_2 and A_4, the latter being the alternative which most decision makers support. Only D_4, D_9 and D_{10} are isolated because of their preferences for alternatives A_1 (D_4 and D_{10}) and A_3 (D_9).

Table 1. Priorities and consistency for each decision maker and each group.

Decision Maker	Priorities				Consistency
	w_1	w_2	w_3	w_4	
D_1	0.1285	0.2860	0.1275	**0.4511**	0.1412
D_2	0.0899	0.2500	0.1418	**0.5110**	0.2042
D_3	0.0900	**0.4759**	0.1582	0.2705	0.1363
D_7	0.0587	0.2501	0.1839	**0.5026**	0.1405
G_1	0.0879	0.3076	0.1518	**0.4502**	0.2790
D_4	**0.5748**	0.1406	0.1086	0.1721	0.1214
D_{10}	**0.6318**	0.1809	0.0797	0.1044	0.1046
G_2	**0.6139**	0.1600	0.0926	0.1318	0.1602
D_5	**0.4021**	0.1570	0.0758	0.3613	0.0888
D_6	0.2131	0.1057	0.0542	**0.6194**	0.2849
D_8	0.1388	0.2763	0.0578	**0.5197**	0.2186
G_3	0.2321	0.1676	0.0617	**0.5338**	0.4071
D_9	0.2846	0.0871	**0.5503**	0.0720	0.2186
G_4	0.2846	0.0871	**0.5503**	0.0720	0.2186
D_{11}	0.1228	**0.4684**	0.2736	0.1294	0.1261
G_5	0.1228	**0.4684**	0.2736	0.1294	0.1261

in **bold** the highest priority.

Table 2. Alpha distributions for each decision maker and each group.

Decision Maker	Alternatives			
	A_1	A_2	A_3	A_4
D_1	0.30	6.11	0.00	**93.59**
D_2	0.19	2.71	0.04	**97.06**
D_3	0.00	**96.75**	0.00	3.25
D_7	0.01	1.75	0.04	**98.20**
G_1	0.00	2.69	0.00	**97.31**
D_4	**99.23**	0.03	0.00	0.74
D_{10}	**99.92**	0.02	0.00	0.06
G_2	**99.96**	0.00	0.00	0.04
D_5	**63.16**	0.01	0.00	**36.83**
D_6	5.24	0.00	0.00	**94.76**
D_8	1.09	3.92	0.00	**94.99**
G_3	2.96	0.00	0.00	**97.04**
D_9	5.49	0.00	**94.51**	0.00
G_4	5.49	0.00	**94.51**	0.00
D_{11}	0.01	**98.75**	1.20	0.04
G_5	0.01	**98.75**	1.20	0.04

in **bold** the probabilities higher than 20%.

Table 3. Gamma distributions for each decision maker (D_i) and each group (G_j) (probabilities higher than 20% are in bold).

Decision Maker	Rankings																							
	1234 [†]	1243	1324	1342	1432	1423	2134	2143	2314	2341	2431	2413	3214	3241	3124	3142	3412	3421	4231	4213	4321	4312	4132	4123
D_1	0.02	0.17	0.00	0.00	0.00	0.11	0.01	0.64	0.00	0.01	0.51	4.94	0.00	0.00	0.00	0.00	0.00	0.00	**49.20**	**43.98**	0.09	0.01	0.01	0.30
D_2	0.05	0.10	0.00	0.00	0.00	0.04	0.01	0.23	0.00	0.00	0.87	1.60	0.00	0.00	0.00	0.00	0.01	0.03	**86.05**	8.68	2.10	0.05	0.03	0.15
D_3	0.00	0.00	0.00	0.00	0.00	0.00	0.43	0.57	0.21	1.05	**91.49**	3.00	0.00	0.00	0.00	0.00	0.00	0.00	3.20	0.02	0.03	0.00	0.00	0.00
D_7	0.01	0.00	0.00	0.00	0.00	0.00	0.00	0.05	0.02	0.04	1.55	0.09	0.00	0.00	0.01	0.00	0.00	0.03	**89.78**	0.07	8.33	0.00	0.01	0.01
G_1	0.00	0.00	0.00	0.00	0.00	0.00	0.00	0.00	0.00	0.00	2.28	0.41	0.00	0.00	0.00	0.00	0.00	0.00	**96.94**	0.36	0.01	0.00	0.00	0.00
D_4	1.75	**20.98**	0.23	0.40	8.64	**67.23**	0.00	0.01	0.00	0.00	0.00	0.02	0.00	0.00	0.00	0.00	0.00	0.00	0.00	0.01	0.01	0.02	0.41	0.29
D_{10}	8.90	**88.84**	0.01	0.01	0.06	2.10	0.00	0.01	0.00	0.00	0.00	0.01	0.00	0.00	0.00	0.00	0.00	0.00	0.00	0.01	0.01	0.00	0.00	0.04
G_2	2.46	**80.42**	0.00	0.02	0.15	16.91	0.00	0.00	0.00	0.00	0.00	0.00	0.00	0.00	0.00	0.00	0.00	0.00	0.00	0.00	0.00	0.00	0.01	0.03
D_5	0.00	0.29	0.00	0.00	0.04	**62.83**	0.00	0.01	0.00	0.00	0.00	0.00	0.00	0.00	0.00	0.00	0.00	0.00	0.00	0.08	0.00	0.00	0.06	**36.69**
D_6	0.01	0.15	0.00	0.00	0.09	4.99	0.00	0.00	0.00	0.00	0.00	0.00	0.00	0.00	0.00	0.00	0.00	0.00	0.36	3.63	0.16	0.16	1.99	**88.46**
D_8	0.01	0.64	0.01	0.00	0.00	0.43	0.00	0.57	0.00	0.00	0.00	3.35	0.00	0.00	0.00	0.00	0.00	0.00	2.27	**90.83**	0.00	0.01	0.01	1.87
G_3	0.00	0.03	0.00	0.00	0.00	2.93	0.00	0.00	0.00	0.00	0.00	0.00	0.00	0.00	0.00	0.00	0.00	0.00	0.00	11.32	0.00	0.00	0.00	**85.72**
D_9	0.00	0.00	5.37	0.12	0.00	0.00	0.00	0.00	0.00	0.00	0.00	0.00	0.05	0.03	**66.57**	**26.61**	1.08	0.17	0.00	0.00	0.00	0.00	0.00	0.00
G_4	0.00	0.00	5.37	0.12	0.00	0.00	0.00	0.00	0.00	0.00	0.00	0.00	0.05	0.03	**66.57**	**26.61**	1.08	0.17	0.00	0.00	0.00	0.00	0.00	0.00
D_{11}	0.01	0.00	0.00	0.00	0.00	0.00	1.14	0.00	**43.39**	**53.94**	0.28	0.00	0.28	0.88	0.00	0.00	0.00	0.04	0.03	0.00	0.01	0.00	0.00	0.00
G_5	0.01	0.00	0.00	0.00	0.00	0.00	1.14	0.00	**43.39**	**53.94**	0.28	0.00	0.28	0.88	0.00	0.00	0.00	0.04	0.03	0.00	0.01	0.00	0.00	0.00

[†] 1234 denotes the ranking $A_1 > A_2 > A_3 > A_4$.

There are 4 homogeneous groups: $\{D_1, D_2, D_7\}$ that prefer alternative A_4 and the ranking 4231; $\{D_5, D_6\}$ position alternatives A_2 and A_3 in the last places and they show a non-negligible preference for A_4; $\{D_4, D_{10}\}$ prefer alternative A_1 and put A_3 in last place; decision maker $\{D_9\}$ prefers A_3. Decision makers D_3, D_8, D_{11} are in more intermediate and ambiguous positions. In the case of D_3, this is due to the greater preference for A_2 (shared with D_{11}) and the non-negligible preference for A_4 (which places them close to the group $\{D_1, D_2, D_7\}$). In the case of D_8, the intermediate position is due to their preferences for A_4 and A_2, in that order, which places them close to the group $\{D_1, D_2, D_7\}$, as well as to the rejection of A_3, which places them close to the group $\{D_5, D_6\}$.

In order to achieve as broad an agreement as possible, alternative A_4 could be suggested, given that a majority of decision makers (in groups G_1 and G_3) showed a preference for it. The negotiation should be aimed at convincing decision makers D_4, D_9, D_{10} and D_{11}.

With regards to the practical implications of the Bayesian procedure proposed in this work, it is worth mentioning that, as in AHP, these applications are numerous, especially in matters of strategic planning where the number of actors is not usually very high.

Table 4. Partitions selected by the Occam window.

Partition	D_1	D_2	D_3	D_4	D_5	D_6	D_7	D_8	D_9	D_{10}	D_{11}	Ratio [†]
1	1	1	1	2	3	3	1	3	4	2	5	1.00
2	1	1	2	3	4	4	1	4	5	3	2	0.48
3	1	1	1	2	3	3	1	1	4	2	5	0.42
4	1	1	2	3	4	4	1	1	5	3	2	0.15

[†] The ratio calculates the quotient of the posterior probability for the most probable model and that for the model corresponding to each partition.

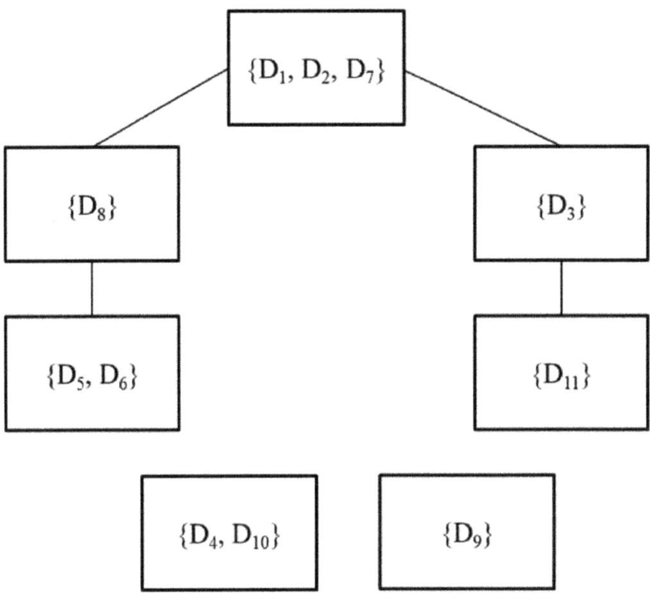

Figure 2. Relations of opinions existing in D.

5. Conclusions

This paper has proposed a methodology for the identification of homogenous opinion groups with AHP in a local context. The methodology is based on the use of Bayesian processes for the selection of hierarchical models that describe the judgments issued by each decision maker in their matrices of pairwise comparisons based on a set of priorities common to each of the members of the group.

Using an exhaustive search method of the most compatible partitions with the judgments issued, the Occam's window of the compared models is defined. From these models it has been shown how it is possible to describe the existing opinions in the groups, information that can be very useful to identify consensus paths among the decision makers that can culminate in a more representative decision-making process.

The search method works in a local AHP context, but has some limitations. First, it functions if the number of decision makers is not very high (≤ 11). The total number of partitions of the set of decision makers is equal to the Bell number $B_K = \sum_{k=0}^{K-1} \binom{K-1}{k} B_k$ with $B_0 = 1$, $B_1 = 1$. The larger the number of decision makers, the more computationally infeasible is the problem. In our case ($K = 11$), the number of possible partitions is 687,570, which is computationally feasible. For instance, if $K = 22$ the number is 4507×10^{15}, then it is necessary to use algorithms that approximately determine Occam's window.

We are currently experimenting with stochastic search algorithms and the results obtained will be published in a future work. A second limitation is that it is necessary that the groups constitute a partition of the set of decision makers and this implies that a decision maker cannot belong to more than one group. Even though this requirement decreases the computational time of the algorithm, it also reduces the flexibility of the method. The development of search strategies that eliminate this unrealistic assumption is worthy of consideration. Finally, it would be interesting to extend the methodology to a global context in which a hierarchy of criteria and sub-criteria is used.

Author Contributions: The paper has been elaborated jointly by the four authors.

Funding: This research was funded by the Spanish Ministry of Economy and Competitiveness and FEDER funding, projects ECO2015-66673-R (first and third authors) and ECO2016-79392-P (second and fourth authors).

Acknowledgments: The authors would like to acknowledge the work of English translation professional David Jones in preparing the final text.

Conflicts of Interest: The authors declare no conflict of interest.

References

1. Moreno-Jiménez, J.M.; Vargas, L.G. Cognitive Multicriteria Decision Making and the Legacy of AHP. *Estud. Econ. Apl.* **2018**, *36*, 67–80.
2. Saaty, T.L. *The Analytic Hierarchy Process*; Mc Graw-Hill: New York, NY, USA, 1980.
3. Saaty, T.L. Group decision making and the AHP. In *The Analytic Hierarchy Process*; Golden, B.L., Wasil, E.A., Harker, P.T., Eds.; Springer: Berlin/Heidelberg, Germany, 1989; pp. 59–67.
4. Ramanathan, R.; Ganesh, L.S. Group preference aggregation methods employed in AHP: An evaluation and intrinsic process for deriving members' weightages. *Eur. J. Oper. Res.* **1994**, *79*, 249–265. [CrossRef]
5. Forman, E.; Peniwati, K. Aggregating individual judgements and priorities with the analytic hierarchy process. *Eur. J. Oper. Res.* **1998**, *108*, 165–169. [CrossRef]
6. Altuzarra, A.; Moreno-Jiménez, J.M.; Salvador, M. A Bayesian priorization procedure for AHP-group decision making. *Eur. J. Oper. Res.* **2007**, *182*, 367–382. [CrossRef]
7. Gelman, A.; Carlin, J.; Stern, H.; Dunson, D.; Vehtary, A.; Rubin, D. *Bayesian Data Analysis*, 3rd ed.; CRC Press: Boca Raton, FL, USA, 2014.
8. Aguarón, J.; Moreno-Jiménez, J.M. The geometric consistency index: Approximate thresholds. *Eur. J. Oper. Res.* **2003**, *147*, 137–145. [CrossRef]
9. Kass, R.E.; Raftery, A.E. Bayes factors. *J. Am. Stat. Assoc.* **1995**, *90*, 773–795. [CrossRef]
10. Madigan, D.; Raftery, A.E. Model Selection and Accounting for Model Uncertainty in Graphical Models Using Occam's Window. *J. Am. Stat. Assoc.* **1994**, *89*, 1535–1546. [CrossRef]
11. Altuzarra, A.; Moreno-Jiménez, J.M.; Salvador, M. A Consensus building in AHP-group decision making: A Bayesian approach. *Oper. Res.* **2010**, *58*, 1755–1773. [CrossRef]

12. Chen, S.M.; Lin, T.E.; Lee, L.W. Group Decision Making using incomplete fuzzy preference relations based on the additive consistency and the order consistency. *Inf. Sci.* **2014**, *259*, 1–15. [CrossRef]
13. Herrera-Viedma, E.; Cabrerizo, F.J.; Kacprzyk, J.; Pedrycz, W. A review of soft consensus models in a fuzzy environment. *Inf. Fusion* **2014**, *17*, 4–13. [CrossRef]
14. Wang, Z.J.; Lin, J. Ratio-based similarity analysis and consensus building for group decision making with interval reciprocal preference relations. *Appl. Soft Comput.* **2016**, *42*, 260–275. [CrossRef]
15. Dong, Y.C.; Ding, Z.G.; Martínez, L.; Herrera, F. Managing consensus based on leadership in opinion dynamics. *Inf. Sci.* **2017**, *397*, 187–205. [CrossRef]
16. Wu, Z.B.; Xu, J. A consensus model for large-scale group decision making with hesitant fuzzy information and changeable clusters. *Inf. Fusion* **2018**, *41*, 217–231. [CrossRef]

Article

Similarity Measures of q-Rung Orthopair Fuzzy Sets Based on Cosine Function and Their Applications

Ping Wang [1], Jie Wang [2], Guiwu Wei [2,*] and Cun Wei [2,3]

[1]   Institute of Technology, Sichuan Normal University, Chengdu 610101, China; WangPing97@163.com
[2]   School of Business, Sichuan Normal University, Chengdu 610101, China; JW970326@163.com (J.W.);
      weicun1990@163.com (C.W.)
[3]   School of Statistics, Southwestern University of Finance and Economics, Chengdu 611130, China
*   Correspondence: weiguiwu1973@sicnu.edu.cn

Received: 23 March 2019; Accepted: 4 April 2019; Published: 9 April 2019

Abstract: In this article, we propose another form of ten similarity measures by considering the function of membership degree, non-membership degree, and indeterminacy membership degree between the q-ROFSs on the basis of the traditional cosine similarity measures and cotangent similarity measures. Then, we utilize our presented ten similarity measures and ten weighted similarity measures between q-ROFSs to deal with multiple attribute decision-making (MADM) problems including pattern recognition and scheme selection. Finally, two numerical examples are provided to illustrate the scientific and effective of the similarity measures for pattern recognition and scheme selection.

Keywords: multiple attribute decision-making (MADM); q-rung orthopair fuzzy sets (q-ROFSs); cosine function; cosine similarity measure; pattern recognition; scheme selection

---

1. Introduction

As an important branch of multiple attribute decision-making (MADM) domains, the similarity measures have been regarded as very useful tools to determine the degree of similarity between two objects. In the previous research literature, an increasing number of researchers pay attention to similarity measures between fuzzy sets (FSs) due to their broad applications in a variety of fields, for instance, pattern recognition, scheme selection, machine learning, image processing, and decision-making, many theories and applications of similarity measures between fuzzy sets (FSs) have been presented and investigated for the past few years. Atanassov [1,2] presented the definition of intuitionistic fuzzy set (IFS), which is an extension form of fuzzy set (FS). Each element contained in IFS was depicted by an ordered pair including the degree of membership μ and non-membership v, and the sum of them is limited to 1. Since IFS theory was proposed, a variety of similarity measures between intuitionistic fuzzy sets (IFSs) have been studied in the document [3–6]. Based on IFS and theories of similarity measures, Li and Cheng [7] presented appropriate similarity measure and gave a numerical example of pattern recognition problems to illustrate the effective of this method. Besides, Mitchell [8] improved Li and Cheng's similarity measures to deal with MADM. According to the extension of the Hamming distance (HD) of fuzzy sets (FSs), Park et al. [9] computed the distance between IFSs based on Hamming distance (HD) and proposed some similarity measures to solve MADM problems [10]. According to the Hausdorff distance, Torra and Narukawa [11] defined some new similarity measures between IFSs. Based on geometric aggregation operators, Xia and Xu [12] proposed the intuitionistic fuzzy geometric distance and intuitionistic fuzzy similarity measures to deal with MADM problems. Ye [13] initially developed the intuitionistic fuzzy cosine similarity measure based on cosine function. Kuo-ChenHung [14] defined the likelihood-based measurement of IFSs for the medical diagnosis and bacteria classification problems. Shi and Ye [15] further modified the cosine similarity measure of IFSs. Based on the cotangent function, Tian [16] presented the intuitionistic fuzzy cotangent similarity measure between IFSs for medical diagnosis. To contain more fuzzy information, Rajarajeswari and Uma [17] further defined the cotangent similarity measure which considered the function of membership degree, non-membership degree, and indeterminacy membership degrees described in IFSs. In addition, Szmidt [18] introduced distances between IFSs and introduced a family of similarity measures which considered the function of membership degree, non-membership degree, and indeterminacy membership degree in IFSs. Ye [19] developed two new cosine similarity measures and weighted cosine similarity measures based on cosine function and the fuzzy information denoted by the function of membership degree, non-membership degree, and indeterminacy membership degree described in intuitionistic fuzzy sets (IFSs). Wei [20] proposed some picture fuzzy similarity measures and applied them in MADM problems. Le Hoang and Pham Hong [21] defined the intuitionistic vector similarity measures for medical diagnosis. Wei and Wei [22] introduced some Pythagorean fuzzy similarity measures based on cosine function and applied them in pattern recognition and medical diagnosis.

More recently years, Pythagorean fuzzy set (PFS) [23] has emerged to describe the indeterminacy and complexity of the evaluation information. Similar to IFS, the PFS also consisted of the function of membership μ and non-membership v; the sum of squares of μ and v is restricted to 1, thus it is clear that the PFS is more widespread than the IFS and can express more decision-making information. For instance, the membership is given as 0.6 and the non-membership is given as 0.8, therefore it is obvious that this problem is only valid for PFS. In other words, all the intuitionistic fuzzy decision-making problems are the special case of Pythagorean fuzzy decision-making problems, which means that PFS can more efficiently deal with MADM problems. In previous literatures, some researching works have been studied by a large amount of investigators [24–28]. Zhang and Xu [29] defined the Pythagorean fuzzy TOPSIS model to deal with the MADM problems. Peng and Yang [30] primarily proposed two Pythagorean fuzzy operations including the division and subtraction operations to better understand PFS. Reformat and Yager [31] handled the collaborative-based recommender system with Pythagorean fuzzy information. Garg [32] defined some new Pythagorean fuzzy aggregation operators including Pythagorean fuzzy Einstein weighted averaging (PFEWA) operator, Pythagorean fuzzy Einstein ordered weighted averaging (PFEOWA) operator, generalized Pythagorean fuzzy Einstein weighted averaging (GPFEWA) operator, and generalized Pythagorean fuzzy Einstein ordered weighted averaging (GPFEOWA) operator. Zeng, et al. [33] utilized the Pythagorean fuzzy ordered weighted averaging weighted average distance (PFOWAWAD) operator to study Pythagorean fuzzy MADM issues. Ren, et al. [34] built the Pythagorean fuzzy TODIM model. Wei and Lu [35] developed some new Maclaurin symmetric mean (MSM) [36] operator based on Pythagorean fuzzy environment. Wei and Wei [22] defined ten cosine similarity measures under Pythagorean fuzzy environment. Liang, et al. [37] investigated some Bonferroni mean operators with Pythagorean fuzzy information. Liang, et al. [38] presented Pythagorean fuzzy Bonferroni mean aggregation operators based on geometric averaging (GA) operations. Combined the PFSs [39–41] and dual hesitant fuzzy sets (DHFSs) [42], Zhao et al. [43] introduced the definition of the dual hesitant Pythagorean fuzzy sets (DHPFSs) and proposed some dual hesitant Pythagorean fuzzy Hamacher aggregation operators.

In spite of this, to express more decision information, Yager [44] initially defined the q-rung orthopair fuzzy sets (q-ROFSs), in which the sum of the qth power of the membership and non-membership is less or equal to 1, that is to say, $\mu^q + v^q \leq 1$. Obviously, q-ROFS are more general for the IFS, and PFSs are special issues of it. Liu and Wang [45] developed the q-rung orthopair fuzzy weighted averaging (q-ROFWA) operator and the q-rung orthopair fuzzy weighted geometric (q-ROFWG) operator. Wei, et al. [46] proposed some q-rung orthopair fuzzy MSM operators, including q-rung orthopair fuzzy MSM (q-ROFMSM) operator, q-rung orthopair fuzzy weighted MSM (q-ROFWMSM) operator, q-rung orthopair fuzzy dual MSM (q-ROFDMSM) operator, and q-rung orthopair fuzzy weighted DMSM (q-ROFWDMSM) operator. Wei, et al. [47] defined some q-rung orthopair fuzzy Heronian mean operators. Yang and Pang [48] presented some new partitioned Bonferroni mean operators under q-rung orthopair fuzzy environment. Liu and Liu [49] provided some power Bonferroni mean operators with linguistic q-rung orthopair fuzzy information. Xu, et al. [50], given the concept of q-rung dual hesitant orthopair fuzzy set (q-RDHOFS), proposed some q-rung dual hesitant orthopair fuzzy Heronian mean operators. Lei and Xu [51] gave some methods for MAGDM with q-rung interval-valued orthopair fuzzy information for green supplier selection.

Although the intuitionistic fuzzy set (IFS) [1,2] and Pythagorean fuzzy set (PFS) [23,39] have been applied in some decision-making areas, for some special cases, such as when the membership degree and non-membership degree are given as 0.7 and 0.8, it is clear that both IFS and PFS theories cannot satisfy this situation. The q-rung orthopair fuzzy set (q-ROFS) is also denoted by the degree of membership and non-membership whose *q-th* power sum of them is restricted to 1. Obviously, the q-ROFS is more general than the q-ROFS and can express more fuzzy information. In other words, the q-ROFS can deal with the MADM problems which IFS cannot, and it is clear that IFS is a part of the q-ROFS, which indicates q-ROFS can be more effective and powerful to deal with fuzzy and uncertain decision-making problems. Thus, to solve such issues, based on the cosine functions and cotangent functions, we shall propose the concept of q-rung orthopair fuzzy cosine similarity measures and q-rung orthopair fuzzy cotangent similarity measures under q-rung orthopair fuzzy environment in this paper, which is a new extension of the similarity measure of IFSs.

To do this, the rest of this article is structured as follows. In the next section, we briefly review some fundamental theories of intuitionistic fuzzy set (IFS) and some intuitionistic fuzzy similarity measures. Some q-rung orthopair fuzzy cosine similarity measures, q-rung orthopair fuzzy weighted cosine similarity measures, q-rung orthopair fuzzy cotangent similarity measures, and q-rung orthopair fuzzy weighted cotangent similarity measures are developed in Section 3. All the above-mentioned similarity measures for q-ROFSs are used to pattern recognition and scheme selection in Section 4. Section 5 concludes the paper with some remarks.

2. Preliminaries

In this part, we shall briefly introduce some basic theories of intuitionistic fuzzy sets (IFSs) and review some similarity measures based on cosine functions and cotangent functions between IFSs.

Definition 1. *Suppose that X is a fixed set, then an intuitionistic fuzzy set (IFS) Q in X [1,2] can be denoted as*

$$Q = \left\{ \left\langle x, \alpha_Q(x), \beta_Q(x) \right\rangle | x \in X \right\} \tag{1}$$

where $\alpha_Q : X \to [0,1]$ means the degree of membership and $\beta_Q(x) : X \to [0,1]$ means the degree of non-membership which satisfies the condition of $0 \le \alpha_Q(x) \le 1, 0 \le \beta_Q(x) \le 1, 0 \le \alpha_Q(x) + \beta_Q(x) \le 1, \forall x \in X.$

Definition 2. *For each intuitionistic fuzzy set (IFS) Q in X [1,2], the degree of indeterminacy membership $\pi_Q(x)$ can be expressed as*

$$\pi_Q(x) = 1 - \alpha_Q(x) - \beta_Q(x), \forall x \in X. \tag{2}$$

The cosine similarity measures and cotangent similarity measures, which can calculate the degree of proximity between any two schemes, have been applied in many practical MADM problems. As we all know, the cosine and cotangent functions are monotone decreasing functions, thus, by considering the distance measures between any two alternatives, the bigger the distance values are, the smaller the calculating results by cosine and cotangent functions are and the lower similarity measures are. Therefore, to select best alternatives in decision-making problems, we always utilize cosine and cotangent similarity measures to obtain the similarity degree between each alternative and the ideal alternative. In what follows, we will briefly review some intuitionistic fuzzy cosine and cotangent similarity measures.

Let $M = \left\{ \left\langle x_j, \alpha_M(x_j), \beta_M(x_j) \right\rangle | x_j \in X \right\}$ and $N = \left\{ \left\langle x_j, \alpha_N(x_j), \beta_N(x_j) \right\rangle | x_j \in X \right\}$ be two intuitionistic fuzzy sets (IFSs), then the intuitionistic fuzzy cosine (*IFC*) measure between M and N proposed by Ye [13] can be shown as

$$IFC^1(M,N) = \frac{1}{n}\sum_{j=1}^{n} \frac{\alpha_M(x_j)\alpha_N(x_j) + \beta_M(x_j)\beta_N(x_j)}{\sqrt{\alpha_M^2(x_j) + \beta_M^2(x_j)}\,\sqrt{\alpha_N^2(x_j) + \beta_N^2(x_j)}} \tag{3}$$

Consider the degree of membership, non-membership and indeterminacy membership, then the intuitionistic fuzzy cosine (*IFC*) measure between M and N proposed by Shi and Ye [15] can be shown as

$$IFC^2(M,N) = \frac{1}{n}\sum_{j=1}^{n} \frac{\alpha_M(x_j)\alpha_N(x_j) + \beta_M(x_j)\beta_N(x_j) + \pi_M(x_j)\pi_N(x_j)}{\sqrt{\alpha_M^2(x_j) + \beta_M^2(x_j) + \pi_M^2(x_j)}\,\sqrt{\alpha_N^2(x_j) + \beta_N^2(x_j) + \pi_N^2(x_j)}} \tag{4}$$

On account of cosine function, Ye [19] developed two intuitionistic fuzzy cosine similarity (*IFCS*) measures between two intuitionistic fuzzy sets (IFSs) M and N.

$$IFCS^1(M,N) = \frac{1}{n}\sum_{j=1}^{n} \cos\left[\frac{\pi}{2}\left(\max\left(\begin{array}{c}|\alpha_M(x_j) - \alpha_N(x_j)|, \\ |\beta_M(x_j) - \beta_N(x_j)|, \\ |\pi_M(x_j) - \pi_N(x_j)|\end{array}\right)\right)\right] \tag{5}$$

$$IFCS^2(M,N) = \frac{1}{n}\sum_{i=1}^{n} \cos\left[\frac{\pi}{4}\left(\begin{array}{c}|\alpha_M(x_j) - \alpha_N(x_j)| + \\ |\beta_M(x_j) - \beta_N(x_j)| + \\ |\pi_M(x_j) - \pi_N(x_j)|\end{array}\right)\right] \tag{6}$$

In addition, the intuitionistic fuzzy cotangent (*IFCot*) similarity measure between any two intuitionistic fuzzy sets (IFSs) M and N proposed by Tian [16] is shown as

$$IFCot^1(M,N) = \frac{1}{n}\sum_{j=1}^{n} \cot\left[\frac{\pi}{4} + \frac{\pi}{4}\left(\max\left(\begin{array}{c}|\alpha_M(x_j) - \alpha_N(x_j)|, \\ |\beta_M(x_j) - \beta_N(x_j)|\end{array}\right)\right)\right] \tag{7}$$

Consider the degree of membership, non-membership, and indeterminacy membership, then the intuitionistic fuzzy cotangent (*IFCot*) similarity measure between any two intuitionistic fuzzy sets (IFSs) M and N proposed by Rajarajeswari and Uma [17] can be shown as

$$IFCot^2(M,N) = \frac{1}{n}\sum_{j=1}^{n} \cot\left[\frac{\pi}{4} + \frac{\pi}{4}\left(\max\left(\begin{array}{c}|\alpha_M(x_j) - \alpha_N(x_j)|, \\ |\beta_M(x_j) - \beta_N(x_j)|, \\ |\pi_M(x_j) - \pi_N(x_j)|\end{array}\right)\right)\right] \tag{8}$$

Consider the weighting vector of the elements in IFS, the weighted intuitionistic fuzzy cosine (*WIFC*) measure, the weighted intuitionistic fuzzy cosine similarity (*WIFCS*) measure, and weighted intuitionistic fuzzy cotangent (*WIFCot*) similarity measure between any two intuitionistic fuzzy sets (IFSs), M and N can be shown as follows [13,15–17,19]

$$WIFC^1(M,N) = \sum_{j=1}^{n} \omega_j \left[\frac{\alpha_M(x_j)\alpha_N(x_j) + \beta_M(x_j)\beta_N(x_j)}{\sqrt{\alpha_M^2(x_j) + \beta_M^2(x_j)}\sqrt{\alpha_N^2(x_j) + \beta_N^2(x_j)}} \right] \tag{9}$$

$$WIFC^2(M,N) = \sum_{j=1}^{n} \omega_j \frac{\alpha_M(x_j)\alpha_N(x_j) + \beta_M(x_j)\beta_N(x_j) + \pi_M(x_j)\pi_N(x_j)}{\sqrt{\alpha_M^2(x_j) + \beta_M^2(x_j) + \pi_M^2(x_j)}\sqrt{\alpha_N^2(x_j) + \beta_N^2(x_j) + \pi_N^2(x_j)}} \tag{10}$$

$$WIFCS^1(M,N) = \sum_{j=1}^{n} \omega_j \cos\left[\frac{\pi}{2} \max\left(\begin{array}{c} |\alpha_M(x_j) - \alpha_N(x_j)|, \\ |\beta_M(x_j) - \beta_N(x_j)|, \\ |\pi_M(x_j) - \pi_N(x_j)| \end{array} \right) \right] \tag{11}$$

$$WIFCS^2(M,N) = \sum_{i=1}^{n} \omega_j \cos\left[\frac{\pi}{4}\left(\begin{array}{c} |\alpha_M(x_j) - \alpha_N(x_j)| + \\ |\beta_M(x_j) - \beta_N(x_j)| + \\ |\pi_M(x_j) - \pi_N(x_j)| \end{array} \right) \right] \tag{12}$$

$$WIFCot^1(M,N) = \sum_{j=1}^{n} \omega_j \cot\left[\frac{\pi}{4} + \frac{\pi}{4}\left(\max\left(\begin{array}{c} |\alpha_M(x_j) - \alpha_N(x_j)|, \\ |\beta_M(x_j) - \beta_N(x_j)| \end{array} \right) \right) \right] \tag{13}$$

$$WIFCot^2(M,N) = \sum_{j=1}^{n} \omega_j \cot\left[\frac{\pi}{4} + \frac{\pi}{4}\left(\max\left(\begin{array}{c} |\alpha_M(x_j) - \alpha_N(x_j)|, \\ |\beta_M(x_j) - \beta_N(x_j)|, \\ |\pi_M(x_j) - \pi_N(x_j)| \end{array} \right) \right) \right] \tag{14}$$

where $\omega_j(j = 1,2,\cdots,n)$ denotes the weighting vector of elements x_j, which satisfies the condition of $\omega_j \in [0,1]$ and $\sum_{j=1}^{n} \omega_j = 1$.

3. Some Similarity Measures Based on Cosine Function for q-ROFSs

Although the intuitionistic fuzzy sets (IFSs) defined by Atanassov's [1,2] have been broadly applied in different areas, for some special cases, such as when membership degree and non-membership degree are given as 0.7 and 0.8, it is clear that IFSs theory cannot satisfy this situation. The q-rung orthopair fuzzy set (q-ROFS) is also denoted by the degree of membership and non-membership, whose *q-th* power sum is restricted to 1, obviously, the q-ROFS is more general than the q-ROFS and can express more fuzzy information. In other words, the q-ROFS can deal with the MADM problems which IFS cannot and it is clear that IFS is a part of the q-ROFS, which indicates q-ROFS can be more effective and powerful to deal with fuzzy and uncertain decision-making problems.

Definition 3. *Suppose P be a fix set, then a q-rung orthopair fuzzy set (q-ROFS) P in X [39,40] can be denoted as*

$$P = \{\langle x, (\alpha_P(x), \beta_P(x))\rangle | x \in X\} \tag{15}$$

where $\alpha_P : X \rightarrow [0,1]$ means the degree of membership and $\beta_P(x) : X \rightarrow [0,1]$ means the degree of non-membership which satisfies the condition of $0 \leq \alpha_P(x) \leq 1, 0 \leq \beta_P(x) \leq 1, 0 \leq (\alpha_P(x))^q + (\beta_P(x))^q \leq 1, q \geq 1, \forall x \in X.$

Definition 4. *For each q-rung orthopair fuzzy set (q-ROFS) P in X [39,40], the degree of indeterminacy membership $\pi_P(x)$ can be expressed as*

$$\pi_P(x) = \sqrt[q]{(\alpha_P(x))^q + (\beta_P(x))^q - (\alpha_P(x))^q (\beta_P(x))^q}, \forall x \in X. \tag{16}$$

Definition 5. *Let $p = (\alpha, \beta)$ be a q-ROFN, a score function can be represented [40] as follows*

$$S(p) = \frac{1}{2}(1 + \alpha^q - \beta^q), S(p) \in [0, 1]. \tag{17}$$

Definition 6. *Let $r_j = (\alpha_j, \beta_j)(j = 1, 2, \cdots, n)$ be a group of q-ROFNs with weighting vector $w = (w_1, w_2, \ldots, w_n)^T$, which satisfies $w_j > 0$, $i = 1, 2, \ldots, n$ and $\sum_{j=1}^{n} w_j = 1$ [40]. Then we can obtain the q-rung orthopair fuzzy weighted averaging (q-ROFWA) operator and the q-rung orthopair fuzzy weighted geometric (q-ROFWG) operator as follows*

$$q - ROFWA(r_1, r_2, \ldots, r_n) = \overset{n}{\underset{j=1}{\oplus}} w_j r_j = \left\langle \left(1 - \prod_{j=1}^{n}\left(1 - \alpha_j^q\right)^{w_j}\right)^{1/q}, \prod_{j=1}^{n} \beta_j^{w_j} \right\rangle \tag{18}$$

$$q - ROFWG(r_1, r_2, \ldots, r_n) = \overset{n}{\underset{j=1}{\otimes}} (r_j)^{w_j} = \left\langle \prod_{j=1}^{n} \alpha_j^{w_j}, \left(1 - \prod_{j=1}^{n}\left(1 - \beta_j^q\right)^{w_j}\right)^{1/q} \right\rangle \tag{19}$$

3.1. Cosine Similarity Measure for q-ROFSs

Suppose that P is a q-rung orthopair fuzzy set (q-ROFS) in a universe of discourse $X = \{x\}$, the elements contained in q-ROFS can be expressed as the function of membership degree $\alpha_P(x)$, the function of non-membership degree $\beta_P(x)$, and the function of indeterminacy membership degree $\pi_P(x)$. Thus, a cosine similarity measure and a weighted cosine similarity measure with q-rung orthopair fuzzy information are presented in an analogous manner to the cosine similarity measure based on Bhattacharya's distance and cosine similarity measure for intuitionistic fuzzy set (IFS) [13].

Let $M = \{\langle x_j, \alpha_M(x_j), \beta_M(x_j)\rangle | x_j \in X\}$ and $N = \{\langle x_j, \alpha_N(x_j), \beta_N(x_j)\rangle | x_j \in X\}$ be two q-rung orthopair fuzzy sets (q-ROFSs), then the q-rung orthopair fuzzy cosine (q-ROFC) measure between M and N can be shown as

$$q - ROFC^1(M, N) = \frac{1}{n} \sum_{j=1}^{n} \frac{\alpha_M^q(x_j)\alpha_N^q(x_j) + \beta_M^q(x_j)\beta_N^q(x_j)}{\sqrt{\left(\alpha_M^q(x_j)\right)^2 + \left(\beta_M^q(x_j)\right)^2}\sqrt{\left(\alpha_N^q(x_j)\right)^2 + \left(\beta_N^q(x_j)\right)^2}} \tag{20}$$

Especially, when we let $n = 1$, the cosine similarity measure between q-ROFSs M and N can be depicted as $C_{q-ROFS}(M, N)$, which will become the correlation coefficient between M and N, which is depicted as $K_{q-ROFS}(M, N)$, i.e., $C_{q-ROFS}(M, N) = K_{q-ROFS}(M, N)$. In addition, the cosine similarity measure between q-ROFSs M and N also satisfies some properties as follows.

(1) $0 \leq q - ROFC^1(M, N) \leq 1$;
(2) $q - ROFC^1(M, N) = q - ROFC^1(N, M)$;
(3) $q - ROFC^1(M, N) = 1$, if $M = N$, $j = 1, 2, \cdots, n$.

Proof.

(1) It is clear that the proposition is true based on the cosine result.

(2) It is clear that the proposition is true.

(3) When $M = N$, it means that $\alpha_M(x_j) = \alpha_N(x_j)$ and $\beta_M(x_j) = \beta_N(x_j)$ for $j = 1, 2, \cdots, n$. So $C^1_{q-ROFS}(M, N) = 1$.

Therefore, we have finished the proofs. $\square$

In what follows, we shall study the distance measure of the angle as $d(M, N) = \arccos\left(C^1_{q-ROFS}(M, N)\right)$. It satisfies some properties as follows.

(1) $d(M, N) \geq 0$, if $0 \leq C^1_{q-ROFS}(M, N) \leq 1$;

(2) $d(M, N) = \arccos(1) = 0$, if $C^1_{q-ROFS}(M, N) = 1$;

(3) $d(M, N) = d(N, M)$, if $C^1_{q-ROFS}(M, N) = C^1_{q-ROFS}(N, M)$,

(4) $d(M, T) \leq d(M, N) + d(N, T)$, if $M \subseteq N \subseteq T$ for any q-ROFS T.

Proof. Clearly the distance measure $d(M, N)$ satisfies properties (1)–(3). In what follows we shall prove that the distance measure $d(M, N)$ satisfies property (4).

For any q-rung orthopair fuzzy set (q-ROFS) $T = \left\{\langle x_j, (\alpha_T(x_j), \beta_T(x_j))\rangle | x_j \in x\right\}$, $M \subseteq N \subseteq T$, let us investigate the distance measures of the angle between the vectors:

$$d_j(M(x_j), N(x_j)) = \arccos(q - ROFC^1_i(M(x_i), N(x_i)))(j = 1, 2, \cdots, n)$$
$$d_j(M(x_j), T(x_j)) = \arccos(q - ROFC^1_i(M(x_i), T(x_i)))(j = 1, 2, \cdots, n)$$
$$d_j(N(x_j), T(x_j)) = \arccos(q - ROFC^1_i(N(x_i), T(x_i)))(j = 1, 2, \cdots, n)$$

where

$$q - ROFC^1_j(M(x_j), N(x_j)) = \frac{\alpha^q_M(x_j)\alpha^q_N(x_j) + \beta^q_M(x_j)\beta^q_N(x_j)}{\sqrt{(\alpha^q_M(x_j))^2 + (\beta^q_M(x_j))^2}\sqrt{(\alpha^q_N(x_j))^2 + (\beta^q_N(x_j))^2}}$$

$$q - ROFC^1_j(M(x_j), T(x_j)) = \frac{\alpha^q_M(x_j)\alpha^q_T(x_j) + \beta^q_M(x_j)\beta^q_T(x_j)}{\sqrt{(\alpha^q_M(x_j))^2 + (\beta^q_M(x_j))^2}\sqrt{(\alpha^q_T(x_j))^2 + (\beta^q_T(x_j))^2}}$$

$$q - ROFC^1_j(N(x_j), T(x_j)) = \frac{\alpha^q_N(x_j)\alpha^q_T(x_j) + \beta^q_N(x_j)\beta^q_T(x_j)}{\sqrt{(\alpha^q_N(x_j))^2 + (\beta^q_N(x_j))^2}\sqrt{(\alpha^q_T(x_j))^2 + (\beta^q_T(x_j))^2}}$$

$M(x_j) = \langle \alpha_M(x_j), \beta_M(x_j)\rangle, N(x_j) = \langle \alpha_N(x_j), \beta_N(x_j)\rangle, T(x_j) = \langle \alpha_T(x_j), \beta_T(x_j)\rangle$ are three vectors in one plane, if $M(x_j) \subseteq N(x_j) \subseteq T(x_j)$, $j = 1, 2, \cdots, n$. Therefore, it is clear that $d_j(M(x_j), T(x_j)) \leq d_j(M(x_j), N(x_j)) + d_j(N(x_j), T(x_j))$ based on the triangle inequality. Combining the inequality $0 \leq (\alpha_P(x_j))^q + (\beta_P(x_j))^q \leq 1$, we can get $d(M, T) \leq d(M, N) + d(N, T)$. Therefore $d(M, N)$ meets the property (4). So we completed the process of proof. $\square$

If we consider three terms—membership degree, non-membership degree, and indeterminacy membership—which are contained in q-ROFSs, assume that there are two q-rung orthopair fuzzy sets, $M = \left\{\langle x_j, \alpha_M(x_j), \beta_M(x_j), \pi_M(x_j)\rangle | x_j \in X\right\}(j = 1, 2, \ldots, n)$ and $N = \left\{\langle x_j, \alpha_N(x_j), \beta_N(x_j), \pi_N(x_j)\rangle | x_j \in X\right\}(j = 1, 2, \ldots, n)$, then the q-rung orthopair fuzzy cosine (q-ROFC) measures between q-ROFSs can be expressed as

$$q - ROFC^2(M, N) = \frac{1}{n}\sum_{j=1}^{n}\left[\frac{\left(\begin{array}{c}\alpha^q_M(x_j)\alpha^q_N(x_j) + \beta^q_M(x_j)\beta^q_N(x_j) \\ +\pi^q_M(x_j)\pi^q_N(x_j)\end{array}\right)}{\left(\begin{array}{c}\sqrt{(\alpha^q_M(x_j))^2 + (\beta^q_M(x_j))^2 + (\pi^q_M(x_j))^2} \\ \times \sqrt{(\alpha^q_N(x_j))^2 + (\beta^q_N(x_j))^2 + (\pi^q_N(x_j))^2}\end{array}\right)}\right] \tag{21}$$

Especially when we let $n = 1$, the cosine similarity measure between q-ROFSs M and N will become the correlation coefficient between q-rung orthopair fuzzy sets (q-ROFSs) M and N. Of course, the cosine similarity measure $q - ROFC^2(M, N)$ also satisfies some properties which are listed as follows.

(1) $0 \leq q - ROFC^2(M, N) \leq 1$;
(2) $q - ROFC^2(M, N) = q - ROFC^2(N, M)$;
(3) $q - ROFC^2(M, N) = 1$, $if\ M = N$, $j = 1, 2, \cdots, n$.

Consider the weighting vector of the elements in q-ROFS, the q-rung orthopair fuzzy weighted cosine (q-ROFWC) measure between two q-rung orthopair fuzzy sets (q-ROFSs) M and N can be shown as follows.

$$q - ROFWC^1(M, N) = \sum_{j=1}^{n} \omega_j \frac{\alpha_M^q(x_j)\alpha_N^q(x_j) + \beta_M^q(x_j)\beta_N^q(x_j)}{\sqrt{\left(\alpha_M^q(x_j)\right)^2 + \left(\beta_M^q(x_j)\right)^2}\sqrt{\left(\alpha_N^q(x_j)\right)^2 + \left(\beta_N^q(x_j)\right)^2}} \tag{22}$$

$$q - ROFWC^2(M, N) = \sum_{j=1}^{n} \omega_j \left[\frac{\left(\begin{array}{c}\alpha_M^q(x_j)\alpha_N^q(x_j) + \beta_M^q(x_j)\beta_N^q(x_j)\\ +\pi_M^q(x_j)\pi_N^q(x_j)\end{array}\right)}{\left(\begin{array}{c}\sqrt{\left(\alpha_M^q(x_j)\right)^2 + \left(\beta_M^q(x_j)\right)^2 + \left(\pi_M^q(x_j)\right)^2}\\ \times\sqrt{\left(\alpha_N^q(x_j)\right)^2 + \left(\beta_N^q(x_j)\right)^2 + \left(\pi_N^q(x_j)\right)^2}\end{array}\right)} \right] \tag{23}$$

where $\omega = (\omega_1, \omega_2, \cdots, \omega_n)^T$ indicates the weighting vector of the elements $x_j(j = 1, 2, \cdots, n)$ contained in q-ROFS and the weighting vector satisfies $\omega_j \in [0, 1]$, $j = 1, 2, \cdots, n$, $\sum_{j=1}^{n} \omega_j = 1$. Especially, when we let weighting vector be $\omega = (1/n, 1/n, \cdots, 1/n)^T$, then the weighted cosine similarity measure will reduce to cosine similarity measure. In other words, when $\omega_j = \frac{1}{n}, j = 1, 2 \cdots, n$, the $q - ROFWC^1(M, N) = q - ROFC^1(M, N)$.

Example 1. *Suppose there are two q-ROFSs $M = \{(x_1, 0.7, 0.4), (x_2, 0.5, 0.6), (x_3, 0.3, 0.8)\}$ and $N = \{(x_1, 0.9, 0.2), (x_2, 0.4, 0.3), (x_3, 0.7, 0.6)\}$, assume $q = 3, \omega_j = (0.2, 0.3, 0.5)$ then according to Equation (19), the weighted cosine similarity measure between M and N can be calculated as*

$$q - ROFWC^1(M, N) = \sum_{j=1}^{n} \omega_j \frac{\alpha_M^q(x_j)\alpha_N^q(x_j) + \beta_M^q(x_j)\beta_N^q(x_j)}{\sqrt{\left(\alpha_M^q(x_j)\right)^2 + \left(\beta_M^q(x_j)\right)^2}\sqrt{\left(\alpha_N^q(x_j)\right)^2 + \left(\beta_N^q(x_j)\right)^2}}$$

$$= \left(\begin{array}{c}\frac{0.2\times\left(0.7^3\times0.9^3 + 0.4^3\times0.2^3\right)}{\sqrt{(0.7^3)^2 + (0.4^3)^2}\times\sqrt{(0.9^3)^2 + (0.2^3)^2}} + \frac{0.3\times\left(0.5^3\times0.4^3 + 0.6^3\times0.3^3\right)}{\sqrt{(0.5^3)^2 + (0.6^3)^2}\times\sqrt{(0.4^3)^2 + (0.3^3)^2}}\\ +\frac{0.5\times\left(0.3^3\times0.7^3 + 0.8^3\times0.6^3\right)}{\sqrt{(0.3^3)^2 + (0.8^3)^2}\times\sqrt{(0.7^3)^2 + (0.6^3)^2}}\end{array}\right)$$

$$= 0.7247$$

Example 2. *Suppose there are two q-ROFSs* $M = \{(x_1, 0.7, 0.4), (x_2, 0.5, 0.6), (x_3, 0.3, 0.8)\}$ *and* $N = \{(x_1, 0.9, 0.2), (x_2, 0.4, 0.3), (x_3, 0.7, 0.6)\}$, *assume* $q = 3, \omega_j = (0.2, 0.3, 0.5)$ *then according to Equation (3) and Equation (20), the weighted cosine similarity measure between M and N can be calculated as*

$$
q - ROFWC^2(M,N) = \sum_{j=1}^{n} \omega_j \left[\frac{\left(\begin{array}{c} \alpha_M^q(x_j)\alpha_N^q(x_j) + \beta_M^q(x_j)\beta_N^q(x_j) \\ + \pi_M^q(x_j)\pi_N^q(x_j) \end{array} \right)}{\begin{array}{c} \sqrt{\left(\alpha_M^q(x_j)\right)^2 + \left(\beta_M^q(x_j)\right)^2 + \left(\pi_M^q(x_j)\right)^2} \\ \times \sqrt{\left(\alpha_N^q(x_j)\right)^2 + \left(\beta_N^q(x_j)\right)^2 + \left(\pi_N^q(x_j)\right)^2} \end{array}} \right]
$$

$$
= \left[\begin{array}{l} 0.2 \times \left(\dfrac{0.7^3 \times 0.9^3 + 0.4^3 \times 0.2^3 + 0.7^3 \times 0.9^3}{\sqrt{(0.7^3)^2 + (0.4^3)^2 + (0.7^3)^2} \times \sqrt{(0.9^3)^2 + (0.2^3)^2 + (0.9^3)^2}} \right) \\[3mm] + 0.3 \times \left(\dfrac{0.5^3 \times 0.4^3 + 0.6^3 \times 0.3^3 + 0.7^3 \times 0.4^3}{\sqrt{(0.5^3)^2 + (0.6^3)^2 + (0.7^3)^2} \times \sqrt{(0.4^3)^2 + (0.3^3)^2 + (0.4^3)^2}} \right) \\[3mm] + 0.5 \times \left(\dfrac{0.3^3 \times 0.7^3 + 0.8^3 \times 0.6^3 + 0.8^3 \times 0.8^3}{\sqrt{(0.3^3)^2 + (0.8^3)^2 + (0.8^3)^2} \times \sqrt{(0.7^3)^2 + (0.6^3)^2 + (0.8^3)^2}} \right) \end{array} \right]
$$

$$
= 0.8789
$$

Evidently, similar to cosine similarity measure $q - ROFC^1(M,N)$, the weighted cosine similarity measure $q - ROFWC^1(M,N)$ also meets three properties as follows.

(1) $0 \leq q - ROFWC^1(M,N) \leq 1$,

(2) $q - ROFWC^1(M,N) = q - ROFWC^1(N,M)$,

(3) $q - ROFWC^1(M,N) = 1, if\ M = N, i = 1, 2, \cdots, n$.

3.2. Similarity Measures of q-ROFSs Based on Cosine Function

In this section, according to the cosine function, we will present some q-rung orthopair fuzzy cosine similarity measures (q-ROFCS) between q-ROFSs and discuss their properties.

Definition 7. *Assume that there are any two q-rung orthopair fuzzy sets (q-ROFSs)* $M = \{\langle x_j, (\alpha_M(x_j), \beta_M(x_j)) \rangle | x_j \in x\}$ *and* $N = \{\langle x_j, (\alpha_N(x_j), \beta_N(x_j)) \rangle | x_j \in x\}$. *Then, we shall propose two q-rung orthopair fuzzy cosine similarity (q-ROFCS) measures between q-ROFSs M and N as follows*

$$
q - ROFCS^1(M,N) = \frac{1}{n} \sum_{j=1}^{n} \cos \left[\frac{\pi}{2} \left(\max \left(\begin{array}{c} \left| \alpha_M^q(x_j) - \alpha_N^q(x_j) \right|, \\ \left| \beta_M^q(x_j) - \beta_N^q(x_j) \right| \end{array} \right) \right) \right] \tag{24}
$$

$$
q - ROFCS^2(M,N) = \frac{1}{n} \sum_{j=1}^{n} \cos \left[\frac{\pi}{4} \left(\begin{array}{c} \left| \alpha_M^q(x_j) - \alpha_N^q(x_j) \right| + \\ \left| \beta_M^q(x_j) - \beta_N^q(x_j) \right| \end{array} \right) \right] \tag{25}
$$

Proposition 1. *Assume that there are any two q-rung orthopair fuzzy sets (q-ROFSs) M and N in* $X = \{x_1, x_2, \cdots, x_n\}$, *the q-rung orthopair fuzzy cosine similarity measures* $q - ROFCS^k(M,N)(k = 1, 2)$ *should satisfy the properties (1)–(4):*

(1) $0 \leq q - ROFCS^k(M,N) \leq 1$;

(2) $q - ROFCS^k(M,N) = 1\ if\ and\ only\ if\ M = N$;

(3) $q - ROFCS^k(M,N) = q - ROFCS^k(N,M)$;

(4)   Let M, N, T be three q-ROFSs in X and $M \subseteq N \subseteq T$, then $q - ROFCS^k(M, T) \leq q - ROFCS^k(M, N)$, $q - ROFCS^k(M, T) \leq q - ROFCS^k(N, T)$.

Proof. (1) Since the calculated results based on the cosine function are within $[0, 1]$, the q-rung orthopair fuzzy cosine similarity measures based on the cosine function are also within $[0, 1]$. Thus $0 \leq q - ROFCS^k(M, N) \leq 1, k = 1, 2$.

(2) For two q-rung orthopair fuzzy sets (q-ROFSs) M and N in $X = \{x_1, x_2, \cdots, x_n\}$, if $M = N$, then $\alpha_M^q(x_j) = \alpha_N^q(x_j), \beta_M^q(x_j) = \beta_N^q(x_j), j = 1, 2, \cdots, n$. Thus, $\left| \alpha_M^q(x_j) - \alpha_N^q(x_j) \right| = 0, \left| \beta_M^q(x_j) - \beta_N^q(x_j) \right| = 0$. So, $q - ROFCS^k(M, N) = 1, k = 1, 2$. If $q - ROFCS^k(M, N) = 1, k = 1, 2$, it implies $\left| \alpha_M^q(x_j) - \alpha_N^q(x_j) \right| = 0$, $j = 1, 2, \cdots, n, \left| \beta_M^q(x_j) - \beta_N^q(x_j) \right| = 0, j = 1, 2, \cdots, n$. Since $\cos(0) = 1$. Then, there are $\alpha_M^q(x_j) = \alpha_N^q(x_j)$, $\beta_M^q(x_j) = \beta_N^q(x_j), j = 1, 2, \cdots, n$. Hence $M = N$.

(3) Proof is straightforward.

(4) If $M \subseteq N \subseteq T$, that means $\alpha_M(x_j) \leq \alpha_N(x_j) \leq \alpha_T(x_j), \beta_M(x_j) \geq \beta_N(x_j) \geq \beta_T(x_j)$, for $j = 1, 2, \cdots, n$. Then $\alpha_M^q(x_j) \leq \alpha_N^q(x_j) \leq \alpha_T^q(x_j), \beta_M^q(x_j) \geq \beta_N^q(x_j) \geq \beta_T^q(x_j)$. Thus, we have

$$\left| \alpha_M^q(x_j) - \alpha_N^q(x_j) \right| \leq \left| \alpha_M^q(x_j) - \alpha_T^q(x_j) \right|, \left| \alpha_N^q(x_j) - \alpha_T^q(x_j) \right| \leq \left| \alpha_M^q(x_j) - \alpha_T^q(x_j) \right|,$$
$$\left| \beta_M^q(x_j) - \beta_N^q(x_j) \right| \leq \left| \beta_M^q(x_j) - \beta_T^q(x_j) \right|, \left| \beta_N^q(x_j) - \beta_C^q(x_j) \right| \leq \left| \beta_M^q(x_j) - \beta_T^q(x_j) \right|.$$

Thus $q - ROFCS^k(M, T) \leq q - ROFCS^k(M, N), q - ROFCS^k(M, T) \leq q - ROFCS^k(M, N)$, as the cosine function is a decreasing function with the interval $[0, \pi/2]$. Then, we finished the process of proofs. $\square$

If we consider three terms including membership degree, non-membership degree, and indeterminacy membership, which are contained in q-ROFSs, assume that there are two q-rung orthopair fuzzy sets $M = \left\{ \left\langle x_j, \alpha_M(x_j), \beta_M(x_j), \pi_M(x_j) \right\rangle \middle| x_j \in X \right\} (j = 1, 2, \ldots, n)$ and $N = \left\{ \left\langle x_j, \alpha_N(x_j), \beta_N(x_j), \pi_N(x_j) \right\rangle \middle| x_j \in X \right\} (j = 1, 2, \ldots, n)$, then the q-rung orthopair fuzzy cosine similarity (q-ROFCS) measures between M and N can be expressed as

$$q - ROFCS^3(M, N) = \frac{1}{n} \sum_{j=1}^{n} \cos \left[\frac{\pi}{2} \left(\max \left(\begin{array}{c} \left| \alpha_M^q(x_j) - \alpha_N^q(x_j) \right|, \\ \left| \beta_M^q(x_j) - \beta_N^q(x_j) \right|, \\ \left| \pi_M^q(x_j) - \pi_N^q(x_j) \right| \end{array} \right) \right) \right] \tag{26}$$

where $q - ROFCS^3(M, N)$ means the q-rung orthopair fuzzy cosine similarity measures between M and N, which consider the maximum distance based on the membership, indeterminacy membership, and non-membership degree.

$$q - ROFCS^4(M, N) = \frac{1}{n} \sum_{j=1}^{n} \cos \left[\frac{\pi}{4} \left(\begin{array}{c} \left| \alpha_M^q(x_j) - \alpha_N^q(x_j) \right| + \\ \left| \beta_M^q(x_j) - \beta_N^q(x_j) \right| + \\ \left| \pi_M^q(x_j) - \pi_N^q(x_j) \right| \end{array} \right) \right] \tag{27}$$

where $q - ROFCS^4(M, N)$ means the q-rung orthopair fuzzy cosine similarity measures between M and N, which consider the sum of distance based on the membership, indeterminacy membership, and non-membership degree.

Consider the weighting vector of the elements in q-ROFS, the q-rung orthopair fuzzy weighted cosine similarity (q-ROFWCS) measure between two q-rung orthopair fuzzy sets (q-ROFSs) M and N can be shown as follows.

$$q-ROFWCS^1(M,N) = \sum_{j=1}^{n} \omega_j \cos\left[\frac{\pi}{2}\left(\max\left(\begin{array}{c}|\alpha_M^q(x_j)-\alpha_N^q(x_j)|, \\ |\beta_M^q(x_j)-\beta_N^q(x_j)|\end{array}\right)\right)\right] \tag{28}$$

where $q-ROFWCS^1(M,N)$ means the q-rung orthopair fuzzy weighted cosine similarity measures between M and N, which consider the maximum distance based on the membership and non-membership degree.

$$q-ROFWCS^2(M,N) = \sum_{j=1}^{n} \omega_j \cos\left[\frac{\pi}{4}\left(\begin{array}{c}|\alpha_M^q(x_j)-\alpha_N^q(x_j)|+ \\ |\beta_M^q(x_j)-\beta_N^q(x_j)|\end{array}\right)\right] \tag{29}$$

where $q-ROFWCS^2(M,N)$ means the q-rung orthopair fuzzy weighted cosine similarity measures between M and N, which consider the sum of distance based on the membership and non-membership degree.

$$q-ROFWCS^3(M,N) = \sum_{j=1}^{n} \omega_j \cos\left[\frac{\pi}{2}\left(\max\left(\begin{array}{c}|\alpha_M^q(x_j)-\alpha_N^q(x_j)|, \\ |\beta_M^q(x_j)-\beta_N^q(x_j)|, \\ |\pi_A^2(x_j)-\pi_B^2(x_j)|\end{array}\right)\right)\right] \tag{30}$$

where $q-ROFWCS^3(M,N)$ means the q-rung orthopair fuzzy weighted cosine similarity measures between M and N, which consider the maximum distance based on the membership, indeterminacy membership, and non-membership degree.

$$q-ROFWCS^4(M,N) = \sum_{j=1}^{n} \omega_j \cos\left[\frac{\pi}{4}\left(\begin{array}{c}|\alpha_M^q(x_j)-\alpha_N^q(x_j)|+ \\ |\beta_M^q(x_j)-\beta_N^q(x_j)|+ \\ |\pi_A^2(x_j)-\pi_B^2(x_j)|\end{array}\right)\right] \tag{31}$$

where $q-ROFWCS^4(M,N)$ means the q-rung orthopair fuzzy weighted cosine similarity measures between M and N, which consider the sum of distance based on the membership, indeterminacy membership, and non-membership degree.

where $\omega = (\omega_1, \omega_2, \cdots, \omega_n)^T$ indicates the weighting vector of the elements $x_j (j = 1, 2, \cdots, n)$ contained in q-ROFS, and the weighting vector satisfies $\omega_j \in [0,1]$, $j = 1, 2, \cdots, n$, $\sum_{j=1}^{n} \omega_j = 1$. Especially, when we let weighting vector be $\omega = (1/n, 1/n, \cdots, 1/n)^T$, then the weighted cosine similarity measure will reduce to cosine similarity measure. In other words, when $\omega_j = \frac{1}{n}, j = 1, 2 \cdots, n$, the $q-ROFWCS^k(M,N) = q-ROFCS^k(M,N)(k = 1, 2, 3, 4)$.

Example 3. *Suppose there are two q-ROFSs, $M = \{(x_1, 0.7, 0.4), (x_2, 0.5, 0.6), (x_3, 0.3, 0.8)\}$ and $N = \{(x_1, 0.9, 0.2), (x_2, 0.4, 0.3), (x_3, 0.7, 0.6)\}$, assume $q = 3, \omega_j = (0.2, 0.3, 0.5)$, then according to Equation (25), the weighted cosine similarity measure between M and N can be calculated as*

$$q-ROFWCS^1(M,N) = \sum_{j=1}^{n} \omega_j \cos\left[\frac{\pi}{2}\left(\max\left(|\alpha_M^q(x_j)-\alpha_N^q(x_j)|, |\beta_M^q(x_j)-\beta_N^q(x_j)|\right)\right)\right]$$

$$= \left(\begin{array}{c} 0.2 \times \cos\left[\frac{\pi}{2}\max\left(|0.7^3 - 0.9^3|, |0.4^3 - 0.2^3|\right)\right] + 0.3 \times \cos\left[\frac{\pi}{2}\max\left(\begin{array}{c}|0.5^3 - 0.4^3|, \\ |0.6^3 - 0.3^3|\end{array}\right)\right] \\ +0.5 \times \cos\left[\frac{\pi}{2}\left(\max\left(|0.3^3 - 0.7^3|, |0.8^3 - 0.6^3|\right)\right)\right]\end{array}\right)$$

$$= 0.8909$$

Evidently, similar to cosine similarity measure $q - ROFCS^k(M, N)(k = 1, 2, 3, 4)$, the weighted cosine similarity measure $q - ROFWCS^k(M, N)(k = 1, 2, 3, 4)$ also meets some properties as follows.

Proposition 2. *Assume that there are any two q-rung orthopair fuzzy sets (q-ROFSs) M and N in $X = \{x_1, x_2, \cdots, x_n\}$, the q-rung orthopair fuzzy weighted cosine similarity measures $q - ROFWCS^k(M, N)(k = 1, 2, 3, 4)$ should satisfy the properties (1)–(4):*

(1) $0 \leq q - ROFWCS^k(M, N) \leq 1$;

(2) $q - ROFWCS^k(M, N) = 1$ if and only if $M = N$;

(3) $q - ROFWCS^k(M, N) = q - ROFWCS^k(N, M)$;

(4) If T is a q-ROFS in X and $M \subseteq N \subseteq T$, then $q - ROFWCS^k(M, T) \leq q - ROFWCS^k(M, N)$, $q - ROFWCS^k(M, T) \leq q - ROFWCS^k(N, T)$.

The proof is similar to Proposition 1, so it is omitted here.

3.3. Similarity Measures of q-ROFSs Based on Cotangent Function

In this section, according to the cotangent function, we will present some q-rung orthopair fuzzy cotangent similarity measures (q-ROFCot) between q-ROFSs and discuss their properties.

Definition 8. *Assume that there are any two q-rung orthopair fuzzy sets (q-ROFSs) $M = \left\{ \left\langle x_j, \left(\alpha_M(x_j), \beta_M(x_j) \right) \right\rangle | x_j \in x \right\}$ and $N = \left\{ \left\langle x_j, \left(\alpha_N(x_j), \beta_N(x_j) \right) \right\rangle | x_j \in x \right\}$. Then, we shall propose two q-rung orthopair fuzzy cotangent (q-ROFCot) measures between q-ROFSs M and N as follows*

$$q - ROFCot^1(M, N) = \frac{1}{n} \sum_{j=1}^{n} \cot \left[\frac{\pi}{4} + \frac{\pi}{4} \left(\max \left(\begin{array}{c} \left| \alpha_M^q(x_j) - \alpha_N^q(x_j) \right|, \\ \left| \beta_M^q(x_j) - \beta_N^q(x_j) \right| \end{array} \right) \right) \right] \tag{32}$$

where $q - ROFCot^1(M, N)$ means the q-rung orthopair fuzzy cotangent similarity measures between M and N, which consider the maximum distance based on the membership and non-membership degree.

$$q - ROFCot^2(M, N) = \frac{1}{n} \sum_{j=1}^{n} \cot \left[\frac{\pi}{4} + \frac{\pi}{8} \left(\begin{array}{c} \left| \alpha_M^q(x_j) - \alpha_N^q(x_j) \right| + \\ \left| \beta_M^q(x_j) - \beta_N^q(x_j) \right| \end{array} \right) \right] \tag{33}$$

where $q - ROFCot^2(M, N)$ means the q-rung orthopair fuzzy cotangent similarity measures between M and N, which consider the sum of distance based on the membership and non-membership degree.

If we consider three terms—membership degree, non-membership degree and indeterminacy membership—which are contained in q-ROFSs, assume that there are two q-rung orthopair fuzzy sets $M = \left\{ \left\langle x_j, \alpha_M(x_j), \beta_M(x_j), \pi_M(x_j) \right\rangle | x_j \in X \right\} (j = 1, 2, \ldots, n)$ and $N = \left\{ \left\langle x_j, \alpha_N(x_j), \beta_N(x_j), \pi_N(x_j) \right\rangle | x_j \in X \right\} (j = 1, 2, \ldots, n)$, then the q-rung orthopair fuzzy cotangent (q-ROFCot) similarity measures between M and N can be expressed as

$$q - ROFCot^3(M, N) = \frac{1}{n} \sum_{j=1}^{n} \cot \left[\frac{\pi}{4} + \frac{\pi}{4} \left(\max \left(\begin{array}{c} \left| \alpha_M^q(x_j) - \alpha_N^q(x_j) \right|, \\ \left| \beta_M^q(x_j) - \beta_N^q(x_j) \right|, \\ \left| \pi_A^2(x_j) - \pi_B^2(x_j) \right| \end{array} \right) \right) \right] \tag{34}$$

where $q - ROFCot^3(M, N)$ means the q-rung orthopair fuzzy cotangent similarity measures between M and N, which consider the maximum distance based on the membership, indeterminacy membership, and non-membership degree.

$$q - ROFCot^4(M, N) = \frac{1}{n} \sum_{j=1}^{n} \cot\left[\frac{\pi}{4} + \frac{\pi}{8}\left(\begin{array}{c} \left|\alpha_M^q(x_j) - \alpha_N^q(x_j)\right| + \\ \left|\beta_M^q(x_j) - \beta_N^q(x_j)\right| + \\ \left|\pi_A^2(x_j) - \pi_B^2(x_j)\right| \end{array}\right)\right] \tag{35}$$

where $q - ROFCot^4(M, N)$ means the q-rung orthopair fuzzy cotangent similarity measures between M and N, which consider the sum of distance based on the membership, indeterminacy membership, and non-membership degree.

Consider the weighting vector of the elements in q-ROFS, the q-rung orthopair fuzzy weighted cotangent (q-ROFWCot) similarity measure between two q-rung orthopair fuzzy sets (q-ROFSs) M and N can be shown as follows.

$$q - ROFWCot^1(M, N) = \sum_{j=1}^{n} \omega_j \cot\left[\frac{\pi}{4} + \frac{\pi}{4}\left(\max\left(\begin{array}{c} \left|\alpha_M^q(x_j) - \alpha_N^q(x_j)\right|, \\ \left|\beta_M^q(x_j) - \beta_N^q(x_j)\right| \end{array}\right)\right)\right] \tag{36}$$

where $q - ROFWCot^1(M, N)$ means the q-rung orthopair fuzzy weighted cotangent similarity measures between M and N, which consider the maximum distance based on the membership and non-membership degree.

$$q - ROFWCot^2(M, N) = \sum_{j=1}^{n} \omega_j \cot\left[\frac{\pi}{4} + \frac{\pi}{8}\left(\begin{array}{c} \left|\alpha_M^q(x_j) - \alpha_N^q(x_j)\right| + \\ \left|\beta_M^q(x_j) - \beta_N^q(x_j)\right| \end{array}\right)\right] \tag{37}$$

where $q - ROFWCot^2(M, N)$ means the q-rung orthopair fuzzy weighted cotangent similarity measures between M and N, which consider the sum of distance based on the membership and non-membership degree.

$$q - ROFWCot^3(M, N) = \sum_{j=1}^{n} \omega_j \cot\left[\frac{\pi}{4} + \frac{\pi}{4}\left(\max\left(\begin{array}{c} \left|\alpha_M^q(x_j) - \alpha_N^q(x_j)\right|, \\ \left|\beta_M^q(x_j) - \beta_N^q(x_j)\right|, \\ \left|\pi_A^2(x_j) - \pi_B^2(x_j)\right| \end{array}\right)\right)\right] \tag{38}$$

where $q - ROFWCot^3(M, N)$ means the q-rung orthopair fuzzy weighted cotangent similarity measures between M and N, which consider the maximum distance based on the membership, indeterminacy membership and non-membership degree.

$$q - ROFWCot^4(M, N) = \sum_{j=1}^{n} \omega_j \cot\left[\frac{\pi}{4} + \frac{\pi}{8}\left(\begin{array}{c} \left|\alpha_M^q(x_j) - \alpha_N^q(x_j)\right| + \\ \left|\beta_M^q(x_j) - \beta_N^q(x_j)\right| + \\ \left|\pi_A^2(x_j) - \pi_B^2(x_j)\right| \end{array}\right)\right] \tag{39}$$

where $q - ROFWCot^4(M, N)$ means the q-rung orthopair fuzzy weighted cotangent similarity measures between M and N, which consider the sum of distance based on the membership, indeterminacy membership and non-membership degree.

Where $\omega = (\omega_1, \omega_2, \cdots, \omega_n)^T$ indicates the weighting vector of the elements $x_j(j = 1, 2, \cdots, n)$ contained in q-ROFS and the weighting vector satisfies $\omega_j \in [0, 1]$, $j = 1, 2, \cdots, n$, $\sum_{j=1}^{n} \omega_j = 1$. Especially, when we let weighting vector be $\omega = (1/n, 1/n, \cdots, 1/n)^T$, then the weighted cotangent similarity measure will reduce to cotangent similarity measure. In other words, when $\omega_j = \frac{1}{n}$, $j = 1, 2 \cdots, n$, the $q - ROFWCot^k(M, N) = q - ROFWCot^k(M, N)(k = 1, 2, 3, 4)$.

Example 4. *Suppose there are two q-ROFSs* $M = \{(x_1, 0.7, 0.4), (x_2, 0.5, 0.6), (x_3, 0.3, 0.8)\}$ *and* $N = \{(x_1, 0.9, 0.2), (x_2, 0.4, 0.3), (x_3, 0.7, 0.6)\}$, *assume* $q = 3, \omega_j = (0.2, 0.3, 0.5)$, *then according to Equation (33), the weighted cotangent similarity measure between M and N can be calculated as*

$$q - ROFWCot^1(M, N) = \sum_{j=1}^{n} \omega_j \cot\left[\frac{\pi}{4} + \frac{\pi}{4}\left(\max\left(\left|\alpha_M^q(x_j) - \alpha_N^q(x_j)\right|, \left|\beta_M^q(x_j) - \beta_N^q(x_j)\right|\right)\right)\right]$$

$$= \left(\begin{array}{c} 0.2 \times \cot\left[\frac{\pi}{4} + \frac{\pi}{4}\max\left(\left|0.7^3 - 0.9^3\right|, \left|0.4^3 - 0.2^3\right|\right)\right] + 0.3 \times \cot\left[\frac{\pi}{4} + \frac{\pi}{4}\max\left(\begin{array}{c}\left|0.5^3 - 00.4^3\right|, \\ \left|0.6^3 - 0.3^3\right|\end{array}\right)\right] \\ +0.5 \times \cot\left[\frac{\pi}{4} + \frac{\pi}{4}\left(\max\left(\left|0.3^3 - 0.7^3\right|, \left|0.8^3 - 0.6^3\right|\right)\right)\right] \end{array} \right)$$

$$= 0.6245$$

4. Applications

In this section, we shall give two applications about the cosine similarity measures and cotangent similarity measures under q-rung orthopair fuzzy environment. The methods proposed in this paper are applied to pattern recognition and scheme selection to demonstrate the effectiveness of these methods.

4.1. Numerical Example 1—Pattern Recognition

There is no doubt that the quantity of construction mainly depends on the quality of building materials. Therefore, building material inspection is the premise of good engineering quality. In the selection of materials must be strictly controlled. Inspection can not only enable the builders to accurately identify qualified materials, but also ensure and improve the quality of the project. Let us consider the pattern recognition problems about classification of building materials, suppose there are five known building materials $A_i(i = 1, 2, 3, 4, 5)$, which are depicted by the q-ROFSs $A_i(i = 1, 2, 3, 4, 5)$ in the feature space $X = \{x_1, x_2, x_3, x_4, x_5\}$ as

$$A_1 = \{(x_1, \, 0.5, 0.8), (x_2, \, 0.6, 0.4), (x_3, \, 0.8, 0.3), (x_4, \, 0.6, 0.9), (x_5, \, 0.1, 0.4)\}$$
$$A_2 = \{(x_1, \, 0.6, 0.7), (x_2, \, 0.7, 0.3), (x_3, \, 0.6, 0.2), (x_4, \, 0.8, 0.6), (x_5, \, 0.3, 0.5)\}$$
$$A_3 = \{(x_1, \, 0.3, 0.4), (x_2, \, 0.7, 0.5), (x_3, \, 0.9, 0.3), (x_4, \, 0.4, 0.8), (x_5, \, 0.2, 0.3)\}$$
$$A_4 = \{(x_1, \, 0.5, 0.3), (x_2, \, 0.4, 0.4), (x_3, \, 0.6, 0.2), (x_4, \, 0.4, 0.7), (x_5, \, 0.2, 0.6)\}$$
$$A_5 = \{(x_1, \, 0.4, 0.7), (x_2, \, 0.2, 0.6), (x_3, \, 0.5, 0.4), (x_4, \, 0.5, 0.3), (x_5, \, 0.4, 0.2)\}$$

Consider an unknown building material $A \in q - ROFSs(X)$ that will be recognized, which is depicted as

$$A = \{(x_1, \, 0.7, 0.6), (x_2, \, 0.8, 0.2), (x_3, \, 0.4, 0.3), (x_4, \, 0.7, 0.8), (x_5, \, 0.4, 0.2)\}$$

The purpose of this problem is classify the pattern A in one of the following classes, $A_1, A_2, A_3, A_4,$ or A_5. For it, the cosine similarity measures and cotangent similarity measures proposed in this paper have been utilized to compute the similarity from A to $A_i(i = 1, 2, 3, 4, 5)$ and the results are listed as follows. (Suppose $q = 3$)

For q-rung orthopair fuzzy cosine ($q\text{-}ROFC^1$) measures, we can obtain

$$
\begin{aligned}
&q - ROFC^1(A_1, A) \\
&= \frac{1}{5}\left(
\begin{array}{c}
\dfrac{\left(0.5^3\times0.7^3+0.8^3\times0.6^3\right)}{\sqrt{(0.5^3)^2+(0.8^3)^2}\times\sqrt{(0.7^3)^2+(0.6^3)^2}} + \dfrac{\left(0.6^3\times0.8^3+0.4^3\times0.2^3\right)}{\sqrt{(0.6^3)^2+(0.4^3)^2}\times\sqrt{(0.8^3)^2+(0.2^3)^2}} \\[2em]
+ \dfrac{\left(0.8^3\times0.4^3+0.3^3\times0.3^3\right)}{\sqrt{(0.8^3)^2+(0.3^3)^2}\times\sqrt{(0.4^3)^2+(0.3^3)^2}} + \dfrac{\left(0.6^3\times0.7^3+0.9^3\times0.8^3\right)}{\sqrt{(0.6^3)^2+(0.9^3)^2}\times\sqrt{(0.7^3)^2+(0.8^3)^2}} \\[2em]
+ \dfrac{\left(0.1^3\times0.4^3+0.4^3\times0.2^3\right)}{\sqrt{(0.1^3)^2+(0.4^3)^2}\times\sqrt{(0.4^3)^2+(0.2^3)^2}}
\end{array}
\right) \\[1em]
&= 0.7443
\end{aligned}
$$

Similarly, we can get

$$
q - ROFC^1(A_2, A) = 0.8033, q - ROFC^1(A_3, A) = 0.7988,
$$
$$
q - ROFC^1(A_4, A) = 0.7345, q - ROFC^1(A_5, A) = 0.6897.
$$

(1) For q-rung orthopair fuzzy cosine ($q\text{-}ROFC^2$) measures we can obtain

$$
q - ROFC^2(A_1, A) = 0.8795, q - ROFC^2(A_2, A) = 0.9116,
$$
$$
q - ROFC^2(A_3, A) = 0.9124, q - ROFC^2(A_4, A) = 0.8766,
$$
$$
q - ROFC^2(A_5, A) = 0.8543.
$$

(2) For q-rung orthopair fuzzy cosine similarity ($q\text{-}ROFCS^1$) measures we can obtain

$$
q - ROFCS^1(A_1, A) = 0.8975, q - ROFCS^1(A_2, A) = 0.9588,
$$
$$
q - ROFCS^1(A_3, A) = 0.8496, q - ROFCS^1(A_4, A) = 0.9057,
$$
$$
q - ROFCS^1(A_5, A) = 0.8654.
$$

(3) For q-rung orthopair fuzzy cosine similarity ($q\text{-}ROFCS^2$) measures we can obtain

$$
q - ROFCS^2(A_1, A) = 0.9559, q - ROFCS^2(A_2, A) = 0.9774,
$$
$$
q - ROFCS^2(A_3, A) = 0.9498, q - ROFCS^2(A_4, A) = 0.9561,
$$
$$
q - ROFCS^2(A_5, A) = 0.9291.
$$

(4) For q-rung orthopair fuzzy cosine similarity ($q\text{-}ROFCS^3$) measures we can obtain

$$
q - ROFCS^3(A_1, A) = 0.8975, q - ROFCS^3(A_2, A) = 0.9588,
$$
$$
q - ROFCS^3(A_3, A) = 0.8364, q - ROFCS^3(A_4, A) = 0.8880,
$$
$$
q - ROFCS^3(A_5, A) = 0.8540.
$$

(5) For q-rung orthopair fuzzy cosine similarity ($q\text{-}ROFCS^4$) measures we can obtain

$$
q - ROFCS^3(A_1, A) = 0.8964, q - ROFCS^3(A_2, A) = 0.9630,
$$
$$
q - ROFCS^3(A_3, A) = 0.8386, q - ROFCS^3(A_4, A) = 0.8701,
$$
$$
q - ROFCS^3(A_5, A) = 0.8362.
$$

(6) For q-rung orthopair fuzzy cotangent similarity ($q\text{-}ROFCot^1$) measures we can obtain

$$
q - ROFCot^1(A_1, A) = 0.6618, q - ROFCot^1(A_2, A) = 0.7633,
$$
$$
q - ROFCot^1(A_3, A) = 0.6362, q - ROFCot^1(A_4, A) = 0.6613,
$$
$$
q - ROFCot^1(A_5, A) = 0.6766.
$$

(7) For q-rung orthopair fuzzy cotangent similarity (q-ROFCot^2) measures we can obtain

$$q - ROFCot^2(A_1, A) = 0.7571, q - ROFCot^2(A_2, A) = 0.8257,$$
$$q - ROFCot^2(A_3, A) = 0.7613, q - ROFCot^2(A_4, A) = 0.7544,$$
$$q - ROFCot^2(A_5, A) = 0.7522.$$

(8) For q-rung orthopair fuzzy cotangent similarity (q-ROFCot^3) measures we can obtain

$$q - ROFCot^3(A_1, A) = 0.6618, q - ROFCot^3(A_2, A) = 0.7633,$$
$$q - ROFCot^3(A_3, A) = 0.6198, q - ROFCot^3(A_4, A) = 0.6318,$$
$$q - ROFCot^3(A_5, A) = 0.6596.$$

(9) For q-rung orthopair fuzzy cotangent similarity (q-ROFCot^4) measures we can obtain

$$q - ROFCot^4(A_1, A) = 0.6588, q - ROFCot^4(A_2, A) = 0.7702,$$
$$q - ROFCot^4(A_3, A) = 0.6259, q - ROFCot^4(A_4, A) = 0.6085,$$
$$q - ROFCot^4(A_5, A) = 0.6496.$$

Thereafter, the above computing results are concluded to list in Table 1 as follows.

Table 1. The similarity measures between $A_i(i = 1, 2, 3, 4, 5)$ and A.

Similarity Measures	(A_1, A)	(A_2, A)	(A_3, A)	(A_4, A)	(A_5, A)
$q - ROFC^1(A_i, A)$	0.7433	0.8003	0.7988	0.7345	0.6897
$q - ROFC^2(A_i, A)$	0.8795	0.9116	0.9124	0.8766	0.8543
$q - ROFCS^1(A_i, A)$	0.8975	0.9588	0.8496	0.9057	0.8654
$q - ROFCS^2(A_i, A)$	0.9559	0.9774	0.9498	0.9561	0.9291
$q - ROFCS^3(A_i, A)$	0.8975	0.9588	0.8364	0.8880	0.8540
$q - ROFCS^4(A_i, A)$	0.8964	0.9630	0.8386	0.8701	0.8362
$q - ROFCot^1(A_i, A)$	0.6618	0.7633	0.6362	0.6613	0.6766
$q - ROFCot^2(A_i, A)$	0.7571	0.8257	0.7613	0.7544	0.7522
$q - ROFCot^3(A_i, A)$	0.6618	0.7633	0.6198	0.6318	0.6596
$q - ROFCot^4(A_i, A)$	0.6588	0.7702	0.6259	0.6085	0.6496

According to the above calculated results listed in Table 1, except for $q - ROFC^2(A_i, A)$, we can easily find that the degree of similarity between A_2 and A is the largest as derived by the other nine similarity measures. This indicates the nine similarity measures allocate the unknown building material A to the known building material A_2 based on the principle of maximum similarity between q-rung orthopair fuzzy sets (q-ROFSs).

In practical decision-making problems, it is important to take the weights of elements into account, if we let the weights of elements $x_i(i = 1, 2, 3, 4, 5)$ be $\omega_i = (0.15, 0.20, 0.25, 0.10, 0.30)$, respectively. Then the weighted cosine similarity measures and weighted cotangent similarity measures proposed in this paper have been utilized to compute the similarity from A to $A_i(i = 1, 2, 3, 4, 5)$ and the results are listed in Table 2 (suppose $q = 3$). (The calculation process is similar to not weighted situation, so it is omitted here.)

According to the above calculated results listed in Table 2, except for $q - ROFC^2(A_i, A)$ and $q - ROFWC^2(A_i, A)$, we can easily find that the degree of similarity between A_2 and A is the largest one as derived by the other eight similarity measures. This indicates the eight similarity measures allocate the unknown building material A to the known building material A_2 based on the principle of maximum similarity between q-rung orthopair fuzzy sets (q-ROFSs).

Table 2. The weighted similarity measures between $A_i (i = 1, 2, 3, 4, 5)$ and A.

Similarity Measures	(A_1, A)	(A_2, A)	(A_3, A)	(A_4, A)	(A_5, A)
$q - ROFWC^1(A_i, A)$	0.6728	0.7515	0.7553	0.6584	0.7336
$q - ROFWC^2(A_i, A)$	0.8457	0.8901	0.8937	0.8406	0.8735
$q - ROFWCS^1(A_i, A)$	0.8962	0.9673	0.8398	0.9114	0.8976
$q - ROFWCS^2(A_i, A)$	0.9601	0.9838	0.9487	0.9621	0.9464
$q - ROFWCS^3(A_i, A)$	0.8962	0.9673	0.8299	0.8986	0.8910
$q - ROFWCS^4(A_i, A)$	0.8961	0.9693	0.8303	0.8883	0.8830
$q - ROFWCot^1(A_i, A)$	0.6740	0.7831	0.6478	0.6735	0.7474
$q - ROFWCot^2(A_i, A)$	0.7740	0.8482	0.7700	0.7733	0.8065
$q - ROFWCot^3(A_i, A)$	0.6740	0.7831	0.6356	0.6522	0.7324
$q - ROFWCot^4(A_i, A)$	0.6727	0.7866	0.6356	0.6389	0.7284

In order to illustrate the effective and scientific of our proposed methods, we shall compare with other decision-making methods, such as the q-rung orthopair fuzzy weighted averaging (q-ROFWA) operator and the q-rung orthopair fuzzy weighted geometric (q-ROFWG) operator proposed by Liu and Wang [40], we obtained the following results in Table 3.

Table 3. The fused values of $A_i (i = 1, 2, 3, 4, 5)$ and A.

	The q-ROFWA Operator	*The q-ROFWG Operator*
A_1	(0.2387,0.4479)	(0.3665,0.2428)
A_2	(0.2300,0.3845)	(0.5173,0.1262)
A_3	(0.3453,0.3827)	(0.4263,0.1162)
A_4	(0.0980,0.3848)	(0.3719,0.1258)
A_5	(0.0751,0.3724)	(0.3766,0.1249)
A	(0.2479,0.2998)	(0.5285,0.1123)

Then, based on distance measure between q-rung orthopair fuzzy numbers (q-ROFNs), we can allocate the unknown building material A to the known building material A_i, the distance measure $d(M, N)$ between q-ROFNs $M = (\alpha_1, \beta_1)$ and $N = (\alpha_2, \beta_2)$ can be depicted as

$$d(M, N) = \frac{\left|(\alpha_1)^q - (\alpha_2)^q\right| + \left|(\beta_1)^q - (\beta_2)^q\right|}{2} \tag{40}$$

For q-ROFWA operator, we can obtain the distance results $d(A_i, A)$ as

$$d(A_1, A) = 0.0323, d(A_2, A) = 0.0165, d(A_3, A) = 0.0275,$$
$$d(A_4, A) = 0.0222, d(A_5, A) = 0.0197$$

For q-ROFWG operator, we can obtain the distance results $d(A_i, A)$ as

$$d(A_1, A) = 0.0556, d(A_2, A) = 0.0049, d(A_3, A) = 0.0352,$$
$$d(A_4, A) = 0.0484, d(A_5, A) = 0.0474$$

From above analysis, for q-ROFWA and q-ROFWG operators, the distance measure between A_2 and A is the minimum one. This indicates that q-ROFWA and q-ROFWG operators allocate the unknown building material A to the known building material A_2. Although, based on the two operators and our developed methods, we can derive the same results, however, the q-ROFWA and q-ROFWG operators have the limitation of considering the interrelationship between attributes; our developed methods can overcome this disadvantage and derive more accuracy and scientific decision-making results.

4.2. Numerical Example 2—Scheme Selection

In this section, we shall present a numerical example to show scheme selection of construction project with q-rung orthopair fuzzy information in order to illustrate the method proposed in this paper. There is a panel with five possible construction projects. $Y_i (i = 1, 2, 3, 4, 5)$ to select. Experts selected five attributes to evaluate from the five possible construction projects: ① G_1 is the capital and technical factors; ② G_2 is the Hoisting construction operation factors; ③ G_3 is the PC component installation factor; ④ G_4 is the internal and external environmental risk factors; and ⑤ G_5 is the professional management level factors. The five possible construction projects $Y_i (i = 1, 2, 3, 4, 5)$ are to be evaluated using the q-rung orthopair fuzzy information by the decision maker under the above five attributes which listed as follows.

$$Y_1 = \{(G_1, 0.6, 0.7), (G_2, 0.5, 0.8), (G_3, 0.6, 0.3), (G_4, 0.7, 0.3), (G_5, 0.4, 0.6)\}$$
$$Y_2 = \{(G_1, 0.7, 0.4), (G_2, 0.8, 0.3), (G_3, 0.5, 0.6), (G_4, 0.2, 0.5), (G_5, 0.6, 0.3)\}$$
$$Y_3 = \{(G_1, 0.6, 0.3), (G_2, 0.4, 0.2), (G_3, 0.7, 0.4), (G_4, 0.5, 0.2), (G_5, 0.9, 0.4)\}$$
$$Y_4 = \{(G_1, 0.8, 0.7), (G_2, 0.5, 0.6), (G_3, 0.4, 0.6), (G_4, 0.6, 0.3), (G_5, 0.4, 0.2)\}$$
$$Y_5 = \{(G_1, 0.7, 0.2), (G_2, 0.4, 0.3), (G_3, 0.5, 0.6), (G_4, 0.3, 0.5), (G_5, 0.7, 0.2)\}$$

Let

$$Y^+ = \left\{ \begin{array}{l} \left(G_1, \max_i(\alpha_{i1}), \min_i(\beta_{i1})\right), \left(G_2, \max_i(\alpha_{i2}), \min_i(\beta_{i2})\right), \\ \left(G_3, \max_i(\alpha_{i3}), \min_i(\beta_{i3})\right), \left(G_4, \max_i(\alpha_{i4}), \min_i(\beta_{i4})\right), \\ \left(G_5, \max_i(\alpha_{i5}), \min_i(\beta_{i5})\right) \end{array} \right\}$$

According to the evaluation results given in Y_1, Y_2, Y_3, Y_4 *and* Y_5, we can easily obtain

$$Y^+ = \{(G_1, 0.8, 0.2), (G_2, 0.8, 0.2), (G_3, 0.7, 0.3), (G_4, 0.7, 0.2), (G_5, 0.9, 0.2)\}$$

Then the weighted cosine similarity measures and weighted cotangent similarity measures proposed in this paper have been utilized to compute the similarity from Y^+ to $Y_i (i = 1, 2, 3, 4, 5)$ and the results are listed in Table 4 (suppose $q = 3$).

Table 4. The similarity measures between $Y_i (i = 1, 2, 3, 4, 5)$ and Y^+.

Similarity Measures	(Y_1, Y)	(Y_2, Y)	(Y_3, Y)	(Y_4, Y)	(Y_5, Y)
$q - ROFC^1(Y_i, Y)$	0.6181	0.9092	0.9958	0.7401	0.7457
$q - ROFC^2(Y_i, Y)$	0.8215	0.7266	0.9975	0.8818	0.8837
$q - ROFCS^1(Y_i, Y)$	0.8099	0.8714	0.9185	0.8147	0.8741
$q - ROFCS^2(Y_i, Y)$	0.8827	0.8927	0.9785	0.9303	0.9542
$q - ROFCS^3(Y_i, Y)$	0.8099	0.9571	0.9185	0.8147	0.8741
$q - ROFCS^4(Y_i, Y)$	0.8161	0.8927	0.9194	0.8249	0.8737
$q - ROFCot^1(Y_i, Y)$	0.6088	0.8912	0.7289	0.5642	0.6103
$q - ROFCot^2(Y_i, Y)$	0.6851	0.6836	0.8494	0.7076	0.7453
$q - ROFCot^3(Y_i, Y)$	0.6088	0.7781	0.7289	0.5642	0.6103
$q - ROFCot^4(Y_i, Y)$	0.6160	0.6836	0.7371	0.5802	0.6089

According to the above calculated results listed in Table 4, we can easily find that the degree of similarity between Y_3 and Y is the largest one as derived by all ten similarity measures. This indicates all ten similarity measures think the alternative Y_3 is closest to be best alternative Y^+ based on the principle of maximum similarity between q-rung orthopair fuzzy sets (q-ROFSs). In other words, Y_3 is the best scheme selection for the construction project.

In practical decision-making problems, it is important to take the weights of elements into account, if we let the weights of elements $x_i (i = 1, 2, 3, 4, 5)$ be $\omega_i = (0.15, 0.20, 0.25, 0.10, 0.30)$, respectively. Then the weighted cosine similarity measures and weighted cotangent similarity measures proposed in this paper have been utilized to compute the similarity from Y to $Y_i(i = 1, 2, 3, 4, 5)$ and the results are listed in Table 5 (suppose $q = 3$).

Table 5. The weighted similarity measures between $Y_i(i = 1, 2, 3, 4, 5)$ and Y^+.

Similarity Measures	(Y_1,Y)	(Y_2,Y)	(Y_3,Y)	(Y_4,Y)	(Y_5,Y)
$q - ROFWC^1(Y_i, Y)$	0.5703	0.7963	0.9956	0.7160	0.8006
$q - ROFWC^2(Y_i, Y)$	0.7994	0.9060	0.9974	0.8711	0.9095
$q - ROFWCS^1(Y_i, Y)$	0.7659	0.8744	0.9292	0.7689	0.8672
$q - ROFWCS^2(Y_i, Y)$	0.8659	0.9529	0.9813	0.9160	0.9533
$q - ROFWCS^3(Y_i, Y)$	0.7659	0.8744	0.9292	0.7689	0.8672
$q - ROFWCS^4(Y_i, Y)$	0.7714	0.8727	0.9303	0.7765	0.8667
$q - ROFWCot^1(Y_i, Y)$	0.5508	0.6633	0.7663	0.5124	0.6001
$q - ROFWCot^2(Y_i, Y)$	0.6494	0.7678	0.8706	0.6723	0.7405
$q - ROFWCot^3(Y_i, Y)$	0.5508	0.6633	0.7663	0.5124	0.6001
$q - ROFWCot^4(Y_i, Y)$	0.5571	0.6607	0.7779	0.5240	0.5983

According to the above calculated results listed in Table 5, we can easily find that the degree of similarity between Y_3 and Y is the largest one as derived by other nine similarity measures. This indicates all ten similarity measures; the alternative Y_3 is closest to be best alternative Y^+ based on the principle of maximum similarity between q-rung orthopair fuzzy sets (q-ROFSs). In other words, Y_3 is the best scheme selection for the construction project.

In order to illustrate the effective and scientific of our proposed methods, we shall compare with other decision-making methods such as the q-rung orthopair fuzzy weighted averaging (q-ROFWA) operator and the q-rung orthopair fuzzy weighted geometric (q-ROFWG) operator proposed by Liu and Wang [40], we can obtain the result which is listed in Table 6.

Table 6. The fused results of $Y_i(i = 1, 2, 3, 4, 5)$.

	The q-ROFWA Operator	The q-ROFWG Operator
Y_1	(0.1697,0.5103)	(0.5203,0.2511)
Y_2	(0.2693,0.3921)	(0.5568,0.0932)
Y_3	(0.4287,0.3112)	(0.6377,0.0420)
Y_4	(0.1772,0.4121)	(0.4833,0.1628)
Y_5	(0.2121,0.3129)	(0.5287,0.0799)

Then according to the score functions of q-rung orthopair fuzzy numbers (q-ROFNs), we can obtain the score values of $Y_i(i = 1, 2, 3, 4, 5)$ which is listed in Table 7.

Table 7. The score values of $Y_i(i = 1, 2, 3, 4, 5)$.

	The q-ROFWA Operator	The q-ROFWG Operator
$s(Y_1)$	0.4360	0.5625
$s(Y_2)$	0.4796	0.5859
$s(Y_3)$	0.5243	0.6296
$s(Y_4)$	0.4678	0.5543
$s(Y_5)$	0.4895	0.5736

Then based on score values, the ordering of $Y_i(i = 1, 2, 3, 4, 5)$ can be determined in Table 8.

From above analysis, based on the two operators and our developed methods, we can obtain that the ordering of alternatives are slightly different and the best results are same, however, the q-ROFWA and q-ROFWG operators have the limitation of considering the interrelationship between attributes, our developed methods can overcome this disadvantage and derive more accuracy and scientific decision-making results.

Table 8. The ordering of $Y_i (i = 1, 2, 3, 4, 5)$.

	Ordering
The q – ROFWA operator	$Y_3 > Y_5 > Y_2 > Y_4 > Y_1$
The q – ROFWG operator	$Y_3 > Y_2 > Y_5 > Y_1 > Y_4$

4.3. Advantages of the Proposed Similarity Measures

Although, the intuitionistic fuzzy sets (IFSs), defined by Atanassov's [1,2], have been broadly applied in different areas, for some special cases, such as when membership degree and non-membership degree are given as 0.7 and 0.8, it is clear that IFSs theory cannot satisfy this situation. The q-rung orthopair fuzzy set (q-ROFS) is also denoted by the degree of membership and non-membership, whose *q-th* power sum of them is restricted to 1; obviously, the q-ROFS is more general than the q-ROFS and can express more fuzzy information. In other words, the q-ROFS can deal with the MADM problems which IFS cannot and it is clear that IFS is a part of the q-ROFS, which indicates q-ROFS can be more effective and powerful to deal with fuzzy and uncertain decision-making problems. Thus, the MADM problem with q-rung orthopair fuzzy information is more effective and suitable for practical scientific and engineering applications.

To date, we can get that the cosine similarity measures and cotangent similarity measures [13,15–17,19] with IFSs have been investigated by a large amount of scholars; however, as mentioned above, there are some special cases that cannot be described by IFS. Therefore, the algorithms based on similarity measures with IFS can't deal with such problems. The cosine similarity measures and cotangent similarity measures with intuitionistic fuzzy information are special case of our proposed similarity measures with q-rung orthopair fuzzy information in this paper. Thus, our defined similarity measures are more suitable and generalized to deal with the real-life problem more accurately than the existing ones.

4.4. Discussion

According to above two numerical examples, we can easily find our proposed methods can express more fuzzy information and apply broadly situations in real MADM problems. Based on the q-rung orthopair fuzzy set (q-ROFS), we developed ten q-rung orthopair fuzzy similarity measures; our research results are more suitable for MADM problems than intuitionistic fuzzy similarity measures and Pythagorean fuzzy similarity measures. For pattern recognition problems, we accurately allocated the unknown building material A to the known building material A_2. For scheme selection, by utilizing our developed ten similarity measures, we obtained the best scheme selection of construction project.

Furthermore, in complicated decision-making environment, the decision-maker's risk attitude is an important factor to think about, our methods can make this come true by altering the parameters whereas other decision-making ways such as q-ROFWA and q-ROFWA operator do not have the ability that dynamic adjust to the parameter according to the decision maker's risk attitude, so it is difficult to solve the risk multiple attribute decision-making in real practice.

5. Conclusions

According to the intuitionistic fuzzy cosine similarity measures and cotangent similarity measures, based on q-rung orthopair fuzzy sets (q-ROFSs), we proposed another form of ten similarity measures by considering the function of membership degree, nonmembership degree and indeterminacy membership degree in q-ROFSs. In addition, we utilized our presented ten similarity measures and ten weighted similarity measures between q-ROFSs to deal MADM problems, including pattern recognition and scheme selection. Finally, two numerical examples and some comparative analysis were provided to illustrate the scientific and effective of the similarity measures for pattern recognition and scheme selection. By utilizing our developed ten similarity measures, we can deal with MADM problems regarding pattern recognition and scheme selection. When comparing our developed ten similarity measures with the q-rung orthopair fuzzy weighted average (q-ROFWA) operator and q-rung orthopair fuzzy weighted geometric (q-ROFWG) operator, our proposed methods can be applied in scheme selection and pattern recognition applications as the q-ROFWA and q-ROFWG operators can be only used to select best alternatives. Moreover, q-ROFWA and q-ROFWG operators have the limitation of considering the interrelationship between fused arguments; our proposed methods can overcome this disadvantage and derive more accuracy and reasonable decision-making results. In the future, works concerning q-ROFSs could focus on dealing with other kinds of decision-making problems such as: staff selection, investment selection, machine selection, project selection, manufacturing systems, etc. [52–59].

Author Contributions: All the authors conceived and worked together to achieve this work; P.W. and J.W. compiled the computing program by Excel and analyzed the data; J.W. and G.W. wrote the paper. Finally, all the authors have read and approved the final manuscript.

Funding: The work was supported by the National Natural Science Foundation of China under Grant No. 61174149 and 71571128, the Humanities and Social Sciences Foundation of Ministry of Education of the People's Republic of China (Grant No. 14YJCZH091, 15XJA630006), and the Construction Plan of Scientific Research Innovation Team for Colleges and Universities in Sichuan Province (15TD0004).

Conflicts of Interest: The authors declare no conflict of interest.

References

1. Atanassov, K. Intuitionistic fuzzy sets. *Fuzzy Sets Syst.* **1986**, *20*, 87–96. [CrossRef]
2. Atanassov, K. More on intuitionistic fuzzy sets. *Fuzzy Sets Syst.* **1989**, *33*, 37–46. [CrossRef]
3. Zeng, W.; Li, D.; Yin, Q. Distance and similarity measures of Pythagorean fuzzy sets and their applications to multiple criteria group decision making. *Int. J. Intell. Syst.* **2018**, *33*, 2236–2254. [CrossRef]
4. Ye, J.; Fang, Z.B.; Cui, W.H. Vector Similarity Measures of Q-Linguistic Neutrosophic Variable Sets and Their Multi-Attribute Decision Making Method. *Symmetry* **2018**, *10*, 531. [CrossRef]
5. Ye, J. Multiple-attribute decision-making method using similarity measures of single-valued neutrosophic hesitant fuzzy sets based on least common multiple cardinality. *J. Intell. Fuzzy Syst.* **2018**, *34*, 4203–4211. [CrossRef]
6. Wu, H.B.; Yuan, Y.; Wei, L.J.; Pei, L.D. On entropy, similarity measure and cross-entropy of single-valued neutrosophic sets and their application in multi-attribute decision making. *Soft Comput.* **2018**, *22*, 7367–7376. [CrossRef]
7. Li, D.F.; Cheng, C.T. New similarity measures of intuitionistic fuzzy sets and application to pattern recognitions. *Pattern Recognit. Lett.* **2002**, *23*, 221–225.
8. Mitchell, H.B. On the Dengfeng-Chuntian similarity measure and its application to pattern recognition. *Pattern Recognit. Lett.* **2003**, *24*, 3101–3104. [CrossRef]
9. Park, J.S.; Kwun, Y.C.; Park, J.H. Some Results and Example for Compatible Maps of Type(beta) on Intuitionistic Fuzzy Metric Space. In *Fuzzy Information and Engineering, Volume 2*; Cao, B., Li, T.F., Zhang, C.Y., Eds.; 2009; Volume 62, pp. 629–638.
10. Li, D.F. Multiattribute decision making method based on generalized OWA operators with intuitionistic fuzzy sets. *Expert Syst. Appl.* **2010**, *37*, 8673–8678. [CrossRef]
11. Torra, V.; Narukawa, Y. On Hesitant Fuzzy Sets and Decision. In Proceedings of the 2009 IEEE International Conference on Fuzzy Systems, Jeju Island, Korea, 20–24 August 2009; pp. 1378–1382.
12. Xia, M.; Xu, Z. Some new similarity measures for intuitionistic fuzzy values and their application in group decision making. *J. Syst. Sci. Syst. Eng.* **2010**, *19*, 430–452. [CrossRef]
13. Ye, J. Cosine similarity measures for intuitionistic fuzzy sets and their applications. *Math. Comput. Model.* **2011**, *53*, 91–97. [CrossRef]
14. Hung, K.-C. Applications of medical information: Using an enhanced likelihood measured approach based on intuitionistic fuzzy sets. *IIE Trans. Healthc. Syst. Eng.* **2012**, *2*, 224–231. [CrossRef]

15. Shi, L.L.; Ye, J. Study on Fault Diagnosis of Turbine Using an Improved Cosine Similarity Measure for Vague Sets. *J. Appl. Sci.* **2013**, *13*, 1781–1786. [CrossRef]

16. Tian, M.Y. A new fuzzy similarity based on cotangent function for medical diagnosis. *Adv. Model. Optim.* **2013**, *15*, 151–156.

17. Rajarajeswari, P.; Uma, N. Intuitionistic Fuzzy Multi Similarity Measure Based on Cotangent Function. *Int. J. Eng. Res. Technool.* **2013**, *2*, 1323–1329.

18. Szmidt, E. *Distances and Similarities in Intuitionistic Fuzzy Sets*; Springer: Berlin/Heidelberg, Germany, 2014.

19. Ye, J. Similarity measures of intuitionistic fuzzy sets based on cosine function for the decision making of mechanical design schemes. *J. Intell. Fuzzy Syst.* **2016**, *30*, 151–158. [CrossRef]

20. Wei, G.W. Some similarity measures for picture fuzzy sets and their applications. *Iran. J. Fuzzy Syst.* **2018**, *15*, 77–89.

21. Le Hoang, S.; Pham Hong, P. On the performance evaluation of intuitionistic vector similarity measures for medical diagnosis. *J. Intell. Fuzzy Syst.* **2016**, *31*, 1597–1608.

22. Wei, G.W.; Wei, Y. Similarity measures of Pythagorean fuzzy sets based on the cosine function and their applications. *Int. J. Intell. Syst.* **2018**, *33*, 634–652. [CrossRef]

23. Yager, R.R. Pythagorean membership grades in multicriteria decision making. *IEEE Trans. Fuzzy Syst.* **2014**, *22*, 958–965. [CrossRef]

24. Nie, R.-X.; Tian, Z.-P.; Wang, J.-Q.; Hu, J.-H. Pythagorean fuzzy multiple criteria decision analysis based on Shapley fuzzy measures and partitioned normalized weighted Bonferroni mean operator. *Int. J. Intell. Syst.* **2019**, *34*, 297–324. [CrossRef]

25. Zhu, L.; Liang, X.F.; Wang, L.; Wu, X.R. Generalized pythagorean fuzzy point operators and their application in multi-attributes decision making. *J. Intell. Fuzzy Syst.* **2018**, *35*, 1407–1418. [CrossRef]

26. Zhao, C.Y.; Tang, X.Y.; Yuan, L.J. MAGDM Method with Pythagorean 2-Tuple Linguistic Information and Applications in the HSE Performance Assessment of Laboratory. *Math. Probl. Eng.* **2018**, *2018*, 1–9. [CrossRef]

27. Zhang, C.; Li, D.Y.; Mu, Y.M.; Song, D. A Pythagorean Fuzzy Multigranulation Probabilistic Model for Mine Ventilator Fault Diagnosis. *Complexity* **2018**, *2018*, 1–19. [CrossRef]

28. Yang, W.; Shi, J.; Liu, Y.; Pang, Y.; Lin, R. Pythagorean Fuzzy Interaction Partitioned Bonferroni Mean Operators and Their Application in Multiple-Attribute Decision-Making. *Complexity* **2018**, *2018*, 1–25. [CrossRef]

29. Zhang, X.L.; Xu, Z.S. Extension of TOPSIS to Multiple Criteria Decision Making with Pythagorean Fuzzy Sets. *Int. J. Intell. Syst.* **2014**, *29*, 1061–1078. [CrossRef]

30. Peng, X.D.; Yang, Y. Some Results for Pythagorean Fuzzy Sets. *Int. J. Intell. Syst.* **2015**, *30*, 1133–1160. [CrossRef]

31. Reformat, M.Z.; Yager, R.R. Suggesting Recommendations Using Pythagorean Fuzzy Sets illustrated Using Netflix Movie Data. In *Information Processing and Management of Uncertainty in Knowledge-Based Systems, Pt I*; Laurent, A., Strauss, O., BouchonMeunier, B., Yager, R.R., Eds.; 2014; Volume 442, pp. 546–556. Available online: https://link.springer.com/chapter/10.1007/978-3-319-08795-5_56 (accessed on 22 March 2019).

32. Garg, H. A New Generalized Pythagorean Fuzzy Information Aggregation Using Einstein Operations and Its Application to Decision Making. *Int. J. Intell. Syst.* **2016**, *31*, 886–920. [CrossRef]

33. Zeng, S.Z.; Chen, J.P.; Li, X.S. A Hybrid Method for Pythagorean Fuzzy Multiple-Criteria Decision Making. *Int. J. Inf. Technol. Decis. Mak.* **2016**, *15*, 403–422. [CrossRef]

34. Ren, P.J.; Xu, Z.S.; Gou, X.J. Pythagorean fuzzy TODIM approach to multi-criteria decision making. *Appl. Soft Comput.* **2016**, *42*, 246–259. [CrossRef]

35. Wei, G.W.; Lu, M. Pythagorean Fuzzy Maclaurin Symmetric Mean Operators in Multiple Attribute Decision Making. *Int. J. Intell. Syst.* **2018**, *33*, 1043–1070. [CrossRef]

36. Maclaurin, C. A second letter to Martin Folkes, Esq.; concerning the roots of equations, with demonstration of other rules of algebra. *Philos. Trans. R. Soc. Lond. Ser. A* **1729**, *36*, 59–96.

37. Liang, D.C.; Zhang, Y.R.J.; Xu, Z.S.; Darko, A.P. Pythagorean fuzzy Bonferroni mean aggregation operator and its accelerative calculating algorithm with the multithreading. *Int. J. Intell. Syst.* **2018**, *33*, 615–633. [CrossRef]

38. Liang, D.C.; Xu, Z.S.; Darko, A.P. Projection Model for Fusing the Information of Pythagorean Fuzzy Multicriteria Group Decision Making Based on Geometric Bonferroni Mean. *Int. J. Intell. Syst.* **2017**, *32*, 966–987. [CrossRef]

39. Yager, R.R. Pythagorean Fuzzy Subsets. In Proceedings of the 2013 Joint IFSA World Congress and NAFIPS Annual Meeting (IFSA/NAFIPS), Edmonton, AB, Canada, 24–28 June 2013; pp. 57–61.
40. Chen, S.M.; Chang, C.H. A new similarity measure between intuitionistic fuzzy sets based on transformation techniques. In Proceedings of the 2014 International Conference on Machine Learning and Cybernetics, Lanzhou, China, 13–16 July 2014; pp. 396–402.
41. Wei, G.W.; Lu, M. Dual hesitant pythagorean fuzzy Hamacher aggregation operators in multiple attribute decision making. *Arch. Control Sci.* **2017**, *27*, 365–395. [CrossRef]
42. Peng, X.D.; Yuan, H.Y.; Yang, Y. Pythagorean Fuzzy Information Measures and Their Applications. *Int. J. Intell. Syst.* **2017**, *32*, 991–1029. [CrossRef]
43. Zhao, H.; Xu, Z.S.; Liu, S.S. Dual hesitant fuzzy information aggregation with Einstein t-conorm and t-norm. *J. Syst. Sci. Syst. Eng.* **2017**, *26*, 240–264. [CrossRef]
44. Yager, R.R. Generalized Orthopair Fuzzy Sets. *IEEE Trans. Fuzzy Syst.* **2017**, *25*, 1222–1230. [CrossRef]
45. Liu, P.D.; Wang, P. Some q-Rung Orthopair Fuzzy Aggregation Operators and their Applications to Multiple-Attribute Decision Making. *Int. J. Intell. Syst.* **2018**, *33*, 259–280. [CrossRef]
46. Wei, G.W.; Wei, C.; Wang, J.; Gao, H.; Wei, Y. Some q-rung orthopair fuzzy maclaurin symmetric mean operators and their applications to potential evaluation of emerging technology commercialization. *Int. J. Intell. Syst.* **2019**, *34*, 50–81. [CrossRef]
47. Wei, G.W.; Gao, H.; Wei, Y. Some q-rung orthopair fuzzy Heronian mean operators in multiple attribute decision making. *Int. J. Intell. Syst.* **2018**, *33*, 1426–1458. [CrossRef]
48. Yang, W.; Pang, Y.F. New q-rung orthopair fuzzy partitioned Bonferroni mean operators and their application in multiple attribute decision making. *Int. J. Intell. Syst.* **2019**, *34*, 439–476. [CrossRef]
49. Liu, P.D.; Liu, W.Q. Multiple-attribute group decision-making based on power Bonferroni operators of linguistic q-rung orthopair fuzzy numbers. *Int. J. Intell. Syst.* **2019**, *34*, 652–689. [CrossRef]
50. Xu, Y.; Shang, X.; Wang, J.; Wu, W.; Huang, H. Some q-Rung Dual Hesitant Fuzzy Heronian Mean Operators with Their Application to Multiple Attribute Group Decision-Making. *Symmetry* **2018**, *10*, 472. [CrossRef]
51. Lei, Q.; Xu, Z.S. Relationships Between Two Types of Intuitionistic Fuzzy Definite Integrals. *IEEE Trans. Fuzzy Syst.* **2016**, *24*, 1410–1425. [CrossRef]
52. Wu, Z.B.; Xu, J.P.; Jiang, X.L.; Zhong, L. Two MAGDM models based on hesitant fuzzy linguistic term sets with possibility distributions: VIKOR and TOPSIS. *Inf. Sci.* **2019**, *473*, 101–120. [CrossRef]
53. Wang, X.G.; Geng, Y.S.; Yao, P.P.; Yang, M.J. Multiple attribute group decision making approach based on extended VIKOR and linguistic neutrosophic Set. *J. Intell. Fuzzy Syst.* **2019**, *36*, 149–160. [CrossRef]
54. Liang, W.; Zhao, G.; Wu, H.; Dai, B. Risk assessment of rockburst via an extended MABAC method under fuzzy environment. *Tunn. Undergr. Space Technol.* **2019**, *83*, 533–544. [CrossRef]
55. Zhou, W.; Chen, J.; Xu, Z.S.; Meng, S. Hesitant fuzzy preference envelopment analysis and alternative improvement. *Inf. Sci.* **2018**, *465*, 105–117. [CrossRef]
56. Zhao, N.; Xu, Z.S. Prioritized Dual Hesitant Fuzzy Aggregation Operators Based on t-Norms and t-Conorms with Their Applications in Decision Making. *Informatica* **2018**, *29*, 581–607. [CrossRef]
57. Zhang, H.Q.; Li, D.W.; Wang, T.; Li, T.R.; Yu, X.; Bouras, A. Hesitant extension of fuzzy-rough set to address uncertainty in classification. *J. Intell. Fuzzy Syst.* **2018**, *34*, 2535–2550. [CrossRef]
58. Zhang, C.; Li, D.Y.; Liang, J.Y. Hesitant fuzzy linguistic rough set over two universes model and its applications. *Int. J. Mach. Learn. Cybern.* **2018**, *9*, 577–588. [CrossRef]
59. Zang, Y.Q.; Zhao, X.D.; Li, S.Y.; Nazir, A. Grey Relational Bidirectional Projection Method for Multicriteria Decision Making with Hesitant Intuitionistic Fuzzy Linguistic Information. *Math. Probl. Eng.* **2018**, *2018*, 1–12. [CrossRef]

Article

Interval-Valued Probabilistic Hesitant Fuzzy Set Based Muirhead Mean for Multi-Attribute Group Decision-Making

R. Krishankumar [1], **K. S. Ravichandran** [1], **M. Ifjaz Ahmed** [1], **Samarjit Kar** [2,*] **and Xindong Peng** [3]

[1]   School of Computing, SASTRA University, Thanjavur 613401, Tamil Nadu, India;
      krishankumar@sastra.ac.in (R.K.); raviks@sastra.edu (K.S.R.); ifjazahmed@it.sastra.edu (M.I.A.)
[2]   Department of Mathematics, National Institute of Technology, Durgapur-713209, West Bengal, India
[3]   School of Information Sciences & Engineering, Shaoguan University, Shaoguan 512005, China;
      pengxindong@sgu.ac.cn
*   Correspondence: samarjit.kar@maths.nitdgp.ac.in

Received: 23 January 2019; Accepted: 4 April 2019; Published: 9 April 2019

Abstract: As a powerful generalization to fuzzy set, hesitant fuzzy set (HFS) was introduced, which provided multiple possible membership values to be associated with a specific instance. But HFS did not consider occurrence probability values, and to circumvent the issue, probabilistic HFS (PHFS) was introduced, which associates an occurrence probability value with each hesitant fuzzy element (HFE). Providing such a precise probability value is an open challenge and as a generalization to PHFS, interval-valued PHFS (IVPHFS) was proposed. IVPHFS provided flexibility to decision makers (DMs) by associating a range of values as an occurrence probability for each HFE. To enrich the usefulness of IVPHFS in multi-attribute group decision-making (MAGDM), in this paper, we extend the Muirhead mean (MM) operator to IVPHFS for aggregating preferences. The MM operator is a generalized operator that can effectively capture the interrelationship between multiple attributes. Some properties of the proposed operator are also discussed. Then, a new programming model is proposed for calculating the weights of attributes using DMs' partial information. Later, a systematic procedure is presented for MAGDM with the proposed operator and the practical use of the operator is demonstrated by using a renewable energy source selection problem. Finally, the strengths and weaknesses of the proposal are discussed in comparison with other methods.

Keywords: group decision-making; hesitant fuzzy set; interval-valued probability; muirhead mean and programming model

---

1. Introduction

Multi-attribute group decision-making (MAGDM) is an interesting and complex day-to-day problem which involves implicit uncertainty and vagueness [1]. Hesitant fuzzy set (HFS) [2] is a powerful extension of the fuzzy set [3] that allows multiple degrees of truth to be associated with each preference information for better handling uncertainty and vagueness. Attracted by the strength of HFS, many researchers used HFS for different MAGDM applications viz., supplier selection [4,5], plant location selection [6], hospital site selection [7] and pattern recognition [8]. Recently, Rodriguez et al. [9] conducted a thorough analysis of HFS and some of its variants and identified its usefulness in MAGDM.

Though HFS is powerful, it lacks the ability to consider occurrence probability for each hesitant fuzzy element (HFE). To alleviate the issue, Xu and Zhou [10] put forward the probabilistic hesitant fuzzy set (PHFS), which associates occurrence probability value for each HFE. Motivated by the power of PHFS, researchers widely explored the idea for multi-attribute decision-making (MADM) [11–17]. Though PHFS alleviates the weakness of HFS to some extent, still the elicitation of occurrence probability is prone to imprecision and inaccuracy. To circumvent the weakness, a generalized model called interval-valued PHFS (IVPHFS) [18] is put forward which associates a range of values as occurrence probability to each HFE with a constraint that the sum of upper limit probability equals unity. As a generalization of Reference [18], Krishankumar et al. [19] proposed an IVPHFS concept which associates a range of values as occurrence probability for each HFE (for flexible elicitation of occurrence probability values). This mitigates the problem of imprecision and inaccuracy in the elicitation of occurrence probability values by providing multiple choices of values as occurrence probability for each HFE.

Based on the literature analysis of PHFS, we can infer that IVPHFS is a recent extension to PHFS which needs to be better explored for effective MAGDM. Group decision-making (GDM) [20] is a widely studied problem which obtains preference information from multiple DMs and aggregates them into single preference information without much loss of information. Mesiar et al. [21] made an interesting analysis of aggregation functions and we can infer that the realization of the interrelationship between attributes is a key factor for aggregation operators. Most of the state-of-the-art operators ignore this theme and are hence, not very suitable for MAGDM.

The MM [22] operator is an aggregation operator that reflects the interrelationship between attributes in a better way by considering risk appetite (refer Section 3 for details)s. MM is a generalized operator which can easily represent other operators viz., arithmetic average, geometric average, generalized arithmetic average, Bonferroni mean [23] and Maclaurin symmetric mean [24] as special cases.

Some challenges that can be encountered from the literature analysis made above are:

1. Elicitation of occurrence probability in a precise manner is difficult and prone to inaccuracies.
2. Following this, aggregation of preference information with better scope to capture the interrelationship between attributes is an open challenge in the IVPHFS context.
3. Further, calculation of weights of attributes by making reasonable use of partial information from DMs is also an open challenge.
4. Understanding the applicability, strengths and weaknesses of the proposed method are also substantial for effective use of the framework in uncertain situations.

Motivated by these challenges and to circumvent the same, in this paper, some key contributions are made:

1. As a generalization to PHFS, IVPHFS [18,19] was proposed which allows the range of occurrence probability values to be associated with each HFE. This mitigates the issue of imprecision and inaccuracy in probability elicitation and addresses challenge (1).
2. MM operator is extended in the IVPHFS context for capturing the interrelationship between attributes in a better way. Also, DMs' preferences are aggregated in a rational manner by considering risk appetite along with the weight of each DM which addresses challenge (2).
3. A new mathematical programming model is put forward in the IVPHFS context for calculation of weights of attributes with the help of partial information from the DMs. The idea is to use this partial information effectively for a reasonable calculation of weights.
4. The applicability of the proposed method is validated by using a green supplier selection problem.
5. Finally, the superiority and weakness of the proposed method are discussed in comparison with other methods.

The rest of the paper is constructed as follows. Section 2 describes some basic concepts of HFS, PHFS and IVPHFS. Section 3 presents the core idea of the research in which the proposed aggregation operator along with some desirable properties is discussed. Further, a new programming model is put forward for calculating attributes' weight values and finally, a systematic procedure is presented for MAGDM using proposed aggregation operator. In Section 4, a numerical example for green supplier selection is put forward to validate the applicability of the proposed method. In Section 5, some superiority and weakness of the proposal are discussed by comparison with other methods and finally, in Section 6, concluding remarks with future research directions are presented.

2. Preliminaries

Let us discuss some basics of HFS, PHFS and IVPHFS concepts.

Definition 1 [2]. *Consider a fixed set T and hesitant fuzzy set (HFS) H on T which is a function h that produces a subset in the interval* $[0,1]$. *Mathematically, it is given by,*

$$H = (t, h_H(t)|t \in T) \tag{1}$$

where $h_H(t)$ *is the set of values in range 0 to 1.*

Remark 1. *For convenience,* $h_H(t) = h(t)$ *is called the hesitant fuzzy element (HFE) and the collection of HFEs is H.*

Definition 2 [10]. *Consider a fixed set T. The PHFS* H_p *on T is given by,*

$$H_P = \left(t, h_{H_p}(\gamma_i, p_i)|t \in T\right) \tag{2}$$

where $h_{H_p}(\gamma_i, p_i)$ *is the probabilistic hesitant fuzzy element with* γ_i *being the membership value of t with its associated occurrence probability* p_i *and* $i = 1, 2, \ldots, \#h_{H_p}$.

Remark 2. *For convenience,* $h_{H_p}(\gamma_i, p_i) = h(\gamma_i, p_i)$ *is called the probabilistic hesitant fuzzy element (PHFE) with* γ_i *and* p_i *in the range 0 to 1.*

Definition 3 [19]. *Consider a fixed set T. The IVPHFS* H_{IP} *on T is given by,*

$$H_{IP} = \left(t, h_{H_p}\left(\gamma_i, \left[p_i^l, p_i^u\right]\right)\right) \tag{3}$$

where $h_{H_p}\left(\gamma_i, \left[p_i^l, p_i^u\right]\right)$ *is the interval-valued probabilistic hesitant fuzzy element with* γ_i *being the membership value of t with its associated occurrence probability value in the interval fashion as* $\left[p_i^l, p_i^u\right]$, *for all* $i = 1, 2, \ldots, \#h_{H_{lp}}$.

Here, γ_i, p_i^l and p_i^u are in the range 0 to 1 and $p_i^l \leq p_i^u$. For simplicity, $h_{H_p}\left(\gamma_i, \left[p_i^l, p_i^u\right]\right) = h\left(\gamma_i, \left[p_i^l, p_i^u\right]\right) = h$ is called interval-valued probabilistic hesitant fuzzy element (IVPHFE).

Consider an example where a DM provides his preference for ice-creams. Initially, he uses PHFS information (refer Definition 2) to rate different ice-creams viz., vanilla, strawberry and chocolate as $H_P = (vennila, \{0.6, (0.4)\}, strawberry, \{0.8, (0.5)\}, chocolate, \{0.4, (0.7)\})$. Later, he uses IVPHFS information (refer Definition 3) to provide probability values in a more flexible manner. $H_{IP} = (vennila, \{0.6, [0.3, 0.45]\}, strawberry, \{0.8, [0.4, 0.6]\}, , chocolate, \{0.4, [0.6, 0.7]\})$. By applying the latter style for preference information, we can increase the flexibility by providing a range of values as probability values.

Definition 4 [19]. *Let* h_1, h_2 *and* h_3 *be three IVPHFEs; then some operations are given by,*

$$h_1 \oplus h_2 = \left(\gamma_1 + \gamma_2 - \gamma_1\gamma_2, \left[p_1^l p_2^l, p_1^u p_2^u\right]\right) \tag{4}$$

$$h_1 \otimes h_2 = \left(\gamma_1\gamma_2, \left[p_1^l p_2^l, p_1^u p_2^u\right]\right) \tag{5}$$

$$\lambda h_1 = \left(1 - (1 - \gamma_1)^\lambda, \left[p_1^l, p_1^u\right]\right) \tag{6}$$

$$h_1^\lambda = \left(\gamma_1^\lambda, \left[p_1^l, p_1^u\right]\right) \tag{7}$$

Here, Equations (4)–(7) represent addition, multiplication, scalar multiplication and power operations. Actually, these equations are algebraic Archimedean T-norm and T-conorm and the additive generators used here are $g(x) = -ln(x)$ for T-norm $T_A(x, y) = xy$ and $h(x) = -ln(1 - x)$ for T-conorm $S_A(x, y) = x + y - xy$.

Property 1: Commutativity

$$h_1 \oplus h_2 = h_2 \oplus h_1$$

$$h_1 \otimes h_2 = h_2 \otimes h_1$$

Property 2: Associativity

$$h_1 \oplus (h_2 \oplus h_3) = (h_1 \oplus h_2) \oplus h_3$$

$$h_1 \otimes (h_2 \otimes h_3) = (h_1 \otimes h_2) \otimes h_3$$

3. Proposed Methodology

3.1. IVPHFS Based MM Operator and Its Properties

This section presents a new extension to the popular and powerful MM operator in the IVPHFS context. The MM operator [22] is a generalized operator which aggregates preferences by properly capturing the interrelationship between attributes. The operator can be used to realize other operators as mentioned above. The MM operator also considers the risk appetite values of DMs along with their relative importance (weights) in its formulation which intuitively produces a rational aggregation of preference information.

Table 1 provides a review on MM operators that are proposed for different fuzzy sets. This provides an idea on the basic concept of MM operator, its practical use in MADM problems. Moreover, the challenges mentioned above are clearly supported by this tabular analysis.

Table 1. Review of MM operator on different fuzzy sets.

Ref.#	Aggregation Operator	Preference Style	Attributes' Weights Calculation	Applications
[25]	MM	Picture fuzzy set	no	Investment risk
[26]	Power MM	T-spherical fuzzy set	no	Air quality evaluation
[27]	MM	The hesitant fuzzy linguistic term set	no	ERP system selection
[28]	MM	Hesitant fuzzy set	no	Evaluation of emergency responses
[29]	Power MM	Neutrosophic cubic fuzzy set	no	Van selection
[30]	MM	Pythagorean fuzzy set	no	Airline evaluation
[31]	MM	q-rung orthopair fuzzy set	no	Investor selection
[32]	MM	The probabilistic linguistic term set	no	Project selection

From Table 1, it can be inferred that (i) preference styles either associate a single value as a probability or ignore probability, which is not reasonable for decision-making; (ii) the MM operator is popularly used for aggregating preference information by effectively capturing the interrelationship between attributes and (iii) finally, attributes' weight values are directly (not calculated) obtained from the DMs which causes inaccuracies in the decision-making process.

Motivated by the power of MM operator and IVPHFS concept, in this paper, we extend the MM operator to IVPHFS and the definition and properties are given below:

Definition 5. *The aggregation of IVPHFEs using IVPHFMM (interval-valued probabilistic hesitant fuzzy Muirhead mean) operator is a mapping from $T^k \to T$ for $k = 1, 2, \ldots, nd$ which is given by,*

$$IVPHFMM^{(\lambda_1, \lambda_2, \ldots, \lambda_{nd})}(h_1, h_2, \ldots, h_{nd})$$

$$= \left(\left(\prod_{k=1}^{nd} \left(\prod_{j=1}^{nd} \gamma_i^{\lambda_j} \right)^{w_k} \right)^{\frac{1}{\Sigma_j \lambda_j}}, \left[\left(\prod_{k=1}^{nd} \left(\prod_{j=1}^{nd} \left(p_i^l\right)^{\lambda_j} \right)^{w_k} \right)^{\frac{1}{\Sigma_j \lambda_j}}, \left(\prod_{k=1}^{nd} \left(\prod_{j=1}^{nd} \left(p_i^u\right)^{\lambda_j} \right)^{w_k} \right)^{\frac{1}{\Sigma_j \lambda_j}} \right] \right) \tag{8}$$

where $\lambda_1, \lambda_2, \ldots, \lambda_k$ is risk appetite values of each DM from the set $\{1, 2, \ldots, nd\}$, nd is the total number of DMs, w_k is the weight of the kth DM.

It must be noted that Equation (8) provides the MM operator in the IVPHFS context. The operator aggregates the membership values, followed by the probability values (in the interval fashion). That is, the lower limit of the probability value is aggregated and then the upper limit of the probability value is aggregated. The square bracket represents the interval values that we obtain upon aggregation of probability values. Further, we present a theorem below to show that aggregation of different IVPHFEs by using IVPHFMM operator produces an IVPHFE.

Remark 3. *The MM operator is initially proposed in Reference [22] and it is given by $\left(\frac{1}{nd!} \sum_{k=1}^{nd} \prod_{j=1}^{nd} a_k^{\lambda_j} \right)^{\frac{1}{\Sigma_j \lambda_j}}$ where nd is the number of DMs and λ_j is the risk appetite values for $j = 1, 2, \ldots, nd$. Risk appetite is defined by ISO 31000 as the amount of risk pursued, retained or taken by an organization. In this case, it is the risk pursued, retained or taken by a DM. The possible values are from the set $\{1, 2, \ldots, nd\}$. The higher the value of λ indicates a higher risk appetite value for the DM.*

Property 1: Idempotent

If IVPHFEs $h_1 = h_2 = \ldots = h_k = h$, then $IVPHFMM^{(\lambda_1, \lambda_2, \ldots, \lambda_k)}(h_1, h_2, \ldots, h_k) = h$.

Proof. From Equation (8),

$$IVPHFMM^{(\lambda_1, \lambda_2, \ldots, \lambda_k)}(h_1, h_2, \ldots, h_k) = \left(\left(\prod_{k=1}^{nd} \left(\prod_{j=1}^{nd} \gamma_i^{\lambda_j} \right)^{w_k} \right)^{\frac{1}{\Sigma_j \lambda_j}}, \left[\left(\prod_{k=1}^{nd} \left(\prod_{j=1}^{nd} \left(p_i^l\right)^{\lambda_j} \right)^{w_k} \right)^{\frac{1}{\Sigma_j \lambda_j}}, \left(\prod_{k=1}^{nd} \left(\prod_{j=1}^{nd} \left(p_i^u\right)^{\lambda_j} \right)^{w_k} \right)^{\frac{1}{\Sigma_j \lambda_j}} \right] \right)$$

$$= \left(\left(\prod_{k=1}^{nd} \left(\gamma_i^{\lambda_1 + \lambda_2 + \ldots + \lambda_{nd}} \right)^{w_k} \right)^{\frac{1}{\Sigma_j \lambda_j}}, \left[\left(\prod_{k=1}^{nd} \left(\left(p_i^l\right)^{\lambda_1 + \lambda_2 + \ldots + \lambda_{nd}} \right)^{w_k} \right)^{\frac{1}{\Sigma_j \lambda_j}}, \left(\prod_{k=1}^{nd} \left(\left(p_i^u\right)^{\lambda_1 + \lambda_2 + \ldots + \lambda_{nd}} \right)^{w_k} \right)^{\frac{1}{\Sigma_j \lambda_j}} \right] \right)$$

By doing the same to DMs' weight values, we obtain,

$$= \left(\left(\gamma_i^{\lambda_1 + \lambda_2 + \ldots + \lambda_{nd}} \right)^{\frac{1}{\Sigma_j \lambda_j}}, \left[\left(\left(p_i^l\right)^{\lambda_1 + \lambda_2 + \ldots + \lambda_{nd}} \right)^{\frac{1}{\Sigma_j \lambda_j}}, \left(\left(p_i^u\right)^{\lambda_1 + \lambda_2 + \ldots + \lambda_{nd}} \right)^{\frac{1}{\Sigma_j \lambda_j}} \right] \right) \quad \text{as } \sum_k w_k = 1. \quad \square$$

$$= \left(\gamma_i, \left[p_i^l, p_i^u\right] \right) = h$$

Property 2: Commutativity

If h_i^* is any permutation of h_i $\forall i = 1, 2, \ldots, k$, then $IVPHFMM^{(\lambda_1, \lambda_2, \ldots, \lambda_k)}(h_1, h_2, \ldots, h_k) = IVPHFMM^{(\lambda_1, \lambda_2, \ldots, \lambda_k)}(h_1^*, h_2^*, \ldots, h_k^*)$.

Proof.

$$
IVPHFMM^{(\lambda_1, \lambda_2, \ldots, \lambda_k)}(h_1^*, h_2^*, \ldots, h_k^*) = \left(
\begin{array}{c}
\left(\prod_{k=1}^{nd} \left(\prod_{j=1}^{nd} \gamma_i^{*\lambda_j} \right)^{w_k} \right)^{\frac{1}{\sum_j \lambda_j}}, \\[4mm]
\left[\left(\prod_{k=1}^{nd} \left(\prod_{j=1}^{nd} \left(p_i^{*l} \right)^{\lambda_j} \right)^{w_k} \right)^{\frac{1}{\sum_j \lambda_j}}, \left(\prod_{k=1}^{nd} \left(\prod_{j=1}^{nd} \left(p_i^{*u} \right)^{\lambda_j} \right)^{w_k} \right)^{\frac{1}{\sum_j \lambda_j}} \right]
\end{array}
\right)
$$

$$
= \left(
\begin{array}{c}
\left(\prod_{k=1}^{nd} \left(\prod_{j=1}^{nd} \gamma_i^{\lambda_j} \right)^{w_k} \right)^{\frac{1}{\sum_j \lambda_j}}, \\[4mm]
\left[\left(\prod_{k=1}^{nd} \left(\prod_{j=1}^{nd} \left(p_i^l \right)^{\lambda_j} \right)^{w_k} \right)^{\frac{1}{\sum_j \lambda_j}}, \left(\prod_{k=1}^{nd} \left(\prod_{j=1}^{nd} \left(p_i^u \right)^{\lambda_j} \right)^{w_k} \right)^{\frac{1}{\sum_j \lambda_j}} \right]
\end{array}
\right) = IVPHFMM^{(\lambda_1, \lambda_2, \ldots, \lambda_k)}(h_1, h_2, \ldots, h_k)
$$

$\square$

Property 3: Monotonicity

Consider IVPHFEs $h' = h_i'$ and $h = h_i$ for $i = 1, 2, \ldots, k$ such that $h_i' \geq h_i$. Then, $IVPHFMM^{(\lambda_1, \lambda_2, \ldots, \lambda_k)}(h_1, h_2, \ldots, h_k) \leq IVPHFMM^{(\lambda_1, \lambda_2, \ldots, \lambda_k)}(h_1', h_2', \ldots, h_k')$.

Proof. Let $\gamma_i' = \left(\prod_{k=1}^{nd} \left(\prod_{j=1}^{nd} \left(\gamma_i' \right)^{\lambda_j} \right)^{w_k} \right)^{\frac{1}{\sum_j \lambda_j}}$ and $\left[p_i^{l'}, p_i^{u'} \right] = \left[\left(\prod_{k=1}^{nd} \left(\prod_{j=1}^{nd} \left(p_i^{l'} \right)^{\lambda_j} \right)^{w_k} \right)^{\frac{1}{\sum_j \lambda_j}}, \left(\prod_{k=1}^{nd} \left(\prod_{j=1}^{nd} \left(p_i^{u'} \right)^{\lambda_j} \right)^{w_k} \right)^{\frac{1}{\sum_j \lambda_j}} \right]$.

Similarly, h_i is defined as above. Now score and deviation measure is adopted from Reference [19] for IVPHFEs. Since $h_i' \geq h_i \ \forall i = 1, 2, \ldots, k$, $s(h') \geq s(h)$ which concludes that $h' \geq h$ and if $s(h') = s(h)$, then calculate deviation and if $\sigma(h) \geq \sigma(h') \ h' \geq h$. Thus, $IVPHFMM^{(\lambda_1, \lambda_2, \ldots, \lambda_k)}(h_1, h_2, \ldots, h_k) \leq IVPHFMM^{(\lambda_1, \lambda_2, \ldots, \lambda_k)}(h_1', h_2', \ldots, h_k')$. $\square$

Property 4: Bounded

For any IVPHFE $h_i \ \forall i = 1, 2, \ldots, k$, $h^- \leq IVPHFMM^{(\lambda_1, \lambda_2, \ldots, \lambda_k)}(h_1, h_2, \ldots, h_k) \leq h^+$. Here, $h^- = min(h_i)$ and $h^+ = max(h_i)$. Initially, $\frac{\gamma_i p_i^l + \gamma_i p_i^u}{2}$ is calculated and the IVPHFE that correspond to minimum and maximum value is considered as h^- and h^+ respectively.

Proof. Based on the monotonic and idempotent property of IVPHFMM operator, we can easily conclude that $IVPHFMM^{(\lambda_1, \lambda_2, \ldots, \lambda_k)}(h_1, h_2, \ldots, h_k) \geq IVPHFMM^{(\lambda_1, \lambda_2, \ldots, \lambda_k)}(h^-, h^-, \ldots, h^-)$. Similarly, $IVPHFMM^{(\lambda_1, \lambda_2, \ldots, \lambda_k)}(h_1, h_2, \ldots, h_k) \leq IVPHFMM^{(\lambda_1, \lambda_2, \ldots, \lambda_k)}(h^+, h^+, \ldots, h^+)$. Combining these two inequalities, we get $h^- \leq IVPHFMM^{(\lambda_1, \lambda_2, \ldots, \lambda_k)}(h_1, h_2, \ldots, h_k) \leq h^+$. $\square$

Theorem 1. *The aggregation of IVPHFEs by IVPHFMM operator produces an IVPHFE.*

Proof. To prove the above theorem, we must show that the aggregated value obeys Definition 3. Now, from Property 4, it is clear that the aggregated value is bounded within the different IVPHFEs taken for aggregation. By extending the idea further, we get $0 \leq \left(\prod_{k=1}^{\#DM} \left(\prod_{j=1}^{\#DM} \gamma_i^{\lambda_j} \right)^{w_k} \right)^{\frac{1}{\sum_j \lambda_j}} \leq 1$ and $0 \leq \left[\left(\prod_{k=1}^{\#DM} \left(\prod_{j=1}^{\#DM} \left(p_i^l \right)^{\lambda_j} \right)^{w_k} \right)^{\frac{1}{\sum_j \lambda_j}}, \left(\prod_{k=1}^{\#DM} \left(\prod_{j=1}^{\#DM} \left(p_i^u \right)^{\lambda_j} \right)^{w_k} \right)^{\frac{1}{\sum_j \lambda_j}} \right] \leq 1$. By combining these two inequalities, we get $0 \leq$

$$\left(\dfrac{\left(\prod\limits_{k=1}^{\#DM}\left(\prod\limits_{j=1}^{\#DM} \gamma_i^{\lambda_j}\right)^{w_k}\right)^{\frac{1}{\Sigma_j \lambda_j}},}{\left[\left(\prod\limits_{k=1}^{\#DM}\left(\prod\limits_{j=1}^{\#DM} \left(p_i^l\right)^{\lambda_j}\right)^{w_k}\right)^{\frac{1}{\Sigma_j \lambda_j}}, \left(\prod\limits_{k=1}^{\#DM}\left(\prod\limits_{j=1}^{\#DM} \left(p_i^u\right)^{\lambda_j}\right)^{w_k}\right)^{\frac{1}{\Sigma_j \lambda_j}}\right]} \right) \leq 1.$$

Thus, $0 \leq IVPHFMM^{(\lambda_1,\lambda_2,\dots,\lambda_k)}(h_1, h_2, \dots, h_k) \leq 1$. Hence, aggregation of IVPHFEs yields an IVPHFE. $\square$

3.2. Weight Calculation for Attributes Using the Proposed Programming Model

This section put forwards a new mathematical programming model in the IVPHFS context for calculating the weights of attributes. There are mainly two types of weight calculation methods. In the first type, weight values are calculated with completely unknown information and some popular examples are analytical hierarchy process (AHP) [33], entropy measure [34] and so forth. In the second type, weight values are calculated with partially known information and this type of weight calculation gives DMs a chance to express their personal preference on each attribute which is considered during the weight calculation process. Whenever partial information is known for each attribute, the effective idea is to use the information for rational calculation of weights.

Motivated by the ability of the second type of weight calculation, in this paper, a new programming model is put forward in the IVPHFS context. The key advantages of the proposed model are (i) it uses the partial information from the DMs in a rational manner in its formulation; (ii) provides flexibility to the DMs to share their personal preferences on each attribute in the form of constraints; (iii) the nature of the attribute (benefit or cost) is also taken into consideration during formulation and (iv) the ideal solution for each attribute is considered for rational calculation of weight values which resemble closely to the human cognition process.

The systematic procedure for attribute weight calculation is presented below:

Step 1: Construct an evaluation matrix of order 3×4 with IVPHFS information. The order of the matrix is DMs by attributes.

Step 2: Calculate the positive ideal solution (PIS) and negative ideal solution (NIS) for each attribute using Equations 9 and 10.

$$h_j^{PIS} = max_{j \in benefit}\left(\sum_{i=1}^{\#h} \gamma_i\left(\frac{p_i^l + p_i^u}{2}\right)\right) \text{ (or) } min_{j \in cost}\left(\sum_{i=1}^{\#h} \gamma_i\left(\frac{p_i^l + p_i^u}{2}\right)\right) \tag{9}$$

$$h_j^{NIS} = min_{j \in benefit}\left(\sum_{i=1}^{\#h} \gamma_i\left(\frac{p_i^l + p_i^u}{2}\right)\right) \text{ (or) } max_{j \in cost}\left(\sum_{i=1}^{\#h} \gamma_i\left(\frac{p_i^l + p_i^u}{2}\right)\right) \tag{10}$$

where h_j^{PIS} and h_j^{NIS} are PIS and NIS values of the jth attribute respectively. The h_j^{PIS} and h_j^{NIS} are calculated for each attribute and they contain IVPHFS information of the corresponding value obtained from Equations 9 and 10.

Step 3: Apply Model 1 to obtain the weights of attributes. Model 1:

$$Min \ Z = \sum_{j=1}^{n} w_j \sum_{i=1}^{m} d\left(h_{ij}, h_j^{PIS}\right) - d\left(h_{ij}, h_j^{NIS}\right)$$

Subject to:

$$0 \leq w_j \leq 1; \sum_{j} w_j = 1.$$

The distance measure is calculated using Equation (11).

$$d(a,b) = \sqrt{\sum_{i=1}^{\#instance}\left(\left(\gamma_{ia}\left(\frac{p_{ia}^l + p_{ia}^u}{2}\right)\right) - \left(\gamma_{ib}\left(\frac{p_{ib}^l + p_{ib}^u}{2}\right)\right)\right)^2} \tag{11}$$

where a and b are any two IVPHFEs.

3.3. Proposed MAGDM Method for Prioritization of Objects

This section develops a ranking procedure for prioritizing objects based on the operational laws and newly proposed IVPHFMM operator. The procedure is presented below:

Step 1: Begin.

Step 2: Construct k decision matrices of order $m \times n$ where m is the number of objects and n is the number of attributes.

Step 3: Aggregate these matrices into a single matrix of order $m \times n$ by using newly proposed IVPHFMM operator (see Section 3.1).

Step 4: Multiply the weight of each attribute with the respective IVPHFE and use Equation (4) attribute-wise to calculate the net value for each object. The resultant value is also an IVPHFE.

Step 5: Prioritize the objects by adopting transformation method given in Equation (12).

$$h_l^{single} = \sum_{i=1}^{\#instance}\left(\frac{\gamma_i p_i^l + \gamma_i p_i^u}{2}\right) \tag{12}$$

where l is the index for the object. Arrange h_l^{single} in the descending order of values to obtain ranking order.

Step 6: End.

Before demonstrating the practical use of the proposed framework, Figure 1 is presented to obtain a clear view of the implementation process of the proposed framework. Initially, DMs' preference information is obtained as IVPHFEs for each object over a specific attribute. These matrices are aggregated using newly proposed IVPHFMM operator (refer Section 3.1) which extends the MM operator in the IVPHFS context. An evaluation matrix is obtained from the DMs for attribute weight calculation. Mathematical programming model (refer to Section 3.2) is proposed for calculating the weights of the attributes with the help of partially known information. Finally, a new systematic procedure (refer to Section 3.3) is developed for prioritizing objects which uses the aggregated matrix and weight vector as input for implementation.

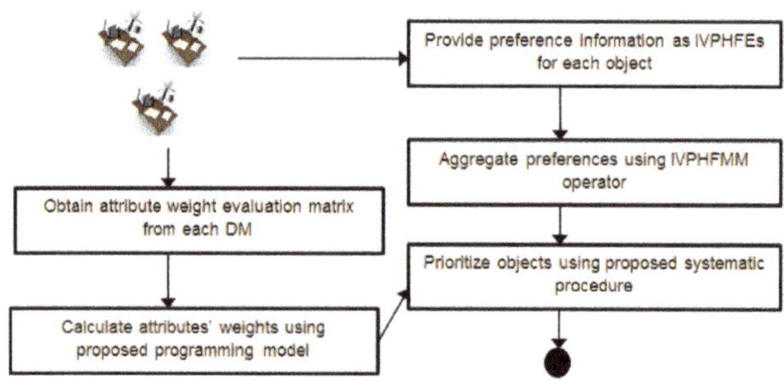

Figure 1. Proposed decision framework in the IVPHFS context.

4. Numerical Example: Renewable Energy Source Selection from the Indian Perspective

This section demonstrates the practicality of the proposed method by solving renewable energy source selection problem from an Indian perspective which is adapted from Reference [35]. India has a great appetite for energy owing to its high technological advancement and opportunities. Around 85,000 MW of energy demand is potentially satisfied with the help of biomass, small hydro, solar and wind energy. Recently, Mardani et al. [36] conducted a thorough survey of energy source selection using MADM methods and projected the key importance of MADM methods for energy selection. Luthra et al. [37] conducted a deep investigation of barriers to energy sources in India and suggested ideas for the mitigation of challenges from the Indian perspective. In a recent study performed by the Economic Times on energy demand and supply, they predicted that by 2040: (i) there will be a 30% increase in energy which crosses the conventional energy charts and forces an urgent need for renewable energy sources; (ii) also by 2040, India will reach 49% renewable energy usage.

Motivated by the investigation, in this paper, we formulate the renewable source selection problem as an MADM problem and present a systematic procedure for the suitable selection of renewable energy sources.

Step 1: Construct a panel of three DMs viz., technical personnel e_1, member of the ministry of energy and natural resource e_2 and senior financial personnel e_3. Make an initial list of renewable energy sources and attributes. By systematic pre-screening, the panel finalizes four renewable energy sources and four attributes. These are adapted from Reference [35] and the panel decides to use IVPHFS information for rating energy sources against the attribute.

Step 2: Form three matrices of order 4×4 where the rows represent the energy sources and columns represent the attributes. Four renewable energy sources considered are geothermal a_1, solar a_2, tidal a_3 and hydro a_4 which are evaluated over four attributes viz., technical aspects c_1, social aspects c_2, financial aspects c_3 and environmental aspects c_4 IVPHFS information (refer Definition 3) is used for rating energy sources and it is depicted in Table 2.

Table 2 presents the preference information by different DMs over renewable energy sources based on a set of attributes. IVPHFS based preference information is adopted.

Step 3: Aggregate the IVPHFEs from Table 2 by using IVPHFMM operator (see Section 3.1 for details). There are three risk appetite values used viz., λ_1, λ_2 and λ_3 which are given by 2, 2 and 1. The weight of each DM is given by 0.3, 0.4 and 0.3. The preference information from Table 2 is aggregated using IVPHFMM operator and it is shown in Table 3. Clearly, from Table 3, we can observe that the aggregated value is also an IVPHFE (refer Definition 3).

Step 4: Calculate the weights of the attributes by using the procedure given in Section 3.2. Table 4 shows an evaluation matrix with IVPHFS information which is used to calculate the PIS and NIS values for each attribute (see Table 4).

By using Tables 4 and 5 the objective function is constructed from Model 1. The constraints are obtained from the DMs and the model is solved using optimization toolbox of MATLAB®. From Model 1, we get the objective function as $MinZ = 0.38w_1 - 0.008w_2 - 0.043w_3 + 0.112w_4$ and the inequality constraints are given by $w_1 \leq 0.3$, $w_2 \leq 0.3$, $w_3 \leq 0.3$ and $w_4 \leq 0.3$. By solving we get, $w_1 = w_2 = w_3 = 0.3$ and $, w_4 = 0.1$.

Table 2. Interval-valued probabilistic hesitant fuzzy set (IVPHFS) based preference information for different DMs.

Energy	Evaluation Attribute			
	c_1	c_2	c_3	c_4
e_1				
a_1	$\{0.85, [0.12, 0.12]$ $0.58, [0.2, 0.48]\}$	$\{0.12, [0.41, 0.41]$ $0.6, [0.29, 0.29]\}$	$\{0.21, [0.74, 0.88]$ $0.8, [0.16, 0.89]\}$	$\{0.79, [0.46, 0.46]$ $0.73, [0.34, 0.83]\}$
a_2	$\{0.86, [0.49, 0.57]$ $0.39, [0.66, 0.66]\}$	$\{0.79, [0.39, 0.39]$ $0.55, [0.54, 0.54]\}$	$\{0.79, [0.51, 0.8]$ $0.39, [0.17, 0.73]\}$	$\{0.28, [0.25, 0.3]$ $0.28, [0.2, 0.2]\}$
a_3	$\{0.21, [0.26, 0.43]$ $0.81, [0.4, 0.76]\}$	$\{0.63, [0.13, 0.85]$ $0.31, [0.56, 0.59]\}$	$\{0.29, [0.21, 0.21]$ $0.61, [0.75, 0.84]\}$	$\{0.2, [0.11, 0.11]$ $0.71, [0.12, 0.12]\}$
a_4	$\{0.71, [0.32, 0.32]$ $0.43, [0.65, 0.88]\}$	$\{0.49, [0.35, 0.71]$ $0.23, [0.2, 0.2]\}$	$\{0.58, [0.64, 0.64]$ $0.19, [0.5, 0.7]\}$	$\{0.28, [0.33, 0.33]$ $0.58, [0.28, 0.43]\}$
e_2				
a_1	$\{0.88, [0.19, 0.19]$ $0.11, [0.17, 0.71]\}$	$\{0.31, [0.39, 0.39]$ $0.56, [0.47, 0.84]\}$	$\{0.89, [0.64, 0.64]$ $0.1, [0.34, 0.73]\}$	$\{0.29, [0.1, 0.11]$ $0.43, [0.1, 0.1]\}$
a_2	$\{0.24, [0.34, 0.44]$ $0.53, [0.71, 0.71]\}$	$\{0.54, [0.45, 0.48]$ $0.8, [0.19, 0.19s]\}$	$\{0.47, [0.15, 0.31]$ $0.79, [0.32, 0.32]\}$	$\{0.35, [0.35, 0.35]$ $0.35, [0.19, 0.23]\}$
a_3	$\{0.65, [0.47, 0.73]$ $0.89, [0.13, 0.77]\}$	$\{0.52, [0.27, 0.51]$ $0.39, [0.18, 0.18]\}$	$\{0.81, [0.34, 0.57]$ $0.46, [0.58, 0.58]\}$	$\{0.4, [0.32, 0.48]$ $0.33, [0.19, 0.23]\}$
a_4	$\{0.85, [0.27, 0.27]$ $0.89, [0.31, 0.79]\}$	$\{0.87, [0.27, 0.27]$ $0.52, [0.29, 0.29]\}$	$\{0.28, [0.39, 0.77]$ $0.43, [0.3, 0.3]\}$	$\{0.11, [0.32, 0.63]$ $0.32, [0.21, 0.46]\}$
e_3				
a_1	$\{0.63, [0.2, 0.2]$ $0.6, [0.33, 0.33]\}$	$\{0.54, [0.45, 0.45]$ $0.68, [0.15, 0.15]\}$	$\{0.85, [0.52, 0.52]$ $0.8, [0.11, 0.11]\}$	$\{0.48, [0.21, 0.22]$ $0.58, [0.22, 0.29]\}$
a_2	$\{0.56, [0.15, 0.15]$ $0.57, [0.65, 0.7]\}$	$\{0.78, [0.57, 0.57]$ $\{0.48, [0.19, 0.29]\}$	$\{0.24, [0.63, 0.63]$ $0.57, [0.38, 0.38]\}$	$\{0.29, [0.31, 0.33]$ $0.33, [0.22, 0.24]\}$
a_3	$\{0.53, [0.44, 0.44]$ $0.16, [0.33, 0.33]\}$	$\{0.84, [0.13, 0.38]$ $0.78, [0.43, 0.63]\}$	$\{0.15, [0.2, 0.86]$ $0.34, [0.35, 0.35]\}$	$\{0.34, [0.22, 0.34]$ $0.31, [0.22, 0.25]\}$
a_4	$\{0.46, [0.83, 0.83]$ $0.64, [0.33, 0.33]\}$	$\{0.2, [0.54, 0.54]$ $0.82, [0.41, 0.41]\}$	$\{0.84, [0.25, 0.25]$ $0.69, [0.23, 0.37]\}$	$\{0.21, [0.31, 0.54]$ $0.51, [0.27, 0.42]\}$

Table 3. Aggregation of preferences by using interval-valued probabilistic hesitant fuzzy Muirhead mean (IVPHFMM) operator.

Energy	Evaluation Attribute			
	c_1	c_2	c_3	c_4
e_{123}				
a_1	$\{0.79, [0.17, 0.17]$ $0.3, [0.22, 0.5]\}$	$\{0.27, [0.41, 0.41]$ $0.61, [0.29, 0.37]\}$	$\{0.57, [0.63, 0.66]$ $0.35, [0.19, 0.44]\}$	$\{0.48, [0.21, 0.22]$ $0.58, [0.22, 0.29]\}$
a_2	$\{0.46, [0.3, 0.34]$ $0.5, [0.68, 0.69]\}$	$\{0.67, [0.46, 0.47]$ $0.61, [0.26, 0.3]\}$	$\{0.45, [0.34, 0.51]$ $0.58, [0.28, 0.43]\}$	$\{0.29, [0.31, 0.33]$ $0.33, [0.22, 0.24]\}$
a_3	$\{0.43, [0.39, 0.54]$ $0.51, [0.24, 0.6]\}$	$\{0.64, [0.17, 0.54]$ $0.45, [0.33, 0.37]\}$	$\{0.36, [0.25, 0.48]$ $0.45, [0.54, 0.56\}$	$\{0.34, [0.22, 0.34]$ $0.31, [0.22, 0.25]\}$
a_4	$\{0.67, [0.4, 0.4]$ $0.49, [0.39, 0.63]\}$	$\{0.47, [0.36, 0.45]$ $0.47, [0.29, 0.29]\}$	$\{0.48, [0.4, 0.52]$ $0.39, [0.32, 0.41\}$	$\{0.21, [0.31, 0.54]$ $0.51, [0.27, 0.42]\}$

Table 4. Evaluation matrix for attribute weight calculation.

DMs	Evaluation Attribute			
	c_1	c_2	c_3	c_4
e_1	$\{0.5, [0.5, 0.7]$ $0.4, [0.35, 0.45]\}$	$\{0.44, [0.45, 0.5]$ $0.54, [0.5, 0.6]\}$	$\{0.55, [0.55, 0.65]$ $0.45, [0.55, 0.7]\}$	$\{0.65, [0.35, 0.4]$ $0.4, [0.4, 0.45]\}$
e_2	$\{0.6, [0.6, 0.65]$ $0.45, [0.45, 0.5]\}$	$\{0.35, [0.6, 0.7]$ $0.5, [0.55, 0.65]\}$	$\{0.45, [0.54, 0.65]$ $0.35, [0.4, 0.6]\}$	$\{0.55, [0.5, 0.7]$ $0.4, [0.6, 0.7]\}$
e_3	$\{0.45, [0.5, 0.6]$ $0.4, [0.55, 0.65]\}$	$\{0.4[0.5, 0.6]$ $0.5, [0.4, 0.55]\}$	$\{0.45, [0.45, 0.55]$ $0.6, [0.35, 0.55]\}$	$\{0.45, [0.55, 0.65]$ $0.6, [0.6, 0.65]\}$

Table 5. PIS & NIS for each attribute.

Ideal Solution	Evaluation Attribute			
	c_1	c_2	c_3	c_4
h^+	$\left\{\begin{array}{l} 0.6,\ [0.6, 0.65] \\ 0.45,\ [0.45, 0.5] \end{array}\right\}$	$\left\{\begin{array}{l} 0.35,\ [0.6, 0.7] \\ 0.5,\ [0.55, 0.65] \end{array}\right\}$	$\left\{\begin{array}{l} 0.45,\ [0.54, 0.65] \\ 0.35,\ [0.4, 0.6] \end{array}\right\}$	$\left\{\begin{array}{l} 0.45,\ [0.55, 0.65] \\ 0.6,\ [0.6, 0.65] \end{array}\right\}$
h^-	$\left\{\begin{array}{l} 0.5,\ [0.5, 0.7] \\ 0.4,\ [0.35, 0.45] \end{array}\right\}$	$\left\{\begin{array}{l} 0.4,\ [0.5, 0.6] \\ 0.5,\ [0.4, 0.55] \end{array}\right\}$	$\left\{\begin{array}{l} 0.55,\ [0.55, 0.65] \\ 0.45,\ [0.55, 0.7] \end{array}\right\}$	$\left\{\begin{array}{l} 0.65,\ [0.35, 0.4] \\ 0.4,\ [0.4, 0.45] \end{array}\right\}$

Step 5: Prioritize the energy sources by using the procedure given in Section 3.3. The cumulative ring sum value for each renewable energy source is given by $\begin{array}{l} 0.97, [0.86, 0.87], 0.93, [0.65, 0.87]; 0.93, [0.83, 0.89], 0.95, [0.87, 0.91]; \\ 0.91, [0.7, 0.93], 0.9, [0.82, 0.92]; 0.93, [0.84, 0.93], 0.92, [0.79, 0.91] \end{array}$ for unbiased attributes' weights and $\begin{array}{l} 0.41, [0.33, 0.34], 0.34, [0.20, 0.33]; 0.37, [0.29, 0.34], 0.39, [0.31, 0.35]; \\ 0.34, [0.23, 0.37], 0.34, [0.30, 0.37]; 0.37, [0.30, 0.36], 0.34, [0.27, 0.34] \end{array}$ for biased attributes' weights. From Equation (12) we get, $a_1 = 1.54$, $a_2 = 1.63$, $a_3 = 1.52$ and $a_4 = 1.6$ for unbiased weight values and $a_1 = 0.23$, $a_2 = 0.25$, $a_3 = 0.23$ and $a_4 = 0.23$ for biased weight values. Thus, the ranking order is $a_2 > a_4 > a_1 > a_3$ and the suitable renewable energy source for the process is solar energy a_2.

Step 6: Perform sensitivity analysis on the risk appetite of each DM by varying the values within the predefined threshold value. Figures 2–4 depict 27 possible risk appetite values and the corresponding prioritized value for each renewable energy source. From the analysis, we can clearly observe that the prioritized order remains unchanged with the final order as $a_2 > a_4 \succcurlyeq a_1 > a_3$ with solar energy as a suitable renewable energy source for the process taken into consideration.

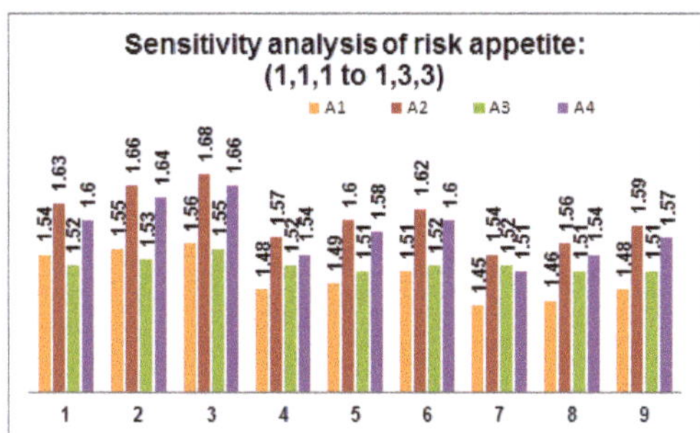

Figure 2. Sensitivity analysis of risk appetite of each DM ((1) 1,1,1 to (9) 1,3,3).

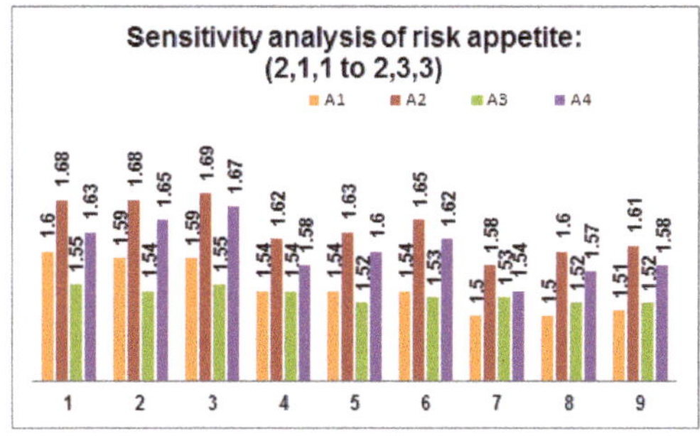

Figure 3. Sensitivity analysis of risk appetite of each DM ((10) 2,1,1 to (18) 2,3,3).

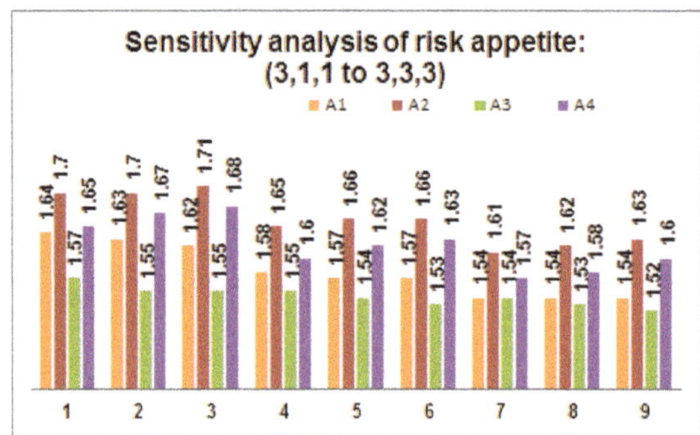

Figure 4. Sensitivity analysis of risk appetite of each DM ((19) 3,1,1 to (27) 3,3,3).

Step 7:   End.

5. Comparative Analysis of Proposed Decision Framework

This section put forwards a comparative study of the proposed aggregation operator with other operators in the IVPHFS context. In order to maintain the homogeneity of the comparison process, we analyse the proposed method with aggregation operators from References [18] and [19] which uses IVPHFEs as preference information.

Table 6 depicts the ranking order from different methods and we can see that the proposed method produces a unique ranking order compared to its counterpart. This is mainly due to the ability of the proposed aggregation operator to capture the interrelationship between multiple attributes. The consistency of the proposed method is realized by calculating the Spearman correlation coefficient [38] for each method. Figure 5 depicts the correlation plot for each method and we can infer that the proposed method is highly consistent with other state-of-the-art methods.

Table 6. The ranking order of renewable energy sources: Proposed vs. Others.

Methods	Renewable Energy Source (s)				Order
	a_1	a_2	a_3	a_4	
[18]	2	1	4	3	$a_2 > a_1 > a_4 > a_3$
[19]	2	1	4	3	$a_2 > a_1 > a_4 > a_3$
Proposed	3	1	4	2	$a_2 > a_4 > a_1 > a_3$

Note: DMs' weight values are considered as 0.3, 0.4 and 0.3 respectively and the attributes' weight values are considered as 0.25.

Table 7 depicts the characteristics analysis of different methods. The characteristics are adapted from the work of Liu and Teng [32].

Some strengths of the proposed method are listed below:

1.   The proposed aggregation operator uses the generalized data structure viz., IVPHFS as preference information. The interval number is associated as an occurrence probability value for each HFE.
2.   The extension of MM operator to IVPHFS context provides DMs with the ability to capture interrelationship between multiple attributes. This property resembles closely with the real-life decision-making problem.
3.   The proposed operator not only obtains weights of DMs but also considers the risk appetite of each DM.
4.   The proposed operator has parameters that are easily customizable for realizing different effects on prioritization of objects.

5. The proposed operator can also realize certain operators as special cases which provide a generalized context for aggregation of preferences that helps DMs is effective management of uncertainty and vagueness.

6. The attributes' weight values are calculated in a rational manner by considering the partial information from each DM.

Some weaknesses of the proposed method are:

1. It is difficult to fix the parameter value for different MAGDM application without a trial and error process.

2. DMs need some training with the data structure for proper elicitation of preferences.

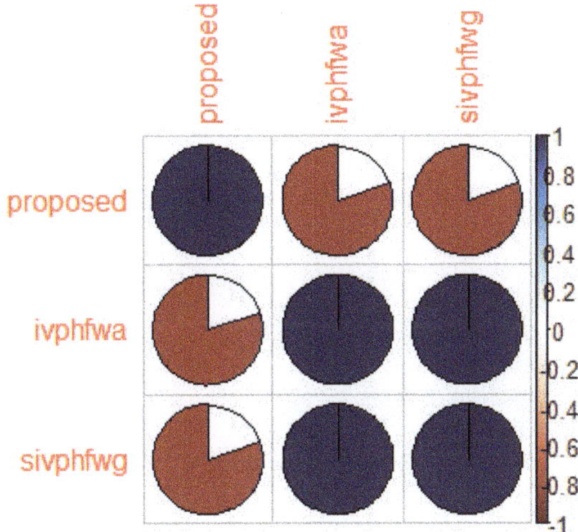

Figure 5. Corrplot for different ranking methods: Proposed vs. Others.

Table 7. Characteristics analysis for proposed vs. other methods.

Characteristics	Methods		
	Proposed	[18]	[19]
Data	IVPHFS based preference information		
Comprehensive data	yes	yes	yes
Interrelationship	yes	no	no
Customizable parameter(s)	yes	no	no
Total pre-order	yes	yes	yes
Generalizability	yes	no	no
The relative importance of DM	yes	yes	yes
Risk appetite of DM	yes	no	no
Information loss	Mitigated in an efficient manner	Mitigated in a moderate manner	

6. Conclusions

This paper presents a new extension to MM operator in the IVPHFS context. Some interesting properties and theorems are also discussed for arriving better theoretical perspective in the field of aggregation and information fusion. The weights of the attributes are also calculated using newly proposed mathematical programming model which uses the DMs' partial information effectively. The data structure used for preference elicitation is a generalization of PHFS concept that mitigates the problem of assigning a precise occurrence probability to each HFE by allowing interval values as occurrence probability for each HFE. The MM operator has the ability to capture interrelationship between multiple attributes and provides a generalized focus on the aggregation of preferences. The customizable parameter in MM operator allows DMs to clearly understand the effect of risk appetite value on the prioritization order. The idea of extending the MM operator to IVPHFS context and the proposing of a new programming model for attribute weight calculation enriches the data structure for better MAGDM. Some desirable properties of the proposed IVPHFMM operator have analyzed for better understanding the operator and applying the same for MAGDM. From the sensitivity analysis of risk appetite values, we can infer that as values increase, the rank values also increase for each renewable energy source, thus allowing rational prioritization of renewable energy sources. Also, sensitivity analysis is carried out for attribute weight values and they infer that the prioritization order is unchanged and the proposed prioritization procedure is stable even after adequate changes are made to the attributes' weights.

As a part of the future direction, plans are made to extend different generalized operators in the IVPHFS context with a discussion on different Archimedean T-norms and T-conorms. Further, plans are made to propose a new decision framework in the IVPHFS context for better MAGDM and large scale group decision-making.

Author Contributions: The individual contribution and responsibilities of the authors were as follows: Author(s) R.K. and M.I.A. designed the research model, collected, pre-processed, analysed the data and the obtained results. They worked on the development of the paper. Authors K.S.R., S.K. and X.P. provided valuable advice throughout the research by sharing insights on model design, methodology and inferences. They also refined the manuscript by providing a valuable suggestion. All the authors have read and approved the final manuscript.

Funding: This research was funded by the University Grants Commission (UGC), India and Department of Science & Technology (DST), India under grant number F./2015-17/RGNF-2015-17-TAM-83 and SR/FST/ETI-349/2013.

Acknowledgments: Author(s) thank the editors and the anonymous reviewers for their insightful comments which improved the quality of the paper.

Conflicts of Interest: The authors declare no conflict of interest.

References

1. Triantaphyllou, E.; Shu, B. Multi-criteria decision making: An operations research approach. *Encycl. Electr. Electron. Eng.* **1998**, *15*, 175–186.
2. Torra, V. Hesitant fuzzy sets. *Int. J. Intell. Syst.* **2010**, *25*, 529–539. [CrossRef]
3. Zadeh, L.A. Fuzzy sets. *Inf. Control* **1965**, *8*, 338–353. [CrossRef]
4. Chai, J.; Ngai, E.W.T. Multi-perspective strategic supplier selection in uncertain environments. *Int. J. Prod. Econ.* **2015**, *166*, 215–225. [CrossRef]
5. Krishankumar, R.; Ravichandran, K.S.; Murthy, K.K.; Saeid, A.B. A scientific decision-making framework for supplier outsourcing using hesitant fuzzy information. *Soft Comput.* **2018**, *22*, 7445–7461. [CrossRef]
6. Aktas, A.; Kabak, M. A hybrid hesitant fuzzy decision-making approach for evaluating solar power plant location sites. *Arab. J. Sci. Eng.* **2018**, 1–13. [CrossRef]
7. Senvar, O.; Otay, I.; Bolturk, E. Hospital site selection via hesitant fuzzy TOPSIS. *IFAC-PapersOnLine* **2016**, *49*, 1140–1145. [CrossRef]
8. Zhang, F.; Chen, J.; Zhu, Y.; Li, J.; Li, Q.; Zhuang, Z. A dual hesitant fuzzy rough pattern recognition approach based on deviation theories and its application in urban traffic modes recognition. *Symmetry.* **2017**, *9*, 262. [CrossRef]
9. Rodríguez, R.M.; Martínez, L.; Torra, V.; Xu, Z.S.; Herrera, F. Hesitant fuzzy sets: State of the art and future directions. *Int. J. Intell. Syst.* **2014**, *29*, 495–524. [CrossRef]
10. Xu, Z.S.; Zhou, W. Consensus building with a group of decision makers under the hesitant probabilistic fuzzy environment. *Fuzzy Optim. Decis. Mak.* **2016**, *16*, 1–23. [CrossRef]
11. Yue, L.; Sun, M.; Shao, Z. The probabilistic hesitant fuzzy weighted average operators and their application in strategic decision making. *J. Inf. Comput. Sci.* **2013**, *10*, 3841–3848. [CrossRef]

12. Li, J.; Wang, Z. Multi-attribute decision making based on prioritized operators under probabilistic hesitant fuzzy environments. *Soft Comput.* **2018**, 1–16. [CrossRef]

13. Hao, Z.; Xu, Z.S.; Zhao, H.; Su, Z. Probabilistic dual hesitant fuzzy set and its application in risk evaluation. *Knowledge-Based Syst.* **2017**, *127*, 16–28. [CrossRef]

14. Zhou, W.; Xu, Z.S. Group consistency and group decision making under uncertain probabilistic hesitant fuzzy preference environment. *Inf. Sci.* **2017**, *414*, 276–288. [CrossRef]

15. Bashir, Z.; Rashid, T.; Watróbski, J.; Salabun, W.; Malik, A. Hesitant probabilistic multiplicative preference relations in group decision making. *Appl. Sci.* **2018**, *8*, 3998. [CrossRef]

16. Gao, J.; Xu, Z.S.; Liao, H. A dynamic reference point method for emergency response under hesitant probabilistic fuzzy environment. *Int. J. Fuzzy Syst.* **2017**, *19*, 1261–1278. [CrossRef]

17. Jiang, F.; Ma, Q. Multi-attribute group decision making under probabilistic hesitant fuzzy environment with application to evaluate the transformation efficiency. *Appl. Intell.* **2017**, *48*, 953–965. [CrossRef]

18. Song, C.; Zhao, H.; Xu, Z.S.; Hao, Z. Interval-valued probabilistic hesitant fuzzy set and its application in the Arctic geopolitical risk evaluation. *Int. J. Intell. Syst.* **2018**, 1–25. [CrossRef]

19. Krishankumar, R.; Ravichandran, K.S.; Kar, S.; Gupta, P.; Mehlawat, M.K. Interval-valued probabilistic hesitant fuzzy set for multi-criteria group decision-making. *Soft Comput.* **2018**, 1–27. [CrossRef]

20. Moscovici, S.; Zavalloni, M. The group as a polarizer of attitudes. *J. Pers. Soc. Psychol.* **1969**, *12*, 125–135. [CrossRef]

21. Mesiar, R.; Calvo, T. Fuzzy sets and their extensions: Representation, aggregation and models. *Stud. Fuzziness Soft Comput.* **2008**, *220*, 1–22.

22. Muirhead, R.F. Some methods applicable to identities and inequalities of symmetric algebraic functions of n letters. *Proc. Edinburgh Math. Soc.* **1902**, *21*, 144–162. [CrossRef]

23. Xia, M.; Xu, Z.S.; Zhu, B. Geometric Bonferroni means with their application in multi-criteria decision making. *Knowledge-Based Syst.* **2013**, *40*, 88–100. [CrossRef]

24. Qin, J.; Liu, X.; Pedrycz, W. Hesitant fuzzy Maclaurin symmetric mean operators and its application to multiple-attribute decision making. *Int. J. Fuzzy Syst.* **2015**, *17*, 509–520. [CrossRef]

25. Wang, R.; Wang, J.; Gao, H.; Wei, G. Methods for MADM with picture fuzzy Muirhead mean operators and their application for evaluating the financial investment risk. *Symmetry* **2019**, *11*, 6. [CrossRef]

26. Liu, P.; Khan, Q.; Mahmood, T.; Hassan, N. T-spherical fuzzy power Muirhead mean operator based on novel operational laws and their application in multi-attribute group decision making. *IEEE Access* **2019**, *7*, 22613–22632. [CrossRef]

27. Liu, P.; Li, Y.; Zhang, M.; Zhang, L.; Zhao, J. Multiple-attribute decision-making method based on hesitant fuzzy linguistic Muirhead mean aggregation operators. *Soft Comput.* **2018**, *22*, 5513–5524. [CrossRef]

28. Hong, Z.; Rong, Y.; Qin, Y.; Liu, Y. Hesitant fuzzy dual Muirhead mean operators and its application to multiple attribute decision making. *J. Intell. Fuzzy Syst.* **2018**, *35*, 2161–2172. [CrossRef]

29. Khan, Q.; Hassan, N.; Mahmood, T. Neutrosophic cubic power Muirhead mean operators with uncertain data for multi-attribute decision-making. *Symmetry* **2018**, *10*, 444. [CrossRef]

30. Xu, Y.; Shang, X.; Wang, J. Pythagorean fuzzy interaction Muirhead means with their application to multi-attribute group decision-making. *Inf.* **2018**, *9*, 157. [CrossRef]

31. Wang, J.; Zhang, R.; Zhu, X.; Zhou, Z.; Shang, X.; Li, W. Some q-rung orthopair fuzzy Muirhead means with their application to multi-attribute group decision-making. *J. Intell. Fuzzy Syst.* **2018**, *36*, 1–19. [CrossRef]

32. Liu, P.; Teng, F. Some Muirhead mean operators for probabilistic linguistic term sets and their applications to multiple attribute decision-making. *Appl. Soft Comput.* **2018**, *68*, 396–431. [CrossRef]

33. Saaty, T.L. Decision making with the analytic hierarchy process. *Int. J. Serv. Sci.* **2008**, *1*, 83. [CrossRef]

34. Gupta, P.; Mehlawat, M.K.; Grover, N. Intuitionistic fuzzy multi-attribute group decision-making with an application to plant location selection based on a new extended VIKOR method. *Inf. Sci.* **2016**, *370*, 184–203. [CrossRef]

35. Chatterjee, K.; Kar, S. A multi-criteria decision making for renewable energy selection using Z-numbers. *Technol. Econ. Dev. Econ.* **2018**, *24*, 739–764. [CrossRef]

36. Mardani, A.; Zavadskas, E.K.; Khalifah, Z.; Zakuan, N.; Jusoh, A.; Nor, K.M.; Khoshnoudi, M. A review of multi-criteria decision-making applications to solve energy management problems: Two decades from 1995 to 2015. *Renew. Sustain. Energy Rev.* **2017**, *71*, 216–256. [CrossRef]

37. Luthra, S.; Kumar, S.; Garg, D.; Haleem, A. Barriers to renewable/sustainable energy technologies adoption: Indian perspective. *Renew. Sustain. Energy Rev.* **2015**, *41*, 762–776. [CrossRef]
38. Spearman, C. The proof and measurement of association between two things. *Am. J. Psychol.* **1904**, *15*, 72–101. [CrossRef]

 *mathematics*

Article

Dual Hesitant Pythagorean Fuzzy Heronian Mean Operators in Multiple Attribute Decision Making

Mei Tang [1], Jie Wang [1], Jianping Lu [1], Guiwu Wei [1], Cun Wei [1,2] and Yu Wei [3,*]

[1] School of Business, Sichuan Normal University, Chengdu 610101, China; 15882583803@163.com (M.T.); JW970326@163.com (J.W.); lujp2002@163.com (J.L.); weiguiwu1973@sicnu.edu.cn (G.W.); weicun1990@163.com (C.W.)
[2] School of Statistics, Southwestern University of Finance and Economics, Chengdu 611130, China
[3] School of Finance, Yunnan University of Finance and Economics, Kunming 650221, China
* Correspondence: weiyusy@126.com

Received: 27 March 2019; Accepted: 6 April 2019; Published: 10 April 2019

Abstract: On account of the indeterminacy and subjectivity of decision makers (DMs) in complexity decision-making environments, the evaluation information over alternatives presented by DMs is usually fuzzy and ambiguous. As the generalization of intuitionistic fuzzy sets (IFSs), the Pythagorean fuzzy set (PFS) is more useful in expressing fuzzy and ambiguous information. Meanwhile, in order to consider human hesitance, dual hesitant Pythagorean fuzzy sets (DHPFSs) are presented, which can be more valid for handling real multiple attribute decision-making (MADM) problems. To fuse the information in DHPFSs more effectively, in this article, some dual hesitant Pythagorean fuzzy Heronian mean operators, which can consider the relationships between arguments being fused, are defined and studied. Evidently, the new proposed operators can obtain more exact results than other existing methods. In addition, some important properties of these Heronian mean (HM) operators are discussed. Subsequently, the defined aggregation operators are used in MADM with dual hesitant Pythagorean fuzzy numbers (DHPFNs), and the MADM model is developed. In accordance with the defined operators and the built model, the dual hesitant Pythagorean fuzzy generalized weighted Heronian mean (DHPFGWHM) operator and dual hesitant Pythagorean fuzzy generalized geometric weighted Heronian mean (DHPFGGWHM) operator are applied to deal with the green supplier selection in supply chain management, and the availability and superiority of the proposed operators are analyzed by comparing them with some existing approaches. The method presented in this paper can effectively solve the MADM problems in which the decision-making information is expressed by DHPFNs and the attributes are interactive.

Keywords: multiple attribute decision-making (MADM); dual hesitant Pythagorean fuzzy sets (DHPFSs); dual hesitant Pythagorean fuzzy generalized weighted Heronian mean (DHPFGWHM) operator; dual hesitant Pythagorean fuzzy generalized geometric weighted Heronian mean (DHPFGGWHM) operator; green supplier selection; supply chain management

---

1. Introduction

In practical decision making environments, it is difficult for decision makers (DMs) to give evaluated information with exact real numbers. To overcome this disadvantage, Zadeh [1] has developed the fuzzy set (FS) theory which utilizes the function of membership degree to express decision making information instead of crisp results between 0 and 1. Based on studies of FS, Atanassov [2] further proposed another function, named the non-membership degree, as a supplementary function. Thus, the intuitionistic fuzzy set (IFS) was constructed; in the IFS each intuitionistic fuzzy set is characterized by the functions of membership degree and non-membership degree between 0 and 1, and the sum of these are limited to 1. Subsequent to these studies, more and more scholars have studied the IFS in relation to many multiple attribute decision making (MADM) problems [3–16]. Xu [17] has defined some intuitionistic fuzzy weighted average operators. Xu and Yager [18] have proposed some aggregation operators, such as the intuitionistic fuzzy weighted geometric (IFWG) operator, the intuitionistic fuzzy ordered weighted geometric (IFOWG) operator, and the intuitionistic fuzzy hybrid geometric (IFHG) operator, based on geometric operation laws and an intuitionistic fuzzy environment. Hung and Yang [19] studied the similarity measures of intuitionistic fuzzy sets based on an Lp-metric. Park et al. [20] have put forward some distance measures of interval-valued intuitionistic fuzzy sets. To handle intuitionistic fuzzy MADM problems with incomplete weight information, Wei [21] utilized the maximizing deviation method to build two intuitionistic fuzzy nonlinear programming models. By considering the entropy weight of an intuitionistic fuzzy set, Hung et al. [22] established a fuzzy TOPSIS decision making model. Luo [23] defined a projection method based on intuitionistic fuzzy information with uncertain attribute weights for MADM. Ye [24] has provided a cross-entropy method with which to handle decision making problems with interval-valued intuitionistic fuzzy information. On account of the indeterminacy of DMs, Zhang [25] has presented some interval-valued hesitant fuzzy aggregation operators and applied them to MADM problems. Liao and Xu [26] have defined some intuitionistic fuzzy hybrid weighted aggregation operators. To express fuzzy information more easily, Liu et al. [27] have developed the concepts of hesitant intuitionistic fuzzy linguistic elements (HLFLEs) and have defined some weighted aggregation operators. Peng et al. [28] have discussed the MADM approach under a hesitant interval-valued intuitionistic fuzzy environment. Chen and Huang [29] have given the definition of hesitant triangular intuitionistic fuzzy set (HTIFS) and investigated its applications in MADM problems.

In addition, some other fuzzy decision making approaches have been proposed by numerous scholars [30–33]. Hu et al. [34] have proposed a novel approach combining fuzzy data envelopment analysis (DEA) and the analytical hierarchical process (AHP) to rank units with multiple fuzzy criteria. Ziemba et al. [35] have studied the online comparison system with certain and uncertain criteria. Diouf and Kwak [36] have studied fuzzy AHP, DEA, and managerial analysis for supplier selection and development from the perspective of open innovation. Dong et al. [37] utilized the modified fuzzy VIKOR and scalable computing method to study the performance evaluation of residential demand responses. Kim and Kim [38] have developed a new model for the optimal LNG import portfolio. Chou et al. [39] have used fuzzy AHP and fuzzy TOPSIS to evaluate the human resource in science and technology (HRST) performance of Southeast Asian countries.

In addition, as an effective MADM tool, the Pythagorean fuzzy set (PFS) [40,41] has emerged as a means to describe the indeterminacy and complexity of evaluation information. Similarly to the IFS, the PFS also consists of a membership degree and non-membership degree, the sum of the squares of which is restricted to 1. Thus, it is clear that the PFS is more widespread than the IFS and can express more decision-making information. For instance, the membership is given as 0.6 and the non-membership is given as 0.8; it is obvious that this problem is only valid for the PFS. In other words, all intuitionistic fuzzy decision-making problems are a special case of Pythagorean fuzzy decision-making problems, which means that the PFS is more efficient in dealing with MADM problems. In previous literature, some research works have been studied by a large number of investigators. Zhang and Xu [42] defined the Pythagorean fuzzy TOPSIS model to deal with MADM problems. Peng and Yang [43] primarily proposed two Pythagorean fuzzy operations including the division and subtraction operations to better understand PFS. Reformat and Yager [44] handled the collaborative-based recommender system with Pythagorean fuzzy information. Gou et al. [45] have studied some important properties of continuous Pythagorean fuzzy information. Garg [46] has defined some new Pythagorean fuzzy aggregation operators, including Pthe ythagorean fuzzy Einstein weighted averaging (PFEWA) operator, the Pythagorean fuzzy Einstein ordered weighted averaging (PFEOWA) operator, the generalized Pythagorean fuzzy Einstein weighted averaging (GPFEWA) operator and the generalized Pythagorean fuzzy Einstein ordered weighted averaging (GPFEOWA) operator. Zeng et al. [47] have utilized the Pythagorean fuzzy ordered weighted averaging weighted average distance (PFOWAWAD) operator to study Pythagorean fuzzy MADM issues. Ren et al. [48] built a Pythagorean fuzzy TODIM model. Liang et al. [49] investigated some Bonferroni mean operators with Pythagorean fuzzy information. Liang et al. [50] have presented Pythagorean fuzzy Bonferroni mean aggregation operators based on geometric averaging (GA) operations. Combining PFSs [40,41] and dual hesitant fuzzy sets (DHFSs) [51,52], Wei and Lu [53] introduced a definition of dual hesitant Pythagorean fuzzy sets (DHPFSs) and proposed some dual hesitant Pythagorean fuzzy Hamacher aggregation operators. Obviously, the DHPFSs have the advantages of considering the hesitance of DMs and expressing fuzzy information more effectively and reasonably.

However, in practical MADM problems, some relationships do exist between arguments being fused, and it is obvious that the dual hesitant Pythagorean fuzzy Hamacher aggregation operators defined by Wei and Lu [53] do not take the relationships between the arguments being fused into consideration. Thus, it is necessary to find another more effective method with which to fuse dual hesitant Pythagorean fuzzy information. To date, the Heronian mean (HM) [54] operator, which can effectively take the interrelationship between arguments into account, has drawn a large quantity of scholars' attention [55–59]. Based on intuitionistic fuzzy information and a geometric operator, Yu [54] developed the intuitionistic fuzzy geometric Heronian mean (IFGHM) operator and the intuitionistic fuzzy geometric weighted Heronian mean (IFGWHM) operator. Liu et al. [60] further proposed some Heronian mean operators under an intuitionistic uncertain linguistic environment for MADM. Yu et al. [61] have defined some linguistic hesitant fuzzy Heronian mean (LHFHM) operators. Li et al. [62] extended the Heronian mean operator to a single valued neutrosophic environment. Wei et al. [63] have presented some q-rung orthopair Heronian mean operators. Considering linguistic variables, Li et al. [64] developed some q-rung orthopair linguistic Heronian mean operators.

In this paper, based on the generalized Heronian mean (GHM) operator and generalized geometric Heronian mean (GGHM) operator, we develop some dual hesitant Pythagorean fuzzy generalized Heronian mean aggregation operators. The remainder of this paper is set out as follows. In the next section, we introduce some basic concepts related to the Pythagorean fuzzy set (PFS), the dual hesitant Pythagorean fuzzy set (DHPFS), and their operational laws. In Section 3, we propose some dual hesitant Pythagorean fuzzy Heronian mean aggregation operators such as: the dual hesitant Pythagorean fuzzy generalized weighted Heronian mean (DHPFGWHM) operator and the dual hesitant Pythagorean fuzzy generalized geometric weighted Heronian mean (DHPFGGWHM) operator. In Section 4, based on the DHPFGWHM and DHPFGGWHM operators, we propose some models for MADM problems with dual hesitant Pythagorean fuzzy information. In Section 5, we present a numerical example for supplier selection in supply chain management with dual hesitant Pythagorean fuzzy information in order to illustrate the method proposed in this paper. Section 6 concludes the paper with some remarks.

2. Preliminaries

2.1. Pythagorean Fuzzy Set

The fundamental definition of PFSs [40,41] are briefly introduced in this section. Then, novel score and accuracy functions of Pythagorean fuzzy numbers (PFNs) are developed. Furthermore, the comparison laws of PFNs are proposed.

Definition 1 [40,41]. *Let X be a fixed set. A Pythagorean fuzzy set (PFS) is an object which can be denoted as*

$$P = \{\langle x, (\alpha_P(x), \beta_P(x))\rangle | x \in X\} \tag{1}$$

where the function $\alpha_P : X \rightarrow [0,1]$ indicates the degree of membership and the function $\beta_P : X \rightarrow [0,1]$ indicates the degree of non-membership of the element $x \in X$ to P, respectively, and, for each $x \in X$, it holds that

$$\left(\alpha_p(x)\right)^2 + \left(\beta_p(x)\right)^2 \leq 1 \tag{2}$$

Definition 2 [42]. *Assume that $p_1 = (\alpha_1, \beta_1)$, $p_2 = (\alpha_2, \beta_2)$, and $p = (\alpha, \beta)$ are three Pythagorean fuzzy numbers (PFNs). Then, some basic operation laws of them can be expressed as:*

$$(1)\, p_1 \oplus p_2 = \left(\sqrt{\left(\alpha_{p_1}\right)^2 + \left(\alpha_{p_2}\right)^2 - \left(\alpha_{p_1}\right)^2 \left(\alpha_{p_1}\right)^2}, \beta_{p_1} \beta_{p_2} \right);$$

$$(2)\, p_1 \otimes p_2 = \left(\alpha_{p_1} \alpha_{p_2}, \sqrt{\left(\beta_{p_1}\right)^2 + \left(\beta_{p_2}\right)^2 - \left(\beta_{p_1}\right)^2 \left(\beta_{p_2}\right)^2} \right);$$

$$(3)\, \lambda p = \left(\sqrt{1 - (1 - \alpha^2)^\lambda}, \beta^\lambda \right), \lambda > 0;$$

$$(4)\, (p)^\lambda = \left(\alpha^\lambda, \sqrt{1 - (1 - \beta^2)^\lambda} \right), \lambda > 0;$$

$$(5)\, p^c = (\beta, \alpha).$$

Example 1. *Assume that $p_1 = (0.5, 0.7)$, $p_2 = (0.3, 0.4)$, and $p = (0.6, 0.3)$ are three Pythagorean fuzzy numbers (PFNs). Suppose $\lambda = 3$. Then, according to the above operation laws, we can obtain:*

$$(1)\, p_1 \oplus p_2 = \left(\sqrt{(0.5)^2 + (0.3)^2 - (0.5)^2 \times (0.3)^2}, 0.7 \times 0.4 \right) = (0.56, 0.28);$$

$$(2)\, p_1 \otimes p_2 = \left(0.5 \times 0.3, \sqrt{(0.7)^2 + (0.4)^2 - (0.7)^2 \times (0.4)^2} \right) = (0.15, 0.64);$$

$$(3)\, 3 \times p = \left(\sqrt{1 - (1 - 0.6^2)^3}, 0.3^3 \right) = (0.8590, 0.0270);$$

$$(4)\, (p)^3 = \left(0.6^3, \sqrt{1 - (1 - 0.3^2)^3} \right) = (0.4964, 0.2160);$$

$$(5)\, p^c = (0.3, 0.6).$$

2.2. Dual Hesitant Pythagorean Fuzzy Set

In this section, we shall introduce the basic definition of the dual hesitant Pythagorean fuzzy set (DHPFS), which is the generalization of the PFS [40,41] and the dual hesitant fuzzy set (DHFS) [51,52]. It is obvious that the DHPFSs consist of two parts, namely, the function of membership hesitancy and the function of non-membership hesitancy, which support more exemplary and flexible access to assigning values for each element in the domain, meaning we have to handle two kinds of hesitancy in this situation.

Definition 3 [53]. *Assume that X is a fixed set. Then, a dual hesitant Pythagorean fuzzy set (DHPFS) on X can be developed as*

$$\widetilde{P} = \left(\langle x, h_{\widetilde{p}}(x), g_{\widetilde{p}}(x) \rangle | x \in X \right) \tag{3}$$

in which $h_{\widetilde{p}}(x)$ and $g_{\widetilde{p}}(x)$ are two sets of some values in $[0,1]$, indicating that the function of membership degrees and non-membership degrees of the element $x \in X$ to the set $\widetilde{P}$, respectively, satisfies the condition

$$\alpha^2 + \beta^2 \leq 1$$

where $\alpha \in h_{\widetilde{p}}(x), \beta \in g_{\widetilde{p}}(x)$, for all $x \in X$. For convenience, the pair $\widetilde{p}(x) = \left(h_{\widetilde{p}}(x), g_{\widetilde{p}}(x) \right)$ is called a dual hesitant Pythagorean fuzzy number (DHPFN) denoted by $\widetilde{p} = (h, g)$, with the conditions $\alpha \in h, \beta \in g$, $0 \leq \alpha, \beta \leq 1, 0 \leq \alpha^2 + \beta^2 \leq 1$.

Definition 4 [53]. *Let $\widetilde{p} = (h, g)$ be a DHPFN. Then, $s(\widetilde{p}) = \frac{1}{2}\left(1 + \frac{1}{\#h}\sum_{\alpha \in h}\alpha^2 - \frac{1}{\#g}\sum_{\beta \in g}\beta^2\right)$ is the score function of $\widetilde{p}$, and $H(\widetilde{p}) = \frac{1}{\#h}\sum_{\alpha \in h}\alpha^2 + \frac{1}{\#g}\sum_{\beta \in g}\beta^2$ is the accuracy function of $\widetilde{p}$, where $\#h$ and $\#g$ are the numbers of the elements in h and g respectively. Then, let $\widetilde{p}_i = (h_i, g_i)(i = 1, 2)$ be any two DHPFNs. Subsequently, we have the following comparison laws:*

- *If $s(\widetilde{p}_1) > s(\widetilde{p}_2)$, then $\widetilde{p}_1$ is superior to $\widetilde{p}_2$, denoted by $\widetilde{p}_1 > \widetilde{p}_2$;*
- *If $s(\widetilde{p}_1) = s(\widetilde{p}_2)$, then:*

  (1)   If $p(\widetilde{p}_1) = p(\widetilde{p}_2)$, then $\widetilde{p}_1$ is equivalent to $\widetilde{p}_2$, denoted by $\widetilde{p}_1 \sim \widetilde{p}_2$;
  (2)   If $p(\widetilde{p}_1) > p(\widetilde{p}_2)$, then $\widetilde{p}_1$ is superior to $\widetilde{p}_2$, denoted by $\widetilde{p}_1 > \widetilde{p}_2$.

Definition 5 [53]. *Assume that $\widetilde{p}_1 = (h_1, g_1)$, $\widetilde{p}_2 = (h_2, g_2)$, and $\widetilde{p} = (h, g)$ are three DHPFNs. Then, some basic operation laws of these can be expressed as:*

$$(1) \quad \widetilde{p}^{\lambda} = \cup_{\alpha \in h, \beta \in g}\left\{\{\alpha^{\lambda}\}, \left\{\sqrt{1 - (1 - \beta^2)^{\lambda}}\right\}\right\}, \lambda > 0;$$

$$(2) \quad \lambda\widetilde{p} = \cup_{\alpha \in h, \beta \in g}\left\{\left\{\sqrt{1 - (1 - \alpha^2)^{\lambda}}\right\}, \{\beta^{\lambda}\}\right\}, \lambda > 0;$$

$$(3) \quad \widetilde{p}_1 \oplus \widetilde{p}_2 = \cup_{\alpha_1 \in h_1, \alpha_2 \in h_2, \beta_1 \in g_1, \beta_2 \in g_2}\left\{\left\{\sqrt{(\alpha_1)^2 + (\alpha_2)^2 - (\alpha_1)^2(\alpha_2)^2}\right\}, \{\beta_1\beta_2\}\right\};$$

$$(4) \quad \widetilde{p}_1 \otimes \widetilde{p}_2 = \cup_{\alpha_1 \in h_1, \alpha_2 \in h_2, \beta_1 \in g_1, \beta_2 \in g_2}\left\{\{\alpha_1\alpha_2\}, \left\{\sqrt{(\beta_1)^2 + (\beta_2)^2 - (\beta_1)^2(\beta_2)^2}\right\}\right\}.$$

Example 2. *Assume that $p_1 = \{\{0.7\}, \{0.3\}\}$, $p_2 = \{\{0.1, 0.2\}, \{0.4\}\}$, and $p = \{\{0.5, 0.6\}, \{0.4\}\}$ are three Pythagorean fuzzy numbers. Suppose $\lambda = 3$. Then, according to the above operation laws, we can obtain*

$$(1) \quad \widetilde{p}^3 = \cup_{\alpha \in h, \beta \in g}\left\{\{0.5^3, 0.6^3\}, \left\{\sqrt{1 - (1 - 0.4^2)^3}\right\}\right\} = \{\{0.125, 0.216\}, \{0.638\}\};$$

$$(2) \quad 3\widetilde{p} = \cup_{\alpha \in h, \beta \in g}\left\{\left\{\begin{array}{c}\sqrt{1 - (1 - 0.5^2)^3}, \\ \sqrt{1 - (1 - 0.6^2)^3}\end{array}\right\}, \{0.4^3\}\right\} = \{\{0.760, 0.859\}, \{0.064\}\};$$

$$(3) \quad \widetilde{p}_1 \oplus \widetilde{p}_2 = \cup_{\alpha_1 \in h_1, \alpha_2 \in h_2, \beta_1 \in g_1, \beta_2 \in g_2}\left\{\left\{\begin{array}{c}\sqrt{0.7^2 + 0.1^2 - 0.7^2 \times 0.1^2}, \\ \sqrt{0.7^2 + 0.2^2 - 0.7^2 \times 0.2^2}\end{array}\right\}, \{0.3 \times 0.4\}\right\}$$
$$= \{\{0.704, 0.714\}, \{0.120\}\}$$

$$(4) \quad \widetilde{p}_1 \otimes \widetilde{p}_2 = \cup_{\alpha_1 \in h_1, \alpha_2 \in h_2, \beta_1 \in g_1, \beta_2 \in g_2}\left\{\{0.7 \times 0.1, 0.7 \times 0.2\}, \left\{\sqrt{0.3^2 + 0.4^2 - 0.3^2 \times 0.4^2}\right\}\right\}$$
$$= \{\{0.070, 0.140\}, \{0.485\}\}$$

2.3. The Heronian Mean Operator

Definition 6 [65]. *Let b_i $(i = 1, 2, \cdots, n)$ be a group of nonnegative real numbers. Then, the Heronian mean (HM) operator can be defined as:*

$$\text{HM}(b_1, b_2, \ldots, b_n) = \frac{2}{n(n+1)}\sum_{i=1}^{n}\sum_{j=i}^{n}(b_i b_j)^{\frac{1}{2}} \tag{4}$$

Definition 7 [54]. *Assume that $\xi, \zeta > 0$, and b_i $(i = 1, 2, \cdots, n)$ are a group of nonnegative real numbers. Then, the GHM operator can be defined as:*

$$\mathrm{GHM}^{\xi,\zeta}(a_1, a_2, \ldots, a_n) = \left(\frac{2}{n(n+1)} \sum_{i=1}^{n} \sum_{j=i}^{n} a_i^{\xi} a_j^{\zeta} \right)^{1/(\xi+\zeta)} \tag{5}$$

When $\xi = \zeta = 1/2$, the GHM operator will reduce to the Heronian mean (HM) operator, which indicates that the HM operator is a special case of the GHM operator.

3. Dual Hesitant Pythagorean Fuzzy Heronian Mean Operators

In the following section Xu et al. [66] proposed the dual hesitant Pythagorean fuzzy generalized Heronian mean (DHPFGHM) operators based on dual hesitant Pythagorean fuzzy numbers (DHPFNs) and GHM operations. In addition, some important properties, such as idempotency, boundedness, and monotonicity are discussed.

3.1. The DHPFGHM Aggregation Operator

Definition 8 [66]. *Let $\xi, \zeta > 0$ and $\widetilde{p}_j = (h_j, g_j)(j = 1, 2, \cdots, n)$ be a group of DHPFNs. Then, we can define the DHPFGHM operator as*

$$\mathrm{DHPFGHM}^{\xi,\zeta}(\widetilde{p}_1, \widetilde{p}_2, \ldots, \widetilde{p}_n) = \left(\frac{2}{n(n+1)} \mathop{\oplus}_{i=1}^{n} \mathop{\oplus}_{j=i}^{n} \left(\widetilde{p}_i^{\xi} \otimes \widetilde{p}_j^{\zeta} \right) \right)^{\frac{1}{\xi+\zeta}} \tag{6}$$

where "$\oplus$" indicates the addition operation law and "$\otimes$" indicates the multiplication operation law of the DHPFNs described in Definition 5. Then, according to these operation laws, Xu et al. [66] obtained Theorem 1.

Theorem 1 [66]. *Let $\xi, \zeta > 0$ and $\widetilde{p}_j = (h_j, g_j)(j = 1, 2, \cdots, n)$ be a group of dual hesitant Pythagorean fuzzy numbers, meaning their fused results by utilizing the DHPFGHM operator is also a DHPFN, and*

$$\mathrm{DHPFGHM}^{\xi,\zeta}(\widetilde{p}_1, \widetilde{p}_2, \ldots, \widetilde{p}_n) = \left(\frac{2}{n(n+1)} \mathop{\oplus}_{i=1}^{n} \mathop{\oplus}_{j=i}^{n} \left(\widetilde{p}_i^{\xi} \otimes \widetilde{p}_j^{\zeta} \right) \right)^{\frac{1}{\xi+\zeta}}$$

$$= \bigcup_{\alpha_i \in h_i, \alpha_j \in h_j, \beta_i \in g_i, \alpha_j \in h_j} \left\{ \left\{ \left(\sqrt{1 - \prod_{i=1,j=i}^{n} \left(1 - \alpha_i^{2\xi} \alpha_j^{2\zeta} \right)^{\frac{2}{n(n+1)}}} \right)^{\frac{1}{\xi+\zeta}} \right\}, \left\{ \sqrt{1 - \left(1 - \prod_{i=1,j=i}^{n} \left(1 - \left(1 - \beta_i^2 \right)^{\xi} \left(1 - \beta_j^2 \right)^{\zeta} \right)^{\frac{2}{n(n+1)}} \right)^{\frac{1}{\xi+\zeta}}} \right\} \right\} \tag{7}$$

Proof. Based on Definition 5:

$$\widetilde{p}_i^{\xi} = \bigcup_{\alpha_i \in h_i, \beta_i \in g_i} \left\{ \{ \alpha_i^{\xi} \}, \left\{ \sqrt{1 - \left(1 - \beta_i^2 \right)^{\xi}} \right\} \right\} \tag{8}$$

$$\widetilde{p}_j^{\zeta} = \bigcup_{\alpha_j \in h_j, \beta_j \in g_j} \left\{ \{ \alpha_j^{\zeta} \}, \left\{ \sqrt{1 - \left(1 - \beta_j^2 \right)^{\zeta}} \right\} \right\} \tag{9}$$

Thus,

$$\widetilde{p}_i^{\xi} \otimes \widetilde{p}_j^{\zeta} = \cup_{\alpha_i \in h_i, \alpha_j \in h_j, \beta_i \in g_i, \alpha_j \in h_j} \left\{ \left\{ \alpha_i^{\xi} \alpha_j^{\zeta} \right\}, \left\{ \sqrt{1 - \left(1 - \beta_i^2\right)^{\xi} \left(1 - \beta_j^2\right)^{\zeta}} \right\} \right\} \tag{10}$$

Therefore,

$$\overset{n}{\underset{i=1}{\oplus}} \overset{n}{\underset{j=i}{\oplus}} \left(\widetilde{p}_i^{\xi} \otimes \widetilde{p}_j^{\zeta} \right)$$

$$= \cup_{\alpha_i \in h_i, \alpha_j \in h_j, \beta_i \in g_i, \alpha_j \in h_j} \left\{ \begin{array}{c} \left\{ \sqrt{1 - \prod_{i=1,j=i}^{n} \left(1 - \alpha_i^{2\xi} \alpha_j^{2\zeta}\right)} \right\}, \\ \left\{ \sqrt{\prod_{i=1,j=i}^{n} \left(1 - \left(1 - \beta_i^2\right)^{\xi} \left(1 - \beta_j^2\right)^{\zeta}\right)} \right\} \end{array} \right\} \tag{11}$$

Furthermore,

$$\frac{2}{n(n+1)} \overset{n}{\underset{i=1}{\oplus}} \overset{n}{\underset{j=i}{\oplus}} \left(\widetilde{p}_i^{\xi} \otimes \widetilde{p}_j^{\zeta} \right)$$

$$= \cup_{\alpha_i \in h_i, \alpha_j \in h_j, \beta_i \in g_i, \alpha_j \in h_j} \left\{ \begin{array}{c} \left\{ \sqrt{1 - \prod_{i=1,j=i}^{n} \left(1 - \alpha_i^{2\xi} \alpha_j^{2\zeta}\right)^{\frac{2}{n(n+1)}}} \right\}, \\ \left\{ \sqrt{\prod_{i=1,j=i}^{n} \left(1 - \left(1 - \beta_i^2\right)^{\xi} \left(1 - \beta_j^2\right)^{\zeta}\right)^{\frac{2}{n(n+1)}}} \right\} \end{array} \right\} \tag{12}$$

Therefore,

$$\mathrm{DHPFGHM}^{\xi,\zeta}(\widetilde{p}_1, \widetilde{p}_2, \ldots, \widetilde{p}_n) = \left(\frac{2}{n(n+1)} \overset{n}{\underset{i=1}{\oplus}} \overset{n}{\underset{j=i}{\oplus}} \left(\widetilde{p}_i^{\xi} \otimes \widetilde{p}_j^{\zeta} \right) \right)^{\frac{1}{\xi+\zeta}}$$

$$= \cup_{\alpha_i \in h_i, \alpha_j \in h_j, \beta_i \in g_i, \alpha_j \in h_j} \left\{ \begin{array}{c} \left\{ \left(\sqrt{1 - \prod_{i=1,j=i}^{n} \left(1 - \alpha_i^{2\xi} \alpha_j^{2\zeta}\right)^{\frac{2}{n(n+1)}}} \right)^{\frac{1}{\xi+\zeta}} \right\}, \\ \left\{ \sqrt{1 - \left(1 - \prod_{i=1,j=i}^{n} \left(1 - \left(1 - \beta_i^2\right)^{\xi} \left(1 - \beta_j^2\right)^{\zeta}\right)^{\frac{2}{n(n+1)}}\right)^{\frac{1}{\xi+\zeta}}} \right\} \end{array} \right\} \tag{13}$$

Thus, the proof has been finished.

Example 3. *Assume that* $\widetilde{p}_1 = \{\{0.7, 0.8\}, \{0.4\}\}$, $\widetilde{p}_2 = \{\{0.3\}, \{0.6, 0.7\}\}$, $\widetilde{p}_3 = \{\{0.1, 0.3\}, \{0.4, 0.6\}\}$ *and* $\widetilde{p}_4 = \{\{0.5\}, \{0.5\}\}$ *are four DHPFNs, and suppose that* $\xi = 2, \zeta = 3$. *Then according to the DHPFGHM operator, we can obtain the fused results as follows. For the membership degree function* α, *the fused results are shown as:*

$$\alpha_1 = \mathrm{DHPFGHM}^{2,3}(0.7, 0.3, 0.1, 0.5) = \left(\sqrt{1 - \prod_{i=1,j=i}^{n} \left(1 - \alpha_i^{2\xi} \alpha_j^{2\zeta}\right)^{\frac{2}{n(n+1)}}} \right)^{\frac{1}{\xi+\zeta}}$$

$$= \left(\sqrt{1 - \left(\begin{array}{c} \left(1 - 0.7^{2\times2} \times 0.7^{2\times3}\right) \times \left(1 - 0.7^{2\times2} \times 0.3^{2\times3}\right) \times \left(1 - 0.7^{2\times2} \times 0.1^{2\times3}\right) \\ \times \left(1 - 0.7^{2\times2} \times 0.5^{2\times3}\right) \times \left(1 - 0.3^{2\times2} \times 0.3^{2\times3}\right) \times \left(1 - 0.3^{2\times2} \times 0.1^{2\times3}\right) \\ \times \left(1 - 0.3^{2\times2} \times 0.5^{2\times3}\right) \times \left(1 - 0.1^{2\times2} \times 0.1^{2\times3}\right) \times \left(1 - 0.1^{2\times2} \times 0.5^{2\times3}\right) \\ \times \left(1 - 0.5^{2\times2} \times 0.5^{2\times3}\right) \end{array} \right)^{\frac{1}{10}}} \right)^{\frac{1}{2+3}}$$

$$= 0.5658$$

Similarly, we can obtain

$$\alpha_2 = \text{DHPFGHM}^{2,3}(0.7, 0.3, 0.3, 0.5) = 0.5664$$
$$\alpha_3 = \text{DHPFGHM}^{2,3}(0.8, 0.3, 0.1, 0.5) = 0.6492$$
$$\alpha_4 = \text{DHPFGHM}^{2,3}(0.8, 0.3, 0.3, 0.5) = 0.6432$$

Hence, we can get $\alpha = \{0.5658, 0.5664, 0.6429, 0.6432\}$.

For the non-membership degree function β, the fused results are shown as:

$$\beta_1 = \text{DHPFGHM}^{2,3}(0.4, 0.6, 0.4, 0.5) = \sqrt{1 - \left(1 - \prod_{i=1,j=i}^{n}\left(1 - \left(1 - \beta_i^2\right)^{\xi}\left(1 - \beta_j^2\right)^{\zeta}\right)^{\frac{2}{n(n+1)}}\right)^{\frac{1}{\xi+\zeta}}}$$

$$= \sqrt{\left(1 - \left(1 - \left(\begin{array}{c}\left(1 - \left(1 - 0.4^2\right)^2 \times \left(1 - 0.4^2\right)^3\right) \times \left(1 - \left(1 - 0.4^2\right)^2 \times \left(1 - 0.6^2\right)^3\right) \\ \times\left(1 - \left(1 - 0.4^2\right)^2 \times \left(1 - 0.4^2\right)^3\right) \times \left(1 - \left(1 - 0.4^2\right)^2 \times \left(1 - 0.5^2\right)^3\right) \\ \times\left(1 - \left(1 - 0.6^2\right)^2 \times \left(1 - 0.6^2\right)^3\right) \times \left(1 - \left(1 - 0.6^2\right)^2 \times \left(1 - 0.4^2\right)^3\right) \\ \times\left(1 - \left(1 - 0.6^2\right)^2 \times \left(1 - 0.5^2\right)^3\right) \times \left(1 - \left(1 - 0.4^2\right)^2 \times \left(1 - 0.4^2\right)^3\right) \\ \times\left(1 - \left(1 - 0.4^2\right)^2 \times \left(1 - 0.5^2\right)^3\right) \times \left(1 - \left(1 - 0.5^2\right)^2 \times \left(1 - 0.5^2\right)^3\right)\end{array}\right)^{\frac{1}{10}}\right)^{\frac{1}{2+3}}\right)}$$

$$= 0.4698$$

Similarly, we can obtain

$$\beta_2 = \text{DHPFGHM}^{2,3}(0.4, 0.6, 0.6, 0.5) = 0.5211$$
$$\beta_3 = \text{DHPFGHM}^{2,3}(0.4, 0.7, 0.4, 0.5) = 0.4856$$
$$\beta_4 = \text{DHPFGHM}^{2,3}(0.4, 0.7, 0.6, 0.5) = 0.5383$$

Hence, we can get $\beta = \{0.4698, 0.5211, 0.4856, 0.5383\}$. Therefore,

$$\text{DHPFGHM}(\widetilde{p}_1, \widetilde{p}_2, \widetilde{p}_3, \widetilde{p}_4) = \left\{ \begin{array}{c} \{0.5658, 0.5664, 0.6429, 0.6432\}, \\ \{0.4698, 0.5211, 0.4856, 0.5383\} \end{array} \right\}.$$

It can be easily proven that the DHPFGHM operator satisfies the following properties.

Property 1. *(Idempotency) If all $\widetilde{p}_j = \left(h_j, g_j\right)(j = 1, 2, \cdots, n)$ are equal, i.e., $\widetilde{p}_j = \widetilde{p}$ for all j, then*

$$\text{DHPFGHM}^{\xi,\zeta}(\widetilde{p}_1, \widetilde{p}_2, \ldots, \widetilde{p}_n) = \widetilde{p} \tag{14}$$

Property 2. *(Boundedness) Let $\widetilde{p}_j = \left(h_j, g_j\right)(j = 1, 2, \cdots, n)$ be a collection of DHPFNs, and let*

$$\widetilde{p}^+ = \cup_{\alpha_j \in h_j, \beta_j \in g_j}\{\{\max_i(\alpha_i)\}, \{\min_i(\beta_i)\}\}, \widetilde{p}^- = \cup_{\alpha_j \in h_j, \beta_j \in g_j}\{\{\min_i(\alpha_i)\}, \{\max_i(\beta_i)\}\}$$

Then

$$\widetilde{p}^- \leq \text{DHPFGHM}^{\xi,\zeta}(\widetilde{p}_1, \widetilde{p}_2, \ldots, \widetilde{p}_n) \leq \widetilde{p}^+ \tag{15}$$

Property 3. *(Monotonicity) Let* $\widetilde{p}_j = \left(h_j, g_j\right)$ *and* $\widetilde{p}'_j = \left(h'_j, g'_j\right), j = 1, 2, \cdots, n,$ *be two sets of DHPFNs. If* $\widetilde{p}_j \leq \widetilde{p}'_j,$ *for all* j, *then*

$$\mathrm{DHPFGHM}^{\xi,\zeta}(\widetilde{p}_1, \widetilde{p}_2, \ldots, \widetilde{p}_n) \leq \mathrm{DHPFGHM}^{\xi,\zeta}(\widetilde{p}'_1, \widetilde{p}'_2, \ldots, \widetilde{p}'_n) \tag{16}$$

3.2. The DHPFGWHM Aggregation Operator

Using Definition 8, we can conclude that the DHPFGHM operator didn't take the importance of arguments being fused into account. However, in many practical MADM problems, we should consider the weights of attributes. To overcome this limitation of the DHPFGHM operator, we propose a novel DHPFGWHM operator as follows.

Definition 9. *Assume that* $\xi, \zeta > 0$ *and* $\widetilde{p}_j = \left(h_j, g_j\right)(j = 1, 2, \cdots, n)$ *are a group of dual hesitant Pythagorean fuzzy numbers (DHPFNs). Then, we define the DHPFGWHM operator as follows:*

$$\mathrm{DHPFGWHM}_w^{\xi,\zeta}(\widetilde{p}_1, \widetilde{p}_2, \ldots, \widetilde{p}_n) = \left(\mathop{\oplus}\limits_{i=1}^{n} \mathop{\oplus}\limits_{j=i}^{n} \left(w_i w_j \left(\widetilde{p}_i^{\xi} \widetilde{p}_j^{\zeta} \right) \right) \right)^{\frac{1}{\xi+\zeta}} \tag{17}$$

According to the operation laws of the DHPFNs described in Definition 5, we can obtain Theorem 2.

Theorem 2. *Assume that* $\xi, \zeta > 0$ *and* $\widetilde{p}_j = \left(h_j, g_j\right)(j = 1, 2, \cdots, n)$ *are a collection of DHPFNs with the weighting vector* $w = (w_1, w_2, \ldots, w_n)^T$, *which satisfies* $w_j > 0, i = 1, 2, \ldots, n$ *and* $\sum\limits_{j=1}^{n} w_j = 1$. *Then, their fused result obtained by utilizing the DHPFGWHM operator is also a DHPFN, and*

$$\mathrm{DHPFGWHM}_w^{\xi,\zeta}(\widetilde{p}_1, \widetilde{p}_2, \ldots, \widetilde{p}_n) = \left(\mathop{\oplus}\limits_{i=1}^{n} \mathop{\oplus}\limits_{j=i}^{n} \left(w_i w_j \left(\widetilde{p}_i^{\xi} \widetilde{p}_j^{\zeta} \right) \right) \right)^{\frac{1}{\xi+\zeta}}$$

$$= \bigcup_{\alpha_i \in h_i, \alpha_j \in h_j, \beta_i \in g_i, \alpha_j \in h_j,} \left\{ \begin{array}{c} \left\{ \left(\sqrt{ 1 - \prod\limits_{i=1, j=i}^{n} \left(1 - \alpha_i^{2\xi} \alpha_j^{2\zeta} \right)^{w_i w_j} } \right)^{\frac{1}{\xi+\zeta}} \right\}, \\[2em] \left\{ \sqrt{ 1 - \left(1 - \prod\limits_{i=1, j=i}^{n} \left(1 - \left(1 - \beta_i^2 \right)^{\xi} \left(1 - \beta_j^2 \right)^{\zeta} \right)^{w_i w_j} \right)^{\frac{1}{\xi+\zeta}} } \right\} \end{array} \right\} \tag{18}$$

Proof. Based on Definition 5, we can obtain:

$$\widetilde{p}_i^{\xi} = \bigcup_{\alpha_i \in h_i, \beta_i \in g_i} \left\{ \left\{ \alpha_i^{\xi} \right\}, \left\{ \sqrt{ 1 - \left(1 - \beta_i^2 \right)^{\xi} } \right\} \right\} \tag{19}$$

$$\widetilde{p}_j^{\zeta} = \bigcup_{\alpha_j \in h_j, \beta_j \in g_j} \left\{ \left\{ \alpha_j^{\zeta} \right\}, \left\{ \sqrt{ 1 - \left(1 - \beta_j^2 \right)^{\zeta} } \right\} \right\} \tag{20}$$

Thus,

$$\widetilde{p}_i^{\xi} \widetilde{p}_j^{\zeta} = \bigcup_{\alpha_i \in h_i, \alpha_j \in h_j, \beta_i \in g_i, \alpha_j \in h_j,} \left\{ \left\{ \alpha_i^{\xi} \alpha_j^{\zeta} \right\}, \left\{ \sqrt{ 1 - \left(1 - \beta_i^2 \right)^{\xi} \left(1 - \beta_j^2 \right)^{\zeta} } \right\} \right\} \tag{21}$$

Therefore,

$$
\begin{aligned}
& w_i w_j \left(\widetilde{p}_i^{\,\xi} \widetilde{p}_j^{\,\zeta} \right) \\
& = \bigcup_{\alpha_i \in h_i, \alpha_j \in h_j, \beta_i \in g_i, \alpha_j \in h_j,}
\left\{
\begin{array}{l}
\left\{ \sqrt{1 - \left(1 - \alpha_i^{2\xi} \alpha_j^{2\zeta}\right)^{w_i w_j}} \right\}, \\[4mm]
\left\{ \left(\sqrt{1 - \left(1 - \beta_i^2\right)^{\xi}\left(1 - \beta_j^2\right)^{\zeta}} \right)^{w_i w_j} \right\}
\end{array}
\right\}
\end{aligned}
\tag{22}
$$

Thereafter,

$$
\begin{aligned}
& \overset{n}{\underset{i=1}{\oplus}} \; \overset{n}{\underset{j=i}{\oplus}} \left(w_i w_j \left(\widetilde{p}_i^{\,\xi} \widetilde{p}_j^{\,\zeta} \right) \right) \\
& = \bigcup_{\alpha_i \in h_i, \alpha_j \in h_j, \beta_i \in g_i, \alpha_j \in h_j,}
\left\{
\begin{array}{l}
\left\{ \sqrt{1 - \prod_{i=1,j=i}^{n} \left(1 - \alpha_i^{2\xi} \alpha_j^{2\zeta}\right)^{w_i w_j}} \right\}, \\[4mm]
\left\{ \left(\sqrt{\prod_{i=1,j=i}^{n} \left(1 - \left(1 - \beta_i^2\right)^{\xi}\left(1 - \beta_j^2\right)^{\zeta}\right)} \right)^{w_i w_j} \right\}
\end{array}
\right\}
\end{aligned}
\tag{23}
$$

Therefore,

$$
\begin{aligned}
& \mathrm{DHPFGWHM}_w^{\xi,\zeta}(\widetilde{p}_1, \widetilde{p}_2, \ldots, \widetilde{p}_n) = \left(\overset{n}{\underset{i=1}{\oplus}} \; \overset{n}{\underset{j=i}{\oplus}} \left(w_i w_j \left(\widetilde{p}_i^{\,\xi} \widetilde{p}_j^{\,\zeta} \right) \right) \right)^{\frac{1}{\xi+\zeta}} \\
& = \bigcup_{\alpha_i \in h_i, \alpha_j \in h_j, \beta_i \in g_i, \alpha_j \in h_j,}
\left\{
\begin{array}{l}
\left\{ \left(\sqrt{1 - \prod_{i=1,j=i}^{n} \left(1 - \alpha_i^{2\xi} \alpha_j^{2\zeta}\right)^{w_i w_j}} \right)^{\frac{1}{\xi+\zeta}} \right\}, \\[4mm]
\left\{ \sqrt{1 - \left(1 - \prod_{i=1,j=i}^{n} \left(1 - \left(1 - \beta_i^2\right)^{\xi}\left(1 - \beta_j^2\right)^{\zeta}\right)^{w_i w_j}\right)^{\frac{1}{\xi+\zeta}}} \right\}
\end{array}
\right\}
\end{aligned}
\tag{24}
$$

Thus, we have finished the proof.

Example 4. *Assume that* $\widetilde{p}_1 = \{\{0.7, 0.8\}, \{0.4\}\}$, $\widetilde{p}_2 = \{\{0.3\}, \{0.6, 0.7\}\}$, $\widetilde{p}_3 = \{\{0.1, 0.3\}, \{0.4, 0.6\}\}$ *and* $\widetilde{p}_4 = \{\{0.5\}, \{0.5\}\}$ *are four DHPFNs, and suppose that* $\xi = 2, \zeta = 3$ *and* $w_j = (0.3, 0.2, 0.1, 0.4)$. *Then, according to the DHPFGWHM operator, we can obtain the fused results as follows. For the membership degree function* α, *the fused results are shown as:*

$$
\alpha_1 = \mathrm{DHPFGWHM}^{2,3}(0.7, 0.3, 0.1, 0.5) = \left(\sqrt{1 - \prod_{i=1,j=i}^{n} \left(1 - \alpha_i^{2\xi} \alpha_j^{2\zeta}\right)^{w_i w_j}} \right)^{\frac{1}{\xi+\zeta}}
$$

$$
= \left(
\sqrt{
\begin{array}{l}
1 - \left(1 - 0.7^{2\times 2} \times 0.7^{2\times 3}\right)^{0.3\times 0.3} \times \left(1 - 0.7^{2\times 2} \times 0.3^{2\times 3}\right)^{0.3\times 0.2} \times \left(1 - 0.7^{2\times 2} \times 0.1^{2\times 3}\right)^{0.3\times 0.1} \\[2mm]
\times \left(1 - 0.7^{2\times 2} \times 0.5^{2\times 3}\right)^{0.3\times 0.4} \times \left(1 - 0.3^{2\times 2} \times 0.3^{2\times 3}\right)^{0.2\times 0.2} \times \left(1 - 0.3^{2\times 2} \times 0.1^{2\times 3}\right)^{0.2\times 0.1} \\[2mm]
\times \left(1 - 0.3^{2\times 2} \times 0.5^{2\times 3}\right)^{0.2\times 0.4} \times \left(1 - 0.1^{2\times 2} \times 0.1^{2\times 3}\right)^{0.1\times 0.1} \times \left(1 - 0.1^{2\times 2} \times 0.5^{2\times 3}\right)^{0.1\times 0.4} \\[2mm]
\times \left(1 - 0.5^{2\times 2} \times 0.5^{2\times 3}\right)^{0.4\times 0.4}
\end{array}
}
\right)^{\frac{1}{5}}
$$

$$
= 0.5630
$$

Similarly, we can obtain

$$\alpha_2 = \text{DHPFGWHM}^{2,3}(0.7, 0.3, 0.3, 0.5) = 0.5632$$
$$\alpha_3 = \text{DHPFGWHM}^{2,3}(0.8, 0.3, 0.1, 0.5) = 0.6376$$
$$\alpha_4 = \text{DHPFGWHM}^{2,3}(0.8, 0.3, 0.3, 0.5) = 0.6377$$

Hence, we can get $\alpha = \{0.5630, 0.5632, 0.6376, 0.6377\}$.

For the non-membership degree function β, the fused results are shown as:

$$\beta_1 = \text{DHPFGWHM}^{2,3}(0.4, 0.6, 0.4, 0.5) = \sqrt{1 - \left(1 - \prod_{i=1,j=i}^{n}\left(1 - \left(1 - \beta_i^2\right)^{\xi}\left(1 - \beta_j^2\right)^{\zeta}\right)^{w_i w_j}\right)^{\frac{1}{\xi+\zeta}}}$$

$$= \left[\begin{array}{l}
\left(1 - \left(1 - \left(1 - 0.4^2\right)^2 \times \left(1 - 0.4^2\right)^3\right)^{0.3\times0.3} \times \left(1 - \left(1 - 0.4^2\right)^2 \times \left(1 - 0.6^2\right)^3\right)^{0.3\times0.2}\right. \\[4pt]
\times \left(1 - \left(1 - 0.4^2\right)^2 \times \left(1 - 0.4^2\right)^3\right)^{0.3\times0.1} \times \left(1 - \left(1 - 0.4^2\right)^2 \times \left(1 - 0.5^2\right)^3\right)^{0.3\times0.4} \\[4pt]
\times \left(1 - \left(1 - 0.6^2\right)^2 \times \left(1 - 0.6^2\right)^3\right)^{0.2\times0.2} \times \left(1 - \left(1 - 0.6^2\right)^2 \times \left(1 - 0.4^2\right)^3\right)^{0.2\times0.1} \\[4pt]
\times \left(1 - \left(1 - 0.6^2\right)^2 \times \left(1 - 0.5^2\right)^3\right)^{0.2\times0.4} \times \left(1 - \left(1 - 0.4^2\right)^2 \times \left(1 - 0.4^2\right)^3\right)^{0.1\times0.1} \\[4pt]
\left. \times \left(1 - \left(1 - 0.4^2\right)^2 \times \left(1 - 0.5^2\right)^3\right)^{0.1\times0.4} \times \left(1 - \left(1 - 0.5^2\right)^2 \times \left(1 - 0.5^2\right)^3\right)^{0.4\times0.4}\right)^{\frac{1}{2+3}}
\end{array} \right]^{1/2}$$

$$= 0.5333$$

Similarly, we can obtain

$$\beta_2 = \text{DHPFGWHM}^{2,3}(0.4, 0.6, 0.6, 0.5) = 0.5480$$
$$\beta_3 = \text{DHPFGWHM}^{2,3}(0.4, 0.7, 0.4, 0.5) = 0.5438$$
$$\beta_4 = \text{DHPFGWHM}^{2,3}(0.4, 0.7, 0.6, 0.5) = 0.5586$$

Hence, we can get $\beta = \{0.5333, 0.5480, 0.5438, 0.5586\}$. Therefore,

$$\text{DHPFGWHM}(\widetilde{p}_1, \widetilde{p}_2, \widetilde{p}_3, \widetilde{p}_4) = \left\{ \begin{array}{l} \{0.5630, 0.5632, 0.6376, 0.6377\}, \\ \{0.5333, 0.5480, 0.5438, 0.5586\} \end{array} \right\}.$$

It can be easily proven that the DHPFGWHM operator satisfies the following properties.

Property 4. *(Idempotency) If all $\widetilde{p}_j = \left(h_j, g_j\right)(j = 1, 2, \cdots, n)$ are equal, i.e., $\widetilde{p}_j = \widetilde{p}$ for all j, then*

$$\text{DHPFGWHM}_w^{\xi,\zeta}(\widetilde{p}_1, \widetilde{p}_2, \ldots, \widetilde{p}_n) = \widetilde{p} \tag{25}$$

Property 5. *(Boundedness) Let $\widetilde{p}_j = \left(h_j, g_j\right)(j = 1, 2, \cdots, n)$ be a collection of DHPFNs, and let*

$$\widetilde{p}^+ = \cup_{\alpha_j \in h_j, \beta_j \in g_j}\{\{\max_i(\alpha_i)\}, \{\min_i(\beta_i)\}\}, \widetilde{p}^- = \cup_{\alpha_j \in h_j, \beta_j \in g_j}\{\{\min_i(\alpha_i)\}, \{\max_i(\beta_i)\}\}$$

Then

$$\widetilde{p}^- \leq \text{DHPFGWHM}_w^{\xi,\zeta}(\widetilde{p}_1, \widetilde{p}_2, \ldots, \widetilde{p}_n) \leq \widetilde{p}^+ \tag{26}$$

Property 6. *(Monotonicity) Let* $\widetilde{p}_j = \left(h_j, g_j\right)$ *and* $\widetilde{p}'_j = \left(h'_j, g'_j\right)$, $j = 1, 2, \cdots, n$, *be two sets of DHPFNs. If* $\widetilde{p}_j \leq \widetilde{p}'_j$, *for all* j, *then*

$$\text{DHPFGWHM}_w^{\xi,\zeta}(\widetilde{p}_1, \widetilde{p}_2, \ldots, \widetilde{p}_n) \leq \text{DHPFGWHM}_w^{\xi,\zeta}\left(\widetilde{p}'_1, \widetilde{p}'_2, \ldots, \widetilde{p}'_n\right) \tag{27}$$

3.3. The DHPFGGHM Aggregation Operator

In the following, based on the geometric mean (GM) operator, Yu [54] extended the GHM operator to a GGHM operator which can be depicted as follows.

Definition 10 [54]**.** *Assume that* $\xi, \zeta > 0$ *and* $b_i (i = 1, 2, \cdots, n)$ *are a group of non-negative real numbers. Then, the generalizeGGHM) operator can be expressed as:*

$$\text{GHM}^{\xi,\zeta}(a_1, a_2, \cdots, a_n) = \frac{1}{\xi + \zeta}\left(\prod_{i=1, j=i}^{n} \left(\xi a_i + \zeta a_j\right)\right)^{\frac{2}{n(n+1)}} \tag{28}$$

In this section, we introduced the GGHM operator with dual hesitant Pythagorean fuzzy information. According to Definition 5, Xu et al. [66] gave the definition of the DHPFGGHM operator as follows.

Definition 11 [66]**.** *Assume that* $\xi, \zeta > 0$ *and* $\widetilde{p}_j = \left(h_j, g_j\right)(j = 1, 2, \cdots, n)$ *are a collection of DHPFNs. Then, the DHPFGGHM operator can be defined as:*

$$\text{DHPFGGHM}^{\xi,\zeta}(\widetilde{p}_1, \widetilde{p}_2, \ldots, \widetilde{p}_n) = \frac{1}{\xi + \zeta}\left(\overset{n}{\underset{i=1}{\otimes}} \overset{n}{\underset{j=i}{\otimes}} \left(\xi \widetilde{p}_i \oplus \zeta \widetilde{p}_j\right)\right)^{\frac{2}{n(n+1)}} \tag{29}$$

According to the operation laws of the DHPFNs described in Definition 5, Xu et al. [66] obtained Theorem 3.

Theorem 3 [66]**.** *Assume that* $\xi, \zeta > 0$ *and* $\widetilde{p}_j = \left(h_j, g_j\right)(j = 1, 2, \cdots, n)$ *are a group of DHPFNs. Then, their fused results obtained by utilizing the DHPFGGHM operator is also a DHPFN, and*

$$\text{DHPFGGHM}^{\xi,\zeta}(\widetilde{p}_1, \widetilde{p}_2, \ldots, \widetilde{p}_n) = \frac{1}{\xi+\zeta}\left(\overset{n}{\underset{i=1}{\otimes}} \overset{n}{\underset{j=i}{\otimes}} \left(\xi \widetilde{p}_i \oplus \zeta \widetilde{p}_j\right)\right)^{\frac{2}{n(n+1)}}$$

$$= \bigcup_{\alpha_i \in h_i, \alpha_j \in h_j, \beta_i \in g_i, \alpha_j \in h_j} \left\{ \begin{array}{c} \left\{\sqrt{1 - \left(1 - \prod_{i=1, j=i}^{n}\left(1 - \left(1 - \alpha_i^2\right)^{\xi}\left(1 - \alpha_j^2\right)^{\zeta}\right)^{\frac{2}{n(n+1)}}\right)^{\frac{1}{\xi+\zeta}}}\right\}, \\[4mm] \left\{\left(\sqrt{1 - \prod_{i=1, j=i}^{n}\left(1 - \beta_i^{2\xi}\beta_j^{2\zeta}\right)^{\frac{2}{n(n+1)}}}\right)^{\frac{1}{\xi+\zeta}}\right\} \end{array} \right\} \tag{30}$$

Proof. Based on Definition 5:

$$\xi \widetilde{p}_i = \bigcup_{\alpha_i \in h_i, \beta_i \in g_i}\left\{\left\{\sqrt{1 - \left(1 - \alpha_i^2\right)^{\xi}}\right\}, \left\{\beta_i^{\xi}\right\}\right\} \tag{31}$$

$$\zeta \widetilde{p}_j = \bigcup_{\alpha_j \in h_j, \beta_j \in g_j}\left\{\left\{\sqrt{1 - \left(1 - \alpha_j^2\right)^{\zeta}}\right\}, \left\{\beta_j^{\zeta}\right\}\right\} \tag{32}$$

Thus,

$$\xi\widetilde{p_i} \oplus \zeta\widetilde{p_j} = \cup_{\alpha_i\in h_i,\alpha_j\in h_j,\beta_i\in g_i,\alpha_j\in h_j,}\left\{\left\{\sqrt{1-\left(1-\alpha_i^2\right)^{\xi}\left(1-\alpha_j^2\right)^{\zeta}}\right\},\left\{\beta_i^{\xi}\beta_j^{\zeta}\right\}\right\} \tag{33}$$

Therefore,

$$\underset{i=1}{\overset{n}{\otimes}}\,\underset{j=i}{\overset{n}{\otimes}}\left(\xi\widetilde{p_i}\oplus\zeta\widetilde{p_j}\right)$$

$$= \cup_{\alpha_i\in h_i,\alpha_j\in h_j,\beta_i\in g_i,\alpha_j\in h_j}\left\{\begin{array}{c}\left\{\sqrt{\prod_{i=1,j=i}^{n}\left(1-\left(1-\alpha_i^2\right)^{\xi}\left(1-\alpha_j^2\right)^{\zeta}\right)}\right\},\\ \left\{\sqrt{1-\prod_{i=1,j=i}^{n}\left(1-\beta_i^{2\xi}\beta_j^{2\zeta}\right)}\right\}\end{array}\right\} \tag{34}$$

Furthermore,

$$\left(\underset{i=1}{\overset{n}{\otimes}}\,\underset{j=i}{\overset{n}{\otimes}}\left(\xi\widetilde{p_i}\oplus\zeta\widetilde{p_j}\right)\right)^{\frac{2}{n(n+1)}}$$

$$= \cup_{\alpha_i\in h_i,\alpha_j\in h_j,\beta_i\in g_i,\alpha_j\in h_j}\left\{\begin{array}{c}\left\{\sqrt{\prod_{i=1,j=i}^{n}\left(1-\left(1-\alpha_i^2\right)^{\xi}\left(1-\alpha_j^2\right)^{\zeta}\right)^{\frac{2}{n(n+1)}}}\right\},\\ \left\{\sqrt{1-\prod_{i=1,j=i}^{n}\left(1-\beta_i^{2\xi}\beta_j^{2\zeta}\right)^{\frac{2}{n(n+1)}}}\right\}\end{array}\right\} \tag{35}$$

Therefore,

$$\text{DHPFGGHM}^{\xi,\zeta}(\widetilde{p_1},\widetilde{p_2},\ldots,\widetilde{p_n}) = \frac{1}{\xi+\zeta}\left(\underset{i=1}{\overset{n}{\otimes}}\,\underset{j=i}{\overset{n}{\otimes}}\left(\xi\widetilde{p_i}\oplus\zeta\widetilde{p_j}\right)\right)^{\frac{2}{n(n+1)}}$$

$$= \cup_{\alpha_i\in h_i,\alpha_j\in h_j,\beta_i\in g_i,\alpha_j\in h_j}\left\{\begin{array}{c}\left\{\sqrt{1-\left(1-\prod_{i=1,j=i}^{n}\left(1-\left(1-\alpha_i^2\right)^{\xi}\left(1-\alpha_j^2\right)^{\zeta}\right)^{\frac{2}{n(n+1)}}\right)^{\frac{1}{\xi+\zeta}}}\right\},\\ \left\{\left(\sqrt{1-\prod_{i=1,j=i}^{n}\left(1-\beta_i^{2\xi}\beta_j^{2\zeta}\right)^{\frac{2}{n(n+1)}}}\right)^{\frac{1}{\xi+\zeta}}\right\}\end{array}\right\} \tag{36}$$

Thus, the proof have been finished.

Example 5. *Assume that* $\widetilde{p_1} = \{\{0.7, 0.8\}, \{0.4\}\}$, $\widetilde{p_2} = \{\{0.3\}, \{0.6, 0.7\}\}$, $\widetilde{p_3} = \{\{0.1, 0.3\}, \{0.4, 0.6\}\}$ *and* $\widetilde{p_4} = \{\{0.5\}, \{0.5\}\}$ *are four DHPFNs, and suppose that* $\xi = 2, \zeta = 3$. *Then, according to the DHPFGGHM operator, we can obtain the fused results as follows. For the membership degree function* α, *the fused results are shown as:*

$$\alpha_1 = \text{DHPFGGHM}^{2,3}(0.7, 0.3, 0.1, 0.5) = \sqrt{1 - \left(1 - \prod_{i=1, j=i}^{n}\left(1 - \left(1 - \alpha_i^2\right)^{\xi}\left(1 - \alpha_j^2\right)^{\zeta}\right)^{\frac{2}{n(n+1)}}\right)^{\frac{1}{\xi+\zeta}}}$$

$$= \sqrt{1 - \left(\left(\begin{array}{l}\left(1 - \left(1 - 0.7^2\right)^2 \times \left(1 - 0.7^2\right)^3\right) \times \left(1 - \left(1 - 0.7^2\right)^2 \times \left(1 - 0.3^2\right)^3\right) \\ \times\left(1 - \left(1 - 0.7^2\right)^2 \times \left(1 - 0.1^2\right)^3\right) \times \left(1 - \left(1 - 0.7^2\right)^2 \times \left(1 - 0.5^2\right)^3\right) \\ \times\left(1 - \left(1 - 0.3^2\right)^2 \times \left(1 - 0.3^2\right)^3\right) \times \left(1 - \left(1 - 0.3^2\right)^2 \times \left(1 - 0.1^2\right)^3\right) \\ \times\left(1 - \left(1 - 0.3^2\right)^2 \times \left(1 - 0.5^2\right)^3\right) \times \left(1 - \left(1 - 0.1^2\right)^2 \times \left(1 - 0.1^2\right)^3\right) \\ \times\left(1 - \left(1 - 0.1^2\right)^2 \times \left(1 - 0.5^2\right)^3\right) \times \left(1 - \left(1 - 0.5^2\right)^2 \times \left(1 - 0.5^2\right)^3\right)\end{array}\right)^{\frac{1}{10}}\right)^{\frac{1}{2+3}}}$$

$$= 0.3461$$

Similarly, we can obtain

$$\alpha_2 = \text{DHPFGGHM}^{2,3}(0.7, 0.3, 0.3, 0.5) = 0.4236$$
$$\alpha_3 = \text{DHPFGGHM}^{2,3}(0.8, 0.3, 0.1, 0.5) = 0.3545$$
$$\alpha_4 = \text{DHPFGGHM}^{2,3}(0.8, 0.3, 0.3, 0.5) = 0.4343$$

Hence, we can get $\alpha = \{0.3461, 0.4236, 0.3545, 0.4343\}$.

For the non-membership degree function β, the fused results are shown as:

$$\beta_1 = \text{DHPFGGHM}^{2,3}(0.4, 0.6, 0.4, 0.5) = \left(\sqrt{1 - \prod_{i=1, j=i}^{n}\left(1 - \beta_i^{2\xi}\beta_j^{2\zeta}\right)^{\frac{2}{n(n+1)}}}\right)^{\frac{1}{\xi+\zeta}}$$

$$= \left(\sqrt{1 - \left(\left(\begin{array}{l}\left(1 - 0.4^{2\times2} \times 0.4^{2\times3}\right) \times \left(1 - 0.4^{2\times2} \times 0.6^{2\times3}\right) \times \left(1 - 0.4^{2\times2} \times 0.4^{2\times3}\right) \\ \times\left(1 - 0.4^{2\times2} \times 0.5^{2\times3}\right) \times \left(1 - 0.6^{2\times2} \times 0.6^{2\times3}\right) \times \left(1 - 0.6^{2\times2} \times 0.4^{2\times3}\right) \\ \times\left(1 - 0.6^{2\times2} \times 0.5^{2\times3}\right) \times \left(1 - 0.4^{2\times2} \times 0.4^{2\times3}\right) \times \left(1 - 0.4^{2\times2} \times 0.5^{2\times3}\right) \\ \times\left(1 - 0.5^{2\times2} \times 0.5^{2\times3}\right)\end{array}\right)^{\frac{1}{10}}\right)^{\frac{1}{2+3}}}\right)$$

$$= 0.5100$$

Similarly, we can obtain

$$\beta_2 = \text{DHPFGGHM}^{2,3}(0.4, 0.6, 0.6, 0.5) = 0.5516$$
$$\beta_3 = \text{DHPFGGHM}^{2,3}(0.4, 0.7, 0.4, 0.5) = 0.5734$$
$$\beta_4 = \text{DHPFGGHM}^{2,3}(0.4, 0.7, 0.6, 0.5) = 0.5968$$

Hence, we can get $\beta = \{0.5100, 0.5516, 0.5734, 0.5968\}$. Therefore,

$$\text{DHPFGGHM}(\widetilde{p}_1, \widetilde{p}_2, \widetilde{p}_3, \widetilde{p}_4) = \left\{\begin{array}{l}\{0.3461, 0.4236, 0.3545, 0.4343\}, \\ \{0.5100, 0.5516, 0.5734, 0.5968\}\end{array}\right\}.$$

It can be easily proven that the DHPFGGHM operator satisfies the following properties.

Property 7. *(Idempotency) If all* $\widetilde{p}_j = \left(h_j, g_j\right)(j = 1, 2, \cdots, n)$ *are equal, i.e.,* $\widetilde{p}_j = \widetilde{p}$ *for all* j, *then*

$$\text{DHPFGGHM}^{\xi,\zeta}(\widetilde{p}_1, \widetilde{p}_2, \ldots, \widetilde{p}_n) = \widetilde{p} \tag{37}$$

Property 8. *(Boundedness) Let* $\widetilde{p}_j = \left(h_j, g_j\right)(j = 1, 2, \cdots, n)$ *be a collection of DHPFNs, and let*

$$\widetilde{p}^{+} = \cup_{\alpha_j \in h_j, \beta_j \in g_j}\{\{\max_i(\alpha_i)\}, \{\min_i(\beta_i)\}\}, \widetilde{p}^{-} = \cup_{\alpha_j \in h_j, \beta_j \in g_j}\{\{\min_i(\alpha_i)\}, \{\max_i(\beta_i)\}\}$$

Then

$$\widetilde{p}^{-} \leq \text{DHPFGGHM}^{\xi,\zeta}(\widetilde{p}_1, \widetilde{p}_2, \ldots, \widetilde{p}_n) \leq \widetilde{p}^{+} \tag{38}$$

Property 9. *(Monotonicity) Let* $\widetilde{p}_j = \left(h_j, g_j\right)$ *and* $\widetilde{p}'_j = \left(h'_j, g'_j\right), j = 1, 2, \cdots, n$, *be two set of DHPFNs. If* $\widetilde{p}_j \leq \widetilde{p}'_j$, *for all* j, *then*

$$\text{DHPFGGHM}^{\xi,\zeta}(\widetilde{p}_1, \widetilde{p}_2, \ldots, \widetilde{p}_n) \leq \text{DHPFGGHM}^{\xi,\zeta}\left(\widetilde{p}'_1, \widetilde{p}'_2, \ldots, \widetilde{p}'_n\right) \tag{39}$$

3.4. The DHPFGGWHM Aggregation Operator

Using Definition 11, we can conclude that the DHPFGGHM operator didn't take the importance of arguments being fused into account. However, in many practical MADM problems, we should consider the weights of attributes. To overcome the limitations of the DHPFGGHM operator, we propose a novel DHPFGGWHM operator as follows.

Definition 12. *Assume that* $\xi, \zeta > 0$ *and* $\widetilde{p}_j = \left(h_j, g_j\right)(j = 1, 2, \cdots, n)$ *are a collection of DHPFNs. Then, the DHPFGGWHM operator can be defined as:*

$$\text{DHPFGGWHM}_w^{\xi,\zeta}(\widetilde{p}_1, \widetilde{p}_2, \ldots, \widetilde{p}_n) = \frac{1}{\xi+\zeta}\left(\overset{n}{\underset{i=1}{\otimes}}\overset{n}{\underset{j=i}{\otimes}}\left(\xi\widetilde{p}_i \oplus \zeta\widetilde{p}_j\right)^{w_i w_j}\right) \tag{40}$$

According to the operation laws of the DHPFNs described in Definition 5, we can obtain Theorem 4.

Theorem 4. *Assume that* $\xi, \zeta > 0$ *and* $\widetilde{p}_j = \left(h_j, g_j\right)(j = 1, 2, \cdots, n)$ *are a collection of DHPFNs with the weighting vector* $w = (w_1, w_2, \ldots, w_n)^T$ *which satisfies* $w_j > 0, i = 1, 2, \ldots, n$ *and* $\sum_{j=1}^{n} w_j = 1$. *Then, their fused result obtained by utilizing the DHPFGGWHM operator is also a DHPFN, and*

$$\text{DHPFGGWHM}_w^{\xi,\zeta}(\widetilde{p}_1, \widetilde{p}_2, \ldots, \widetilde{p}_n) = \frac{1}{\xi+\zeta}\left(\overset{n}{\underset{i=1}{\otimes}}\overset{n}{\underset{j=i}{\otimes}}\left(\xi\widetilde{p}_i \oplus \zeta\widetilde{p}_j\right)^{w_i w_j}\right)$$

$$= \cup_{\alpha_i \in h_i, \alpha_j \in h_j, \beta_i \in g_i, \alpha_j \in h_j,}\left\{\left\{\sqrt{1 - \left(1 - \prod_{i=1,j=i}^{n}\left(1 - \left(1-\alpha_i^2\right)^{\xi}\left(1-\alpha_j^2\right)^{\zeta}\right)^{w_i w_j}\right)^{\frac{1}{\xi+\zeta}}}\right\}, \left\{\left(\sqrt{1 - \prod_{i=1,j=i}^{n}\left(1 - \beta_i^{2\xi}\beta_j^{2\zeta}\right)^{w_i w_j}}\right)^{\frac{1}{\xi+\zeta}}\right\}\right\} \tag{41}$$

Proof. Based on Definition 5, we can obtain:

$$\xi\widetilde{p}_i = \cup_{\alpha_i\in h_i,\beta_i\in g_i}\left\{\left\{\sqrt{1-\left(1-\alpha_i^2\right)^\xi}\right\},\left\{\beta_i^\xi\right\}\right\} \tag{42}$$

$$\zeta\widetilde{p}_j = \cup_{\alpha_j\in h_j,\beta_j\in g_j}\left\{\left\{\sqrt{1-\left(1-\alpha_j^2\right)^\zeta}\right\},\left\{\beta_j^\zeta\right\}\right\} \tag{43}$$

Thus,

$$\xi\widetilde{p}_i \oplus \zeta\widetilde{p}_j = \cup_{\alpha_i\in h_i,\alpha_j\in h_j,\beta_i\in g_i,\alpha_j\in h_j,}\left\{\left\{\sqrt{1-\left(1-\alpha_i^2\right)^\xi\left(1-\alpha_j^2\right)^\zeta}\right\},\left\{\beta_i^\xi\beta_j^\zeta\right\}\right\} \tag{44}$$

Therefore,

$$\left(\xi\widetilde{p}_i \oplus \zeta\widetilde{p}_j\right)^{w_iw_j}$$

$$= \cup_{\alpha_i\in h_i,\alpha_j\in h_j,\beta_i\in g_i,\alpha_j\in h_j,}\left\{\begin{array}{c}\left\{\left(\sqrt{1-\left(1-\alpha_i^2\right)^\xi\left(1-\alpha_j^2\right)^\zeta}\right)^{w_iw_j}\right\}, \\ \left\{\sqrt{1-\left(1-\beta_i^{2\xi}\beta_j^{2\zeta}\right)^{w_iw_j}}\right\}\end{array}\right\} \tag{45}$$

Thereafter,

$$\mathop{\otimes}_{i=1}^{n}\mathop{\otimes}_{j=i}^{n}\left(\xi\widetilde{p}_i \oplus \zeta\widetilde{p}_j\right)^{w_iw_j}$$

$$= \cup_{\alpha_i\in h_i,\alpha_j\in h_j,\beta_i\in g_i,\alpha_j\in h_j,}\left\{\begin{array}{c}\left\{\left(\sqrt{\prod_{i=1,j=i}^{n}\left(1-\left(1-\alpha_i^2\right)^\xi\left(1-\alpha_j^2\right)^\zeta\right)}\right)^{w_iw_j}\right\}, \\ \left\{\sqrt{1-\prod_{i=1,j=i}^{n}\left(1-\beta_i^{2\xi}\beta_j^{2\zeta}\right)^{w_iw_j}}\right\}\end{array}\right\} \tag{46}$$

Therefore,

$$\text{DHPFGGWHM}_w^{\xi,\zeta}(\widetilde{p}_1,\widetilde{p}_2,\ldots,\widetilde{p}_n) = \frac{1}{\xi+\zeta}\left(\mathop{\otimes}_{i=1}^{n}\mathop{\otimes}_{j=i}^{n}\left(\xi\widetilde{p}_i \oplus \zeta\widetilde{p}_j\right)^{w_iw_j}\right)$$

$$= \cup_{\alpha_i\in h_i,\alpha_j\in h_j,\beta_i\in g_i,\alpha_j\in h_j,}\left\{\begin{array}{c}\left\{\sqrt{1-\left(1-\prod_{i=1,j=i}^{n}\left(1-\left(1-\alpha_i^2\right)^\xi\left(1-\alpha_j^2\right)^\zeta\right)^{w_iw_j}\right)^{\frac{1}{\xi+\zeta}}}\right\}, \\ \left\{\left(\sqrt{1-\prod_{i=1,j=i}^{n}\left(1-\beta_i^{2\xi}\beta_j^{2\zeta}\right)^{w_iw_j}}\right)^{\frac{1}{\xi+\zeta}}\right\}\end{array}\right\} \tag{47}$$

Thus, we have finished the proof.

Example 6. *Assume that* $\widetilde{p}_1 = \{\{0.7,0.8\},\{0.4\}\}$, $\widetilde{p}_2 = \{\{0.3\},\{0.6,0.7\}\}$, $\widetilde{p}_3 = \{\{0.1,0.3\},\{0.4,0.6\}\}$ *and* $\widetilde{p}_4 = \{\{0.5\},\{0.5\}\}$ *are four DHPFNs, and suppose that* $\xi = 2, \zeta = 3$ *and* $w_j = (0.3,0.2,0.1,0.4)$. *Then, according to the DHPFGGWHM operator, we can obtain the fused results as follows. For the membership degree function* α, *the fused results are shown as:*

$$\alpha_1 = \text{DHPFGGWHM}^{2,3}(0.7, 0.3, 0.1, 0.5) = \sqrt{1 - \left(1 - \prod_{i=1,j=i}^{n} \left(1 - \left(1 - \alpha_i^2\right)^{\xi}\left(1 - \alpha_j^2\right)^{\zeta}\right)^{w_i w_j}\right)^{\frac{1}{\xi+\zeta}}}$$

$$= \sqrt{ \left(\begin{array}{l} \left(1 - \left(1 - \left(1 - 0.7^2\right)^2 \times \left(1 - 0.7^2\right)^3\right)^{0.3\times0.3} \times \left(1 - \left(1 - 0.7^2\right)^2 \times \left(1 - 0.3^2\right)^3\right)^{0.3\times0.2}\right. \\ \times\left(1 - \left(1 - 0.7^2\right)^2 \times \left(1 - 0.1^2\right)^3\right)^{0.3\times0.1} \times \left(1 - \left(1 - 0.7^2\right)^2 \times \left(1 - 0.5^2\right)^3\right)^{0.3\times0.4} \\ \times\left(1 - \left(1 - 0.3^2\right)^2 \times \left(1 - 0.3^2\right)^3\right)^{0.2\times0.2} \times \left(1 - \left(1 - 0.3^2\right)^2 \times \left(1 - 0.1^2\right)^3\right)^{0.2\times0.1} \\ \times\left(1 - \left(1 - 0.3^2\right)^2 \times \left(1 - 0.5^2\right)^3\right)^{0.2\times0.4} \times \left(1 - \left(1 - 0.1^2\right)^2 \times \left(1 - 0.1^2\right)^3\right)^{0.1\times0.1} \\ \left.\times\left(1 - \left(1 - 0.1^2\right)^2 \times \left(1 - 0.5^2\right)^3\right)^{0.1\times0.4} \times \left(1 - \left(1 - 0.5^2\right)^2 \times \left(1 - 0.5^2\right)^3\right)^{0.4\times0.4}\right) \end{array}\right)^{\frac{1}{2+3}} }$$

$$= 0.5156$$

Similarly, we can obtain

$$\alpha_2 = \text{DHPFGGWHM}^{2,3}(0.7, 0.3, 0.3, 0.5) = 0.5378$$
$$\alpha_3 = \text{DHPFGGWHM}^{2,3}(0.8, 0.3, 0.1, 0.5) = 0.5273$$
$$\alpha_4 = \text{DHPFGGWHM}^{2,3}(0.8, 0.3, 0.3, 0.5) = 0.5503$$

Hence, we can get $\alpha = \{0.5156, 0.5378, 0.5273, 0.5503\}$.

For the non-membership degree function β, the fused results are shown as:

$$\beta_1 = \text{DHPFGGWHM}^{2,3}(0.4, 0.6, 0.4, 0.5) = \left(\sqrt{1 - \prod_{i=1,j=i}^{n} \left(1 - \beta_i^{2\xi}\beta_j^{2\zeta}\right)^{w_i w_j}}\right)^{\frac{1}{\xi+\zeta}}$$

$$= \left(\begin{array}{l} \sqrt{ \begin{array}{l} \left(1 - 0.4^{2\times2} \times 0.4^{2\times3}\right)^{0.3\times0.3} \times \left(1 - 0.4^{2\times2} \times 0.6^{2\times3}\right)^{0.3\times0.2} \times \left(1 - 0.4^{2\times2} \times 0.4^{2\times3}\right)^{0.3\times0.1} \\ \times\left(1 - 0.4^{2\times2} \times 0.5^{2\times3}\right)^{0.3\times0.4} \times \left(1 - 0.6^{2\times2} \times 0.6^{2\times3}\right)^{0.2\times0.2} \times \left(1 - 0.6^{2\times2} \times 0.4^{2\times3}\right)^{0.2\times0.1} \\ \times\left(1 - 0.6^{2\times2} \times 0.5^{2\times3}\right)^{0.2\times0.4} \times \left(1 - 0.4^{2\times2} \times 0.4^{2\times3}\right)^{0.1\times0.1} \times \left(1 - 0.4^{2\times2} \times 0.5^{2\times3}\right)^{0.1\times0.4} \\ \times\left(1 - 0.5^{2\times2} \times 0.5^{2\times3}\right)^{0.4\times0.4} \end{array} } \end{array} \right)^{\frac{1}{5}}$$

$$= 0.4850$$

Similarly, we can obtain

$$\beta_2 = \text{DHPFGGWHM}^{2,3}(0.4, 0.6, 0.6, 0.5) = 0.5006$$
$$\beta_3 = \text{DHPFGGWHM}^{2,3}(0.4, 0.7, 0.4, 0.5) = 0.5338$$
$$\beta_4 = \text{DHPFGGWHM}^{2,3}(0.4, 0.7, 0.6, 0.5) = 0.5433$$

Hence, we can get $\beta = \{0.4850, 0.5006, 0.5338, 0.5433\}$. Therefore,

$$\text{DHPFGGWHM}(\widetilde{p}_1, \widetilde{p}_2, \widetilde{p}_3, \widetilde{p}_4) = \left\{ \begin{array}{l} \{0.5156, 0.5378, 0.5273, 0.5503\}, \\ \{0.4850, 0.5006, 0.5338, 0.5433\} \end{array} \right\}.$$

It can be easily proven that the DHPFGGWHM operator satisfies the following properties.

Property 10. *(Idempotency) If all* $\widetilde{p}_j = \left(h_j, g_j\right)(j = 1, 2, \cdots, n)$ *are equal, i.e.,* $\widetilde{p}_j = \widetilde{p}$ *for all* j, *then*

$$\text{DHPFGGWHM}_w^{\xi,\zeta}(\widetilde{p}_1, \widetilde{p}_2, \ldots, \widetilde{p}_n) = \widetilde{p} \tag{48}$$

Property 11. *(Boundedness) Let* $\widetilde{p}_j = \left(h_j, g_j\right)(j = 1, 2, \cdots, n)$ *be a collection of DHPFNs, and let*

$$\widetilde{p}^+ = \cup_{\alpha_j \in h_j, \beta_j \in g_j}\{\{\max_i(\alpha_i)\}, \{\min_i(\beta_i)\}\}, \widetilde{p}^- = \cup_{\alpha_j \in h_j, \beta_j \in g_j}\{\{\min_i(\alpha_i)\}, \{\max_i(\beta_i)\}\}$$

Then

$$\widetilde{p}^- \leq \text{DHPFGGWHM}_w^{\xi,\zeta}(\widetilde{p}_1, \widetilde{p}_2, \ldots, \widetilde{p}_n) \leq \widetilde{p}^+ \tag{49}$$

Property 12. *(Monotonicity) Let* $\widetilde{p}_j = \left(h_j, g_j\right)$ *and* $\widetilde{p}_j' = \left(h_j', g_j'\right), j = 1, 2, \cdots, n$, *be two sets of DHPFNs. If* $\widetilde{p}_j \leq \widetilde{p}_{j'}'$, *for all* j, *then*

$$\text{DHPFGGWHM}_w^{\xi,\zeta}(\widetilde{p}_1, \widetilde{p}_2, \ldots, \widetilde{p}_n) \leq \text{DHPFGGWHM}_w^{\xi,\zeta}(\widetilde{p}_1', \widetilde{p}_2', \ldots, \widetilde{p}_n') \tag{50}$$

4. An Approach to MADM with DHPFNs Information

In this section, we shall use the DHPFGWHM and DHPFGGWHM operators to deal with MADM problems with dual hesitant Pythagorean fuzzy information. Suppose that there are m alternatives $\eta = \{\eta_1, \eta_2, \cdots, \eta_m\}$, and each alternative is characterized by n attributes $\delta = \{\delta_1, \delta_2, \cdots, \delta_n\}$ with the weighting vector being $w_j = \{w_1, w_2, \cdots, w_n\}$. Then, the dual hesitant Pythagorean fuzzy matrix can be constructed as $\widetilde{P} = \left(\widetilde{p}_{ij}\right)_{m \times n}$, with each element $\widetilde{p}_{ij} = \left(h_{ij}, g_{ij}\right)(i = 1, 2, \cdots, m, j = 1, 2, \cdots, n)$ indicating a dual hesitant Pythagorean fuzzy number, where h_{ij} means the membership degree set with several values in $[0, 1]$, and g_{ij} means the no-membership degree set with several values in $[0, 1]$.

In what follows, we apply the DHPFGWHM or DHPFGGWHM operator to MADM problems for supplier selection in supply chain management with dual hesitant Pythagorean fuzzy information.

Step 1. In order to derive the fused results of each alternative, for alternatives $\eta = \{\eta_1, \eta_2, \cdots, \eta_m\}$, based on the weighting vector $w_j = \{w_1, w_2, \cdots, w_n\}$ and dual hesitant Pythagorean fuzzy information $\widetilde{p}_{ij} = \left(h_{ij}, g_{ij}\right)(i = 1, 2, \cdots, m, j = 1, 2, \cdots, n)$ given in matrix $\widetilde{P} = \left(\widetilde{p}_{ij}\right)_{m \times n}$, we can aggregate all the DHPFNs by the DHPFGWHM operator

$$\widetilde{p}_i = \text{DHPFGWHM}_w^{\xi,\zeta}(\widetilde{p}_{i1}, \widetilde{p}_{i2}, \ldots, \widetilde{p}_{in}) = \left(\overset{n}{\underset{k=1}{\oplus}} \overset{n}{\underset{j=k}{\oplus}} \left(w_{ik}w_{ij}\left(\widetilde{p}_{ik}^\xi \widetilde{p}_{ij}^\zeta\right)\right)\right)^{\frac{1}{\xi+\zeta}}$$

$$= \cup_{\alpha_{ik} \in h_{ik}, \alpha_{ij} \in h_{ij}, \beta_{ij} \in g_{ij}, \alpha_{ij} \in h_{ij},} \left\{ \begin{array}{c} \left\{\left(\sqrt{1 - \prod_{k=1,j=k}^{n}\left(1 - \alpha_{ik}^{2\xi}\alpha_{ij}^{2\zeta}\right)^{w_iw_j}}\right)^{\frac{1}{\xi+\zeta}}\right\}, \\ \left\{\left(\sqrt{1 - \left(1 - \prod_{k=1,j=k}^{n}\left(1 - \left(1 - \beta_{ik}^2\right)^\xi\left(1 - \beta_{ij}^2\right)^\zeta\right)^{w_iw_j}\right)^{\frac{1}{\xi+\zeta}}}\right)\right\} \end{array} \right\} \tag{51}$$

or the DHPFGGWHM operator

$$\widetilde{p}_i = \mathrm{DHPFGGWHM}_w^{\xi,\zeta}(\widetilde{p}_{i1},\widetilde{p}_{i2},\ldots,\widetilde{p}_{in}) = \frac{1}{\xi+\zeta}\left(\mathop{\otimes}_{k=1}^{n} \mathop{\otimes}_{j=k}^{n} \left(\xi\widetilde{p}_{ik} \oplus \zeta\widetilde{p}_{ij} \right)^{w_{ik}w_{ij}} \right)$$

$$= \cup_{\alpha_{ik}\in h_{ik},\alpha_{ij}\in h_{ij},\beta_{ij}\in g_{ij},\alpha_{ij}\in h_{ij},} \left\{ \begin{array}{c} \left\{ \left(\sqrt{1-\left(1-\prod_{k=1,j=k}^{n}\left(1-\left(1-\alpha_{ik}^2\right)^{\xi}\left(1-\alpha_{ij}^2\right)^{\zeta}\right)^{w_iw_j}\right)^{\frac{1}{\xi+\zeta}}} \right) \right\}, \\[4mm] \left\{ \left(\sqrt{1-\prod_{k=1,j=k}^{n}\left(1-\beta_{ik}^{2\xi}\beta_{ij}^{2\zeta}\right)^{w_iw_j}} \right)^{\frac{1}{\xi+\zeta}} \right\} \end{array} \right\} \tag{52}$$

to obtain the overall fused results $\widetilde{p}_i(i=1,2,\cdots,m)$.

Step 2. To obtain the rank of all the alternatives, we need to adapt the score function and accuracy function described in Definition 4. Firstly, based on the score function equation, we can compute the score values $S(\widetilde{p}_i)$ $(i=1,2,\cdots,m)$ of $\widetilde{p}_i(i=1,2,\cdots,m)$. If all the score values of $\widetilde{p}_i(i=1,2,\cdots,m)$ are different, we can easily obtain the ordering of alternatives. Then, if there is no difference between any two scores $S(\widetilde{p}_i)$ and $S(\widetilde{p}_j)$, we need to compute the accuracy values $H(\widetilde{p}_i)$ and $H(\widetilde{p}_j)$ of $\widetilde{p}_i$ and $\widetilde{p}_j$, respectively, and then determine the ordering of all the alternatives η_i and η_j based on the accuracy results $H(\widetilde{p}_i)$ and $H(\widetilde{p}_j)$.

Step 3. Determine the ordering of all the alternatives $\eta_i(i=1,2,\cdots,m)$ and select the best one(s) according to the scores values $S(\widetilde{p}_i)(i=1,2,\cdots,m)$ and accuracy results $H(\widetilde{p}_i)$. Thus, we have finished the decision making process by using the DHPFGWHM operator or the DHPFGGWHM operator.

5. Numerical Example and Comparative Analysis

5.1. Numerical Example

In this section we present a numerical example for supplier selection in supply chain management with dual hesitant Pythagorean fuzzy information in order to demonstrate the method proposed in this paper. Suppose there is a problem to do with the supplier selection in supply chain management which is a classical MADM problem. There are five prospective suppliers $\eta_i(i=1,2,3,4,5)$ for four attributes $\delta_j(j=1,2,3,4)$. The four attributes include product quality (δ_1), service (δ_2), delivery, (δ_3) and price (δ_4), respectively. In order to avoid influencing each other, the decision makers are required to evaluate the five suppliers $\eta_i(i=1,2,3,4,5)$ under the above four attributes in anonymity. The decision matrix $\widetilde{P} = (\widetilde{p}_{ij})_{5\times4}$ is presented in Table 1, where $\widetilde{p}_{ij}(i=1,2,3,4,5,j=1,2,3,4)$ are in the form of DHPFNs. (Suppose the weighting vector is $w_j = (0.25,0.34,0.27,0.14)$)

Table 1. Dual hesitant Pythagorean fuzzy decision matrix.

Alternatives	δ_1	δ_2	δ_3	δ_4
η_1	{{0.4,0.5},{0.7}}	{{0.5,0.6},{0.4,0.5}}	{{0.3,0.4},{0.8}}	{{0.5,0.6},{0.6}}
η_2	{{0.7},{0.5}}	{{0.3,0.5,0.6},{0.5}}	{{0.3},{0.7,0.8,0.9}}	{{0.6},{0.5,0.6}}
η_3	{{0.6,0.8},{0.3}}	{{0.3},{0.8,0.9}}	{{0.3,0.4,0.5},{0.7}}	{{0.6,0.7,0.8},{0.4}}
η_4	{{0.8},{0.4}}	{{0.7,0.8,0.9},{0.3}}	{{0.2,0.3},{0.4}}	{{0.2},{0.7,0.8,0.9}}
η_5	{{0.1,0.2},{0.3}}	{{0.3,0.4,0.5},{0.6}}	{{0.5,0.6},{0.3}}	{{0.3,0.4,0.5},{0.6}}

In what follows, we can utilize our developed methods to deal with the supplier selection in supply chain management with dual hesitant Pythagorean fuzzy information.

Step 1. We aggregate the dual hesitant Pythagorean fuzzy information given in the matrix by utilizing the DHPFGWHM operator to obtain the overall preference values $\widetilde{p}_i$ of the supplier in supply chain management $\eta_i (i = 1, 2, 3, 4, 5)$. Taking the alternative η_1 as an example (here, we take $\xi = \zeta = 2$), we have

$$
\begin{aligned}
\widetilde{p}_1 &= \text{DHPFGWHM}_w^{\xi,\zeta}(\widetilde{p}_{11}, \widetilde{p}_{12}, \widetilde{p}_{13}, \widetilde{p}_{14}) = \left(\mathop{\oplus}\limits_{k=1}^{4} \mathop{\oplus}\limits_{j=k}^{4} \left(w_{ik} w_{ij} \left(\widetilde{p}_{ik}^{\xi} \widetilde{p}_{ij}^{\zeta} \right) \right) \right)^{\frac{1}{\xi+\zeta}} \\
&= \bigcup_{\alpha_{ik} \in h_{ik}, \alpha_{ij} \in h_{ij}, \beta_{ij} \in g_{ij}, \alpha_{ij} \in h_{ij},} \left\{ \begin{array}{l} \left\{ \left(\sqrt{1 - \prod\limits_{k=1, j=k}^{4} \left(1 - \alpha_{ik}^{2\xi} \alpha_{ij}^{2\zeta}\right)^{w_i w_j}} \right)^{\frac{1}{\xi+\zeta}} \right\}, \\[2em] \left\{ \left(\sqrt{1 - \left(1 - \prod\limits_{k=1, j=k}^{4} \left(1 - \left(1 - \beta_{ik}^2\right)^{\xi} \left(1 - \beta_{ij}^2\right)^{\zeta}\right)^{w_i w_j}\right)^{\frac{1}{\xi+\zeta}}} \right) \right\} \end{array} \right\} \\
&= \{ \{\{0.4, 0.5\}, \{0.7\}\}, \{\{0.5, 0.6\}, \{0.4, 0.5\}\}, \{\{0.3, 0.4\}, \{0.8\}\}, \{\{0.5, 0.6\}, \{0.6\}\}\} \\
&= \{\{0.4234, 0.4461, 0.4335, 0.4547, 0.4824, 0.4964, 0.4887, 0.5023, 0.4448, 0.4642, \\
&\qquad 0.4536, 0.4719, 0.4957, 0.5087, 0.5016, 0.5143\}, \{0.6319, 0.6725\}\}
\end{aligned}
$$

Step 2. Compute the scores results $s(\widetilde{p}_i)$ $(i = 1, 2, 3, 4, 5)$ of the overall dual hesitant Pythagorean fuzzy preference values $\widetilde{p}_i$ $(i = 1, 2, 3, 4, 5)$:

$$
\begin{aligned}
s(\widetilde{p}_1) &= 0.3998, s(\widetilde{p}_2) = 0.4536, s(\widetilde{p}_3) = 0.4669 \\
s(\widetilde{p}_4) &= 0.6255, s(\widetilde{p}_5) = 0.4674
\end{aligned}
$$

Step 3. Determine the ordering of all the suppliers $\eta_i (i = 1, 2, 3, 4, 5)$ based on the scores values $s(\widetilde{p}_i)$ $(i = 1, 2, 3, 4, 5)$: $\eta_4 \succ \eta_5 \succ \eta_3 \succ \eta_2 \succ \eta_1$, and it is clear that the most desirable supplier is η_4.

Similarly, if we utilize the DHPFGGWHM operator to solve this MADM, the decision making steps can be described as follows.

Step 1'. Aggregate all dual hesitant Pythagorean fuzzy values $\widetilde{p}_{ij}(j = 1, 2, 3, 4)$ by using the DHPFGGWHM operator to derive the overall dual hesitant Pythagorean fuzzy values $\widetilde{p}_i (i = 1, 2, \cdots, 5)$ of the supplier η_i. Taking supplier η_1 for an example (here, we take $\xi = \zeta = 2$), we have

$$
\begin{aligned}
\widetilde{p}_1 &= \text{DHPFGGWHM}_w^{\xi,\zeta}(\widetilde{p}_{11}, \widetilde{p}_{12}, \widetilde{p}_{13}, \widetilde{p}_{14}) = \frac{1}{\xi+\zeta} \left(\mathop{\otimes}\limits_{k=1}^{4} \mathop{\otimes}\limits_{j=k}^{4} \left(\xi \widetilde{p}_{ik} \oplus \zeta \widetilde{p}_{ij} \right)^{w_{ik} w_{ij}} \right) \\
&= \bigcup_{\alpha_{ik} \in h_{ik}, \alpha_{ij} \in h_{ij}, \beta_{ij} \in g_{ij}, \alpha_{ij} \in h_{ij},} \left\{ \begin{array}{l} \left\{ \left(\sqrt{1 - \left(1 - \prod\limits_{k=1, j=k}^{4} \left(1 - \left(1 - \alpha_{ik}^2\right)^{\xi} \left(1 - \alpha_{ij}^2\right)^{\zeta}\right)^{w_i w_j}\right)^{\frac{1}{\xi+\zeta}}} \right) \right\}, \\[2em] \left\{ \left(\sqrt{1 - \prod\limits_{k=1, j=k}^{4} \left(1 - \beta_{ik}^{2\xi} \beta_{ij}^{2\zeta}\right)^{w_i w_j}} \right)^{\frac{1}{\xi+\zeta}} \right\} \end{array} \right\} \\
&= \{ \{\{0.4, 0.5\}, \{0.7\}\}, \{\{0.5, 0.6\}, \{0.4, 0.5\}\}, \{\{0.3, 0.4\}, \{0.8\}\}, \{\{0.5, 0.6\}, \{0.6\}\}\} \\
&= \{\{0.4929, 0.5022, 0.5190, 0.5287, 0.5186, 0.5283, 0.5460, 0.5560, 0.5140, 0.5236, \\
&\qquad 0.5406, 0.5506, 0.5406, 0.5505, 0.5686, 0.5790\}, \{0.6421, 0.6493\}\}
\end{aligned}
$$

Step 2'. Compute the scores results $s(\widetilde{p}_i)$ $(i = 1, 2, 3, 4, 5)$ of the overall dual hesitant Pythagorean fuzzy values $\widetilde{p}_i (i = 1, 2, 3, 4, 5)$ of the supplier $\widetilde{p}_i$:

$$
\begin{aligned}
s(\widetilde{p}_1) &= 0.4349, s(\widetilde{p}_2) = 0.4549, s(\widetilde{p}_3) = 0.3976 \\
s(\widetilde{p}_4) &= 0.5240, s(\widetilde{p}_5) = 0.4780
\end{aligned}
$$

Step 3′. Determine the ordering of all the suppliers $\eta_i (i = 1, 2, 3, 4, 5)$ based on the score results $s(\widetilde{p}_i)$ $(i = 1, 2, 3, 4, 5)$ of $\widetilde{p}_i (i = 1, 2, \cdots, 5)$: $\eta_4 > \eta_5 > \eta_2 > \eta_1 > \eta_3$ and it is clear that the most desirable supplier in supply chain management is η_4.

According to the above analysis, we can easily find that although the overall rating values of the alternatives are slightly different by using two operators respectively, the most desirable supplier in supply chain management is η_4.

5.2. Influence of Parameters on the Final Result

The parameters ξ and ζ play an important role in the final ranking of alternatives. We may obtain different ordering results by assigning different values to ξ and ζ. By altering the values of ξ and ζ, different ranking results are obtained, as shown in Tables 2 and 3. Therefore, the DHPFGWHM and DHPFGGWHM operators are shown to be considerably flexible by using a parameter vector. Tables 2 and 3 show that the ranking results increase and become steady with the increase of values in the parameter vector. That is, the final results become increasingly objective by considering the interrelationship among the attribute values. These features of the DHPFGWHM and DHPFGGWHM operators are crucial in real MADM problems.

Table 2. Ordering by the DHPFGWHM operators.

Parameter	$s(\eta_1)$	$s(\eta_2)$	$s(\eta_3)$	$s(\eta_4)$	$s(\eta_5)$	Ordering
$\xi = \zeta = 0.5$	0.2989	0.3322	0.3126	0.4804	0.3732	$\eta_4 > \eta_5 > \eta_2 > \eta_3 > \eta_1$
$\xi = \zeta = 1$	0.3504	0.3921	0.3834	0.5540	0.4208	$\eta_4 > \eta_5 > \eta_2 > \eta_3 > \eta_1$
$\xi = \zeta = 2$	0.3998	0.4536	0.4669	0.6255	0.4674	$\eta_4 > \eta_5 > \eta_3 > \eta_2 > \eta_1$
$\xi = \zeta = 3$	0.4288	0.4887	0.5197	0.6621	0.4933	$\eta_4 > \eta_3 > \eta_5 > \eta_2 > \eta_1$
$\xi = \zeta = 4$	0.4493	0.5117	0.5559	0.6846	0.5108	$\eta_4 > \eta_3 > \eta_2 > \eta_5 > \eta_1$
$\xi = \zeta = 5$	0.4647	0.5280	0.5822	0.7001	0.5235	$\eta_4 > \eta_3 > \eta_2 > \eta_5 > \eta_1$

Table 3. Ordering by the DHPFGGWHM operators.

Parameter	$s(\eta_1)$	$s(\eta_2)$	$s(\eta_3)$	$s(\eta_4)$	$s(\eta_5)$	Ordering
$\xi = \zeta = 0.5$	0.5484	0.5775	0.5431	0.6850	0.5728	$\eta_4 > \eta_2 > \eta_5 > \eta_1 > \eta_3$
$\xi = \zeta = 1$	0.4901	0.5176	0.4710	0.6165	0.5256	$\eta_4 > \eta_5 > \eta_2 > \eta_1 > \eta_3$
$\xi = \zeta = 2$	0.4349	0.4549	0.3976	0.5240	0.4780	$\eta_4 > \eta_5 > \eta_2 > \eta_1 > \eta_3$
$\xi = \zeta = 3$	0.4033	0.4146	0.3563	0.4591	0.4515	$\eta_4 > \eta_5 > \eta_2 > \eta_1 > \eta_3$
$\xi = \zeta = 4$	0.3815	0.3854	0.3290	0.4139	0.4344	$\eta_5 > \eta_4 > \eta_2 > \eta_1 > \eta_3$
$\xi = \zeta = 5$	0.3653	0.3633	0.3094	0.3817	0.4223	$\eta_5 > \eta_4 > \eta_1 > \eta_2 > \eta_3$

5.3. Comparative Analysis

The prominent characteristic of the DHPFGWHM and DHPFGGWHM operators is that they can consider the interrelationship among the DHFNs. Next, we shall compare our developed methods with the dual hesitant Pythagorean fuzzy weighted average (DHPFWA) and dual hesitant Pythagorean fuzzy weighted geometric (DHPFWG) operators [53], with the comparative analysis results listed as follows.

According to Table 1 and attribute weights, the fused values obtained by the DHPFGWA operator are:

$$\widetilde{p}_1 = \text{DHPFWA}(\widetilde{p}_{11}, \widetilde{p}_{12}, \widetilde{p}_{13}, \widetilde{p}_{14}) = \overset{4}{\underset{j=1}{\oplus}} w_j \widetilde{p}_{1j}$$

$$= \left\{ \left\{ \begin{array}{l} 0.4325, 0.4527, 0.4522, 0.4711, 0.4793, 0.4967, 0.4962, 0.5126, \\ 0.4580, 0.4766, 0.4761, 0.4936, 0.5013, 0.5174, 0.5170, 0.5323 \end{array} \right\}, \left\{ \begin{array}{l} 0.5872, \\ 0.6334 \end{array} \right\} \right\};$$

$$\widetilde{p}_2 = \text{DHPFWA}(\widetilde{p}_{21}, \widetilde{p}_{22}, \widetilde{p}_{23}, \widetilde{p}_{24}) = \overset{4}{\underset{j=1}{\oplus}} w_j \widetilde{p}_{2j}$$
$$= \{\{0.5005, 0.5461, 0.5788\}, \{0.5476, 0.5617, 0.5677, 0.5823, 0.5860, 0.6011\}\};$$

$$\widetilde{p}_3 = \text{DHPFWA}(\widetilde{p}_{31}, \widetilde{p}_{32}, \widetilde{p}_{33}, \widetilde{p}_{34}) = \overset{4}{\underset{j=1}{\oplus}} w_j \widetilde{p}_{3j}$$
$$= \left\{ \left\{ \begin{array}{l} 0.4547, 0.4812, 0.5178, 0.4775, 0.5022, 0.5364, 0.5071, 0.5296, 0.5610, \\ 0.5595, 0.5784, 0.6051, 0.5757, 0.5936, 0.6190, 0.5973, 0.6139, 0.6376 \end{array} \right\}, \left\{ \begin{array}{l} 0.5480, \\ 0.5704 \end{array} \right\} \right\};$$

$$\widetilde{p}_4 = \text{DHPFWA}(\widetilde{p}_{41}, \widetilde{p}_{42}, \widetilde{p}_{43}, \widetilde{p}_{44}) = \overset{4}{\underset{j=1}{\oplus}} w_j \widetilde{p}_{4j}$$
$$= \{\{0.6278, 0.6347, 0.6796, 0.6852, 0.7529, 0.7570\}, \{0.3923, 0.3997, 0.4063\}\};$$

$$\widetilde{p}_5 = \text{DHPFWA}(\widetilde{p}_{51}, \widetilde{p}_{52}, \widetilde{p}_{53}, \widetilde{p}_{54}) = \overset{4}{\underset{j=1}{\oplus}} w_j \widetilde{p}_{5j}$$
$$= \left\{ \left\{ \begin{array}{l} 0.3434, 0.3574, 0.3761, 0.3936, 0.4053, 0.4213, 0.3763, 0.3888, 0.4056, \\ 0.4214, 0.4321, 0.4467, 0.4172, 0.4281, 0.4428, 0.4568, 0.4663, 0.4794, \\ 0.3531, 0.3666, 0.3848, 0.4017, 0.4132\ 0.4287, 0.3849, 0.3971, 0.4134, \\ 0.4288, 0.4393, 0.4535, 0.4247, 0.4353, 0.4497, 0.4634, 0.4727, 0.4855 \end{array} \right\}, \{0.4184\} \right\}.$$

Then, based on the score function of the dual hesitant Pythagorean fuzzy elements (DHPFEs), we can obtain the score results of $\widetilde{p}_i$ as:

$$s(\widetilde{p}_1) = 0.4317, s(\widetilde{p}_2) = 0.4822, s(\widetilde{p}_3) = 0.4977$$
$$s(\widetilde{p}_4) = 0.6592, s(\widetilde{p}_5) = 0.5005$$

Then, we rank all the suppliers in supply chain management $\eta_i (i = 1, 2, 3, 4, 5)$ in accordance with the scores $s(\widetilde{p}_i)$ $(i = 1, 2, 3, 4, 5)$ of the overall dual hesitant Pythagorean fuzzy values $\widetilde{p}_i (i = 1, 2, \cdots, 5)$: $\eta_4 > \eta_5 > \eta_3 > \eta_2 > \eta_1$ and thus the most desirable supplier in supply chain management is obtained, which is η_4.

According to Table 1 and attribute weights, the fused values by the DHPFWG operator are:

$$\widetilde{p}_1 = \text{DHPFWG}(\widetilde{p}_{11}, \widetilde{p}_{12}, \widetilde{p}_{13}, \widetilde{p}_{14}) = \overset{4}{\underset{j=1}{\otimes}} \left(\widetilde{p}_{1j}\right)^{w_j}$$
$$= \left\{ \left\{ \begin{array}{l} 0.4119, 0.4226, 0.4452, 0.4567, 0.4383, 0.4496, 0.4737, 0.4859, \\ 0.4356, 0.4468, 0.4708, 0.4829, 0.4634, 0.4754, 0.5009, 0.5138 \end{array} \right\}, \left\{ \begin{array}{l} 0.6574, \\ 0.6735 \end{array} \right\} \right\};$$

$$\widetilde{p}_2 = \text{DHPFWG}(\widetilde{p}_{21}, \widetilde{p}_{22}, \widetilde{p}_{23}, \widetilde{p}_{24}) = \overset{4}{\underset{j=1}{\otimes}} \left(\widetilde{p}_{2j}\right)^{w_j}$$
$$= \{\{0.4086, 0.4861, 0.5171\}, \{0.5694, 0.5822, 0.6203, 0.6311, 0.6945, 0.7026\}\};$$

$$\widetilde{p}_3 = \text{DHPFWG}(\widetilde{p}_{31}, \widetilde{p}_{32}, \widetilde{p}_{33}, \widetilde{p}_{34}) = \overset{4}{\underset{j=1}{\otimes}} \left(\widetilde{p}_{3j}\right)^{w_j}$$
$$= \left\{ \left\{ \begin{array}{l} 0.3931, 0.4017, 0.4093, 0.4335, 0.4430, 0.4513, 0.4677, 0.4779, 0.4869, \\ 0.4224, 0.4316, 0.4398, 0.4658, 0.4760, 0.4850, 0.5026, 0.5135, 0.5232 \end{array} \right\}, \left\{ \begin{array}{l} 0.6622, \\ 0.7404 \end{array} \right\} \right\};$$

$$\widetilde{p}_4 = \text{DHPFWG}(\widetilde{p}_{41}, \widetilde{p}_{42}, \widetilde{p}_{43}, \widetilde{p}_{44}) = \overset{4}{\underset{j=1}{\otimes}} \left(\widetilde{p}_{4j}\right)^{w_j}$$
$$= \{\{0.6278, 0.6347, 0.6796, 0.6852, 0.7529, 0.7570\}, \{0.3923, 0.3997, 0.4063\}\};$$

$$\widetilde{p}_5 = \text{DHPFWG}(\widetilde{p}_{51}, \widetilde{p}_{52}, \widetilde{p}_{53}, \widetilde{p}_{54}) = \overset{4}{\underset{j=1}{\otimes}} \left(\widetilde{p}_{5j}\right)^{w_j}$$
$$= \left\{ \left\{ \begin{array}{l} 0.2617, 0.2724, 0.2811, 0.2749, 0.2862, 0.2952, 0.2885, 0.3004, 0.3099, \\ 0.3031, 0.3156, 0.3256, 0.3113, 0.3241, 0.3344, 0.3270, 0.3404, 0.3512, \\ 0.3112, 0.3240, 0.3342, 0.3269, 0.3403, 0.3511, 0.3431, 0.3572, 0.3686, \\ 0.3605, 0.3753, 0.3872, 0.3702, 0.3854, 0.3976, 0.3889, 0.4049, 0.4177 \end{array} \right\}, \{0.4811\} \right\}.$$

Then, based on the score function of the DHPFEs, we can obtain the score results of $\widetilde{p}_i$ as:

$$s(\widetilde{p}_1) = 0.3851, s(\widetilde{p}_2) = 0.4099, s(\widetilde{p}_3) = 0.3584$$
$$s(\widetilde{p}_4) = 0.4939, s(\widetilde{p}_5) = 0.4410$$

Then, we rank all the suppliers in supply chain management $\eta_i(i = 1, 2, 3, 4, 5)$ in accordance with the scores $s(\widetilde{p}_i)$ $(i = 1, 2, 3, 4, 5)$ of the overall dual hesitant Pythagorean fuzzy values $\widetilde{p}_i(i = 1, 2, \cdots, 5)$: $\eta_4 > \eta_5 > \eta_2 > \eta_1 > \eta_3$ and thus the most desirable supplier in supply chain management is obtained, which is η_4.

According to Table 4, we can easily conclude that the ordering is slightly different and that these are some of the best alternatives. However, our defined operators are mainly characteristic of the advantages that can consider the interrelationship between the arguments being fused into consideration and consider the human hesitance in practical MADM problems. Obviously, the DHPFWA and DHPFWG operators defined by Wei and Lu [53] cannot consider the interrelationship between the arguments being fused. In addition, in a complicated decision-making environment, the decision maker's risk attitude is an important factor to think about, and our methods can do this by altering the parameters ξ and ζ, whereas the DHPFWA and DHPFWG operators presented by Wei and Lu [53] do not have the ability to dynamically adjust to the parameters according to the decision maker's risk attitude, meaning it is difficult to solve risk multiple attribute decision making in real practice.

Table 4. Ordering of the suppliers by the DHPFGGWHM operators.

Methods	Ordering
The DHPFWA operator [53]	$\eta_4 > \eta_5 > \eta_3 > \eta_2 > \eta_1$
The DHPFWG operator [53]	$\eta_4 > \eta_5 > \eta_2 > \eta_1 > \eta_3$
The DHPFGWHM operator	$\eta_4 > \eta_5 > \eta_3 > \eta_2 > \eta_1$
The DHPFGGWHM operator	$\eta_4 > \eta_5 > \eta_2 > \eta_1 > \eta_3$

6. Conclusions

Dual hesitant Pythagorean fuzzy numbers have applied the advantages of DHFSs and PFSs. They can flexibly denote decision-making information as well as effectively characterize the reliability of information. Thus, it is meaningful to study MADM problems with DHPFNs. In this paper, based on the generalized Heronian mean operator and generalized geometric Heronian mean operator, we developed some dual hesitant Pythagorean fuzzy Heronian mean aggregation operators: dual hesitant Pythagorean fuzzy generalized weighted Heronian mean (DHPFGWHM) operator and dual hesitant Pythagorean fuzzy generalized geometric weighted Heronian mean (DHPFGGWHM) operator. The significant merits of these defined operators are investigated. Moreover, we have adopted DHPFGWHM and DHPFGGWHM operators to build a decision-making model for MADM problems. In the end, we utilize a concrete instance for suppliers selection in supply chain management to demonstrate our defined model and to testify its accuracy and scientific ability. However, our developed methods can only deal with MADMs with dual hesitant Pythagorean fuzzy information, and it is clear that these operators cannot handle more complicated decision making problems, such as when the sum square of the membership and non-membership is more than 1. In the future, we shall continue studying MADM problems with the application and extension of the developed operators to other domains [67,68] and proposed more suitable methods [69–75].

Author Contributions: M.T., J.W., J.L., G.W., C.W. and Y.W. conceived and worked together to achieve this work, M.T. and J.W. compiled the computing program by Excel and analyzed the data, J.W. and G.W. wrote the paper. Finally, all the authors have read and approved the final manuscript.

Funding: This work was supported by the University Students' Innovation and Entrepreneurship Training Program Project of Sichuan Normal University under Grant No. 201810636122 and the National Natural Science Foundation of China under Grant No. 71571128.

Conflicts of Interest: The authors declare no conflict of interest.

References

1. Zadeh, L.A. Fuzzy Sets. *Inf. Control* **1965**, *8*, 338–356.
2. Atanassov, K. Intuitionistic fuzzy sets. *Fuzzy Sets Syst.* **1986**, *20*, 87–96. [CrossRef]
3. Zhou, W.; Xu, Z.S. Extended Intuitionistic Fuzzy Sets Based on the Hesitant Fuzzy Membership and their Application in Decision Making with Risk Preference. *Int. J. Intell. Syst.* **2018**, *33*, 417–443. [CrossRef]
4. Zhao, F.; Liu, H.; Fan, J.; Chen, C.W.; Lan, R.; Li, N. Intuitionistic fuzzy set approach to multi-objective evolutionary clustering with multiple spatial information for image segmentation. *Neurocomput.* **2018**, *312*, 296–309. [CrossRef]
5. Zhang, Z.M. Geometric Bonferroni means of interval-valued intuitionistic fuzzy numbers and their application to multiple attribute group decision making. *Neural Comput. Appl.* **2018**, *29*, 1139–1154. [CrossRef]
6. Zhang, G.; Zhang, Z.; Kong, H. Some Normal Intuitionistic Fuzzy Heronian Mean Operators Using Hamacher Operation and Their Application. *Symmetry* **2018**, *10*, 199. [CrossRef]
7. Li, Z.; Gao, H.; Wei, G. Methods for Multiple Attribute Group Decision Making Based on Intuitionistic Fuzzy Dombi Hamy Mean Operators. *Symmetry* **2018**, *10*, 574. [CrossRef]
8. Wei, G.W. TODIM Method for Picture Fuzzy Multiple Attribute Decision Making. *Informatica* **2018**, *29*, 555–566. [CrossRef]
9. Wang, J.; Wei, G.W.; Lu, M. TODIM Method for Multiple Attribute Group Decision Making under 2-Tuple Linguistic Neutrosophic Environment. *Symmetry* **2018**, *10*, 486. [CrossRef]
10. Zhai, Y.L.; Xu, Z.S.; Liao, H.C. Measures of Probabilistic Interval-Valued Intuitionistic Hesitant Fuzzy Sets and the Application in Reducing Excessive Medical Examinations. *IEEE Trans. Fuzzy Syst.* **2018**, *26*, 1651–1670.
11. Li, Z.; Wei, G.; Lu, M. Pythagorean Fuzzy Hamy Mean Operators in Multiple Attribute Group Decision Making and Their Application to Supplier Selection. *Symmetry* **2018**, *10*, 505. [CrossRef]
12. Wu, L.; Wei, G.; Gao, H.; Wei, Y. Some Interval-Valued Intuitionistic Fuzzy Dombi Hamy Mean Operators and Their Application for Evaluating the Elderly Tourism Service Quality in Tourism Destination. *Mathematics* **2018**, *6*, 294. [CrossRef]
13. Wang, J.; Gao, H.; Wei, G.; Wei, Y. Methods for Multiple-Attribute Group Decision Making with q-Rung Interval-Valued Orthopair Fuzzy Information and Their Applications to the Selection of Green Suppliers. *Symmetry* **2019**, *11*, 56. [CrossRef]
14. Wei, G.W.; Zhang, Z.P. Some single-valued neutrosophic Bonferroni power aggregation operators in multiple attribute decision making. *J. Ambient. Intell. Humaniz. Comput.* **2019**, *10*, 863–882. [CrossRef]
15. Zhang, S.; Gao, H.; Wei, G.; Wei, Y.; Wei, C. Evaluation Based on Distance from Average Solution Method for Multiple Criteria Group Decision Making under Picture 2-Tuple Linguistic Environment. *Mathematics* **2019**, *7*, 243. [CrossRef]
16. Wei, G.-W. Pythagorean Fuzzy Hamacher Power Aggregation Operators in Multiple Attribute Decision Making. *Fundam. Inform.* **2019**, *166*, 57–85. [CrossRef]
17. Xu, Z.S. Intuitionistic fuzzy aggregation operators. *IEEE Trans. Fuzzy Syst.* **2007**, *15*, 1179–1187.
18. Xu, Z.; Yager, R.R. Some geometric aggregation operators based on intuitionistic fuzzy sets. *Int. J. Gen. Syst.* **2006**, *35*, 417–433. [CrossRef]
19. Hung, W.L.; Yang, M.S. Similarity measures of intuitionistic fuzzy sets based on L-p metric. *Int. J. Approx. Reason.* **2007**, *46*, 120–136. [CrossRef]
20. Park, J.H.; Lim, K.M.; Park, J.S.; Kwun, Y.C. Distances between Interval-valued Intuitionistic Fuzzy Sets. In Proceedings of the International Symposium on Nonlinear Dynamics, Shanghai, China, 27–30 October 2007.
21. Wei, G.-W. Maximizing deviation method for multiple attribute decision making in intuitionistic fuzzy setting. *Knowl. Based Syst.* **2008**, *21*, 833–836. [CrossRef]
22. Hung, C.C.; Chen, L.H.; Ao, S.I. A Fuzzy TOPSIS Decision Making Model with Entropy Weight under Intuitionistic Fuzzy Environment. In Proceedings of the International of Multi Conference of Engineers and Computer Scientist (IMECS), Hong Kong, China, 18–20 March 2009.
23. Luo, Y.J. IEEE, Projection Method for Multiple Attribute Decision Making with Uncertain Attribute Weights under Intuitionistic Fuzzy Environment. In Proceedings of the Chinese Control and Decision Conference, Guilin, China, 17–19 June 2009; pp. 2945–2948.

24. Ye, J. Fuzzy cross entropy of interval-valued intuitionistic fuzzy sets and its optimal decision-making method based on the weights of alternatives. *Syst. Appl.* **2011**, *38*, 6179–6183. [CrossRef]

25. Zhang, Z.M. Interval-Valued Intuitionistic Hesitant Fuzzy Aggregation Operators and Their Application in Group Decision-Making. *J. Appl. Math.* **2013**, *2013*. [CrossRef]

26. Liao, H.C.; Xu, Z.S. Intuitionistic Fuzzy Hybrid Weighted Aggregation Operators. *Int. J. Intell. Syst.* **2014**, *29*, 971–993. [CrossRef]

27. Liu, X.Y.; Ju, Y.B.; Yang, S.H. Hesitant intuitionistic fuzzy linguistic aggregation operators and their applications to multiple attribute decision making. *J. Intell. Fuzzy Syst.* **2014**, *27*, 1187–1201.

28. Peng, J.-J.; Wang, J.-Q.; Wang, J.; Chen, X.-H. Multicriteria Decision-Making Approach with Hesitant Interval-Valued Intuitionistic Fuzzy Sets. *Sci. J.* **2014**, *2014*, 1–22. [CrossRef] [PubMed]

29. Chen, J.J.; Huang, X.J. Hesitant triangular intuitionistic fuzzy information and its application to multi-attribute decision making problem. *J. Nonlinear Sci. Appl.* **2017**, *10*, 1012–1029. [CrossRef]

30. Leśniak, A.; Kubek, D.; Plebankiewicz, E.; Zima, K.; Belniak, S. Fuzzy AHP Application for Supporting Contractors' Bidding Decision. *Symmetry* **2018**, *10*, 642. [CrossRef]

31. Adeel, A.; Akram, M.; Ahmed, I.; Nazar, K. Novel m-Polar Fuzzy Linguistic ELECTRE-I Method for Group Decision-Making. *Symmetry* **2019**, *11*, 471. [CrossRef]

32. Turskis, Z.; Goranin, N.; Nurusheva, A.; Boranbayev, S. A Fuzzy WASPAS-Based Approach to Determine Critical Information Infrastructures of EU Sustainable Development. *Sustainability* **2019**, *11*, 424. [CrossRef]

33. Ziemba, P.; Becker, J. Analysis of the Digital Divide Using Fuzzy Forecasting. *Symmetry* **2019**, *11*, 166. [CrossRef]

34. Hu, C.-K.; Liu, F.-B. A Hybrid Fuzzy DEA/AHP Methodology for Ranking Units in a Fuzzy Environment. *Symmetry* **2017**, *9*, 273. [CrossRef]

35. Ziemba, P.; Jankowski, J.; Wątróbski, J. Online Comparison System with Certain and Uncertain Criteria Based on Multi-criteria Decision Analysis Method. In *Text, Speech and Dialogue*; Springer Nature: Heidelberger, Germany, 2017; Volume 10449, pp. 579–589.

36. Diouf, M.; Kwak, C. Fuzzy AHP, DEA, and Managerial Analysis for Supplier Selection and Development; From the Perspective of Open Innovation. *Sustainability* **2018**, *10*, 3779. [CrossRef]

37. Dong, J.; Li, R.; Huang, H. Performance Evaluation of Residential Demand Response Based on a Modified Fuzzy VIKOR and Scalable Computing Method. *Energies* **2018**, *11*, 1097. [CrossRef]

38. Kim, J.; Kim, J. Optimal Portfolio for LNG Importation in Korea Using a Two-Step Portfolio Model and a Fuzzy Analytic Hierarchy Process. *Energies* **2018**, *11*, 3049. [CrossRef]

39. Chou, Y.C.; Yen, H.Y.; Dang, V.T.; Sun, C.C. Assessing the Human Resource in Science and Technology for Asian Countries: Application of Fuzzy AHP and Fuzzy TOPSIS. *Symmetry* **2019**, *11*, 251. [CrossRef]

40. Yager, R.R. Pythagorean Fuzzy Subsets. In Proceedings of the Joint IFSA World Congress and NAFIPS Annual Meeting (IFSA/NAFIPS), Edmonton, Alberta, Canada, 24–28 June 2013; pp. 57–61.

41. Yager, R.R. Pythagorean Membership Grades in Multicriteria Decision Making. *IEEE Trans. Syst.* **2014**, *22*, 958–965. [CrossRef]

42. Zhang, X.L.; Xu, Z.S. Extension of TOPSIS to Multiple Criteria Decision Making with Pythagorean Fuzzy Sets. *Int. J. Intell. Syst.* **2014**, *29*, 1061–1078. [CrossRef]

43. Peng, X.; Yang, Y. Some Results for Pythagorean Fuzzy Sets. *Int. J. Intell. Syst.* **2015**, *30*, 1133–1160. [CrossRef]

44. Reformat, M.Z.; Yager, R.R. Suggesting Recommendations Using Pythagorean Fuzzy Sets illustrated Using Netflix Movie Data. In *Communications in Computer and Information Science*; Springer Nature: Basingstoke, UK, 2014; Volume 442, pp. 546–556.

45. Gou, X.J.; Xu, Z.S.; Ren, P.J. The Properties of Continuous Pythagorean Fuzzy Information. *Int. J. Intell. Syst.* **2016**, *31*, 401–424. [CrossRef]

46. Garg, H. A New Generalized Pythagorean Fuzzy Information Aggregation Using Einstein Operations and Its Application to Decision Making. *Int. J. Intell. Syst.* **2016**, *31*, 886–920. [CrossRef]

47. Zeng, S.Z.; Chen, J.P.; Li, X.S. A Hybrid Method for Pythagorean Fuzzy Multiple-Criteria Decision Making. *Int. J. Inf. Technol. Decis. Mak.* **2016**, *15*, 403–422. [CrossRef]

48. Ren, P.J.; Xu, Z.S.; Gou, X.J. Pythagorean fuzzy TODIM approach to multi-criteria decision making. *Appl. Soft Comput.* **2016**, *42*, 246–259. [CrossRef]

49. Liang, D.; Zhang, Y.; Xu, Z.; Darko, A.P. Pythagorean fuzzy Bonferroni mean aggregation operator and its accelerative calculating algorithm with the multithreading. *Int. J. Intell. Syst.* **2018**, *33*, 615–633. [CrossRef]

50. Liang, D.; Xu, Z.; Darko, A.P. Projection Model for Fusing the Information of Pythagorean Fuzzy Multicriteria Group Decision Making Based on Geometric Bonferroni Mean. *Int. J. Intell. Syst.* **2017**, *32*, 966–987. [CrossRef]
51. Zhu, B.; Xu, Z.S.; Xia, M.M. Dual Hesitant Fuzzy Sets. *J. Appl. Math.* **2012**, *2012*. [CrossRef]
52. Wang, H.J.; Zhao, X.F.; Wei, G.W. Dual hesitant fuzzy aggregation operators in multiple attribute decision making. *J. Intell. Fuzzy Syst.* **2014**, *26*, 2281–2290.
53. Wei, G.; Lu, M. Dual hesitant pythagorean fuzzy Hamacher aggregation operators in multiple attribute decision making. *Arch. Control. Sci.* **2017**, *27*, 365–395. [CrossRef]
54. Yu, D.J. Intuitionistic fuzzy geometric Heronian mean aggregation operators. *Appl. Soft Comput.* **2013**, *13*, 1235–1246. [CrossRef]
55. Fan, C.; Ye, J.; Feng, S.; Fan, E.; Hu, K. Multi-Criteria Decision-Making Method Using Heronian Mean Operators under a Bipolar Neutrosophic Environment. *Mathematics* **2019**, *7*, 97. [CrossRef]
56. Zang, Y.Q.; Zhao, X.D.; Li, S.Y. Interval-Valued Dual Hesitant Fuzzy Heronian Mean Aggregation Operators and their Application to Multi-Attribute Decision Making. *Int. J. Comput. Intell. Appl.* **2018**, *17*, 1850005. [CrossRef]
57. Wei, G.W.; Lu, M.; Gao, H. Picture fuzzy heronian mean aggregation operators in multiple attribute decision making. *Int. J. Knowl. Intell. Eng. Syst.* **2018**, *22*, 167–175. [CrossRef]
58. Wang, H.H.; Ju, Y.B.; Liu, P.D.; Ju, D.W.; Liu, Z.M. Some trapezoidal interval type-2 fuzzy Heronian mean operators and their application in multiple attribute group decision making. *J. Intell. Fuzzy Syst.* **2018**, *35*, 2323–2337. [CrossRef]
59. Fan, C.; Ye, J. Heronian Mean Operator of Linguistic Neutrosophic Cubic Numbers and Their Multiple Attribute Decision-Making Methods. *Math. Probl. Eng.* **2018**, *2018*, 1–13. [CrossRef]
60. Liu, P.; Liu, Z.; Zhang, X. Some intuitionistic uncertain linguistic Heronian mean operators and their application to group decision making. *Appl. Math. Comput.* **2014**, *230*, 570–586. [CrossRef]
61. Yu, S.-M.; Zhou, H.; Chen, X.-H.; Wang, J.-Q. A Multi-Criteria Decision-Making Method Based on Heronian Mean Operators Under a Linguistic Hesitant Fuzzy Environment. *Asia-Pacific J. Oper. Res.* **2015**, *32*, 1550035. [CrossRef]
62. Li, Y.H.; Liu, P.D.; Chen, Y.B. Some Single Valued Neutrosophic Number Heronian Mean Operators and Their Application in Multiple Attribute Group Decision Making. *Informatica* **2016**, *27*, 85–110. [CrossRef]
63. Wei, G.; Gao, H.; Wei, Y. Some q-rung orthopair fuzzy Heronian mean operators in multiple attribute decision making. *Int. J. Intell. Syst.* **2018**, *33*, 1426–1458. [CrossRef]
64. Li, L.; Zhang, R.T.; Wang, J.; Shang, X.P. Some q-rung orthopair linguistic Heronian mean operators with their application to multi-attribute group decision making. *Arch. Control Sci.* **2018**, *28*, 551–583.
65. Beliakov, G.P.; Calvo, A.T. *Aggregation Functions: A Guide for Practitioners*; Springer: Heidelberg, Germany, 2007.
66. Xu, Y.; Shang, X.; Wang, J.; Wu, W.; Huang, H. Some q-Rung Dual Hesitant Fuzzy Heronian Mean Operators with Their Application to Multiple Attribute Group Decision-Making. *Symmetry* **2018**, *10*, 472. [CrossRef]
67. Wei, Y.; Yu, Q.; Liu, J.; Cao, Y. Hot money and China's stock market volatility: Further evidence using the GARCH-MIDAS model. *Phys. A: Stat. Mech. Appl.* **2018**, *492*, 923–930. [CrossRef]
68. Wei, Y.; Liu, J.; Lai, X.; Hu, Y. Which determinant is the most informative in forecasting crude oil market volatility: Fundamental, speculation, or uncertainty? *Energy Economics* **2017**, *68*, 141–150. [CrossRef]
69. Wang, R.; Wang, J.; Gao, H.; Wei, G.W. Methods for MADM with Picture Fuzzy Muirhead Mean Operators and Their Application for Evaluating the Financial Investment Risk. *Symmetry* **2019**, *11*, 6. [CrossRef]
70. Wang, J.; Wei, G.W.; Wei, Y. Models for Green Supplier Selection with Some 2-Tuple Linguistic Neutrosophic Number Bonferroni Mean Operators. *Symmetry* **2018**, *10*, 131. [CrossRef]
71. Gao, H. Pythagorean fuzzy Hamacher Prioritized aggregation operators in multiple attribute decision making. *J. Intell. Fuzzy Syst.* **2018**, *35*, 2229–2245. [CrossRef]
72. Deng, X.M.; Wei, G.W.; Gao, H.; Wang, J. Models for Safety Assessment of Construction Project With Some 2-Tuple Linguistic Pythagorean Fuzzy Bonferroni Mean Operators. *IEEE Access* **2018**, *6*, 52105–52137. [CrossRef]
73. Deng, X.M.; Wang, J.; Wei, G.W.; Lu, M. Models for Multiple Attribute Decision Making with Some 2-Tuple Linguistic Pythagorean Fuzzy Hamy Mean Operators. *Mathematics* **2018**, *6*, 11. [CrossRef]

74. Wei, G.W.; Garg, H.; Gao, H.; Wei, C. Interval-Valued Pythagorean Fuzzy Maclaurin Symmetric Mean Operators in Multiple Attribute Decision Making. *IEEE Access* **2018**, *6*, 67866–67884. [CrossRef]
75. Li, Z.X.; Wei, G.W.; Gao, H. Methods for Multiple Attribute Decision Making with Interval-Valued Pythagorean Fuzzy Information. *Mathematics* **2018**, *6*, 228. [CrossRef]

 *mathematics*

Article

Some Hesitant Fuzzy Hamacher Power-Aggregation Operators for Multiple-Attribute Decision-Making

Mi Jung Son [1], Jin Han Park [2,*] and Ka Hyun Ko [2]

[1]   Department of Data Information, Korea Maritime and Ocean University, Busan 606-791, Korea
[2]   Department of Applied Mathematics, Pukyong National University, Busan 608-737, Korea
*   Correspondence: jihpark@pknu.ac.kr; Tel.: +82-51-629-5530

Received: 31 May 2019; Accepted: 26 June 2019; Published: 2 July 2019

Abstract: As an extension of the fuzzy set, the hesitant fuzzy set is used to effectively solve the hesitation of decision-makers in group decision-making and to rigorously express the decision information. In this paper, we first introduce some new hesitant fuzzy Hamacher power-aggregation operators for hesitant fuzzy information based on Hamacher t-norm and t-conorm. Some desirable properties of these operators is shown, and the interrelationships between them are given. Furthermore, the relationships between the proposed aggregation operators and the existing hesitant fuzzy power-aggregation operators are discussed. Based on the proposed aggregation operators, we develop a new approach for multiple-attribute decision-making problems. Finally, a practical example is provided to illustrate the effectiveness of the developed approach, and the advantages of our approach are analyzed by comparison with other existing approaches.

Keywords: hesitant fuzzy element (HFE); Hamacher operations; hesitant fuzzy Hamacher power-aggregation operators; multiple-attribute decision-making (MADM)

---

1. Introduction

Since the fuzzy set (FS) was introduced by Zadeh [1], it has received much attention for its applicability. Some classical extensions of the FS, such as the interval-valued fuzzy set (IVFS) [2], intuitionistic fuzzy set (IFS) [3], interval-valued intuitionistic fuzzy set (IVIFS) [4], type-2 fuzzy set (T2FS) [5], type-n fuzzy set (TnFS) [5], and fuzzy multiset (FMS) [6], were then developed. However, it is often faced with the fact that the difficulty of setting membership degree for an element in a set arises not from the possibility distribution of possible values (as in T2FS) or the margin of error (as in IVFS or IFS), but from the hesitation between several different values. The concept of hesitant fuzzy set (HFS) was introduced by Narukawa and Torra [7,8] to deal with such cases. The HFS has the advantage of representing the membership degree of one element to a set by a set of possible values between 0 and 1, so it is an effective tool to represent a decision-maker's hesitation in expressing his/her preferences for objects than the FS or its classical extensions. In this regard, the HFS theory has been applied to many practical applications such as decision-making [9–18].

The goal of multiple-attribute decision-making (MADM), based on preferences provided by the decision-makers, is to select the most desirable alternative(s) from a given set of feasible alternatives. MADM methods classified as conventional and fuzzy. The conventional MADM methods are seen inadequate to handle uncertainty in linguistic terms [19]. Hence, it is proposed to apply MADM methods with the FS and its extensions to cope with vagueness in a decision-making process. Furthermore, these fuzzy methods enable more concrete results. Besides, the FS and its extensions helps to decision-makers to express their opinions by means of linguistic terms. Therefore, more sensitive results can be obtained by applying fuzzy MADM methods to various science and engineering fields such as supplier selection and forecasting [20–23].

The aggregation operators are most commonly used as tools to combine each individual preference information into the overall preference information and to elicit collective preference value for each alternative. The power average (PA) and power-ordered weighted average (POWA), introduced by Yager [24], are the nonlinear weighted average aggregation tools whose weight vectors depend on the input arguments and allow the argument values to support each other [25]. In particular, compared to most aggregation operators, the PA and POWA operators have the advantage of incorporating information about the relationship between argument values that are combined. So these operators have received a lot of attention from researchers in recent years, particularly Xu and Yager [25], Zhou et al. [26] and Zhang [15] introduced some new power-aggregation operators, including the weighted generalizations of these operators. However, these power-aggregation operators only deal with arguments, which are exact numerical values.

In real life, we often face situations where input arguments are expressed not as exact numerical values but as interval numbers [27], intuitionistic fuzzy numbers [28–30], interval-valued intuitionistic fuzzy numbers [31], linguistic variables [32–34], uncertain linguistic variables [35–37], or hesitant fuzzy elements (HFEs) [10,11]. Many extensions of power-aggregation operators have been proposed to address these situations: the uncertain power-aggregation operators [25,38,39], intuitionistic fuzzy power-aggregation operators [26,40], interval-valued intuitionistic fuzzy power-aggregation operators [40], linguistic power-aggregation operators [41–44] and hesitant fuzzy power-aggregation operators [15,18]. In particular, with respect to HFEs, Zhang [15] proposed a family of hesitant fuzzy power-aggregation operators, including the hesitant fuzzy power-weighted average/geometric (HFPWA or HFPWG), generalized hesitant fuzzy power-weighted average/geometric (GHFPWA or GHFPWG), hesitant fuzzy power-ordered weighted average/geometric (HFPOWA or HFPOWG), and generalized hesitant fuzzy power-ordered weighted average/geometric (GHFPOWA or GHFPOWG) operators, and applied them to solve multiple criteria group decision-making problems under hesitant fuzzy environment.

It is worthwhile to mention that operational rules play a key role in integrating information using power-aggregation operators. A lot of research about power-aggregation operators for the FS and its extensions has been done by operational rules using various pairs of triangular norm (shortly *t*-norm) and triangular conorm (shortly *t*-conorm) [45] in recent years. The aforementioned hesitant fuzzy power-aggregation operators, such as the HFPWA, HFPWG, GHFPWA, GHFPWG, HFPOWA, HFPOWG, GHFPOWA, and GHFPOWG operators, are based on the algebraic product and algebraic sum operational rules on HFEs, which are a pair of the special dual *t*-norm and *t*-conorm [45]. The algebraic product and algebraic sum are the basic operations on HFEs, they are not the only ones. The Einstein *t*-norm and *t*-conorm, as another pair of special *t*-norm and dual *t*-conorm, are alternatives to the algebraic product and algebraic sum, respectively, for operational rules on HFEs. Yu [16] extended the Einstein *t*-norm and *t*-conorm to HFEs, and developed some hesitant fuzzy Einstein aggregation operators based on the Einstein product and Einstein sum operational rules on HFEs. By mean of these operational rules on HFEs, Yu et al. [18] proposed a wide range of hesitant fuzzy power-aggregation operators, such as the hesitant fuzzy Einstein power-weighted average/geometric (HFEPWA or HFEPWG), generalized hesitant fuzzy Einstein power-weighted average/geometric (GHFEPWA or GHFEPWG), hesitant fuzzy Einstein power-ordered weighted average/geometric (HFEPOWA or HFEPOWG), and generalized hesitant fuzzy Einstein power-ordered weighted average/geometric (GHFEPOWA or GHFEPOWG) operators, and applied them to deal with MADM with hesitant fuzzy information. Hamacher [46] proposed a more generalized *t*-norm and *t*-conorm, called the Hamacher *t*-norm and *t*-conorm. These Hamacher *t*-norm and *t*-conorm are more general and flexible because they are a generalization of the algebraic *t*-norm and *t*-conorm and the Einstein *t*-norm and *t*-conorm [46]. Tan et al. [17] gave some operations on HFEs based on Hamacher *t*-norm and *t*-conorm, and developed some hesitant fuzzy Hamacher aggregation operators. In this paper, by means of Hamacher operations on HFEs, we propose a family of hesitant fuzzy Hamacher power-aggregation operators that allow decision-makers to have more choices in MADM problems. This study is very necessary because it is an integrated treatment of works by Zhang [15] and Yu et al. [18].

To do so, this paper is organized as follows: In Section 2, some basic concepts and notions of HFSs and Hamacher operations of HFEs based on the Hamacher *t*-norm and *t*-conorm are reviewed. Some results of Hamacher operations of HFEs are investigated. In Section 3, we present a wide range of hesitant fuzzy Hamacher power-aggregation operators for hesitant fuzzy information, some of their basic properties are discussed, and the relationships between the proposed operators and the existing hesitant fuzzy aggregation operators are investigated. In Section 4, we apply the proposed operators to develop an approach to MADM with hesitant fuzzy information. An example application of the new approach is provided, and a comparison with other hesitant fuzzy MADM approaches is performed. Some concluding remarks is given in Section 5.

2. Basic Concepts and Operations

2.1. Triangular Norms and Conorms

The operators play an important role in the beginning of FS theory. The *t*-norm and *t*-conorm used to define generalized union and intersection, respectively, is one of the important concepts in FS theory. They are defined as follows.

Definition 1. *[45] A triangular norm (t-norm) is a binary operation T on the unit interval $[0,1]$, i.e., a function $T : [0,1] \times [0,1] \rightarrow [0,1]$, such that for all $x, y, z \in [0,1]$, the following four axioms are satisfied:*
(1) (Boundary condition) $T(1,x) = x$;
(2) (Commutativity) $T(x,y) = T(y,x)$;
(3) (Associativity) $T(x, T(y,z)) = T(T(x,y),z)$;
(4) (Monotonicity) $T(x_1, y_1) \leq T(x_2, y_2)$ if $x_1 \leq x_2$ and $y_1 \leq y_2$.
The corresponding triangular conorm (t-conorm) of T (or the dual of T) is the function $S : [0,1] \times [0,1] \rightarrow [0,1]$ defined by $S(x,y) = 1 - T(1-x, 1-y)$ for each $x,y \in [0,1]$.

Among many *t*-norms and *t*-conorms, there are the following basic *t*-norms and *t*-conorms: minimum T_M and maximum S_M, algebraic product T_A and algebraic sum S_A, Einstein product T_E and Einstein sum S_E, bounded difference T_B and bounded sum S_B, and drastic product T_D and drastic sum S_D, given respectively as follows:

- $T_M(x,y) = \min(x,y)$, $S_M(x,y) = \max(x,y)$;
- $T_A(x,y) = xy$, $S_A(x,y) = x + y - xy$;
- $T_E(x,y) = \dfrac{xy}{1 + (1-x)(1-y)}$, $S_E(x,y) = \dfrac{x+y}{1+xy}$;
- $T_B(x,y) = \max(0, x + y - 1)$, $S_B(x,y) = \min(1, x + y)$;
- $T_D(x,y) = \begin{cases} 0, & \text{if } (x,y) \in [0,1)^2 \\ \min(x,y), & \text{otherwise} \end{cases}$, $S_D(x,y) = \begin{cases} 1, & \text{if } (x,y) \in (0,1]^2 \\ \max(x,y), & \text{otherwise.} \end{cases}$

These *t*-norms and *t*-conorms are ordered as follows:

$$T_D \leq T_B \leq T_E \leq T_A \leq T_M, \text{ and } S_M \leq S_A \leq S_E \leq S_B \leq S_D. \tag{1}$$

From (1), since the drastic product T_D and minimum T_M are the smallest and the largest *t*-norms, respectively, we know that $T_D \leq T \leq T_M$ for any *t*-norm T. In particular, the algebraic product T_A and the Einstein product T_E are two prototypic examples of the class of strict Archimedean *t*-norms [45].

Hamacher [46] proposed, as more generalized t-norm and t-conorm, the Hamacher t-norm and t-conorm as follows:

$$T_H^\zeta(x,y) = \frac{xy}{\zeta + (1-\zeta)(x+y-xy)}, \quad S_H^\zeta(x,y) = \frac{x+y-xy-(1-\zeta)xy}{\zeta+(1-\zeta)xy}, \quad \zeta > 0. \tag{2}$$

From (2), when $\zeta = 1$, then the Hamacher t-norm and t-conorm reduce to the algebraic t-norm T_A and t-conorm S_A, respectively; when $\zeta = 2$, then then the Hamacher t-norm and t-conorm reduce to the Einstein t-norm T_E and t-conorm S_E, respectively.

2.2. Hesitant Fuzzy Sets and Hesitant Fuzzy Elements

In the following, some basic concepts of hesitant fuzzy set and hesitant fuzzy element are briefly reviewed [7,8,10].

Definition 2. *[7,8] Let X be a fixed set, a hesitant fuzzy set (HFS) on X is defined in terms of function h that returns a subset of $[0,1]$ when applied to X. The HFS can be represented as the following mathematical symbol:*

$$E = \{\langle x, h_E(x)\rangle | x \in X\}, \tag{3}$$

where $h_E(x)$ is a set of values in $[0,1]$ that denote the possible membership degrees of the element $x \in X$ to the set E. For convenience, we refer to $h = h_E(x)$ as a hesitant fuzzy element (HFE) and to H the set of all HFEs.

Given three HFEs h, h_1 and h_2, Torra and Narukawa [7,8] and Xia and Xu [10] defined the following HFE operations:

(1) $h^c = \cup_{\gamma \in h}\{1 - \gamma\}$;
(2) $h_1 \cup h_2 = \cup_{\gamma_1 \in h_1, \gamma_2 \in h_2}\{\gamma_1 \vee \gamma_2\}$;
(3) $h_1 \cap h_2 = \cup_{\gamma_1 \in h_1, \gamma_2 \in h_2}\{\gamma_1 \wedge \gamma_2\}$;
(4) $h^\lambda = \cup_{\gamma \in h}\{\gamma^\lambda\}, \lambda > 0$;
(5) $\lambda h = \cup_{\gamma \in h}\{1 - (1-\gamma)^\lambda\}, \lambda > 0$;
(6) $h_1 \oplus h_2 = \cup_{\gamma_1 \in h_1, \gamma_2 \in h_2}\{\gamma_1 + \gamma_2 - \gamma_1\gamma_2\}$;
(7) $h_1 \otimes h_2 = \cup_{\gamma_1 \in h_1, \gamma_2 \in h_2}\{\gamma_1\gamma_2\}$.

Xia and Xu [10] also defined the following comparison rules for HFEs:

Definition 3. *[10] For a HFE h, $s(h) = \frac{\sum_{\gamma \in h}\gamma}{l(h)}$ is called the score function of h, where $l(h)$ is the number of elements in h. For two HFEs h_1 and h_2,*

- *if $s(h_1) > s(h_2)$, then h_1 is superior to h_2, denoted by $h_1 > h_2$;*
- *if $s(h_1) = s(h_2)$, then h_1 is indifferent to h_2, denoted by $h_1 = h_2$.*

Let h_1 and h_2 be two HFEs. In the most case, $l(h_1) \neq l(h_2)$; for convenience, let $l = \max\{l(h_1), l(h_2)\}$. To compare h_1 and h_2, Xu and Xia [11] extended the shorter HFE until the length of both HFEs was the same. The simplest way to extend the shorter HFE is to add the same value repeatedly. In fact, we can extend the shorter ones by adding any values in them. The selection of these values mainly depends on the decision-makers' risk preferences. Optimists anticipate desirable outcomes and may add the maximum value, while pessimists expect unfavorable outcomes and may add the minimum value [11]. In this paper, we assume that the decision-makers are all pessimistic (other situation can also be studied similarly).

Xu and Xia [11] proposed various distance measures for HFEs, including the hesitant normalized Hamming distance defined as follows:

$$d(h_1, h_2) = \frac{1}{l}\sum_{i=1}^{l}\left|h_1^{\sigma(i)} - h_2^{\sigma(i)}\right|, \tag{4}$$

where $h_1^{\sigma(i)}$ and $h_2^{\sigma(i)}$ are the ith largest values in h_1 and h_2, respectively.

Intrinsically, the addition and multiplication operators proposed by Xia and Xu [10] are algebraic sum and algebraic product operational rules on HFEs, respectively, and are a special pair of dual t-norm and t-conorm. Recently, Tan et al. [17] extended these operations to obtain more general operations on HFEs by means of the Hamacher t-norm and t-conorm as follows:

Definition 4. [17] *For any given three HFEs h, h_1, h_2, and $\zeta > 0$, the Hamacher operations on HFEs are defined as follows:*

(1) $h_1 \oplus_H h_2 = \bigcup_{\gamma_1 \in h_1, \gamma_2 \in h_2} \left\{ \frac{\gamma_1 + \gamma_2 - \gamma_1 \gamma_2 - (1-\zeta)\gamma_1\gamma_2}{1 - (1-\zeta)\gamma_1\gamma_2} \right\}$;

(2) $h_1 \otimes_H h_2 = \bigcup_{\gamma_1 \in h_1, \gamma_2 \in h_2} \left\{ \frac{\gamma_1\gamma_2}{\zeta + (1-\zeta)(\gamma_1 + \gamma_2 - \gamma_1\gamma_2)} \right\}$;

(3) $\lambda \cdot_H h = \bigcup_{\gamma \in h} \left\{ \frac{(1+(\zeta-1)\gamma)^\lambda - (1-\gamma)^\lambda}{(1+(\zeta-1)\gamma)^\lambda + (\zeta-1)(1-\gamma)^\lambda} \right\}, \lambda > 0$;

(4) $h^{\wedge_H \lambda} = \bigcup_{\gamma \in h} \left\{ \frac{\zeta \gamma^\lambda}{(1+(\zeta-1)(1-\gamma))^\lambda + (\zeta-1)\gamma^\lambda} \right\}, \lambda > 0$.

In particular, if $\zeta = 1$, then these operations on HFEs reduce to those proposed by Xia and Xu [10]; if $\zeta = 2$, then these operations on HFEs reduce to the following:

(1) $h_1 \oplus_\varepsilon h_2 = \bigcup_{\gamma_1 \in h_1, \gamma_2 \in h_2} \left\{ \frac{\gamma_1 + \gamma_2}{1 - \gamma_1\gamma_2} \right\}$;

(2) $h_1 \otimes_\varepsilon h_2 = \bigcup_{\gamma_1 \in h_1, \gamma_2 \in h_2} \left\{ \frac{\gamma_1\gamma_2}{1 + (1-\gamma_1)(1-\gamma_2)} \right\}$;

(3) $\lambda \cdot_\varepsilon h = \bigcup_{\gamma \in h} \left\{ \frac{(1+\gamma)^\lambda - (1-\gamma)^\lambda}{(1+\gamma)^\lambda + (1-\gamma)^\lambda} \right\}, \lambda > 0$;

(4) $h^{\wedge_\varepsilon \lambda} = \bigcup_{\gamma \in h} \left\{ \frac{2\gamma^\lambda}{(2-\gamma)^\lambda + \gamma^\lambda} \right\}, \lambda > 0$,

which are defined as Einstein operations on HFEs by Yu [16].

Theorem 1. *Let h, h_1 and h_2 be three HFEs, $\lambda > 0$, $\lambda_1 > 0$ and $\lambda_2 > 0$, then*

(1) $h_1 \oplus_H h_2 = h_2 \oplus_H h_1$;

(2) $h \oplus_H (h_1 \oplus_H h_2) = (h \oplus_H h_1) \oplus_H h_2$;

(3) $\lambda_1 \cdot_H (\lambda_2 \cdot_H h) = (\lambda_1\lambda_2) \cdot_H h$;

(4) $\lambda \cdot_H (h_1 \oplus_H h_2) = (\lambda \cdot_H h_1) \oplus_H (\lambda \cdot_H h_2)$;

(5) $h_1 \otimes_H h_2 = h_2 \otimes_H h_1$;

(6) $h \otimes_H (h_1 \otimes_H h_2) = (h \otimes_H h_1) \otimes_H h_2$;

(7) $(h_1 \otimes_H h_2)^{\wedge_H \lambda} = h_1^{\wedge_H \lambda} \otimes_H h_2^{\wedge_H \lambda}$;

(8) $(h^{\wedge_H \lambda_1})^{\wedge_H \lambda_2} = h^{\wedge_H (\lambda_1\lambda_2)}$.

Proof. Since (1), (2), (5) and (6) are trivial, we prove (3), (4), (7) and (8).

(3) Since $\lambda_2 \cdot_H h = \bigcup_{\gamma \in h} \left\{ \frac{(1+(\zeta-1)\gamma)^{\lambda_2} - (1-\gamma)^{\lambda_2}}{(1+(\zeta-1)\gamma)^{\lambda_2} + (\zeta-1)(1-\gamma)^{\lambda_2}} \right\}$, then we have

$$\lambda_1 \cdot_H (\lambda_2 \cdot_H h)$$

$$= \bigcup_{\gamma \in h} \left\{ \frac{\left(1 + (\zeta-1)\frac{(1+(\zeta-1)\gamma)^{\lambda_2} - (1-\gamma)^{\lambda_2}}{(1+(\zeta-1)\gamma)^{\lambda_2} + (\zeta-1)(1-\gamma)^{\lambda_2}}\right)^{\lambda_1} - \left(1 - \frac{(1+(\zeta-1)\gamma)^{\lambda_2} - (1-\gamma)^{\lambda_2}}{(1+(\zeta-1)\gamma)^{\lambda_2} + (\zeta-1)(1-\gamma)^{\lambda_2}}\right)^{\lambda_1}}{\left(1 + (\zeta-1)\frac{(1+(\zeta-1)\gamma)^{\lambda_2} - (1-\gamma)^{\lambda_2}}{(1+(\zeta-1)\gamma)^{\lambda_2} + (\zeta-1)(1-\gamma)^{\lambda_2}}\right)^{\lambda_1} + (\zeta-1)\left(1 - \frac{(1+(\zeta-1)\gamma)^{\lambda_2} - (1-\gamma)^{\lambda_2}}{(1+(\zeta-1)\gamma)^{\lambda_2} + (\zeta-1)(1-\gamma)^{\lambda_2}}\right)^{\lambda_1}} \right\}$$

$$= \bigcup_{\gamma \in h} \left\{ \frac{(1 + (\zeta-1)\gamma)^{(\lambda_1\lambda_2)} - (1-\gamma)^{(\lambda_1\lambda_2)}}{(1 + (\zeta-1)\gamma)^{(\lambda_1\lambda_2)} + (\zeta-1)(1-\gamma)^{(\lambda_1\lambda_2)}} \right\}$$

$$= (\lambda_1\lambda_2) \cdot_H h.$$

(4) Since $h_1 \oplus_H h_2 = \cup_{\gamma_1 \in h_1, \gamma_2 \in h_2} \left\{ \frac{\gamma_1 + \gamma_2 - \gamma_1 \gamma_2 - (1-\zeta)\gamma_1 \gamma_2}{1 - (1-\zeta)\gamma_1 \gamma_2} \right\}$, by the operational law (3) in Definition 4, we have

$$\lambda \cdot_H (h_1 \oplus_H h_2)$$

$$= \cup_{\gamma_1 \in h_1, \gamma_2 \in h_2} \left\{ \frac{(1 + (\zeta - 1)\frac{\gamma_1 + \gamma_2 - \gamma_1 \gamma_2 - (1-\zeta)\gamma_1 \gamma_2}{1 - (1-\zeta)\gamma_1 \gamma_2})^\lambda - (1 - \frac{\gamma_1 + \gamma_2 - \gamma_1 \gamma_2 - (1-\zeta)\gamma_1 \gamma_2}{1 - (1-\zeta)\gamma_1 \gamma_2})^\lambda}{(1 + (\zeta - 1)\frac{\gamma_1 + \gamma_2 - \gamma_1 \gamma_2 - (1-\zeta)\gamma_1 \gamma_2}{1 - (1-\zeta)\gamma_1 \gamma_2})^\lambda + (\zeta - 1)(1 - \frac{\gamma_1 + \gamma_2 - \gamma_1 \gamma_2 - (1-\zeta)\gamma_1 \gamma_2}{1 - (1-\zeta)\gamma_1 \gamma_2})^\lambda} \right\}$$

$$= \cup_{\gamma_1 \in h_1, \gamma_2 \in h_2} \left\{ \frac{((1 + (\zeta - 1)\gamma_1)(1 + (\zeta - 1)\gamma_2))^\lambda - ((1 - \gamma_1)(1 - \gamma_2))^\lambda}{((1 + (\zeta - 1)\gamma_1)(1 + (\zeta - 1)\gamma_2))^\lambda + (\zeta - 1)((1 - \gamma_1)(1 - \gamma_2))^\lambda} \right\}.$$

Since $\lambda \cdot_H h_1 = \cup_{\gamma_1 \in h_1} \left\{ \frac{(1 + (\zeta - 1)\gamma_1)^\lambda - (1 - \gamma_1)^\lambda}{(1 + (\zeta - 1)\gamma_1)^\lambda + (\zeta - 1)(1 - \gamma_1)^\lambda} \right\}$ and $\lambda \cdot_H h_2 = \cup_{\gamma_2 \in h_2} \left\{ \frac{(1 + (\zeta - 1)\gamma_2)^\lambda - (1 - \gamma_2)^\lambda}{(1 + (\zeta - 1)\gamma_2)^\lambda + (\zeta - 1)(1 - \gamma_2)^\lambda} \right\}$, we have

$$(\lambda \cdot_H h_1) \oplus_H (\lambda \cdot_H h_2)$$

$$= \cup_{\gamma_1 \in h_1, \gamma_2 \in h_2} \left\{ \frac{\left[\begin{array}{c} \frac{(1+(\zeta-1)\gamma_1)^\lambda - (1-\gamma_1)^\lambda}{(1+(\zeta-1)\gamma_1)^\lambda + (\zeta-1)(1-\gamma_1)^\lambda} + \frac{(1+(\zeta-1)\gamma_2)^\lambda - (1-\gamma_2)^\lambda}{(1+(\zeta-1)\gamma_2)^\lambda + (\zeta-1)(1-\gamma_2)^\lambda} \\ - \frac{(1+(\zeta-1)\gamma_1)^\lambda - (1-\gamma_1)^\lambda}{(1+(\zeta-1)\gamma_1)^\lambda + (\zeta-1)(1-\gamma_1)^\lambda} \frac{(1+(\zeta-1)\gamma_2)^\lambda - (1-\gamma_2)^\lambda}{(1+(\zeta-1)\gamma_2)^\lambda + (\zeta-1)(1-\gamma_2)^\lambda} \\ -(1-\zeta)\frac{(1+(\zeta-1)\gamma_1)^\lambda - (1-\gamma_1)^\lambda}{(1+(\zeta-1)\gamma_1)^\lambda + (\zeta-1)(1-\gamma_1)^\lambda} \cdot \frac{(1+(\zeta-1)\gamma_2)^\lambda - (1-\gamma_2)^\lambda}{(1+(\zeta-1)\gamma_2)^\lambda + (\zeta-1)(1-\gamma_2)^\lambda} \end{array} \right]}{1 - (1-\zeta)\frac{(1+(\zeta-1)\gamma_1)^\lambda - (1-\gamma_1)^\lambda}{(1+(\zeta-1)\gamma_1)^\lambda + (\zeta-1)(1-\gamma_1)^\lambda} \cdot \frac{(1+(\zeta-1)\gamma_2)^\lambda - (1-\gamma_2)^\lambda}{(1+(\zeta-1)\gamma_2)^\lambda + (\zeta-1)(1-\gamma_2)^\lambda}} \right\}$$

$$= \cup_{\gamma_1 \in h_1, \gamma_2 \in h_2} \left\{ \frac{((1 + (\zeta - 1)\gamma_1)(1 + (\zeta - 1)\gamma_2))^\lambda - ((1 - \gamma_1)(1 - \gamma_2))^\lambda}{((1 + (\zeta - 1)\gamma_1)(1 + (\zeta - 1)\gamma_2))^\lambda + (\zeta - 1)((1 - \gamma_1)(1 - \gamma_2))^\lambda} \right\}.$$

Hence $\lambda \cdot_H (h_1 \oplus_H h_2) = (\lambda \cdot_H h_1) \oplus_H (\lambda \cdot_H h_2)$.

(7) Since $h_1 \otimes_H h_2 = \cup_{\gamma_1 \in h_1, \gamma_2 \in h_2} \left\{ \frac{\gamma_1 \gamma_2}{\zeta + (1-\zeta)(\gamma_1 + \gamma_2 - \gamma_1 \gamma_2)} \right\}$, by the operational law (4) in Definition 4, we have

$$(h_1 \otimes_H h_2)^{\wedge_H \lambda}$$

$$= \cup_{\gamma_1 \in h_1, \gamma_2 \in h_2} \left\{ \frac{\zeta \left(\frac{\gamma_1 \gamma_2}{\zeta + (1-\zeta)(\gamma_1 + \gamma_2 - \gamma_1 \gamma_2)} \right)^\lambda}{\left(1 + (\zeta - 1)(1 - \frac{\gamma_1 \gamma_2}{\zeta + (1-\zeta)(\gamma_1 + \gamma_2 - \gamma_1 \gamma_2)})\right)^\lambda + (\zeta - 1) \left(\frac{\gamma_1 \gamma_2}{\zeta + (1-\zeta)(\gamma_1 + \gamma_2 - \gamma_1 \gamma_2)} \right)^\lambda} \right\}$$

$$= \cup_{\gamma_1 \in h_1, \gamma_2 \in h_2} \left\{ \frac{\zeta \gamma_1^\lambda \gamma_2^\lambda}{((1 + (\zeta - 1)(1 - \gamma_1))(1 + (\zeta - 1)(1 - \gamma_2)))^\lambda + (\zeta - 1)\gamma_1^\lambda \gamma_2^\lambda} \right\}.$$

Since $h_1^{\wedge_H \lambda} = \cup_{\gamma_1 \in h_1} \left\{ \frac{\zeta \gamma_1^\lambda}{(1+(\zeta-1)(1-\gamma_1))^\lambda + (\zeta-1)\gamma_1^\lambda} \right\}$ and $h_2^{\wedge_H \lambda} = \cup_{\gamma_2 \in h_2} \left\{ \frac{\zeta \gamma_2^\lambda}{(1+(\zeta-1)(1-\gamma_2))^\lambda + (\zeta-1)\gamma_2^\lambda} \right\}$,
we have

$$h_1^{\wedge_H \lambda} \otimes_H h_2^{\wedge_H \lambda}$$

$$= \cup_{\gamma_1 \in h_1, \gamma_2 \in h_2} \left\{ \frac{\frac{\zeta \gamma_1^\lambda}{(1+(\zeta-1)(1-\gamma_1))^\lambda + (\zeta-1)\gamma_1^\lambda} \cdot \frac{\zeta \gamma_2^\lambda}{(1+(\zeta-1)(1-\gamma_2))^\lambda + (\zeta-1)\gamma_2^\lambda}}{\left[\zeta + (1-\zeta) \left(\frac{\zeta \gamma_1^\lambda}{(1+(\zeta-1)(1-\gamma_1))^\lambda + (\zeta-1)\gamma_1^\lambda} + \frac{\zeta \gamma_2^\lambda}{(1+(\zeta-1)(1-\gamma_2))^\lambda + (\zeta-1)\gamma_2^\lambda} - \frac{\zeta \gamma_1^\lambda}{(1+(\zeta-1)(1-\gamma_1))^\lambda + (\zeta-1)\gamma_1^\lambda} \cdot \frac{\zeta \gamma_2^\lambda}{(1+(\zeta-1)(1-\gamma_2))^\lambda + (\zeta-1)\gamma_2^\lambda} \right) \right]} \right\}$$

$$= \cup_{\gamma_1 \in h_1, \gamma_2 \in h_2} \left\{ \frac{\zeta \gamma_1^\lambda \gamma_2^\lambda}{((1+(\zeta-1)(1-\gamma_1))(1+(\zeta-1)(1-\gamma_2)))^\lambda + (\zeta-1)\gamma_1^\lambda \gamma_2^\lambda} \right\}.$$

Hence $(h_1 \otimes_H h_2)^{\wedge_H \lambda} = h_1^{\wedge_H \lambda} \otimes_H h_2^{\wedge_H \lambda}$.

(8) Since $h^{\wedge_H \lambda_1} = \cup_{\gamma \in h} \left\{ \frac{\zeta \gamma^{\lambda_1}}{(1+(\zeta-1)(1-\gamma))^{\lambda_1} + (\zeta-1)\gamma^{\lambda_1}} \right\}$, we have

$$(h^{\wedge_H \lambda_1})^{\wedge_H \lambda_2}$$

$$= \cup_{\gamma \in h} \left\{ \frac{\zeta \left(\frac{\zeta \gamma^{\lambda_1}}{(1+(\zeta-1)(1-\gamma))^{\lambda_1} + (\zeta-1)\gamma^{\lambda_1}} \right)^{\lambda_2}}{\left(1 + (\zeta-1)(1 - \frac{\zeta \gamma^{\lambda_1}}{(1+(\zeta-1)(1-\gamma))^{\lambda_1} + (\zeta-1)\gamma^{\lambda_1}}) \right)^{\lambda_2} + (\zeta-1) \left(\frac{\zeta \gamma^{\lambda_1}}{(1+(\zeta-1)(1-\gamma))^{\lambda_1} + (\zeta-1)\gamma^{\lambda_1}} \right)^{\lambda_2}} \right\}$$

$$= \cup_{\gamma \in h} \left\{ \frac{\zeta \gamma^{(\lambda_1 \lambda_2)}}{(1+(\zeta-1)(1-\gamma))^{(\lambda_1 \lambda_2)} + (\zeta-1)\gamma^{(\lambda_1 \lambda_2)}} \right\}$$

$$= h^{\wedge_H (\lambda_1 \lambda_2)}.$$

$\square$

However, the operational laws $(\lambda_1 \cdot_H h) \oplus_H (\lambda_2 \cdot_H h) = (\lambda_1 + \lambda_2) \cdot_H h$ and $h^{\wedge_H \lambda_1} \otimes_H h^{\wedge_H \lambda_2} = h^{\wedge_H (\lambda_1 + \lambda_2)}$ do not hold in general. To illustrate these, we give an example as follows:

Example 1. *Let* $h = \{0.3, 0.5\}$, $\lambda_1 = \lambda_2 = 1$ *and* $\zeta = 3$, *then*

$$(\lambda_1 \cdot_H h) \oplus_H (\lambda_2 \cdot_H h) = h \oplus_H h = \cup_{i,j=1,2} \left\{ \frac{\gamma_i + \gamma_j + \gamma_i \gamma_j}{1 + 2\gamma_i \gamma_j} \right\}$$

$$= \{0.5874, 0.7308, 0.7308, 0.8333\},$$

$$(\lambda_1 + \lambda_2) \cdot_H h = 2 \cdot_H h = \cup_{i=1,2} \left\{ \frac{(1+2\gamma_i)^2 - (1-\gamma_i)^2}{(1+2\gamma_i)^2 + 2(1-\gamma_i)^2} \right\} = \{0.5874, 0.8333\}.$$

From Definition 3, *we have* $s((\lambda_1 \cdot_H h) \oplus_H (\lambda_2 \cdot_H h)) = 0.7199 > 0.7104 = s((\lambda_1 + \lambda_2) \cdot_H h)$ *and thus* $(\lambda_1 \cdot_H h) \oplus_H (\lambda_2 \cdot_H h) \neq (\lambda_1 + \lambda_2) \cdot_H h$. *Furthermore, we have* $s(h^{\wedge_H \lambda_1} \otimes_H h^{\wedge_H \lambda_2}) = 0.0971 < 0.1060 = s(h^{\wedge_H (\lambda_1 + \lambda_2)})$ *and thus* $h^{\wedge_H \lambda_1} \otimes_H h^{\wedge_H \lambda_2} \neq h^{\wedge_H (\lambda_1 + \lambda_2)}$.

3. Hesitant Fuzzy Hamacher Power-Weighted Aggregation Operators

In this section, based on the Hamacher operation, we shall extend the power-aggregation operators to accommodate the situations where the input arguments are HFEs.

3.1. Hesitant Fuzzy Hamacher Power-Weighted Average/geometric Operators

Based on the PA operator [24] and hesitant fuzzy Hamacher weighted average (HFHWA) operator [17], we firstly define the hesitant fuzzy Hamacher power-weighted average (HFHPWA) operator as follows.

Definition 5. *Let h_i $(i = 1, 2, \ldots, n)$ be a collection of HFEs and $w = (w_1, w_2, \ldots, w_n)^T$ be the weight vector of h_i $(i = 1, 2, \ldots, n)$ such that $w_i \in [0, 1]$ and $\sum_{i=1}^{n} w_i = 1$. A hesitant fuzzy Hamacher power-weighted average (HFHPWA) operator is a function $H^n \to H$ such that*

$$\text{HFHPWA}_\zeta(h_1, h_2, \ldots, h_n) = \oplus_{H_{i=1}^{n}} \left(\frac{w_i(1 + T(h_i)) \cdot_H h_i}{\sum_{i=1}^{n} w_i(1 + T(h_i))} \right), \tag{5}$$

where parameter $\zeta > 0$, $T(h_i) = \sum_{j=1, j \neq i}^{n} w_j \text{Sup}(h_i, h_j)$ and $\text{Sup}(h_i, h_j)$ is the support for h_i from h_j, satisfying the following conditions:

(1) $\text{Sup}(h_i, h_j) \in [0, 1]$;

(2) $\text{Sup}(h_i, h_j) = \text{Sup}(h_j, h_i)$;

(3) $\text{Sup}(h_i, h_j) \geq \text{Sup}(h_s, h_t)$ if $d(h_i, h_j) \leq d(h_s, h_t)$, where d is the hesitant normalized Hamming distance measure between two HFEs given in Equation (4).

Here, the support measure (Sup) can be used to measure the closeness of a preference value with other preference value because it is essentially similarity measure.

Theorem 2. *Let h_i $(i = 1, 2, \ldots, n)$ be a collection of HFEs and $w = (w_1, w_2, \ldots, w_n)^T$ be the weight vector of h_i $(i = 1, 2, \ldots, n)$ such that $w_i \in [0, 1]$ and $\sum_{i=1}^{n} w_i = 1$, then the aggregated value by HFHPWA operator is also a HFE, and*

$$\text{HFHPWA}_\zeta(h_1, h_2, \ldots, h_n)$$

$$= \bigcup_{\gamma_1 \in h_1, \gamma_2 \in h_2, \ldots, \gamma_n \in h_n} \left\{ \frac{\prod_{i=1}^{n}(1 + (\zeta - 1)\gamma_i)^{\frac{w_i(1+T(h_i))}{\sum_{i=1}^{n} w_i(1+T(h_i))}} - \prod_{i=1}^{n}(1 - \gamma_i)^{\frac{w_i(1+T(h_i))}{\sum_{i=1}^{n} w_i(1+T(h_i))}}}{\prod_{i=1}^{n}(1 + (\zeta - 1)\gamma_i)^{\frac{w_i(1+T(h_i))}{\sum_{i=1}^{n} w_i(1+T(h_i))}} + (\zeta - 1)\prod_{i=1}^{n}(1 - \gamma_i)^{\frac{w_i(1+T(h_i))}{\sum_{i=1}^{n} w_i(1+T(h_i))}}} \right\}. \tag{6}$$

Proof. Equation (6) can be proved by mathematical induction on n as follows.

For $n = 1$, the result of Equation (6) is clear.

Suppose that Equation (6) holds for $n = k$, that is

$$\text{HFHPWA}_\zeta(h_1, h_2, \ldots, h_k)$$

$$= \bigcup_{\gamma_1 \in h_1, \gamma_2 \in h_2, \ldots, \gamma_k \in h_k} \left\{ \frac{\prod_{i=1}^{k}(1 + (\zeta - 1)\gamma_i)^{\frac{w_i(1+T(h_i))}{\sum_{i=1}^{k} w_i(1+T(h_i))}} - \prod_{i=1}^{k}(1 - \gamma_i)^{\frac{w_i(1+T(h_i))}{\sum_{i=1}^{k} w_i(1+T(h_i))}}}{\prod_{i=1}^{k}(1 + (\zeta - 1)\gamma_i)^{\frac{w_i(1+T(h_i))}{\sum_{i=1}^{k} w_i(1+T(h_i))}} + (\zeta - 1)\prod_{i=1}^{k}(1 - \gamma_i)^{\frac{w_i(1+T(h_i))}{\sum_{i=1}^{k} w_i(1+T(h_i))}}} \right\}.$$

Then, when $n = k + 1$, by Definitions 4 and 5, we have

$$\text{HFHPWA}_\zeta(h_1, h_2, \ldots, h_{k+1}) = \oplus_{H\,i=1}^{k} \left(\frac{w_i(1 + T(h_i)) \cdot_H h_i}{\sum_{i=1}^{k+1} w_i(1 + T(h_i))} \right) \oplus_H \left(\frac{w_{k+1}(1 + T(h_{k+1})) \cdot_H h_{k+1}}{\sum_{i=1}^{k+1} w_i(1 + T(h_i))} \right)$$

$$= \cup_{\gamma_1 \in h_1, \gamma_2 \in h_2, \ldots, \gamma_k \in h_k} \left\{ \frac{\prod_{i=1}^{k} (1 + (\zeta - 1)\gamma_i)^{\frac{w_i(1+T(h_i))}{\sum_{i=1}^{k+1} w_i(1+T(h_i))}} - \prod_{i=1}^{k} (1 - \gamma_i)^{\frac{w_i(1+T(h_i))}{\sum_{i=1}^{k+1} w_i(1+T(h_i))}}}{\prod_{i=1}^{k} (1 + (\zeta - 1)\gamma_i)^{\frac{w_i(1+T(h_i))}{\sum_{i=1}^{k+1} w_i(1+T(h_i))}} + (\zeta - 1)\prod_{i=1}^{k} (1 - \gamma_i)^{\frac{w_i(1+T(h_i))}{\sum_{i=1}^{k+1} w_i(1+T(h_i))}}} \right\}$$

$$\oplus_H \cup_{\gamma_{k+1} \in h_{k+1}} \left\{ \frac{(1 + (\zeta - 1)\gamma_{k+1})^{\frac{w_{k+1}(1+T(h_{k+1}))}{\sum_{i=1}^{k+1} w_i(1+T(h_i))}} - (1 - \gamma_{k+1})^{\frac{w_{k+1}(1+T(h_{k+1}))}{\sum_{i=1}^{k+1} w_i(1+T(h_i))}}}{(1 + (\zeta - 1)\gamma_{k+1})^{\frac{w_{k+1}(1+T(h_{k+1}))}{\sum_{i=1}^{k+1} w_i(1+T(h_i))}} + (\zeta - 1)(1 - \gamma_{k+1})^{\frac{w_{k+1}(1+T(h_{k+1}))}{\sum_{i=1}^{k+1} w_i(1+T(h_i))}}} \right\}.$$

Let $a_1 = \prod_{i=1}^{k} (1 + (\zeta - 1)\gamma_i)^{\frac{w_i(1+T(h_i))}{\sum_{i=1}^{k+1} w_i(1+T(h_i))}}$, $b_1 = \prod_{i=1}^{k} (1 - \gamma_i)^{\frac{w_i(1+T(h_i))}{\sum_{i=1}^{k+1} w_i(1+T(h_i))}}$, $a_2 = (1 + (\zeta - 1)\gamma_{k+1})^{\frac{w_{k+1}(1+T(h_{k+1}))}{\sum_{i=1}^{k+1} w_i(1+T(h_i))}}$ and $b_2 = (1 - \gamma_{k+1})^{\frac{w_{k+1}(1+T(h_{k+1}))}{\sum_{i=1}^{k+1} w_i(1+T(h_i))}}$, then

$$\text{HFHPWA}_\zeta(h_1, h_2, \ldots, h_{k+1})$$

$$= \cup_{\gamma_1 \in h_1, \gamma_2 \in h_2, \ldots, \gamma_k \in h_k} \left\{ \frac{a_1 - b_1}{a_1 + (\zeta - 1)b_1} \right\} \oplus_H \cup_{\gamma_{k+1} \in h_{k+1}} \left\{ \frac{a_2 - b_2}{a_2 + (\zeta - 1)b_2} \right\}$$

$$= \cup_{\gamma_1 \in h_1, \ldots, \gamma_k \in h_k, \gamma_{k+1} \in h_{k+1}} \left\{ \frac{\left[\frac{a_1 - b_1}{a_1 + (\zeta - 1)b_1} + \frac{a_2 - b_2}{a_2 + (\zeta - 1)b_2} - \frac{a_1 - b_1}{a_1 + (\zeta - 1)b_1} \cdot \frac{a_2 - b_2}{a_2 + (\zeta - 1)b_2} \right]}{1 - (1 - \zeta)\frac{a_1 - b_1}{a_1 + (\zeta - 1)b_1} \cdot \frac{a_2 - b_2}{a_2 + (\zeta - 1)b_2}} \right\}$$

$$= \cup_{\gamma_1 \in h_1, \gamma_2 \in h_2, \ldots, \gamma_{k+1} \in h_{k+1}} \left\{ \frac{a_1 a_2 - b_1 b_2}{a_1 a_2 + (\zeta - 1)b_1 b_2} \right\}$$

$$= \cup_{\gamma_1 \in h_1, \gamma_2 \in h_2, \ldots, \gamma_{k+1} \in h_{k+1}} \left\{ \frac{\prod_{i=1}^{k+1} (1 + (\zeta - 1)\gamma_i)^{\frac{w_i(1+T(h_i))}{\sum_{i=1}^{k+1} w_i(1+T(h_i))}} - \prod_{i=1}^{k+1} (1 - \gamma_i)^{\frac{w_i(1+T(h_i))}{\sum_{i=1}^{k+1} w_i(1+T(h_i))}}}{\prod_{i=1}^{k+1} (1 + (\zeta - 1)\gamma_i)^{\frac{w_i(1+T(h_i))}{\sum_{i=1}^{k+1} w_i(1+T(h_i))}} + (\zeta - 1)\prod_{i=1}^{k+1} (1 - \gamma_i)^{\frac{w_i(1+T(h_i))}{\sum_{i=1}^{k+1} w_i(1+T(h_i))}}} \right\},$$

i.e., Equation (6) holds for $n = k + 1$. Thus, Equation (6) holds for all n. $\square$

Remark 1. *(1) If* $\text{Sup}(h_i, h_j) = k$, *for all* $i \neq j$, *then*

$$\text{HFHPWA}_\zeta(h_1, h_2, \ldots, h_n) = \oplus_{H\,i=1}^{n} (w_i \cdot_H h_i)$$

$$= \cup_{\gamma_1 \in h_1, \gamma_2 \in h_2, \ldots, \gamma_n \in h_n} \left\{ \frac{\prod_{i=1}^{n} (1 + (\zeta - 1)\gamma_i)^{w_i} - \prod_{i=1}^{n} (1 - \gamma_i)^{w_i}}{\prod_{i=1}^{n} (1 + (\zeta - 1)\gamma_i)^{w_i} + (\zeta - 1)\prod_{i=1}^{n} (1 - \gamma_i)^{w_i}} \right\}, \tag{7}$$

which indicates that when all supports are the same, the HFHPWA operator reduces to the hesitant fuzzy Hamacher weighted average (HFHWA) operator [17].

(2) For the HFHPWA operator, if $\zeta = 1$, *then the HFHPWA operator reduces to the following:*

$$\text{HFHPWA}_1(h_1, h_2, \ldots, h_n) = \cup_{\gamma_1 \in h_1, \gamma_2 \in h_2, \ldots, \gamma_n \in h_n} \left\{ 1 - \prod_{i=1}^{n} (1 - \gamma_i)^{\frac{w_i(1+T(h_i))}{\sum_{i=1}^{n} w_i(1+T(h_i))}} \right\} \tag{8}$$

which is called the hesitant fuzzy power-weighted average (HFPWA) operator and if $\zeta = 2$, then the HFHPWA operator reduces to the hesitant fuzzy Einstein power-weighted average (HFEPWA) operator [18]:

$$\text{HFHPWA}_2(h_1, h_2, \ldots, h_n)$$
$$= \bigcup_{\gamma_1 \in h_1, \gamma_2 \in h_2, \ldots, \gamma_n \in h_n} \left\{ \frac{\prod_{i=1}^n (1 + \gamma_i)^{\frac{w_i(1+T(h_i))}{\sum_{i=1}^n w_i(1+T(h_i))}} - \prod_{i=1}^n (1 - \gamma_i)^{\frac{w_i(1+T(h_i))}{\sum_{i=1}^n w_i(1+T(h_i))}}}{\prod_{i=1}^n (1 + \gamma_i)^{\frac{w_i(1+T(h_i))}{\sum_{i=1}^n w_i(1+T(h_i))}} + \prod_{i=1}^n (1 - \gamma_i)^{\frac{w_i(1+T(h_i))}{\sum_{i=1}^n w_i(1+T(h_i))}}} \right\}. \tag{9}$$

To analyze the relationship between the HFHPWA operator and the HFPWA operator, we introduce the following lemma.

Lemma 1. *[47,48] Let $x_i > 0$, $w_i > 0$, $i = 1, 2, \ldots, n$, and $\sum_{i=1}^n w_i = 1$, then $\prod_{i=1}^n x_i^{w_i} \leq \sum_{i=1}^n w_i x_i$, with equality if and only if $x_1 = x_2 = \cdots = x_n$.*

Theorem 3. *Let h_i $(i = 1, 2, \ldots, n)$ be a collection of HFEs and $w = (w_1, w_2, \ldots, w_n)^T$ be the weight vector of h_i such that $w_i \in [0, 1]$ and $\sum_{i=1}^n w_i = 1$, then*

$$\text{HFHPWA}_\zeta(h_1, h_2, \ldots, h_n) \leq \text{HFPWA}(h_1, h_2, \ldots, h_n).$$

Proof. For any $\gamma_i \in h_i$ $(i = 1, 2, \ldots, n)$, by Lemma 1, we have

$$\prod_{i=1}^n (1 + (\zeta - 1)\gamma_i)^{\frac{w_i(1+T(h_i))}{\sum_{i=1}^n w_i(1+T(h_i))}} + (\zeta - 1)\prod_{i=1}^n (1 - \gamma_i)^{\frac{w_i(1+T(h_i))}{\sum_{i=1}^n w_i(1+T(h_i))}}$$
$$\leq \sum_{i=1}^n \frac{w_i(1 + T(h_i))}{\sum_{i=1}^n w_i(1 + T(h_i))} (1 + (\zeta - 1)\gamma_i) + (\zeta - 1)\sum_{i=1}^n \frac{w_i(1 + T(h_i))}{\sum_{i=1}^n w_i(1 + T(h_i))} (1 - \gamma_i) = \zeta.$$

Then,

$$\frac{\prod_{i=1}^n (1 + (\zeta - 1)\gamma_i)^{\frac{w_i(1+T(h_i))}{\sum_{i=1}^n w_i(1+T(h_i))}} - \prod_{i=1}^n (1 - \gamma_i)^{\frac{w_i(1+T(h_i))}{\sum_{i=1}^n w_i(1+T(h_i))}}}{\prod_{i=1}^n (1 + (\zeta - 1)\gamma_i)^{\frac{w_i(1+T(h_i))}{\sum_{i=1}^n w_i(1+T(h_i))}} + (\zeta - 1)\prod_{i=1}^n (1 - \gamma_i)^{\frac{w_i(1+T(h_i))}{\sum_{i=1}^n w_i(1+T(h_i))}}}$$
$$= 1 - \frac{\zeta \prod_{i=1}^n (1 - \gamma_i)^{\frac{w_i(1+T(h_i))}{\sum_{i=1}^n w_i(1+T(h_i))}}}{\prod_{i=1}^n (1 + (\zeta - 1)\gamma_i)^{\frac{w_i(1+T(h_i))}{\sum_{i=1}^n w_i(1+T(h_i))}} + (\zeta - 1)\prod_{i=1}^n (1 - \gamma_i)^{\frac{w_i(1+T(h_i))}{\sum_{i=1}^n w_i(1+T(h_i))}}}$$
$$\leq 1 - \frac{\zeta \prod_{i=1}^n (1 - \gamma_i)^{\frac{w_i(1+T(h_i))}{\sum_{i=1}^n w_i(1+T(h_i))}}}{\zeta} = 1 - \prod_{i=1}^n (1 - \gamma_i)^{\frac{w_i(1+T(h_i))}{\sum_{i=1}^n w_i(1+T(h_i))}},$$

which implies that $\oplus_H{}_{i=1}^n \left(\frac{w_i(1+T(h_i)) \cdot_H h_i}{\sum_{i=1}^n w_i(1+T(h_i))} \right) \leq \oplus_{i=1}^n \left(\frac{w_i(1+T(h_i))h_i}{\sum_{i=1}^n w_i(1+T(h_i))} \right).$ Thus, we obtain $\text{HFHPWA}_\zeta(h_1, h_2, \ldots, h_n) \leq \text{HFPWA}(h_1, h_2, \ldots, h_n).$ $\square$

Theorem 3 shows that the values aggregated by the HFHPWA operator are not larger than those obtained by the HFPWA operator. That is to say, the HFHPWA operator reflects the decision-maker's pessimistic attitude than the HFPWA operator in aggregation process. Furthermore, based on Theorem 2, we have the properties of the HFHPWA operator as follows.

Theorem 4. *Let h_i $(i = 1, 2, \ldots, n)$ be a collection of HFEs and $w = (w_1, w_2, \ldots, w_n)^T$ be the weight vector of h_i such that $w_i \in [0, 1]$ and $\sum_{i=1}^n w_i = 1$, then we have the followings:*

(1) Boundedness: If $h^- = \min\{\gamma_i | \gamma_i \in h_i\}$ and $h^+ = \max\{\gamma_i | \gamma_i \in h_i\}$, then

$$h^- \leq \text{HFHPWA}_\zeta(h_1, h_2, \ldots, h_n) \leq h^+.$$

(2) *Monotonicity:* Let h'_i $(i = 1, 2, \ldots, n)$ be a collection of HFEs, if $w = (w_1, w_2, \ldots, w_n)^T$ is also the weight vector of h'_i, and $\gamma_i \leq \gamma'_i$ for any h_i and h'_i $(i = 1, 2, \ldots, n)$, then

$$\mathrm{HFHPWA}_\zeta(h_1, h_2, \ldots, h_n) \leq \mathrm{HFHPWA}_\zeta(h'_1, h'_2, \ldots, h'_n).$$

Proof. (1) Let $f(x) = \frac{1+(\zeta-1)x}{1-x}$, $x \in [0,1)$, then $f'(x) = \frac{\zeta}{(1-x)^2} > 0$ and thus $f(x)$ is an increasing function. Since $h^- \leq \gamma_i \leq h^+$ for all i, then $f(h^-) \leq f(\gamma_i) \leq f(h^+)$, i.e., $\frac{1+(\zeta-1)h^-}{1-h^-} \leq \frac{1+(\zeta-1)\gamma_i}{1-\gamma_i} \leq \frac{1+(\zeta-1)h^+}{1-h^+}$. For convenience, let $t_i = \frac{w_i(1+T(h_i))}{\sum_{i=1}^n w_i(1+T(h_i))}$. Since $w = (w_1, w_2, \ldots, w_n)^T$ is the weight vector of h_i satisfying $w_i \in [0,1]$ and $\sum_{i=1}^n w_i = 1$, then for all i, we have

$$\left(\frac{1+(\zeta-1)h^-}{1-h^-}\right)^{t_i} \leq \left(\frac{1+(\zeta-1)\gamma_i}{1-\gamma_i}\right)^{t_i} \leq \left(\frac{1+(\zeta-1)h^+}{1-h^+}\right)^{t_i}$$

$$\Leftrightarrow \left(1+\frac{\zeta h^-}{1-h^-}\right)^{t_i} \leq \left(\frac{1+(\zeta-1)\gamma_i}{1-\gamma_i}\right)^{t_i} \leq \left(1+\frac{\zeta h^+}{1-h^+}\right)^{t_i}$$

$$\Leftrightarrow \prod_{i=1}^n \left(1+\frac{\zeta h^-}{1-h^-}\right)^{t_i} \leq \prod_{i=1}^n \left(\frac{1+(\zeta-1)\gamma_i}{1-\gamma_i}\right)^{t_i} \leq \prod_{i=1}^n \left(1+\frac{\zeta h^+}{1-h^+}\right)^{t_i}$$

$$\Leftrightarrow 1+\frac{\zeta h^-}{1-h^-} \leq \prod_{i=1}^n \left(\frac{1+(\zeta-1)\gamma_i}{1-\gamma_i}\right)^{t_i} \leq 1+\frac{\zeta h^+}{1-h^+}$$

$$\Leftrightarrow \zeta+\frac{\zeta h^-}{1-h^-} \leq \prod_{i=1}^n \left(\frac{1+(\zeta-1)\gamma_i}{1-\gamma_i}\right)^{t_i} + (\zeta-1) \leq \zeta+\frac{\zeta h^+}{1-h^+}$$

$$\Leftrightarrow \frac{1}{\zeta+\frac{\zeta h^+}{1-h^+}} \leq \frac{1}{\prod_{i=1}^n \left(\frac{1+(\zeta-1)\gamma_i}{1-\gamma_i}\right)^{t_i} + (\zeta-1)} \leq \frac{1}{\zeta+\frac{\zeta h^-}{1-h^-}}$$

$$\Leftrightarrow \frac{1-h^+}{\zeta} \leq \frac{\prod_{i=1}^n (1-\gamma_i)^{t_i}}{\prod_{i=1}^n (1+(\zeta-1)\gamma_i)^{t_i} + (\zeta-1)\prod_{i=1}^n (1-\gamma_i)^{t_i}} \leq \frac{1-h^-}{\zeta}$$

$$\Leftrightarrow 1-h^+ \leq \frac{\zeta\prod_{i=1}^n (1-\gamma_i)^{t_i}}{\prod_{i=1}^n (1+(\zeta-1)\gamma_i)^{t_i} + (\zeta-1)\prod_{i=1}^n (1-\gamma_i)^{t_i}} \leq 1-h^-$$

$$\Leftrightarrow h^- \leq 1 - \frac{\zeta\prod_{i=1}^n (1-\gamma_i)^{t_i}}{\prod_{i=1}^n (1+(\zeta-1)\gamma_i)^{t_i} + (\zeta-1)\prod_{i=1}^n (1-\gamma_i)^{t_i}} \leq h^+$$

$$\Leftrightarrow h^- \leq \frac{\prod_{i=1}^n (1+(\zeta-1)\gamma_i)^{t_i} - \prod_{i=1}^n (1-\gamma_i)^{t_i}}{\prod_{i=1}^n (1+(\zeta-1)\gamma_i)^{t_i} + (\zeta-1)\prod_{i=1}^n (1-\gamma_i)^{t_i}} \leq h^+.$$

Thus, we have $h^- \leq \mathrm{HFHPWA}_\zeta(h_1, h_2, \ldots, h_n) \leq h^+$.

(2) Let $f(x) = \frac{1+(\zeta-1)x}{1-x}$, $x \in [0,1)$, then by (1), $f(x)$ is an increasing function. If for all h_i and h_i', $\gamma_i \leq \gamma_i'$, then $\frac{1+(\zeta-1)\gamma_i}{1-\gamma_i} \leq \frac{1+(\zeta-1)\gamma_i'}{1-\gamma_i'}$. For convenience, let $t_i = \frac{w_i(1+T(h_i))}{\sum_{i=1}^n w_i(1+T(h_i))}$, then we have

$$\left(\frac{1+(\zeta-1)\gamma_i}{1-\gamma_i}\right)^{t_i} \leq \left(\frac{1+(\zeta-1)\gamma_i'}{1-\gamma_i'}\right)^{t_i}$$

$$\Leftrightarrow \prod_{i=1}^n \left(\frac{1+(\zeta-1)\gamma_i}{1-\gamma_i}\right)^{t_i} + (\zeta-1) \leq \prod_{i=1}^n \left(\frac{1+(\zeta-1)\gamma_i'}{1-\gamma_i'}\right)^{t_i} + (\zeta-1)$$

$$\Leftrightarrow \frac{1}{\prod_{i=1}^n \left(\frac{1+(\zeta-1)\gamma_i}{1-\gamma_i}\right)^{t_i} + (\zeta-1)} \geq \frac{1}{\prod_{i=1}^n \left(\frac{1+(\zeta-1)\gamma_i'}{1-\gamma_i'}\right)^{t_i} + (\zeta-1)}$$

$$\Leftrightarrow \frac{\zeta \prod_{i=1}^n (1-\gamma_i)^{t_i}}{\prod_{i=1}^n (1+(\zeta-1)\gamma_i)^{t_i} + (\zeta-1)\prod_{i=1}^n (1-\gamma_i)^{t_i}} \geq \frac{\zeta \prod_{i=1}^n (1-\gamma_i')^{t_i}}{\prod_{i=1}^n (1+(\zeta-1)\gamma_i')^{t_i} + (\zeta-1)\prod_{i=1}^n (1-\gamma_i')^{t_i}}$$

$$\Leftrightarrow 1 - \frac{\zeta \prod_{i=1}^n (1-\gamma_i)^{t_i}}{\left[\begin{array}{c}\prod_{i=1}^n (1+(\zeta-1)\gamma_i)^{t_i} \\ +(\zeta-1)\prod_{i=1}^n (1-\gamma_i)^{t_i}\end{array}\right]} \leq 1 - \frac{\zeta \prod_{i=1}^n (1-\gamma_i')^{t_i}}{\left[\begin{array}{c}\prod_{i=1}^n (1+(\zeta-1)\gamma_i')^{t_i} \\ +(\zeta-1)\prod_{i=1}^n (1-\gamma_i')^{t_i}\end{array}\right]}$$

$$\Leftrightarrow \frac{\prod_{i=1}^n (1+(\zeta-1)\gamma_i)^{t_i} - \prod_{i=1}^n (1-\gamma_i)^{t_i}}{\left[\begin{array}{c}\prod_{i=1}^n (1+(\zeta-1)\gamma_i)^{t_i} \\ +(\zeta-1)\prod_{i=1}^n (1-\gamma_i)^{t_i}\end{array}\right]} \leq \frac{\prod_{i=1}^n (1+(\zeta-1)\gamma_i')^{t_i} - \prod_{i=1}^n (1-\gamma_i')^{t_i}}{\left[\begin{array}{c}\prod_{i=1}^n (1+(\zeta-1)\gamma_i')^{t_i} \\ +(\zeta-1)\prod_{i=1}^n (1-\gamma_i')^{t_i}\end{array}\right]}.$$

Thus, by Theorem 2, $\text{HFHPWA}_\zeta(h_1, h_2, \ldots, h_n) \leq \text{HFHPWA}_\zeta(h_1', h_2', \ldots, h_n')$. $\square$

However, the HFHPWA operator is neither idempotent nor commutative, as illustrated by the following example.

Example 2. *Let $h_1 = \{0.8, 0.6\}$, $h_2 = \{0.9, 0.5\}$ and $h_3 = \{0.7, 0.6\}$ be three HFEs, $w = (0.3, 0.5, 0.2)^T$ be the weight vector of h_1, h_2 and h_3. Assume that $\text{Sup}(h_i, h_j)$ $(i, j = 1, 2, 3, i \neq j)$ is the support for h_i from h_j given by $\text{Sup}(h_i, h_j) = 1 - d(h_i, h_j)$, where $d(h_i, h_j)$ is the hesitant Hamming distance between h_i and h_j. Then by Theorem 2, we have*

$$\text{HFHPWA}_5(h_1, h_2, h_3) = \{0.8388, 0.6555, 0.7942, 0.5797, 0.8261, 0.6332, 0.7786, 0.5549\},$$

$$\text{HFHPWA}_5(h_2, h_3, h_1) = \{0, 8013, 0.7683, 0.6723, 0.6255, 0.7650, 0.7275, 0.6208, 0.5701\},$$

$$\text{HFHPWA}_5(h_3, h_3, h_3) = \{0.7000, 0.6559, 0.6704, 0.6237, 0.6793, 0.6334, 0.6485, 0.6000\}.$$

From Definition 3, we have $s(h_1) = s(h_2) = 0, 7$, $s(h_3) = 0.65$, $s(\text{HFHPWA}_5(h_1, h_2, h_3)) = 0.7076$, $s(\text{HFHPWA}_5(h_2, h_3, h_1)) = 0.6938$ and $s(\text{HFHPWA}_5(h_3, h_3, h_3)) = 0.6514$. Then $s(\text{HFHPWA}_5(h_3, h_3, h_3)) \neq s(h_3)$ and thus $\text{HFHPWA}_5(h_3, h_3, h_3) \neq h_3$, which implies that the HFHPWA operator is not idempotent. Furthermore, since $s(\text{HFHPWA}_5(h_1, h_2, h_3)) \neq s(\text{HFHPWA}_5(h_2, h_3, h_1))$, we have $\text{HFHPWA}_5(h_1, h_2, h_3) \neq \text{HFHPWA}_5(h_2, h_3, h_1)$. Thus, the HFHPWA operator is not commutative.

Based on the power geometric (PG) operator [25] and hesitant fuzzy Hamacher weighted geometric (HFHWG) operator [17], we also define the hesitant fuzzy Hamacher power-weighted geometric operator as follows.

Definition 6. *Let h_i $(i = 1, 2, \ldots, n)$ be a collection of HFEs and $w = (w_1, w_2, \ldots, w_n)^T$ be the weight vector of h_i $(i = 1, 2, \ldots, n)$ such that $w_i \in [0, 1]$ and $\sum_{i=1}^n w_i = 1$. A hesitant fuzzy Hamacher power-weighted geometric (HFHPWG) operator is a function $H^n \to H$ such that*

$$\text{HFHPWG}_\zeta(h_1, h_2, \ldots, h_n) = \otimes_{H i=1}^n \left(h_i^{\wedge_H \frac{w_i(1+T(h_i))}{\sum_{i=1}^n w_i(1+T(h_i))}}\right), \tag{10}$$

where parameter $\zeta > 0$, $T(h_i) = \sum_{j=1, j\neq i}^{n} w_j \mathrm{Sup}(h_i, h_j)$ and $\mathrm{Sup}(h_i, h_j)$ is the support for h_i from h_j.

Theorem 5. *Let h_i $(i = 1, 2, \ldots, n)$ be a collection of HFEs and $w = (w_1, w_2, \ldots, w_n)^T$ be the weight vector of h_i $(i = 1, 2, \ldots, n)$ such that $w_i \in [0,1]$ and $\sum_{i=1}^{n} w_i = 1$, then the aggregated value by HFHPWG operator is also a HFE, and*

$$\mathrm{HFHPWG}_{\zeta}(h_1, h_2, \ldots, h_n) = \cup_{\gamma_1 \in h_1, \gamma_2 \in h_2, \ldots, \gamma_n \in h_n} \left\{ \frac{\zeta \prod_{i=1}^{n} (\gamma_i)^{\frac{w_i(1+T(h_i))}{\sum_{i=1}^{n} w_i(1+T(h_i))}}}{\left[\begin{array}{c} \prod_{i=1}^{n} (1 + (\zeta-1)(1-\gamma_i))^{\frac{w_i(1+T(h_i))}{\sum_{i=1}^{n} w_i(1+T(h_i))}} \\ + (\zeta-1) \prod_{i=1}^{n} (\gamma_i)^{\frac{w_i(1+T(h_i))}{\sum_{i=1}^{n} w_i(1+T(h_i))}} \end{array} \right]} \right\}. \tag{11}$$

Proof. Similar to the proof of Theorem 2, Equation (11) can be proved by mathematical induction on n. $\square$

Remark 2. *(1) If $\mathrm{Sup}(h_i, h_j) = k$, for all $i \neq j$, then*

$$\mathrm{HFHPWG}_{\zeta}(h_1, h_2, \ldots, h_n) = \otimes_{H \, i=1}^{n} \left(h_i^{\wedge_H w_i} \right) \tag{12}$$

which indicates that when all supports are the same, the HFHPWG operator reduces to the hesitant fuzzy Hamacher weighted geometric (HFHWG) operator [17].

(2) If $\zeta = 1$, then then the HFHPWG operator reduces to the following:

$$\mathrm{HFHPWG}_1(h_1, h_2, \ldots, h_n) = \cup_{\gamma_1 \in h_1, \gamma_2 \in h_2, \ldots, \gamma_n \in h_n} \left\{ \prod_{i=1}^{n} (\gamma_i)^{\frac{w_i(1+T(h_i))}{\sum_{i=1}^{n} w_i(1+T(h_i))}} \right\} \tag{13}$$

which is called the hesitant fuzzy power-weighted geometric (HFPWG) operator and if $\zeta = 2$, then the HFHPWG operator reduces to the hesitant fuzzy Einstein power-weighted geometric (HFEPWG) operator [18]:

$$\mathrm{HFHPWG}_2(h_1, h_2, \ldots, h_n)$$
$$= \cup_{\gamma_1 \in h_1, \gamma_2 \in h_2, \ldots, \gamma_n \in h_n} \left\{ \frac{2 \prod_{i=1}^{n} (\gamma_i)^{\frac{w_i(1+T(h_i))}{\sum_{i=1}^{n} w_i(1+T(h_i))}}}{\prod_{i=1}^{n} (2 - \gamma_i)^{\frac{w_i(1+T(h_i))}{\sum_{i=1}^{n} w_i(1+T(h_i))}} + \prod_{i=1}^{n} (\gamma_i)^{\frac{w_i(1+T(h_i))}{\sum_{i=1}^{n} w_i(1+T(h_i))}}} \right\}. \tag{14}$$

Theorem 6. *Let h_i $(i = 1, 2, \ldots, n)$ be a collection of HFEs and $w = (w_1, w_2, \ldots, w_n)^T$ be the weight vector of h_i such that $w_i \in [0,1]$ and $\sum_{i=1}^{n} w_i = 1$, then*

$$\mathrm{HFHPWG}_{\zeta}(h_1, h_2, \ldots, h_n) \geq \mathrm{HFPWG}(h_1, h_2, \ldots, h_n).$$

Proof. For any $\gamma_i \in h_i$ $(i = 1, 2, \ldots, n)$, by Lemma 1, we have

$$\prod_{i=1}^{n} (1 + (\zeta-1)(1-\gamma_i))^{\frac{w_i(1+T(h_i))}{\sum_{i=1}^{n} w_i(1+T(h_i))}} + (\zeta-1) \prod_{i=1}^{n} (\gamma_i)^{\frac{w_i(1+T(h_i))}{\sum_{i=1}^{n} w_i(1+T(h_i))}}$$
$$\leq \sum_{i=1}^{n} \frac{w_i(1 + T(h_i))}{\sum_{i=1}^{n} w_i(1 + T(h_i))} (1 + (\zeta-1)(1-\gamma_i)) + (\zeta-1) \sum_{i=1}^{n} \frac{w_i(1 + T(h_i))}{\sum_{i=1}^{n} w_i(1 + T(h_i))} \gamma_i = \zeta.$$

Then,

$$\frac{\zeta \prod_{i=1}^{n} (\gamma_i)^{\frac{w_i(1+T(h_i))}{\sum_{i=1}^{n} w_i(1+T(h_i))}}}{\prod_{i=1}^{n} (1 + (\zeta-1)(1-\gamma_i))^{\frac{w_i(1+T(h_i))}{\sum_{i=1}^{n} w_i(1+T(h_i))}} + (\zeta-1) \prod_{i=1}^{n} (\gamma_i)^{\frac{w_i(1+T(h_i))}{\sum_{i=1}^{n} w_i(1+T(h_i))}}} \geq \prod_{i=1}^{n} (\gamma_i)^{\frac{w_i(1+T(h_i))}{\sum_{i=1}^{n} w_i(1+T(h_i))}},$$

which implies that $\otimes_{H\,i=1}^{n}\left(h_i^{\wedge_H\frac{w_i(1+T(h_i))}{\sum_{i=1}^{n}w_i(1+T(h_i))}}\right) \geq \otimes_{i=1}^{n}\left(h_i^{\frac{w_i(1+T(h_i))h_i}{\sum_{i=1}^{n}w_i(1+T(h_i))}}\right)$, i.e.,

$\mathrm{HFHPWG}_\zeta(h_1,h_2,\ldots,h_n) \geq \mathrm{HFPWG}(h_1,h_2,\ldots,h_n)$. $\square$

Theorem 6 shows that the HFHPWG operator reflects the decision-maker's more optimistic attitude than the HFPWG operator in aggregation process. Furthermore, similar to Theorem 4, we have the properties of the HFHPWG operator as follows.

Theorem 7. *bLet h_i $(i = 1,2,\ldots,n)$ be a collection of HFEs and $w = (w_1,w_2,\ldots,w_n)^T$ be the weight vector of h_i such that $w_i \in [0,1]$ and $\sum_{i=1}^{n}w_i = 1$, then we have the followings:*

(1) Boundedness: If $h^- = \min\{\gamma_i|\gamma_i \in h_i\}$ and $h^+ = \max\{\gamma_i|\gamma_i \in h_i\}$, then

$$h^- \leq \mathrm{HFHPWG}_\zeta(h_1,h_2,\ldots,h_n) \leq h^+.$$

(2) Monotonicity: Let h'_i $(i = 1,2,\ldots,n)$ be a collection of HFEs, if $w = (w_1,w_2,\ldots,w_n)^T$ is also the weight vector of h'_i, and $\gamma_i \leq \gamma'_i$ for any h_i and h'_i $(i = 1,2,\ldots,n)$, then

$$\mathrm{HFHPWG}_\zeta(h_1,h_2,\ldots,h_n) \leq \mathrm{HFHPWG}_\zeta(h'_1,h'_2,\ldots,h'_n).$$

Proof. (1) Let $g(x) = \frac{1+(\zeta-1)(1-x)}{x}$, $x \in (0,1]$, then $g'(x) = \frac{-\zeta}{x^2} < 0$, thus $g(x)$ is a decreasing function. Since $h^- \leq \gamma_i \leq h^+$ for all i, then $g(h^-) \geq g(\gamma_i) \geq g(h^+)$, i.e., $\frac{1+(\zeta-1)(1-h^+)}{h^+} \leq \frac{1+(\zeta-1)(1-\gamma_i)}{\gamma_i} \leq \frac{1+(\zeta-1)(1-h^-)}{h^-}$. Since $w = (w_1,w_2,\ldots,w_n)^T$ is the weight vector of h_i satisfying $w_i \in [0,1]$ and $\sum_{i=1}^{n}w_i = 1$, then for all i, let $t_i = \frac{w_i(1+T(h_i))}{\sum_{i=1}^{n}w_i(1+T(h_i))}$, we have

$$\left(\frac{1+(\zeta-1)(1-h^+)}{h^+}\right)^{t_i} \leq \left(\frac{1+(\zeta-1)(1-\gamma_i)}{\gamma_i}\right)^{t_i} \leq \left(\frac{1+(\zeta-1)(1-h^-)}{h^-}\right)^{t_i}$$

$$\Leftrightarrow \prod_{i=1}^{n}\left(\frac{1+(\zeta-1)(1-h^+)}{h^+}\right)^{t_i} \leq \prod_{i=1}^{n}\left(\frac{1+(\zeta-1)(1-\gamma_i)}{\gamma_i}\right)^{t_i} \leq \prod_{i=1}^{n}\left(\frac{1+(\zeta-1)(1-h^-)}{h^-}\right)^{t_i}$$

$$\Leftrightarrow \frac{\zeta}{h^+} - (\zeta-1) \leq \prod_{i=1}^{n}\left(\frac{1+(\zeta-1)(1-\gamma_i)}{\gamma_i}\right)^{t_i} \leq \frac{\zeta}{h^-} - (\zeta-1)$$

$$\Leftrightarrow \frac{\zeta}{h^+} \leq \prod_{i=1}^{n}\left(\frac{1+(\zeta-1)(1-\gamma_i)}{\gamma_i}\right)^{t_i} + (\zeta-1) \leq \frac{\zeta}{h^-}$$

$$\Leftrightarrow \frac{h^-}{\zeta} \leq \frac{1}{\prod_{i=1}^{n}\left(\frac{1+(\zeta-1)(1-\gamma_i)}{\gamma_i}\right)^{t_i} + (\zeta-1)} \leq \frac{h^+}{\zeta}$$

$$\Leftrightarrow \frac{h^-}{\zeta} \leq \frac{\prod_{i=1}^{n}(\gamma_i)^{t_i}}{\prod_{i=1}^{n}(1+(\zeta-1)(1-\gamma_i))^{t_i} + (\zeta-1)\prod_{i=1}^{n}(\gamma_i)^{t_i}} \leq \frac{h^+}{\zeta}$$

$$\Leftrightarrow h^- \leq \frac{\zeta\prod_{i=1}^{n}(\gamma_i)^{t_i}}{\prod_{i=1}^{n}(1+(\zeta-1)(1-\gamma_i))^{t_i} + (\zeta-1)\prod_{i=1}^{n}(\gamma_i)^{t_i}} \leq h^+$$

Thus, we have $h^- \leq \mathrm{HFHPWG}_\zeta(h_1,h_2,\ldots,h_n) \leq h^+$.

(2) Let $g(x) = \frac{1+(\zeta-1)(1-x)}{x}$, $x \in (0,1]$, then by (1), $g(x)$ is a decreasing function. Then for all i, $\gamma_i \leq \gamma_i'$, we have $\frac{1+(\zeta-1)(1-\gamma_i)}{\gamma_i} \geq \frac{1+(\zeta-1)(1-\gamma_i')}{\gamma_i'}$. For convenience, let $t_i = \frac{w_i(1+T(h_i))}{\sum_{i=1}^{n} w_i(1+T(h_i))}$, then we have

$$\left(\frac{1+(\zeta-1)(1-\gamma_i)}{\gamma_i}\right)^{t_i} \geq \left(\frac{1+(\zeta-1)(1-\gamma_i')}{\gamma_i'}\right)^{t_i}$$

$$\Leftrightarrow \prod_{i=1}^{n}\left(\frac{1+(\zeta-1)(1-\gamma_i)}{\gamma_i}\right)^{t_i} \geq \prod_{i=1}^{n}\left(\frac{1+(\zeta-1)(1-\gamma_i')}{\gamma_i'}\right)^{t_i}$$

$$\Leftrightarrow \prod_{i=1}^{n}\left(\frac{1+(\zeta-1)(1-\gamma_i)}{\gamma_i}\right)^{t_i} + (\zeta-1) \geq \prod_{i=1}^{n}\left(\frac{1+(\zeta-1)(1-\gamma_i')}{\gamma_i'}\right)^{t_i} + (\zeta-1)$$

$$\Leftrightarrow \frac{1}{\prod_{i=1}^{n}\left(\frac{1+(\zeta-1)(1-\gamma_i)}{\gamma_i}\right)^{t_i} + (\zeta-1)} \leq \frac{1}{\prod_{i=1}^{n}\left(\frac{1+(\zeta-1)(1-\gamma_i')}{\gamma_i'}\right)^{t_i} + (\zeta-1)}$$

$$\Leftrightarrow \frac{\zeta \prod_{i=1}^{n}(\gamma_i)^{t_i}}{\left[\begin{array}{c}\prod_{i=1}^{n}(1+(\zeta-1)(1-\gamma_i))^{t_i}\\ +(\zeta-1)\prod_{i=1}^{n}(\gamma_i)^{t_i}\end{array}\right]} \leq \frac{\zeta \prod_{i=1}^{n}(\gamma_i')^{t_i}}{\left[\begin{array}{c}\prod_{i=1}^{n}(1+(\zeta-1)(1-\gamma_i'))^{t_i}\\ +(\zeta-1)\prod_{i=1}^{n}(\gamma_i')^{t_i}\end{array}\right]}.$$

Thus, by Theorem 5, $\mathrm{HFHPWG}_\zeta(h_1, h_2, \ldots, h_n) \leq \mathrm{HFHPWG}_\zeta(h_1', h_2', \ldots, h_n')$. $\square$

However, the HFHPWG operator is also neither idempotent nor commutative, as illustrated by the following example.

Example 3. *Let h_1, h_2 and h_3 be three HFEs, w be the weight vector of them, and $\mathrm{Sup}(h_i, h_j)$ $(i,j = 1,2,3,$ $i \neq j)$ be the support for h_i from h_j given in Example 2. Then by Theorem 5, we have*

$$\mathrm{HFHPWG}_5(h_1, h_1, h_1) = \{0.8000, 0.7095, 0.7390, 0.6455, 0.7573, 0.6644, 0.6945, 0.6000\},$$

$$\mathrm{HFHPWG}_5(h_1, h_2, h_3) = \{0.8277, 0.6398, 0.7678, 0.5751, 0.8077, 0.6177, 0.7467, 0.5532\},$$

$$\mathrm{HFHPWG}_5(h_2, h_3, h_1) = \{0,7881, 0.7438, 0.6611, 0.6145, 0.7445, 0.6989, 0.6152, 0.5686\}.$$

According to Definition 3, we have $s(\mathrm{HFHPWG}_5(h_1, h_1, h_1)) = 0.7013$, $s(\mathrm{HFHPWG}_5(h_1, h_2, h_3)) = 0.6920$ and $s(\mathrm{HFHPWG}_5(h_2, h_3, h_1)) = 0.6793$. Then $s(\mathrm{HFHPWG}_5(h_1, h_1, h_1)) \neq s(h_1)$ and thus $\mathrm{HFHPWG}_5(h_1, h_1, h_1) \neq h_1$, which implies that the HFHPWG operator is not idempotent. Furthermore, since $s(\mathrm{HFHPWG}_5(h_1, h_2, h_3)) \neq s(\mathrm{HFHPWG}_5(h_2, h_3, h_1))$, we have $\mathrm{HFHPWG}_5(h_1, h_2, h_3) \neq \mathrm{HFHPW}_5(h_2, h_3, h_1)$. Thus, the HFHPWG operator is not commutative.

Theorem 8. *Let h_i $(i = 1, 2, \ldots, n)$ be a collection of HFEs and $w = (w_1, w_2, \ldots, w_n)^T$ be the weight vector of h_i such that $w_i \in [0,1]$ and $\sum_{i=1}^{n} w_i = 1$, then we have*
(1) $\mathrm{HFHPWA}_\zeta(h_1^c, h_2^c, \ldots, h_n^c) = (\mathrm{HFHPWG}_\zeta(h_1, h_2, \ldots, h_n))^c$;
(2) $\mathrm{HFHPWG}_\zeta(h_1^c, h_2^c, \ldots, h_n^c) = (\mathrm{HFHPWA}_\zeta(h_1, h_2, \ldots, h_n))^c$.

Proof. Since (2) is similar (1), we only prove (1).

$$\text{HFHPWA}_\zeta\left(h_1^c, h_2^c, \ldots, h_n^c\right)$$

$$= \bigcup_{\gamma_1 \in h_1, \gamma_2 \in h_2, \ldots, \gamma_n \in h_n} \left\{ \frac{\prod_{i=1}^n \left(1 + (\zeta-1)(1-\gamma_i)\right)^{\frac{w_i(1+T(h_i))}{\sum_{i=1}^n w_i(1+T(h_i))}} - \prod_{i=1}^n \left(\gamma_i\right)^{\frac{w_i(1+T(h_i))}{\sum_{i=1}^n w_i(1+T(h_i))}}}{\prod_{i=1}^n \left(1 + (\zeta-1)(1-\gamma_i)\right)^{\frac{w_i(1+T(h_i))}{\sum_{i=1}^n w_i(1+T(h_i))}} + (\zeta-1)\prod_{i=1}^n \left(\gamma_i\right)^{\frac{w_i(1+T(h_i))}{\sum_{i=1}^n w_i(1+T(h_i))}}} \right\}$$

$$= \bigcup_{\gamma_1 \in h_1, \gamma_2 \in h_2, \ldots, \gamma_n \in h_n} \left\{ 1 - \frac{\zeta\prod_{i=1}^n \left(\gamma_i\right)^{\frac{w_i(1+T(h_i))}{\sum_{i=1}^n w_i(1+T(h_i))}}}{\left[\begin{array}{c} \prod_{i=1}^n \left(1 + (\zeta-1)(1-\gamma_i)\right)^{\frac{w_i(1+T(h_i))}{\sum_{i=1}^n w_i(1+T(h_i))}} \\ + (\zeta-1)\prod_{i=1}^n \left(\gamma_i\right)^{\frac{w_i(1+T(h_i))}{\sum_{i=1}^n w_i(1+T(h_i))}} \end{array}\right]} \right\}$$

$$= \left(\text{HFHPWG}_\zeta\left(h_1, h_2, \ldots, h_n\right)\right)^c.$$

$\square$

3.2. Generalized Hesitant Fuzzy Hamacher Power-Weighted Average/Geometric Operators

Definition 7. *Let h_i $(i = 1, 2, \ldots, n)$ be a collection of HFEs, $w = (w_1, w_2, \ldots, w_n)^T$ be the weight vector of h_i $(i = 1, 2, \ldots, n)$ such that $w_i \in [0,1]$ and $\sum_{i=1}^n w_i = 1$. For a parameter $\lambda > 0$, a generalized hesitant fuzzy Hamacher power-weighted average (GHFHPWA) operator is a function $H^n \to H$ such that*

$$\text{GHFHPWA}_\zeta(h_1, h_2, \ldots, h_n) = \left(\oplus_{H i=1}^n \left(\frac{w_i(1 + T(h_i)) \cdot_H h_i^{\wedge_H \lambda}}{\sum_{i=1}^n w_i(1 + T(h_i))}\right)\right)^{\wedge_H \frac{1}{\lambda}}, \tag{15}$$

where parameter $\zeta > 0$, $T(h_i) = \sum_{j=1, j \neq i}^n w_j \text{Sup}(h_i, h_j)$ and $\text{Sup}(h_i, h_j)$ is the support for h_i from h_j.

Theorem 9. *Let h_i $(i = 1, 2, \ldots, n)$ be a collection of HFEs and $w = (w_1, w_2, \ldots, w_n)^T$ be the weight vector of h_i $(i = 1, 2, \ldots, n)$ such that $w_i \in [0,1]$ and $\sum_{i=1}^n w_i = 1$, then the aggregated value by GHFHPWA operator is also a HFE, and*

$$\text{GHFHPWA}_\zeta(h_1, h_2, \ldots, h_n)$$

$$= \bigcup_{\gamma_1 \in h_1, \gamma_2 \in h_2, \ldots, \gamma_n \in h_n} \left\{ \frac{\zeta\left(\prod_{i=1}^n a_i^{\frac{w_i(1+T(h_i))}{\sum_{i=1}^n w_i(1+T(h_i))}} - \prod_{i=1}^n b_i^{\frac{w_i(1+T(h_i))}{\sum_{i=1}^n w_i(1+T(h_i))}}\right)^{\frac{1}{\lambda}}}{\left[\begin{array}{c} \left(\prod_{i=1}^n a_i^{\frac{w_i(1+T(h_i))}{\sum_{i=1}^n w_i(1+T(h_i))}} + (\zeta^2-1)\prod_{i=1}^n b_i^{\frac{w_i(1+T(h_i))}{\sum_{i=1}^n w_i(1+T(h_i))}}\right)^{\frac{1}{\lambda}} \\ + (\zeta-1)\left(\prod_{i=1}^n a_i^{\frac{w_i(1+T(h_i))}{\sum_{i=1}^n w_i(1+T(h_i))}} - \prod_{i=1}^n b_i^{\frac{w_i(1+T(h_i))}{\sum_{i=1}^n w_i(1+T(h_i))}}\right)^{\frac{1}{\lambda}} \end{array}\right]} \right\}, \tag{16}$$

where $a_i = (1 + (\zeta-1)(1-\gamma_i))^\lambda + (\zeta^2-1)\gamma_i^\lambda$ and $b_i = (1 + (\zeta-1)(1-\gamma_i))^\lambda - \gamma_i^\lambda$.

Proof. We first use the mathematical induction on n to prove

$$\oplus_H{}_{i=1}^n \left(\frac{w_i(1+T(h_i)) \cdot_H h_i^{\wedge_H \lambda}}{\sum_{i=1}^n w_i(1+T(h_i))} \right)$$

$$= \cup_{\gamma_1 \in h_1, \gamma_2 \in h_2, \ldots, \gamma_n \in h_n} \left\{ \frac{\prod_{i=1}^n a_i^{\frac{w_i(1+T(h_i))}{\sum_{i=1}^n w_i(1+T(h_i))}} - \prod_{i=1}^n b_i^{\frac{w_i(1+T(h_i))}{\sum_{i=1}^n w_i(1+T(h_i))}}}{\prod_{i=1}^n a_i^{\frac{w_i(1+T(h_i))}{\sum_{i=1}^n w_i(1+T(h_i))}} + (\zeta-1)\prod_{i=1}^n b_i^{\frac{w_i(1+T(h_i))}{\sum_{i=1}^n w_i(1+T(h_i))}}} \right\}. \tag{17}$$

(1) When $n = 1$, since $\frac{w_i(1+T(h_i))}{\sum_{i=1}^n w_i(1+T(h_i))} = 1$, we have

$$\oplus_H{}_{i=1}^n \left(\frac{w_i(1+T(h_i)) \cdot_H h_i^{\wedge_H \lambda}}{\sum_{i=1}^n w_i(1+T(h_i))} \right) = h_1^{\wedge_H \lambda}$$

$$= \cup_{\gamma_1 \in h_1} \left\{ \frac{\zeta \gamma_1^\lambda}{(1+(\zeta-1)(1-\gamma_1))^\lambda + (\zeta-1)\gamma_1^\lambda} \right\}$$

$$= \cup_{\gamma_1 \in h_1} \left\{ \frac{a_1 - b_1}{a_1 + (\zeta-1)b_1} \right\}.$$

Thus, Equation (17) holds for $n = 1$.

(2) Suppose that Equation (17) holds for $n = k$, that is

$$\oplus_H{}_{i=1}^k \left(\frac{w_i(1+T(h_i)) \cdot_H h_i^{\wedge_H \lambda}}{\sum_{i=1}^k w_i(1+T(h_i))} \right)$$

$$= \cup_{\gamma_1 \in h_1, \gamma_2 \in h_2, \ldots, \gamma_k \in h_k} \left\{ \frac{\prod_{i=1}^k a_i^{\frac{w_i(1+T(h_i))}{\sum_{i=1}^k w_i(1+T(h_i))}} - \prod_{i=1}^k b_i^{\frac{w_i(1+T(h_i))}{\sum_{i=1}^k w_i(1+T(h_i))}}}{\prod_{i=1}^k a_i^{\frac{w_i(1+T(h_i))}{\sum_{i=1}^k w_i(1+T(h_i))}} + (\zeta-1)\prod_{i=1}^k b_i^{\frac{w_i(1+T(h_i))}{\sum_{i=1}^k w_i(1+T(h_i))}}} \right\},$$

then, when $n = k + 1$, by the operational laws in Definition 4, we have

$$\oplus_H{}_{i=1}^{k+1} \left(\frac{w_i(1+T(h_i)) \cdot_H h_i^{\wedge_H \lambda}}{\sum_{i=1}^{k+1} w_i(1+T(h_i))} \right)$$

$$= \oplus_H{}_{i=1}^k \left(\frac{w_i(1+T(h_i)) \cdot_H h_i^{\wedge_H \lambda}}{\sum_{i=1}^{k+1} w_i(1+T(h_i))} \right) \oplus_H \left(\frac{w_{k+1}(1+T(h_{k+1})) \cdot_H h_{k+1}^{\wedge_H \lambda}}{\sum_{i=1}^{k+1} w_i(1+T(h_i))} \right)$$

$$= \cup_{\gamma_1 \in h_1, \gamma_2 \in h_2, \ldots, \gamma_k \in h_k} \left\{ \frac{\prod_{i=1}^k a_i^{\frac{w_i(1+T(h_i))}{\sum_{i=1}^{k+1} w_i(1+T(h_i))}} - \prod_{i=1}^k b_i^{\frac{w_i(1+T(h_i))}{\sum_{i=1}^{k+1} w_i(1+T(h_i))}}}{\prod_{i=1}^k a_i^{\frac{w_i(1+T(h_i))}{\sum_{i=1}^{k+1} w_i(1+T(h_i))}} + (\zeta-1)\prod_{i=1}^k b_i^{\frac{w_i(1+T(h_i))}{\sum_{i=1}^{k+1} w_i(1+T(h_i))}}} \right\}$$

$$\oplus_H \cup_{\gamma_{k+1} \in h_{k+1}} \left\{ \frac{a_{k+1}^{\frac{w_{k+1}(1+T(h_{k+1}))}{\sum_{i=1}^{k+1} w_i(1+T(h_i))}} - b_{k+1}^{\frac{w_{k+1}(1+T(h_{k+1}))}{\sum_{i=1}^{k+1} w_i(1+T(h_i))}}}{a_{k+1}^{\frac{w_{k+1}(1+T(h_{k+1}))}{\sum_{i=1}^{k+1} w_i(1+T(h_i))}} + (\zeta-1)b_{k+1}^{\frac{w_{k+1}(1+T(h_{k+1}))}{\sum_{i=1}^{k+1} w_i(1+T(h_i))}}} \right\}$$

$$= \cup_{\gamma_1 \in h_1, \gamma_2 \in h_2, \ldots, \gamma_k \in h_k, \gamma_{k+1} \in h_{k+1}} \left\{ \frac{\prod_{i=1}^{k+1} a_i^{\frac{w_i(1+T(h_i))}{\sum_{i=1}^{k+1} w_i(1+T(h_i))}} - \prod_{i=1}^{k+1} b_i^{\frac{w_i(1+T(h_i))}{\sum_{i=1}^{k+1} w_i(1+T(h_i))}}}{\prod_{i=1}^{k+1} a_i^{\frac{w_i(1+T(h_i))}{\sum_{i=1}^{k+1} w_i(1+T(h_i))}} + (\zeta-1)\prod_{i=1}^{k+1} b_i^{\frac{w_i(1+T(h_i))}{\sum_{i=1}^{k+1} w_i(1+T(h_i))}}} \right\},$$

i.e., Equation (17) holds for $n = k + 1$. Thus, Equation (17) holds for all n.

Hence, by the operational laws in Definition 4, we have

$$\text{GHFHPWA}_\zeta(h_1, h_2, \ldots, h_n) = \left(\oplus_{H_{i=1}^n} \left(\frac{w_i(1 + T(h_i)) \cdot_H h_i^{\wedge_H \lambda}}{\sum_{i=1}^n w_i(1 + T(h_i))} \right) \right)^{\wedge_H \frac{1}{\lambda}}$$

$$= \cup_{\gamma_1 \in h_1, \gamma_2 \in h_2, \ldots, \gamma_n \in h_n} \left\{ \frac{\zeta \left(\frac{\prod_{i=1}^n a_i^{\frac{w_i(1+T(h_i))}{\sum_{i=1}^n w_i(1+T(h_i))}} - \prod_{i=1}^n b_i^{\frac{w_i(1+T(h_i))}{\sum_{i=1}^n w_i(1+T(h_i))}}}{\prod_{i=1}^n a_i^{\frac{w_i(1+T(h_i))}{\sum_{i=1}^n w_i(1+T(h_i))}} + (\zeta-1) \prod_{i=1}^n b_i^{\frac{w_i(1+T(h_i))}{\sum_{i=1}^n w_i(1+T(h_i))}}} \right)^{\frac{1}{\lambda}}}{\left[\left(1 + (\zeta-1)(1 - \frac{\prod_{i=1}^n a_i^{\frac{w_i(1+T(h_i))}{\sum_{i=1}^n w_i(1+T(h_i))}} - \prod_{i=1}^n b_i^{\frac{w_i(1+T(h_i))}{\sum_{i=1}^n w_i(1+T(h_i))}}}{\prod_{i=1}^n a_i^{\frac{w_i(1+T(h_i))}{\sum_{i=1}^n w_i(1+T(h_i))}} + (\zeta-1) \prod_{i=1}^n b_i^{\frac{w_i(1+T(h_i))}{\sum_{i=1}^n w_i(1+T(h_i))}}}) \right)^{\frac{1}{\lambda}} + (\zeta-1) \left(\frac{\prod_{i=1}^n a_i^{\frac{w_i(1+T(h_i))}{\sum_{i=1}^n w_i(1+T(h_i))}} - \prod_{i=1}^n b_i^{\frac{w_i(1+T(h_i))}{\sum_{i=1}^n w_i(1+T(h_i))}}}{\prod_{i=1}^n a_i^{\frac{w_i(1+T(h_i))}{\sum_{i=1}^n w_i(1+T(h_i))}} + (\zeta-1) \prod_{i=1}^n b_i^{\frac{w_i(1+T(h_i))}{\sum_{i=1}^n w_i(1+T(h_i))}}} \right)^{\frac{1}{\lambda}} \right]} \right\}$$

$$= \cup_{\gamma_1 \in h_1, \gamma_2 \in h_2, \ldots, \gamma_n \in h_n} \left\{ \frac{\zeta \left(\prod_{i=1}^n a_i^{\frac{w_i(1+T(h_i))}{\sum_{i=1}^n w_i(1+T(h_i))}} - \prod_{i=1}^n b_i^{\frac{w_i(1+T(h_i))}{\sum_{i=1}^n w_i(1+T(h_i))}} \right)^{\frac{1}{\lambda}}}{\left[\left(\prod_{i=1}^n a_i^{\frac{w_i(1+T(h_i))}{\sum_{i=1}^n w_i(1+T(h_i))}} + (\zeta^2 - 1) \prod_{i=1}^n b_i^{\frac{w_i(1+T(h_i))}{\sum_{i=1}^n w_i(1+T(h_i))}} \right)^{\frac{1}{\lambda}} + (\zeta-1) \left(\prod_{i=1}^n a_i^{\frac{w_i(1+T(h_i))}{\sum_{i=1}^n w_i(1+T(h_i))}} - \prod_{i=1}^n b_i^{\frac{w_i(1+T(h_i))}{\sum_{i=1}^n w_i(1+T(h_i))}} \right)^{\frac{1}{\lambda}} \right]} \right\},$$

which completes the proof of the theorem. $\square$

Remark 3. *(1) If* $\text{Sup}(h_i, h_j) = k$, *for all* $i \neq j$, *then*

$$\text{GHFHPWA}_\zeta(h_1, h_2, \ldots, h_n) = \left(\oplus_{H_{i=1}^n} \left(w_i \cdot_H h_i^{\wedge_H \lambda} \right) \right)^{\wedge_H \frac{1}{\lambda}} \tag{18}$$

and thus, the GHFHPWA operator reduces to the generalized hesitant fuzzy Hamacher weighted average (GHFHWA) operator [17].

(2) If $\zeta = 1$, *then the GHFHPWA operator reduces to the generalized hesitant fuzzy power-weighted average (GHFPWA) operator [15]:*

$$\text{GHFHPWA}_1(h_1, h_2, \ldots, h_n) = \cup_{\gamma_1 \in h_1, \gamma_2 \in h_2, \ldots, \gamma_n \in h_n} \left\{ \left(1 - \prod_{i=1}^n \left(1 - \gamma_i^\lambda \right)^{\frac{w_i(1+T(h_i))}{\sum_{i=1}^n w_i(1+T(h_i))}} \right)^{\frac{1}{\lambda}} \right\} \tag{19}$$

and if $\zeta = 2$, then the GHFHPWA operator reduces to the generalized hesitant fuzzy Einstein power-weighted average (GHFEPWA) operator [18]:

$$\text{GHFHPWA}_2(h_1, h_2, \ldots, h_n)$$

$$= \cup_{\gamma_1 \in h_1, \gamma_2 \in h_2, \ldots, \gamma_n \in h_n} \left\{ \frac{2\left(\prod_{i=1}^n a_i^{\frac{w_i(1+T(h_i))}{\sum_{i=1}^n w_i(1+T(h_i))}} - \prod_{i=1}^n b_i^{\frac{w_i(1+T(h_i))}{\sum_{i=1}^n w_i(1+T(h_i))}} \right)^{\frac{1}{\lambda}}}{\left[\left(\prod_{i=1}^n a_i^{\frac{w_i(1+T(h_i))}{\sum_{i=1}^n w_i(1+T(h_i))}} + 3\prod_{i=1}^n b_i^{\frac{w_i(1+T(h_i))}{\sum_{i=1}^n w_i(1+T(h_i))}} \right)^{\frac{1}{\lambda}} + \left(\prod_{i=1}^n a_i^{\frac{w_i(1+T(h_i))}{\sum_{i=1}^n w_i(1+T(h_i))}} - \prod_{i=1}^n b_i^{\frac{w_i(1+T(h_i))}{\sum_{i=1}^n w_i(1+T(h_i))}} \right)^{\frac{1}{\lambda}} \right]} \right\}, \quad (20)$$

where $a_i = (2 - \gamma_i)^\lambda + 3\gamma_i^\lambda$, $b_i = (2 - \gamma_i)^\lambda - \gamma_i^\lambda$.

(3) *If $\lambda = 1$, then $a_i = \zeta(1 + (\zeta - 1)\gamma_i)$ and $b_i = \zeta(1 - \gamma_i)$, and thus the GHFHPWA operator reduces to the HFHPWA operator.*

In particular, if $w = (\frac{1}{n}, \frac{1}{n}, \ldots, \frac{1}{n})^T$, then the GHFHPWA operator reduces to the generalized hesitant fuzzy Hamacher power average (GHFHPA) operator:

$$\text{GHFHPA}_\zeta(h_1, h_2, \ldots, h_n) = \left(\oplus_{H_{i=1}}^n \left(\frac{(1 + T'(h_i)) \cdot_H h_i^{\wedge_H \lambda}}{\sum_{i=1}^n (1 + T'(h_i))} \right) \right)^{\wedge_H \frac{1}{\lambda}}$$

$$= \cup_{\gamma_1 \in h_1, \gamma_2 \in h_2, \ldots, \gamma_n \in h_n} \left\{ \frac{\zeta\left(\prod_{i=1}^n a_i^{\frac{(1+T'(h_i))}{\sum_{i=1}^n (1+T'(h_i))}} - \prod_{i=1}^n b_i^{\frac{(1+T'(h_i))}{\sum_{i=1}^n (1+T'(h_i))}} \right)^{\frac{1}{\lambda}}}{\left[\left(\prod_{i=1}^n a_i^{\frac{(1+T'(h_i))}{\sum_{i=1}^n (1+T'(h_i))}} + (\zeta^2 - 1)\prod_{i=1}^n b_i^{\frac{(1+T'(h_i))}{\sum_{i=1}^n (1+T'(h_i))}} \right)^{\frac{1}{\lambda}} + (\zeta - 1)\left(\prod_{i=1}^n a_i^{\frac{w_i(1+T(h_i))}{\sum_{i=1}^n w_i(1+T(h_i))}} - \prod_{i=1}^n b_i^{\frac{w_i(1+T(h_i))}{\sum_{i=1}^n w_i(1+T(h_i))}} \right)^{\frac{1}{\lambda}} \right]} \right\}, \quad (21)$$

where $a_i = (1 + (\zeta - 1)(1 - \gamma_i))^\lambda + (\zeta^2 - 1)\gamma_i^\lambda$, $b_i = (1 + (\zeta - 1)(1 - \gamma_i))^\lambda - \gamma_i^\lambda$ and $T'(h_i) = \frac{1}{n}\sum_{j=1, j \neq i}^n \text{Sup}(h_i, h_j)$.

Definition 8. *Let h_i $(i = 1, 2, \ldots, n)$ be a collection of HFEs and $w = (w_1, w_2, \ldots, w_n)^T$ be the weight vector of h_i $(i = 1, 2, \ldots, n)$ such that $w_i \in [0, 1]$ and $\sum_{i=1}^n w_i = 1$. For $\lambda > 0$, a generalized hesitant fuzzy Hamacher power-weighted geometric (GHFHPWG) operator is a function $H^n \to H$ such that*

$$\text{GHFHPWG}_\zeta(h_1, h_2, \ldots, h_n) = \frac{1}{\lambda} \cdot_H \left(\otimes_{H_{i=1}}^n \left((\lambda \cdot_H h_i)^{\wedge_H \frac{w_i(1+T(h_i))}{\sum_{i=1}^n w_i(1+T(h_i))}} \right) \right), \quad (22)$$

where $\zeta > 0$, $T(h_i) = \sum_{j=1, j \neq i}^n w_j \text{Sup}(h_i, h_j)$ and $\text{Sup}(h_i, h_j)$ is the support for h_i from h_j.

Theorem 10. *Let h_i ($i = 1, 2, \ldots, n$) be a collection of HFEs and $w = (w_1, w_2, \ldots, w_n)^T$ be the weight vector of h_i ($i = 1, 2, \ldots, n$) such that $w_i \in [0, 1]$ and $\sum_{i=1}^{n} w_i = 1$, then the aggregated value by GHFHPWG operator is also a HFE, and*

$$\text{GHFHPWG}_\zeta(h_1, h_2, \ldots, h_n)$$

$$= \bigcup_{\gamma_1 \in h_1, \gamma_2 \in h_2, \ldots, \gamma_n \in h_n} \left\{ \frac{\left[\left(\prod_{i=1}^{n} c_i^{\frac{w_i(1+T(h_i))}{\sum_{i=1}^{n} w_i(1+T(h_i))}} + (\zeta^2 - 1) \prod_{i=1}^{n} d_i^{\frac{w_i(1+T(h_i))}{\sum_{i=1}^{n} w_i(1+T(h_i))}} \right)^{\frac{1}{\lambda}} - \left(\prod_{i=1}^{n} c_i^{\frac{w_i(1+T(h_i))}{\sum_{i=1}^{n} w_i(1+T(h_i))}} - \prod_{i=1}^{n} d_i^{\frac{w_i(1+T(h_i))}{\sum_{i=1}^{n} w_i(1+T(h_i))}} \right)^{\frac{1}{\lambda}} \right]}{\left[\left(\prod_{i=1}^{n} c_i^{\frac{w_i(1+T(h_i))}{\sum_{i=1}^{n} w_i(1+T(h_i))}} + (\zeta^2 - 1) \prod_{i=1}^{n} d_i^{\frac{w_i(1+T(h_i))}{\sum_{i=1}^{n} w_i(1+T(h_i))}} \right)^{\frac{1}{\lambda}} + (\zeta - 1) \left(\prod_{i=1}^{n} c_i^{\frac{w_i(1+T(h_i))}{\sum_{i=1}^{n} w_i(1+T(h_i))}} - \prod_{i=1}^{n} d_i^{\frac{w_i(1+T(h_i))}{\sum_{i=1}^{n} w_i(1+T(h_i))}} \right)^{\frac{1}{\lambda}} \right]} \right\}, \quad (23)$$

where $c_i = (1 + (\zeta - 1)\gamma_i)^\lambda + (\zeta^2 - 1)(1 - \gamma_i)^\lambda$ and $d_i = (1 + (\zeta - 1)\gamma_i)^\lambda - (1 - \gamma_i)^\lambda$.

Proof. Similar to the proof of Theorem 9, Equation (23) can be proved by mathematical induction on n. $\square$

In particular, if $w = (\frac{1}{n}, \frac{1}{n}, \ldots, \frac{1}{n})^T$, then the GHFHPWG operator reduces to the generalized hesitant fuzzy Hamacher power geometric (GHFHPG) operator:

$$\text{GHFHPA}_\zeta(h_1, h_2, \ldots, h_n) = \frac{1}{\lambda} \cdot_H \left(\otimes_{H_{i=1}}^{n} \left((\lambda \cdot_H h_i)^{\wedge_H \frac{(1+T'(h_i))}{\sum_{i=1}^{n}(1+T'(h_i))}} \right) \right)$$

$$= \bigcup_{\gamma_1 \in h_1, \gamma_2 \in h_2, \ldots, \gamma_n \in h_n} \left\{ \frac{\left[\left(\prod_{i=1}^{n} c_i^{\frac{(1+T'(h_i))}{\sum_{i=1}^{n}(1+T'(h_i))}} + (\zeta^2 - 1) \prod_{i=1}^{n} d_i^{\frac{(1+T'(h_i))}{\sum_{i=1}^{n}(1+T'(h_i))}} \right)^{\frac{1}{\lambda}} - \left(\prod_{i=1}^{n} c_i^{\frac{(1+T'(h_i))}{\sum_{i=1}^{n}(1+T'(h_i))}} - \prod_{i=1}^{n} d_i^{\frac{(1+T'(h_i))}{\sum_{i=1}^{n}(1+T'(h_i))}} \right)^{\frac{1}{\lambda}} \right]}{\left[\left(\prod_{i=1}^{n} c_i^{\frac{(1+T'(h_i))}{\sum_{i=1}^{n}(1+T'(h_i))}} + (\zeta^2 - 1) \prod_{i=1}^{n} d_i^{\frac{(1+T'(h_i))}{\sum_{i=1}^{n}(1+T'(h_i))}} \right)^{\frac{1}{\lambda}} + (\zeta - 1) \left(\prod_{i=1}^{n} c_i^{\frac{(1+T'(h_i))}{\sum_{i=1}^{n}(1+T'(h_i))}} - \prod_{i=1}^{n} d_i^{\frac{(1+T'(h_i))}{\sum_{i=1}^{n}(1+T'(h_i))}} \right)^{\frac{1}{\lambda}} \right]} \right\}, \quad (24)$$

where $c_i = (1 + (\zeta - 1)\gamma_i)^\lambda + (\zeta^2 - 1)(1 - \gamma_i)^\lambda$, $d_i = (1 + (\zeta - 1)\gamma_i)^\lambda - (1 - \gamma_i)^\lambda$ and $T'(h_i) = \frac{1}{n} \sum_{j=1, j \neq i}^{n} \text{Sup}(h_i, h_j)$.

Remark 4. *(1) If $\zeta = 1$, then the GHFHPWG operator reduces to the generalized hesitant fuzzy power-weighted geometric (GHFPWG) operator [15]:*

$$\text{GHFHPWG}_1(h_1, h_2, \ldots, h_n) = \bigcup_{\gamma_1 \in h_1, \gamma_2 \in h_2, \ldots, \gamma_n \in h_n} \left\{ \left(\prod_{i=1}^{n} \left(\gamma_i^\lambda \right)^{\frac{w_i(1+T(h_i))}{\sum_{i=1}^{n} w_i(1+T(h_i))}} \right)^{\frac{1}{\lambda}} \right\}. \quad (25)$$

and if $\zeta = 2$, then the GHFHPWG operator reduces to the generalized hesitant fuzzy Einstein power-weighted geometric (GHFEPWG) operator [18]:

$$\text{GHFHPWG}_2(h_1, h_2, \ldots, h_n)$$

$$= \bigcup_{\gamma_1 \in h_1, \gamma_2 \in h_2, \ldots, \gamma_n \in h_n} \left\{ \frac{\left[\left(\prod_{i=1}^n c_i^{\frac{w_i(1+T(h_i))}{\sum_{i=1}^n w_i(1+T(h_i))}} + 3\prod_{i=1}^n d_i^{\frac{w_i(1+T(h_i))}{\sum_{i=1}^n w_i(1+T(h_i))}} \right)^{\frac{1}{\lambda}} - \left(\prod_{i=1}^n c_i^{\frac{w_i(1+T(h_i))}{\sum_{i=1}^n w_i(1+T(h_i))}} - \prod_{i=1}^n d_i^{\frac{w_i(1+T(h_i))}{\sum_{i=1}^n w_i(1+T(h_i))}} \right)^{\frac{1}{\lambda}} \right]}{\left[\left(\prod_{i=1}^n c_i^{\frac{w_i(1+T(h_i))}{\sum_{i=1}^n w_i(1+T(h_i))}} + 3\prod_{i=1}^n d_i^{\frac{w_i(1+T(h_i))}{\sum_{i=1}^n w_i(1+T(h_i))}} \right)^{\frac{1}{\lambda}} + (\zeta - 1)\left(\prod_{i=1}^n c_i^{\frac{w_i(1+T(h_i))}{\sum_{i=1}^n w_i(1+T(h_i))}} - \prod_{i=1}^n d_i^{\frac{w_i(1+T(h_i))}{\sum_{i=1}^n w_i(1+T(h_i))}} \right)^{\frac{1}{\lambda}} \right]} \right\}, \quad (26)$$

where $c_i = (1 + \gamma_i)^\lambda + 3(1 - \gamma_i)^\lambda$, $d_i = (1 + \gamma_i)^\lambda - (1 - \gamma_i)^\lambda$. (2) If $\text{Sup}(h_i, h_j) = k$, for all $i \neq j$, then

$$\text{GHFHPWG}_\zeta(h_1, h_2, \ldots, h_n) = \frac{1}{\lambda} \cdot_H \left(\otimes_H{}_{i=1}^n (\lambda \cdot_H h_i)^{\wedge_H w_i} \right) \quad (27)$$

and thus, the GHFHPWG operator reduces to the generalized hesitant fuzzy Hamacher weighted geometric (GHFHWG) operator [17].

(3) If $\lambda = 1$, then $c_i = \zeta(1 + (\zeta - 1)(1 - \gamma_i))$ and $d_i = \zeta\gamma_i$, and so the GHFHPWG operator reduces to the HFHPWG operator.

Theorem 11. *Let h_i ($i = 1, 2, \ldots, n$) be a collection of HFEs and $w = (w_1, w_2, \ldots, w_n)^T$ be the weight vector of h_i such that $w_i \in [0, 1]$ and $\sum_{i=1}^n w_i = 1$, then we have*
(1) $\text{GHFHPWA}_\zeta(h_1^c, h_2^c, \ldots, h_n^c) = (\text{GHFHPWG}_\zeta(h_1, h_2, \ldots, h_n))^c$;
(2) $\text{GHFHPWG}_\zeta(h_1^c, h_2^c, \ldots, h_n^c) = (\text{GHFHPWA}_\zeta(h_1, h_2, \ldots, h_n))^c$.

Proof. Similar to the proof of Theorem 8. □

3.3. Hesitant Fuzzy Hamacher Power-Ordered Weighted Average/Geometric Operators

Motivated by the idea of the POWA operator [24], POWG operator [25] and Hamacher operations, we define the hesitant fuzzy Hamacher power-ordered weighted average (HFHPOWA) operator and hesitant fuzzy Hamacher power-ordered weighted geometric (HFHPOWG) operator as follows.

Definition 9. *Let h_i ($i = 1, 2, \ldots, n$) be a collection of HFEs. A hesitant fuzzy Hamacher power-ordered weighted average (HFHPOWA) operator is a function $H^n \to H$ such that*

$$\text{HFHPOWA}_\zeta(h_1, h_2, \ldots, h_n) = \oplus_H{}_{i=1}^n \left(u_i \cdot_H h_{\sigma(i)} \right), \quad (28)$$

where parameter $\zeta > 0$, $h_{\sigma(i)}$ is the ith largest HFE of h_j ($j = 1, 2, \ldots, n$), and u_i ($i = 1, 2, \ldots, n$) is a collection of weights such that

$$u_i = g\left(\frac{R_i}{TV}\right) - g\left(\frac{R_{i-1}}{TV}\right), \quad R_i = \sum_{j=1}^i V_{\sigma(j)}, \quad TV = \sum_{i=1}^n V_{\sigma(i)},$$

$$V_{\sigma(i)} = 1 + T(h_{\sigma(i)}), \quad T(h_{\sigma(i)}) = \sum_{j=1, j \neq i}^n \text{Sup}(h_{\sigma(i)}, h_{\sigma(j)}), \quad (29)$$

where $T(h_{\sigma(i)})$ denotes the support of ith largest HFE by all of the other HFEs, $\text{Sup}(h_{\sigma(i)}, h_{\sigma(j)})$ indicates the support of the ith largest HFE for the jth largest HFE, and $g : [0,1] \to [0,1]$ is a basic unit-interval monotone (BUM) function with the following properties: (1) $g(0) = 0$; (2) $g(1) = 1$; and (3) $g(x) \geq g(y)$ if $x > y$.

Theorem 12. *Let h_i $(i = 1, 2, \ldots, n)$ be a collection of HFEs, then the aggregated value by HFHPOWA operator is also an HFE, and*

$$\text{HFHPOWA}_\zeta(h_1, h_2, \ldots, h_n)$$

$$= \bigcup_{\gamma_{\sigma(1)} \in h_{\sigma(1)}, \gamma_{\sigma(2)} \in h_{\sigma(2)}, \ldots, \gamma_{\sigma(n)} \in h_{\sigma(n)}} \left\{ \frac{\prod_{i=1}^{n}\left(1 + (\zeta-1)\gamma_{\sigma(i)}\right)^{u_i} - \prod_{i=1}^{n}\left(1 - \gamma_{\sigma(i)}\right)^{u_i}}{\prod_{i=1}^{n}\left(1 + (\zeta-1)\gamma_{\sigma(i)}\right)^{u_i} + (\zeta-1)\prod_{i=1}^{n}\left(1 - \gamma_{\sigma(i)}\right)^{u_i}} \right\}, \quad (30)$$

where u_i $(i = 1, 2, \ldots, n)$ is a collection of weights satisfying the condition (29).

Remark 5. *(1) If $\text{Sup}(h_i, h_j) = k$ for all $i \neq j$ and $g(x) = x$, then $u_i = \frac{1}{n}$, $i - 1, 2, \ldots, n$, and so*

$$\text{HFHPOWA}_\zeta(h_1, h_2, \ldots, h_n) = \oplus_H{}_{i=1}^{n}\left(\frac{1}{n} \cdot_H h_i\right)$$

which indicates that the HFHPOWA operator reduces to the hesitant fuzzy Hamacher average (HFHA) operator [17].

(2) If $\zeta = 1$, then the HFHPOWA operator reduces to the hesitant fuzzy power-ordered weighted average (HFPOWA) operator [15]:

$$\text{HFHPOWA}_1(h_1, h_2, \ldots, h_n) = \oplus_{i=1}^{n}\left(u_i h_{\sigma(i)}\right)$$

$$= \bigcup_{\gamma_{\sigma(1)} \in h_{\sigma(1)}, \gamma_{\sigma(2)} \in h_{\sigma(2)}, \ldots, \gamma_{\sigma(n)} \in h_{\sigma(n)}} \left\{1 - \prod_{i=1}^{n}\left(1 - \gamma_{\sigma(i)}\right)^{u_i}\right\}, \quad (31)$$

where u_i $(i = 1, 2, \ldots, n)$ is a collection of weights satisfying the condition (29). *If $\zeta = 2$, then the HFHPOWA operator* (30) *reduces to the hesitant fuzzy Einstein power-ordered weighted average (HFEPOWA) operator* [18]:

$$\text{HFHPOWA}_2(h_1, h_2, \ldots, h_n)$$

$$= \bigcup_{\gamma_{\sigma(1)} \in h_{\sigma(1)}, \gamma_{\sigma(2)} \in h_{\sigma(2)}, \ldots, \gamma_{\sigma(n)} \in h_{\sigma(n)}} \left\{ \frac{\prod_{i=1}^{n}\left(1 + \gamma_{\sigma(i)}\right)^{u_i} - \prod_{i=1}^{n}\left(1 - \gamma_{\sigma(i)}\right)^{u_i}}{\prod_{i=1}^{n}\left(1 + \gamma_{\sigma(i)}\right)^{u_i} + \prod_{i=1}^{n}\left(1 - \gamma_{\sigma(i)}\right)^{u_i}} \right\}, \quad (32)$$

where u_i $(i = 1, 2, \ldots, n)$ is a collection of weights satisfying the condition (29).

Similar to Theorems 3 and 4, we have the properties of HFHPOWA operator as follows.

Theorem 13. *If h_i $(i = 1, 2, \ldots, n)$ is a collection of HFEs and u_i $(i = 1, 2, \ldots, n)$ is the collection of the weights which satisfies the condition* (29), *then*

$$\text{HFHPOWA}_\zeta(h_1, h_2, \ldots, h_n) \leq \text{HFPOWA}(h_1, h_2, \ldots, h_n).$$

Theorem 14. *If h_i $(i = 1, 2, \ldots, n)$ is a collection of HFEs and u_i $(i = 1, 2, \ldots, n)$ is the collection of the weights which satisfies the condition* (29), *then we have the followings:*

(1) Boundedness: If $h^- = \min\{\gamma_i | \gamma_i \in h_i\}$ and $h^+ = \max\{\gamma_i | \gamma_i \in h_i\}$, then

$$h^- \leq \text{HFHPOWA}_\zeta(h_1, h_2, \ldots, h_n) \leq h^+.$$

(2) Monotonicity: Let h_i' $(i = 1, 2, \ldots, n)$ be a collection of HFEs, if for any $h_{\sigma(i)}$ and $h'_{\sigma(i)}$ $(i = 1, 2, \ldots, n)$, $\gamma_{\sigma(i)} \leq \gamma'_{\sigma(i)}$, then

$$\text{HFHPOWA}_\zeta(h_1, h_2, \ldots, h_n) \leq \text{HFHPWA}_\zeta(h_1', h_2', \ldots, h_n').$$

Definition 10. *Let h_i $(i = 1, 2, \ldots, n)$ be a collection of HFEs. A hesitant fuzzy Hamacher power-ordered weighted geometric (HFHPOWG) operator is a function $H^n \to H$ such that*

$$\text{HFHPOWG}_\zeta(h_1, h_2, \ldots, h_n) = \otimes_{H\,i=1}^{n} \left(h_{\sigma(i)}^{\wedge_H u_i} \right), \tag{33}$$

where parameter $\zeta > 0$, $h_{\sigma(i)}$ is the ith largest HFE of h_j $(j = 1, 2, \ldots, n)$, and u_i $(i = 1, 2, \ldots, n)$ is a collection of weights satisfying the condition (29).

Theorem 15. *If h_i $(i = 1, 2, \ldots, n)$ is a collection of HFEs, then the aggregated value by HFHPOWG operator is also an HFE, and*

$$\text{HFHPOWG}_\zeta(h_1, h_2, \ldots, h_n)$$

$$= \cup_{\gamma_{\sigma(1)} \in h_{\sigma(1)}, \gamma_{\sigma(2)} \in h_{\sigma(2)}, \ldots, \gamma_{\sigma(n)} \in h_{\sigma(n)}} \left\{ \frac{\zeta \prod_{i=1}^{n} \gamma_{\sigma(i)}^{u_i}}{\prod_{i=1}^{n} \left(1 + (\zeta - 1)(1 - \gamma_{\sigma(i)}) \right)^{u_i} + (\zeta - 1) \prod_{i=1}^{n} \gamma_{\sigma(i)}^{u_i}} \right\}, \tag{34}$$

where $h_{\sigma(i)}$ is the ith largest HFE of h_j $(j = 1, 2, \ldots, n)$ and u_i $(i = 1, 2, \ldots, n)$ is the collection of the weights satisfying the condition (29).

Remark 6. *(1) If $\text{Sup}(h_i, h_j) = k$, for all $i \neq j$, and $g(x) = x$, then*

$$\text{HFHPOWG}_\zeta(h_1, h_2, \ldots, h_n) = \otimes_{H\,i=1}^{n} \left(h_i^{\wedge_H \frac{1}{n}} \right)$$

which indicates that the HFHPOWG operator reduces to the hesitant fuzzy Hamacher geometric (HFHG) operator [17].

(2) If $\zeta = 1$, then the HFHPOWG operator (34) reduces to the hesitant fuzzy power-ordered weighted geometric (HFPOWG) operator [15]:

$$\text{HFHPOWG}_1(h_1, h_2, \ldots, h_n) = \otimes_{i=1}^{n} \left(h_{\sigma(i)} \right)^{u_i}$$

$$= \cup_{\gamma_{\sigma(1)} \in h_{\sigma(1)}, \gamma_{\sigma(2)} \in h_{\sigma(2)}, \ldots, \gamma_{\sigma(n)} \in h_{\sigma(n)}} \left\{ \prod_{i=1}^{n} \gamma_{\sigma(i)}^{u_i} \right\}, \tag{35}$$

where u_i $(i = 1, 2, \ldots, n)$ is a collection of weights satisfying the condition (29). If $\zeta = 2$, then the HFHPOWG operator (34) reduces to the hesitant fuzzy Einstein power-ordered weighted geometric (HFEPOWG) operator [18]:

$$\text{HFHPOWG}_2(h_1, h_2, \ldots, h_n)$$

$$= \cup_{\gamma_{\sigma(1)} \in h_{\sigma(1)}, \gamma_{\sigma(2)} \in h_{\sigma(2)}, \ldots, \gamma_{\sigma(n)} \in h_{\sigma(n)}} \left\{ \frac{2 \prod_{i=1}^{n} \gamma_{\sigma(i)}^{u_i}}{\prod_{i=1}^{n} \left(2 - \gamma_{\sigma(i)} \right)^{u_i} + \prod_{i=1}^{n} \gamma_{\sigma(i)}^{u_i}} \right\}, \tag{36}$$

where u_i $(i = 1, 2, \ldots, n)$ is a collection of weights satisfying the condition (29).

Similar to Theorems 6–8, we have the properties of HFHPOWG operator as follows.

Theorem 16. *If h_i ($i = 1, 2, \ldots, n$) is a collection of HFEs and u_i ($i = 1, 2, \ldots, n$) is the collection of the weights which satisfies the condition (29), then*

$$\mathrm{HFHPOWG}_\zeta(h_1, h_2, \ldots, h_n) \geq \mathrm{HFPOWG}(h_1, h_2, \ldots, h_n).$$

Theorem 17. *If h_i ($i = 1, 2, \ldots, n$) is a collection of HFEs and u_i ($i = 1, 2, \ldots, n$) is the collection of the weights which satisfies the condition (29), then we have the followings:*

(1) Boundedness: If $h^- = \min\{\gamma_i | \gamma_i \in h_i\}$ and $h^+ = \max\{\gamma_i | \gamma_i \in h_i\}$, then

$$h^- \leq \mathrm{HFHPOWG}_\zeta(h_1, h_2, \ldots, h_n) \leq h^+.$$

(2) Monotonicity: Let h_i' ($i = 1, 2, \ldots, n$) be a collection of HFEs, if for any $h_{\sigma(i)}$ and $h'_{\sigma(i)}$ ($i = 1, 2, \ldots, n$), $\gamma_{\sigma(i)} \leq \gamma'_{\sigma(i)}$, then

$$\mathrm{HFHPOWG}_\zeta(h_1, h_2, \ldots, h_n) \leq \mathrm{HFHPOWG}_\zeta(h_1', h_2', \ldots, h_n').$$

Theorem 18. *If h_i ($i = 1, 2, \ldots, n$) is a collection of HFEs and u_i ($i = 1, 2, \ldots, n$) is the collection of the weights satisfying the condition (29), then we have*

(1) $\mathrm{HFHPOWA}_\zeta(h_1^c, h_2^c, \ldots, h_n^c) = (\mathrm{HFHPOWG}_\zeta(h_1, h_2, \ldots, h_n))^c$;

(2) $\mathrm{HFHPOWG}_\zeta(h_1^c, h_2^c, \ldots, h_n^c) = (\mathrm{HFHPOWA}_\zeta(h_1, h_2, \ldots, h_n))^c$.

In what follows, we define the generalized hesitant fuzzy Hamacher power-ordered weighted average (GHFHPOWA) operator and generalized hesitant fuzzy Hamacher power-ordered weighted geometric (GHFHPOWG) operator.

Definition 11. *Let h_i ($i = 1, 2, \ldots, n$) be a collection of HFEs. For $\lambda > 0$, a generalized hesitant fuzzy Hamacher power-ordered weighted average (GHFHPOWA) operator is a function $H^n \to H$ such that*

$$\mathrm{GHFHPOWA}_\zeta(h_1, h_2, \ldots, h_n) = \left(\oplus_{H i=1}^{n} \left(u_i \cdot_H h_{\sigma(i)}^{\wedge_H \lambda} \right) \right)^{\wedge_H \frac{1}{\lambda}}, \tag{37}$$

where $\zeta > 0$, $h_{\sigma(i)}$ is the ith largest HFE of h_j ($j = 1, 2, \ldots, n$), and u_i ($i = 1, 2, \ldots, n$) is a collection of weights satisfying the condition (29).

Theorem 19. *Let h_i ($i = 1, 2, \ldots, n$) be a collection of HFEs, then the aggregated value by GHFHPOWA operator is also an HFE, and*

$$\mathrm{GHFHPOWA}_\zeta(h_1, h_2, \ldots, h_n)$$

$$= \bigcup_{\gamma_{\sigma(1)} \in h_{\sigma(1)}, \gamma_{\sigma(2)} \in h_{\sigma(2)}, \ldots, \gamma_{\sigma(n)} \in h_{\sigma(n)}} \left\{ \frac{\zeta \left(\prod_{i=1}^{n} a_i^{u_i} - \prod_{i=1}^{n} b_i^{u_i} \right)^{\frac{1}{\lambda}}}{\left[\begin{array}{c} \left(\prod_{i=1}^{n} a_i^{u_i} + (\zeta^2 - 1) \prod_{i=1}^{n} b_i^{u_i} \right)^{\frac{1}{\lambda}} \\ + (\zeta - 1) \left(\prod_{i=1}^{n} a_i^{u_i} - \prod_{i=1}^{n} b_i^{u_i} \right)^{\frac{1}{\lambda}} \end{array} \right]} \right\}, \tag{38}$$

where $a_i = (1 + (\zeta - 1)(1 - \gamma_{\sigma(i)})^\lambda + (\zeta^2 - 1)\gamma_{\sigma(i)}^\lambda$, $b_i = (1 + (\zeta - 1)(1 - \gamma_{\sigma(i)}))^\lambda - \gamma_{\sigma(i)}^\lambda$ and u_i ($i = 1, 2, \ldots, n$) is a collection of weights satisfying the condition (29).

Remark 7. *(1) If $\lambda = 1$, then $a_i = \zeta(1 + (\zeta - 1)\gamma_{\sigma(i)})$ and $b_i = \zeta(1 - \gamma_{\sigma(i)})$, and thus the GHFHPOWA operator reduces to the HFHPOWA operator. In fact, by Equation (38), we have*

$$\text{GHFHPOWA}_\zeta(h_1, h_2, \ldots, h_n)$$

$$= \bigcup_{\gamma_{\sigma(1)} \in h_{\sigma(1)}, \gamma_{\sigma(2)} \in h_{\sigma(2)}, \ldots, \gamma_{\sigma(n)} \in h_{\sigma(n)}} \left\{ \frac{\zeta \left(\prod_{i=1}^n a_i^{u_i} - \prod_{i=1}^n b_i^{u_i} \right)^{\frac{1}{\lambda}}}{\left[\begin{array}{c} \left(\prod_{i=1}^n a_i^{u_i} + (\zeta^2 - 1) \prod_{i=1}^n b_i^{u_i} \right)^{\frac{1}{\lambda}} \\ + (\zeta - 1) \left(\prod_{i=1}^n a_i^{u_i} - \prod_{i=1}^n b_i^{u_i} \right)^{\frac{1}{\lambda}} \end{array} \right]} \right\}$$

$$= \bigcup_{\gamma_{\sigma(1)} \in h_{\sigma(1)}, \gamma_{\sigma(2)} \in h_{\sigma(2)}, \ldots, \gamma_{\sigma(n)} \in h_{\sigma(n)}} \left\{ \frac{\prod_{i=1}^n a_i^{u_i} - \prod_{i=1}^n b_i^{u_i}}{\prod_{i=1}^n a_i^{u_i} + (\zeta - 1) \prod_{i=1}^n b_i^{u_i}} \right\}$$

$$= \bigcup_{\gamma_{\sigma(1)} \in h_{\sigma(1)}, \gamma_{\sigma(2)} \in h_{\sigma(2)}, \ldots, \gamma_{\sigma(n)} \in h_{\sigma(n)}} \left\{ \frac{\prod_{i=1}^n \left(1 + (\zeta - 1)\gamma_{\sigma(i)}\right)^{u_i} - \prod_{i=1}^n \left(1 - \gamma_{\sigma(i)}\right)^{u_i}}{\prod_{i=1}^n \left(1 + (\zeta - 1)\gamma_{\sigma(i)}\right)^{u_i} + (\zeta - 1) \prod_{i=1}^n \left(1 - \gamma_{\sigma(i)}\right)^{u_i}} \right\}$$

$$= \text{HFHPOWA}_\zeta(h_1, h_2, \ldots, h_n).$$

(2) If $\zeta = 1$, then the GHFHPOWA operator reduces to the generalized hesitant fuzzy power-ordered weighted average (GHFPOWA) operator [15]:

$$\text{GHFHPOWA}_1(h_1, h_2, \ldots, h_n) = \bigcup_{\gamma_{\sigma(1)} \in h_{\sigma(1)}, \gamma_{\sigma(2)} \in h_{\sigma(2)}, \ldots, \gamma_{\sigma(n)} \in h_{\sigma(n)}} \left\{ \left(1 - \prod_{i=1}^n \left(1 - \gamma_{\sigma(i)}^\lambda \right)^{u_i} \right)^{\frac{1}{\lambda}} \right\}, \quad (39)$$

where u_i ($i = 1, 2, \ldots, n$) is a collection of weights satisfying the condition (29), and if $\zeta = 2$, then the GHFHPOWA operator reduces to the generalized hesitant fuzzy Einstein power-ordered weighted average (GHFEPOWA) operator [18]:

$$\text{GHFHPOWA}_2(h_1, h_2, \ldots, h_n)$$

$$= \bigcup_{\gamma_{\sigma(1)} \in h_{\sigma(1)}, \gamma_{\sigma(2)} \in h_{\sigma(2)}, \ldots, \gamma_{\sigma(n)} \in h_{\sigma(n)}} \left\{ \frac{2 \left(\prod_{i=1}^n a_i^{u_i} - \prod_{i=1}^n b_i^{u_i} \right)^{\frac{1}{\lambda}}}{\left(\prod_{i=1}^n a_i^{u_i} + 3 \prod_{i=1}^n b_i^{u_i} \right)^{\frac{1}{\lambda}} + \left(\prod_{i=1}^n a_i^{u_i} - \prod_{i=1}^n b_i^{u_i} \right)^{\frac{1}{\lambda}}} \right\}, \quad (40)$$

where $a_i = (2 - \gamma_{\sigma(i)})^\lambda + 3\gamma_{\sigma(i)}^\lambda$, $b_i = (2 - \gamma_{\sigma(i)})^\lambda - \gamma_{\sigma(i)}^\lambda$ and u_i ($i = 1, 2, \ldots, n$) is a collection of weights satisfying the condition (29).

Definition 12. *Let h_i ($i = 1, 2, \ldots, n$) be a collection of HFEs. For $\lambda > 0$, a generalized hesitant fuzzy Hamacher power-ordered weighted geometric (GHFHPOWG) operator is a function $H^n \to H$ such that*

$$\text{GHFHPOWG}_\zeta(h_1, h_2, \ldots, h_n) = \frac{1}{\lambda} \cdot_H \left(\otimes_{H_{i=1}^n} \left(\left(\lambda \cdot_H h_{\sigma(i)} \right)^{\wedge_H u_i} \right) \right), \quad (41)$$

where $\zeta > 0$, $h_{\sigma(i)}$ is the ith largest HFE of h_j ($j = 1, 2, \ldots, n$), and u_i ($i = 1, 2, \ldots, n$) is a collection of weights satisfying the condition (29).

Theorem 20. *Let h_i ($i = 1, 2, \ldots, n$) be a collection of HFEs, then the aggregated value by GHFHPOWG operator is also an HFE, and*

$$\text{GHFHPOWG}_\zeta(h_1, h_2, \ldots, h_n)$$

$$= \bigcup_{\gamma_{\sigma(1)} \in h_{\sigma(1)}, \gamma_{\sigma(2)} \in h_{\sigma(2)}, \ldots, \gamma_{\sigma(n)} \in h_{\sigma(n)}} \left\{ \frac{\begin{bmatrix} \left(\prod_{i=1}^n c_i^{u_i} + (\zeta^2 - 1) \prod_{i=1}^n d_i^{u_i}\right)^{\frac{1}{\lambda}} \\ - \left(\prod_{i=1}^n c_i^{u_i} - \prod_{i=1}^n d_i^{u_i}\right)^{\frac{1}{\lambda}} \end{bmatrix}}{\begin{bmatrix} \left(\prod_{i=1}^n c_i^{u_i} + (\zeta^2 - 1) \prod_{i=1}^n d_i^{u_i}\right)^{\frac{1}{\lambda}} \\ + (\zeta - 1)\left(\prod_{i=1}^n c_i^{u_i} - \prod_{i=1}^n d_i^{u_i}\right)^{\frac{1}{\lambda}} \end{bmatrix}} \right\}, \qquad (42)$$

where $c_i = (1 + (\zeta - 1)\gamma_{\sigma(i)})^\lambda + (\zeta^2 - 1)(1 - \gamma_{\sigma(i)})^\lambda$, $d_i = (1 + (\zeta - 1)\gamma_{\sigma(i)})^\lambda - (1 - \gamma_{\sigma(i)})^\lambda$ and u_i ($i = 1, 2, \ldots, n$) is a collection of weights satisfying the condition (29).

Remark 8. *(1) If $\lambda = 1$, then $c_i = \zeta(1 + (\zeta - 1)(1 - \gamma_{\sigma(i)}))$ and $d_i = \zeta\gamma_{\sigma(i)}$, and the GHFHPOWG operator reduces to the HFHPOWG operator.*

(2) If $\zeta = 1$, then the GHFHPOWG operator reduces to the generalized hesitant fuzzy power-ordered weighted geometric (GHFPOWG) operator [15]:

$$\text{GHFHPOWG}_1(h_1, h_2, \ldots, h_n) = \bigcup_{\gamma_{\sigma(1)} \in h_{\sigma(1)}, \gamma_{\sigma(2)} \in h_{\sigma(2)}, \ldots, \gamma_{\sigma(n)} \in h_{\sigma(n)}} \left\{ \left(\prod_{i=1}^n \left(\gamma_{\sigma(i)}^\lambda\right)^{u_i}\right)^{\frac{1}{\lambda}} \right\}, \qquad (43)$$

where u_i ($i = 1, 2, \ldots, n$) is a collection of weights satisfying the condition (29), and if $\zeta = 2$, then the GHFHPOWG operator reduces to the generalized hesitant fuzzy Einstein power-ordered weighted geometric (GHFEPOWG) operator [18]:

$$\text{GHFHPOWG}_2(h_1, h_2, \ldots, h_n)$$

$$= \bigcup_{\gamma_{\sigma(1)} \in h_{\sigma(1)}, \gamma_{\sigma(2)} \in h_{\sigma(2)}, \ldots, \gamma_{\sigma(n)} \in h_{\sigma(n)}} \left\{ \frac{\left(\prod_{i=1}^n c_i^{u_i} + 3\prod_{i=1}^n d_i^{u_i}\right)^{\frac{1}{\lambda}} - \left(\prod_{i=1}^n c_i^{u_i} - \prod_{i=1}^n d_i^{u_i}\right)^{\frac{1}{\lambda}}}{\left(\prod_{i=1}^n c_i^{u_i} + 3\prod_{i=1}^n d_i^{u_i}\right)^{\frac{1}{\lambda}} + \left(\prod_{i=1}^n c_i^{u_i} - \prod_{i=1}^n d_i^{u_i}\right)^{\frac{1}{\lambda}}} \right\}, \qquad (44)$$

where $c_i = (1 + \gamma_{\sigma(i)})^\lambda + 3(1 - \gamma_{\sigma(i)})^\lambda$, $d_i = (1 + \gamma_{\sigma(i)})^\lambda - (1 - \gamma_{\sigma(i)})^\lambda$, and u_i ($i = 1, 2, \ldots, n$) is a collection of weights satisfying the condition (29).

4. Method for Multiple-Attribute Decision-Making Based on Hesitant Fuzzy Hamacher Power-Aggregation Operators

In this section, we use hesitant fuzzy Hamacher power-aggregation operators to develop an approach to MADM with hesitant fuzzy information.

Let $X = \{x_1, x_2, \ldots, x_n\}$ be a set of n alternatives, and $G = \{g_1, g_2, \ldots, g_m\}$ be a set of m attributes, whose weight vector is $w = (w_1, w_2, \ldots, w_m)^T$, satisfying $w_i > 0$ ($i = 1, 2, \ldots, m$) and $\sum_{i=1}^m w_i = 1$, where w_i denotes the importance degree of the attribute g_i. Suppose the group of decision-makers provides the evaluating value that the alternative x_j ($i = 1, 2, \ldots, n$) satisfies the attribute g_i ($j = 1, 2, \ldots, m$) represented by the HFEs h_{ij} ($i = 1, 2, \ldots, m; j = 1, 2, \ldots, n$). All these HFEs are contained in the hesitant fuzzy decision matrix $D = \left(h_{ij}\right)_{m \times n}$.

The following steps can be used to solve the MADM problem under the hesitant fuzzy environment, and obtain an optimal alternative:

Step 1: Obtain the normalized hesitant fuzzy decision matrix. In general, the attribute set G can be divided two subsets: G_1 and G_2, where G_1 and G_2 are the set of benefit attributes and cost attributes, respectively. If all the attributes are of the same type, then the evaluation values do not need normalization, whereas if there are benefit attributes and cost attributes in MADM, in such cases, we may transform the evaluation values of cost type into the evaluation values of the benefit type by the following normalization formula:

$$r_{ij} = \begin{cases} h_{ij}, & j \in G_1 \\ h_{ij}^c, & j \in G_2, \end{cases} \tag{45}$$

where $h_{ij}^c = \cup_{\gamma_{ij} \in \bar{h}_{ij}} \{1 - \gamma_{ij}\}$ is the complement of h_{ij}. Then we obtain the normalized hesitant fuzzy decision matrix $H = (r_{ij})_{m \times n}$.

Step 2: Calculate the supports

$$\text{Sup}(r_{ij}, r_{kj}) = 1 - d(r_{ij}, r_{kj}), \ j = 1, 2, \ldots, n, \ i, k = 1, 2, \ldots, m, \tag{46}$$

which satisfy conditions (1)–(3) in Definition 5. Here we assume that $d(r_{ij}, r_{kj})$ is the hesitant normalized Hamming distance between r_{ij} and r_{kj} given in Equation (4).

Step 3: Calculate the weights of evaluating values. Use the weights w_i ($i = 1, 2, \ldots, m$) of attributes g_i ($i = 1, 2, \ldots, m$) to calculate the weighted support $T(r_{ij})$ of the HFE r_{ij} by the other HFEs r_{kj} ($k = 1, 2, \ldots, m$, and $k \neq i$):

$$T(r_{ij}) = \sum_{k=1, k \neq i}^{m} w_k \text{Sup}(r_{ij}, r_{kj}) \tag{47}$$

and then use the weights w_i ($i = 1, 2, \ldots, m$) of attributes g_i ($i = 1, 2, \ldots, m$) to calculate the weights ρ_{ij} ($i = 1, 2, \ldots, m$) that are associated with HFEs r_{ij} ($i = 1, 2, \ldots, m$):

$$\rho_{ij} = \frac{w_i(1 + T(r_{ij}))}{\sum_{i=1}^{m} w_i(1 + T(r_{ij}))}, \ i = 1, 2, \ldots, m, \tag{48}$$

where $\rho_{ij} \geq 0$, $i = 1, 2, \ldots, m$, and $\sum_{i=1}^{m} \rho_{ij} = 1$.

Step 4: Compute overall assessments of alternatives. Use the HFHPWA operator (Equation (6)):

$$r_j = \text{HFHPWA}_\zeta(h_1, h_2, \ldots, h_n)$$
$$= \cup_{\gamma_{1j} \in r_{1j}, \gamma_{2j} \in r_{2j}, \ldots, \gamma_{mj} \in r_{mj}} \left\{ \frac{\prod_{i=1}^{m}(1 + (\zeta - 1)\gamma_{ij})^{\rho_{ij}} - \prod_{i=1}^{m}(1 - \gamma_{ij})^{\rho_{ij}}}{\prod_{i=1}^{m}(1 + (\zeta - 1)\gamma_{ij})^{\rho_{ij}} + (\zeta - 1)\prod_{i=1}^{m}(1 - \gamma_{ij})^{\rho_{ij}}} \right\}, \tag{49}$$

or the HFHPWG operator (Equation (11)):

$$r_j = \text{HFHPWG}_\zeta(h_1, h_2, \ldots, h_n)$$
$$= \cup_{\gamma_{1j} \in r_{1j}, \gamma_{2j} \in r_{2j}, \ldots, \gamma_{mj} \in r_{mj}} \left\{ \frac{\zeta \prod_{i=1}^{m}(\gamma_{ij})^{\rho_{ij}}}{\prod_{i=1}^{m}(1 + (\zeta - 1)(1 - \gamma_{ij}))^{\rho_{ij}} + (\zeta - 1)\prod_{i=1}^{m}(\gamma_{ij})^{\rho_{ij}}} \right\}. \tag{50}$$

to aggregate all the evaluating values $\bar{r}_{ij}$ ($1 = 1, 2, \ldots, m$) of the jth column and get the overall rating value $\bar{r}_j$ corresponding to the alternative x_j ($j = 1, 2, \ldots, n$).

Step 5: Rank the order of all alternatives. Use the method in Definition 3 to rank the overall rating values r_j ($j = 1, 2, \ldots, n$), rank all the alternatives x_j ($j = 1, 2, \ldots, n$) in accordance with r_j ($j = 1, 2, \ldots, n$) in descending order, and finally select the most desirable alternative(s) with the largest overall evaluation value.

Step 6: End.

Remark 9. *As previously discussed, a family of hesitant fuzzy Hamacher power-aggregation operators, including the HFHPWA, HFHPWG, GHFHPWA, GHFHPWG, HFHPOWA, HFHPOWG, GHFHPOWA, and GHFHPOWG operators, is proposed for aggregating hesitant fuzzy information. This family is composed* *of two kinds: the HFHPWA operator and HFHPWG operator, and other aggregation operators are developed based on them. Therefore, in Step 3, the HFHPWA and HFHPWG operators are chosen to aggregate hesitant fuzzy information.*

Example 4. *There is a five-member board of directors of a company. They plan to invest their money in a suitable project with a lot of potential over the next five years [17]. Assume that the board of directors will evaluate the four possible projects $X = \{x_1, x_2, x_3, x_4\}$. To evaluate and rank these projects, four attributes $G = \{g_1, g_2, g_3, g_4\}$ are suggested by the Balances Score Card Methodology, i.e., (1) g_1 is the financial perspective; (2) g_2 is the customer satisfaction; (3) g_3 is the internal business process perspective; and (4) g_4 is the learning and growth perspective. Please note that these attributes are all benefit attributes and the corresponding weight vector is $w = (0.2, 0.3, 0.15, 0.35)^T$. The five members of the board of directors provide the evaluating values of the projects x_j ($j = 1, 2, 3, 4$) with respect to attributes g_i ($i = 1, 2, 3, 4$) and construct their hesitant fuzzy decision matrix $D = (h_{ij})_{4 \times 4}$ (see Table 1), where $h_{ij} \in H$ is a HFE that denotes all of the possible values for alternative x_j under the attribute g_i.*

Table 1. Hesitant fuzzy decision matrix D.

	x_1	x_2	x_3	x_4
g_1	$\{0.2, 0.4, 0.7\}$	$\{0.2, 0.4, 0.7, 0.9\}$	$\{0.3, 0.5, 0.6, 0.7\}$	$\{0.3, 0.5, 0.6\}$
g_2	$\{0.2, 0.6, 0.8\}$	$\{0.1, 0.2, 0.4, 0.5\}$	$\{0.2, 0.4, 0.5, 0.6\}$	$\{0.2, 0.4\}$
g_3	$\{0.2, 0.3, 0.6, 0.7, 0.9\}$	$\{0.3, 0.4, 0.6, 0.9\}$	$\{0.3, 0.5, 0.7, 0.8\}$	$\{0.5, 0.6, 0.7\}$
g_4	$\{0.3, 0.4, 0.5, 0.7, 0.8\}$	$\{0.5, 0.6, 0.8, 0.9\}$	$\{0.2, 0.5, 0.6, 0.7\}$	$\{0.8, 0.9\}$

Then we use the above proposed approach to choose the optimal project.

Step 1: Since these attributes are all benefit attributes, it is not necessary to normalize the decision matrix D.

Step 2: Use Equation(46) to calculate the supports $\mathrm{Sup}(h_{ij}, h_{kj})$ ($j = 1, 2, 3, 4$, $i, k = 1, 2, 3, 4$, $i \neq k$). For simplicity, we denote $(\mathrm{Sup}(h_{ij}, h_{kj}))_{1 \times 4}$ by Sup_{ik}, which refers to the supports between the ith and kth columns of D:

$$\mathrm{Sup}_{12} = \mathrm{Sup}_{21} = (0.900, 0.750, 0.900, 0.800), \quad \mathrm{Sup}_{13} = \mathrm{Sup}_{31} = (0.800, 0.950, 0.950, 0.867),$$
$$\mathrm{Sup}_{14} = \mathrm{Sup}_{41} = (0.800, 0.850, 0.975, 0.633), \quad \mathrm{Sup}_{23} = \mathrm{Sup}_{32} = (0.860, 0.750, 0.850, 0.667),$$
$$\mathrm{Sup}_{24} = \mathrm{Sup}_{42} = (0.860, 0.600, 0.925, 0.495), \quad \mathrm{Sup}_{34} = \mathrm{Sup}_{43} = (0.920, 0.850, 0.925, 0.767).$$

Step 3: Use Equation (47) to calculate the weighted support $T(h_{ij})$ of HFE h_{ij} by the other HFEs h_{kj} ($k = 1, 2, 3, 4$, $k \neq i$), which are contained in the matrix $T = (T(h_{ij}))_{4 \times 4}$:

$$T = \begin{pmatrix} 0.6700 & 0.6650 & 0.7538 & 0.5916 \\ 0.6100 & 0.4725 & 0.6313 & 0.4333 \\ 0.7400 & 0.7125 & 0.7688 & 0.6420 \\ 0.5560 & 0.4775 & 0.6113 & 0.3902 \end{pmatrix}$$

and use Equation (48) to calculate the weights ρ_{ij} of HFEs h_{ij} ($i = 1, 2, 3, 4$), which are contained in the matrix $V = (\rho_{ij})_{4 \times 4}$:

$$V = \begin{pmatrix} 0.2058 & 0.2150 & 0.2101 & 0.2149 \\ 0.2977 & 0.2852 & 0.2932 & 0.2903 \\ 0.1609 & 0.1659 & 0.1589 & 0.1663 \\ 0.3356 & 0.3339 & 0.3378 & 0.3285 \end{pmatrix}.$$

Step 4: Let $\zeta = 0.5$ and use the HFHPWA operator (Equation (49)) to aggregate all of the evaluating values h_{ij} ($i = 1, 2, 3, 4$) in the jth column of D and then, derive the overall rating value h_j ($j = 1, 2, 3, 4$) of the alternative x_j $(j = 1, 2, 3, 4)$. The overall rating values h_j are not listed here because of limited space. Using Definition 3, we calculate the score functions $s(h_j)$ of h_j $(j = 1, 2, 3, 4)$ as follows:

$$s(h_1) = 0.5741, s(h_2) = 0.6195, s(h_3) = 0.5256, s(h_4) = 0.6646.$$

Then we rank the h_j $(j = 1, 2, 3, 4)$ in descending order of $s(h_j)$:

$$h_4 > h_2 > h_1 > h_3.$$

Step 5: Rank all the alternatives x_j $(j = 1, 2, 3, 4)$ as follows:

$$x_4 \succ x_2 \succ x_1 \succ x_3.$$

Thus, the best alternative is x_4.

Furthermore, let $\zeta = 0.1, 0.3, 0.5, 3, 5, 7$, respectively, which represents different preferences for decision-makers on decision information. We can obtain the corresponding score values and rankings of the alternatives (listed in Table 2).

Table 2. Score values obtained with the HFHPWA operator and rankings of alternatives.

Aggregation Operator	Score Values	Rankings
$\text{HFHPWA}_{0.1}$	$s(h_1) = 0.5957, s(h_2) = 0.6501, s(h_3) = 0.5318, s(h_4) = 0.7000$	$x_4 \succ x_2 \succ x_1 \succ x_3$
$\text{HFHPWA}_{0.3}$	$s(h_1) = 0.5826, s(h_2) = 0.6314, s(h_3) = 0.5298, s(h_4) = 0.6780$	$x_4 \succ x_2 \succ x_1 \succ x_3$
$\text{HFHPWA}_{0.5}$	$s(h_1) = 0.5741, s(h_2) = 0.6195, s(h_3) = 0.5256, s(h_4) = 0.6646$	$x_4 \succ x_2 \succ x_1 \succ x_3$
HFHPWA_{3}	$s(h_1) = 0.5419, s(h_2) = 0.5752, s(h_3) = 0.5063, s(h_4) = 0.6186$	$x_4 \succ x_2 \succ x_1 \succ x_3$
HFHPWA_{5}	$s(h_1) = 0.5350, s(h_2) = 0.5653, s(h_3) = 0.5017, s(h_4) = 0.6095$	$x_4 \succ x_2 \succ x_1 \succ x_3$
HFHPWA_{7}	$s(h_1) = 0.5313, s(h_2) = 0.5598, s(h_3) = 0.4993, s(h_4) = 0.6047$	$x_4 \succ x_2 \succ x_1 \succ x_3$

To explain how the different parameter value ζ plays a role in the aggregation operator, we use the different values ζ, given by decision-makers. As shown in Table 2, the score values obtained by the HPHPWA operator become smaller as the parameter value ζ increases. Thus, decision-makers can choose the parameter value ζ according to their preferences.

Table 3. Score values obtained with the HFHPWG operator and rankings of alternatives.

Aggregation Operator	Score Values	Rankings
$\text{HFHPWG}_{0.1}$	$s(h_1) = 0.4397, s(h_2) = 0.4118, s(h_3) = 0.4187, s(h_4) = 0.4673$	$x_4 \succ x_1 \succ x_3 \succ x_2$
$\text{HFHPWG}_{0.3}$	$s(h_1) = 0.4520, s(h_2) = 0.4321, s(h_3) = 0.4373, s(h_4) = 0.4824$	$x_4 \succ x_1 \succ x_3 \succ x_2$
$\text{HFHPWG}_{0.5}$	$s(h_1) = 0.4603, s(h_2) = 0.4444, s(h_3) = 0.4483, s(h_4) = 0.4930$	$x_4 \succ x_1 \succ x_3 \succ x_2$
HFHPWG_{3}	$s(h_1) = 0.4935, s(h_2) = 0.4936, s(h_3) = 0.4904, s(h_4) = 0.5410$	$x_4 \succ x_2 \succ x_1 \succ x_3$
HFHPWG_{5}	$s(h_1) = 0.5009, s(h_2) = 0.5027, s(h_3) = 0.5005, s(h_4) = 0.5536$	$x_4 \succ x_2 \succ x_1 \succ x_3$
HFHPWG_{7}	$s(h_1) = 0.5049, s(h_2) = 0.5126, s(h_3) = 0.5061, s(h_4) = 0.5608$	$x_4 \succ x_2 \succ x_3 \succ x_1$

If the HFHPWA operator is replaced by HFHPWG operator in the above Step 4, Table 3 lists the score values of overall rating values and rankings of alternatives. The score values obtained by the HFHPWG operator become larger as parameter ζ increases. Comparing Table 2 with Table 3, we can observe that the score value obtained by the HFHPWA operator greater than the score value obtained by the HFHPWG operator for the same parameter value ζ and the same aggregation value. For HFHPWA operators, the parameter value does not affect the final rankings of the alternatives, but for the HFHPWG operators it is shown that the choice of parameter value has a greater impact on the score values and thus the rankings of the alternatives. In the above analysis, we see that while the best alternative obtained by the HFHPWA operator are the same as that obtained by the HFHPWG operator, the rankings of the alternatives differs between the HFHPWA and HFHPWG operators.

To compare our approach with some other approaches, we apply Xia and Xu's approach [10], Zhu et al.'s approach [13], and Tan et al.'s approach [17] to above example. The results of the rankings of the alternatives are shown in Table 4 below.

Table 4. Rankings of alternatives by other approaches.

Approach	Use Tool	Rankings
Xia and Xu [10]	HFWA operator	$x_4 \succ x_2 \succ x_1 \succ x_3$
	HFWG operator	$x_4 \succ x_1 \succ x_3 \succ x_2$
Zhu et al. [13]	Hesitant fuzzy TOPSIS	$x_4 \succ x_2 \succ x_1 \succ x_3$
Tan et al. [17]	HFHWA operator	$x_4 \succ x_2 \succ x_1 \succ x_3$
	HFHWG operator	$x_4 \succ x_3 \succ x_1 \succ x_2 \ (\zeta \in (0, 0.36])$
		$x_4 \succ x_1 \succ x_3 \succ x_2 \ (\zeta \in (0.36, 1.39])$
		$x_4 \succ x_1 \succ x_2 \succ x_3 \ (\zeta \in (1.39, 3.50])$
		$x_4 \succ x_2 \succ x_1 \succ x_3 \ (\zeta \in (3.50, 10.0])$

From this analysis, we can see that the best alternative is the same for the both HFHPWA and HFHPWG operators, or both HFWA and HFWG operators, or both HFHWA and HFHWG operators, but the ranking of alternatives is different between the HFHPWA and HFHPWG operators, or HFWA and HFWG operators, or HFHWA and HFHWG operators. It reflects that the final results may be different by different types of hesitant fuzzy aggregation operators. Also, the ranking of alternatives obtained by the hesitant fuzzy TOPSIS method is the same those by the HFHPWA, HFWA, and HFHWA operators. This result shows the validity of the proposed approach in this paper.

Compared with the existing hesitant fuzzy MADM approaches, our proposed approach has two advantages: First, decision-makers often have an optimistic or pessimistic attitude in the face of decision information. In this case, optimistic attitude often leads to a preference for risk-seeking, and pessimistic one results in a preference for avoiding risk. The parameter ζ takes into account the decision-maker's subjective attitude to decision-making problem and are therefore useful in obtaining a better decision result. Second, different parameter values clearly indicate changes in the ranking of alternatives. Compared to a fixed evaluated result obtained by existing aggregation operators such as the HFWA and HFWG operators, our evaluated result can better reflect the variety.

5. Conclusions

Hesitant fuzzy information aggregation is one of key issues in the hesitant fuzzy MADM, an important field of research in decision science in an uncertain environment as well as HFS theory. Based on Hamacher operations of HFEs, in this paper, we have developed a family of hesitant fuzzy Hamacher power-aggregation operators, including the HFHWPA, HFHPWG, GHFHPWA, GHFHPWG, HFHPOWA, HFHPOWG, GHFHPOWA, and GHFHPOWG operators. Some basic properties of the proposed aggregation operators, such as boundedness and monotonicity, and the relationships between them have been investigated and discussed. We compared the proposed aggregation operators with the hesitant fuzzy aggregation operators developed by Yu et al. [18] and Zhang [15] and represented their corresponding relations. These proposed hesitant fuzzy Hamacher power-aggregation operators are integrated treatment of operators proposed by Yu et al. [18] and Zhang [15], and provide a complement to the existing work on HFSs. An approach of the hesitant fuzzy MADM based on the HFHPWA and HFHPWG operators has been developed and an example of money investment selection has been provided to describe the hesitant fuzzy MADM process. Some advantages of our proposed approach are shown by comparison with those previously proposed by Xia and Xu [10], Zhu et al. [13] and Tan et al. [17].

In future work, we will present a series of hesitant fuzzy power-aggregation operators using Frank t-norm and t-conorm and apply them to develop approaches for multiple-attribute group decision-making. Furthermore, we will discuss the extension of power-aggregation operators to probabilistic hesitant fuzzy environment.

Author Contributions: J.H.P. drafted the initial manuscript and conceived the MADM framework. M.J.S. provided the relevant literature review and illustrated example. M.J.S. and K.H.K. revised the manuscript and analyzed the data.

Funding: This work was supported by a Research Grant of Pukyong National University (2019).

Conflicts of Interest: The authors declare no conflict of interest.

References

1. Zadeh, L.A. Fuzzy sets. *Inf. Control* **1965**, *8*, 338–353. [CrossRef]
2. Zadeh, L.A. The concept of a linguistic variable and its application to approximate reasoning-I. *Inf. Sci.* **1975**, *8*, 199–249. [CrossRef]
3. Atanassov, K. Intuitionistic fuzzy sets. *Fuzzy Sets Syst.* **1986**, *20*, 87–96. [CrossRef]
4. Atanassov, K.; Gargov, G. Interval-valued intuitionistic fuzzy sets. *Fuzzy Sets Syst.* **1989**, *31*, 343–349. [CrossRef]
5. Dubois, D.; Prade, H. *Fuzzy Sets and Systems: Theory and Applications*; Academic Press: Cambridge, MA, USA, 1980.
6. Miyamoto, S. Fuzzy multisets and their generalizations. In Proceedings of the International Conference on Membrane Computing, Curtea de Arges, Romania, 21–25 August 2000; pp. 225–236.
7. Torra, V. Hesitant fuzzy sets. *Int. J. Intell. Syst.* **2010**, *25*, 529–539. [CrossRef]
8. Torra, V.; Narukawa, Y. On hesitant fuzzy sets and decision. In Proceedings of the 18th IEEE International Conference on Fuzzy Systems, Jeju Island, Korea, 20–24 August 2009; pp. 1378–1382.
9. Rodríguez, R.; Martínez, L.; Herrera, E. Hesitant fuzzy linguistic term sets for decision making. *IEEE Trans. Fuzzy Syst.* **2012**, *20*, 109–119. [CrossRef]
10. Xia, M.M.; Xu, Z.S. Hesitant fuzzy information aggregation in decision making. *Int. J. Approx. Reason.* **2011**, *52*, 395–407. [CrossRef]
11. Xu, Z.S.; Xia, M.M. Distance and similarity measures for hesitant fuzzy sets. *Inf. Sci.* **2011**, *181*, 2128–2138. [CrossRef]
12. Wei, G.W. Hesitant fuzzy prioritized operatros and their application to multiple attribute group decision making. *Knowl.-Based Syst.* **2012**, *31*, 176–182. [CrossRef]
13. Zhu, B.; Xu, Z.S.; Xia, M.M. Hesitant fuzzy geometric Bonferroni means. *Inf. Sci.* **2012**, *205*, 72–85. [CrossRef]
14. Xia, M.M.; Xu, Z.S.; Chen, N. Some hesitant fuzzy aggregation operators with their application in group decision making. *Group Dec. Negoit.* **2013**, *22*, 259–279. [CrossRef]
15. Zhang, Z.M. Hesitant fuzzy power aggregation operators and their application to multiple attribute group decision making. *Inf. Sci.* **2013**, *234*, 150–181. [CrossRef]
16. Yu, D.J. Some hesitant fuzzy infromation aggregation operators based on Einstein operational laws. *Int. J. Intell. Syst.* **2014**, *29*, 320–340. [CrossRef]
17. Tan, C.; Yi, W.; Chen, X. Hesitant fuzzy Hamacher aggregation operators for multicriteria decision making. *Appl. Soft Comput.* **2015**, *26*, 325–349. [CrossRef]
18. Yu, Q.; Hou, F.; Zhai, Y.; Du, Y. Some hesitant fuzzy Einstein aggregation operators and their application to multiple attribute group decision making. *Int. J. Intell. Syst.* **2016**, *31*, 722–746. [CrossRef]
19. Kahraman, C.; Oztaysi, B.; Ucal Sarı, I.; Turanoglu, E. Fuzzy analytic hierarchy process with interval type-2 fuzzy sets. *Knowl. Base Syst.* **2016**, *59*, 48–57. [CrossRef]
20. Wang, C.N.; Yang, C.Y.; Cheng, H.C. Fuzzy multi-criteria decision-making model for supplier evaluation and selection in wind power plant project. *Mathematics* **2019**, *7*, 417. [CrossRef]
21. Ziemba, P.; Becker, J. Analysis of the digital divide using fuzzy forecasting. *Symmetry* **2019**, *11*, 166. [CrossRef]
22. Mari, S.I.; Memon, M.S.; Ramzan, M.B.; Qureshi, S.M.; Iqbal, M.W. Interactive fuzzy multi criteria decision making approach for supplier selection and order allocation in a resilient supply China. *Mathematics* **2019**, *7*, 137. [CrossRef]

23. Wang, T.C.; Tsai, S.Y. Solar panel supplier selection for the photovoltaic system design by using fuzzy multi-criteria decision making (MCDM) approaches. *Energies* **2018**, *11*, 1989. [CrossRef]
24. Yager, R.R. The power average operator. *IEEE Trans. Syst. Man Cybern.* **2001**, *31*, 724–731. [CrossRef]
25. Xu, Z.S.; Yager, R.R. Power-geometric operators and their use in group decision making. *IEEE Trans. Fuzzy Syst.* **2010**, *18*, 94–105.
26. Zhou, L.G.; Chen, H.Y.; Liu, J.P. Generalized power aggregation operators and their applications in group decision making. *Comput. Ind. Eng.* **2012**, *62*, 989–999. [CrossRef]
27. Xu, R.N.; Zhai, X.Y. Extensions of the analytic hierarchy process in fuzzy environment. *Fuzzy Sets Syst.* **1992**, *52*, 251–257.
28. Chen, T.Y. A comparative analysis of score functions for multiple criteria decision making in intuitionistic fuzzy settings. *Inf. Sci.* **2011**, *181*, 3652–3676. [CrossRef]
29. Guo, K.H.; Li, W.L. An attitudinal-based method for constructing intuitionistic fuzzy information in hybrid MADM under uncertainty. *Inf. Sci.* **2012**, *208*, 28–38. [CrossRef]
30. Wu, J.Z.; Chen, F.; Nie, C.P.; Zhang, Q. Intuitionistic fuzzy-valued Choquet integral and its application in multicriteria decision making. *Inf. Sci.* **2013**, *222*, 509–527. [CrossRef]
31. Xu, Z.S. Methods for aggregating interval-valued intuitionistic fuzzy information and their application to decision making. *Control Decis.* **2007**, *22*, 215–219.
32. Chou, J.R. A linguistic evaluation approach for unversal design. *Inf. Sci.* **2012**, *190*, 76–94. [CrossRef]
33. Delgado, M.; Herrera, F.; Herrera-Viedma, E.; Martínez, L. Combining numerical and linguistic information in group decision making. *Inf. Sci.* **1998**, *107*, 177–194. [CrossRef]
34. Delgado, M.; Herrera, F.; Herrera-Viedma, E.; Martínez, L. A fusion approach for managing multigranularity linguistic term sets in decision making. *Fuzzy Sets Syst.* **2000**, *114*, 43–58.
35. Xu, Z.S. Uncertain linguistic aggregation opweerators based on approach to multiple attribute group decision making under uncertain linguistic environment. *Inf. Sci.* **2004**, *168*, 171–184. [CrossRef]
36. Xu, Z.S. Induced uncertain linguistic OWA operators applied to group decision making. *Inf. Fusion* **2006**, *7*, 231–238. [CrossRef]
37. Xu, Z.S. An approach based on the uncertain LOWG and induced uncertain LOWG operators to group decision making with uncertain multiplicative linguistic preference relations. *Decis. Support Syst.* **2006**, *41*, 488–499. [CrossRef]
38. Xu, Z.S.; Cai, X.Q. Uncertain power average operators for aggregating interval fuzzy preference relations. *Group Dec. Negoit.* **2012**, *21*, 381–397. [CrossRef]
39. Park, J.H.; Park, J.M.; Seo, J.J.; Kwun, Y.C. Power harmonic operators and their applications in group decision making. *J. Comput. Anal. Appl.* **2013**, *15*, 1120–1137.
40. Xu, Z.S. Approaches to multiple attribute group decision making based on intuitionistic fuzzy power aggregation operators. *Knowl.-Based Syst.* **2011**, *24*, 749–760. [CrossRef]
41. Xu, Y.J.; Merigó, J.M.; Wang, H.M. Linguistic power aggregation operators and their application to muktiple attribute group decision making. *Appl. Math. Model.* **2012**, *36*, 5427–5444. [CrossRef]
42. Xu, Y.J.; Wang, H.M. Approaches based on 2-tuple linguistic power aggregation operators for multiple attribute group decision making under linguistic environment. *Appl. Soft Comput.* **2011**, *11*, 3988–3997. [CrossRef]
43. Xu, Y.J.; Wang, H.M. Power geometric operators for group decision making under multiplicative linguistic preference relations. *Int. J. Uncertain. Fuzz. Knowl. Based Syst.* **2012**, *20*, 139–159. [CrossRef]
44. Zhou, L.G.; Chen, H.Y. A generalization of the power aggregation operators for linguistic environment and its application in group decision making. *Knowl.-Based Syst.* **2012**, *26*, 216–224. [CrossRef]
45. Klement, E.; Mesiar, R.; Pap, E. *Triangular Norms*; Kluwer Academic Publishers: Dordrecht, The Netherlands, 2000.
46. Hamacher, H. *Uber Logische Verknunpfungenn Unssharfer Aussagen undderen Zugenhorige Bewertungsfunktione*; Trappl, R., Klir, G.J., Riccardi, L., Eds.; Progress in Cybernatics and Systems Research: Washington, DC, USA, 1978; Volume 3, pp. 276–288.

47. Xu, Z.S. On consistency of the weighted geometric mean complex judgement matrix in AHP. *Eur. J. Oper. Res.* **2000**, *126*, 683–687. [CrossRef]

48. Torra, V.; Narukawa, Y. *Modeling Decisions: Information Fusion and Aggregation Operators*; Springer: Berlin, Germany, 2007.

 *mathematics*

Article

A Comparative Analysis of Simulated Annealing and Variable Neighborhood Search in the ATCo Work-Shift Scheduling Problem

Faustino Tello *, Antonio Jiménez-Martín *, Alfonso Mateos * and Pablo Lozano *

Departamento de Inteligencia Artificial, E.T.S.I. Informáticos, Universidad Politécnica de Madrid, Campus de Montegancedo S/N, 28660 Boadilla del Monte, Spain

* Correspondence: faustino.tello@upm.es (F.T.); antonio.jimenez-martin@upm.es (A.J.-M.); alfonso.mateos@upm.es (A.M.); plozano94@gmail.com (P.L.)

Received: 25 June 2019; Accepted: 16 July 2019; Published: 17 July 2019

Abstract: This paper deals with the air traffic controller (ATCo) work shift scheduling problem. This is a multi-objective optimization problem, as it involves identifying the best possible distribution of ATCo work and rest periods and positions, ATCo workload and control center changes in order to cover an airspace sector configuration, while, at the same time, complying with ATCo working conditions. We propose a three-phase problem-solving methodology based on the variable neighborhood search (VNS) to tackle this problem. The solution structure should resemble the previous template-based solution. Initial infeasible solutions are built using a template-based heuristic in Phase 1. Then, VNS is conducted in Phase 2 in order to arrive at a feasible solution. This constitutes the starting point of a new search process carried out in Phase 3 to derive an optimal solution based on a weighted sum fitness function. We analyzed the performance in the proposed methodology of VNS against simulated annealing, as well as the use of regular expressions compared with the implementation in the code to verify the feasibility of the analyzed solutions, taking into account four representative and complex instances of the problem corresponding to different airspace sectorings.

Keywords: air traffic management; work-shift scheduling problem; variable neighborhood search; performance analysis

---

1. Introduction

The key concept at the heart of air traffic management (ATM) network operations is air traffic flow and capacity management (ATFCM). ATFCM should optimize traffic flows so that airlines can operate safe and efficient flights depending on air traffic control capacity. In Europe, the network manager operations center (NMOC) constantly monitors the balance between the airspace capacity and traffic load. NMOC activities are divided into four –strategic, pre-tactical, tactical and post-operational– phases [1].

The strategic phase is related to capacity prediction at ATC centers by air navigation service providers (ANSPs). ANSPs prepare a routing scheme with the help of NMOC seven days ahead of operations.

The pre-tactical phase is related to the definition of the initial network plan. The NMOC publishes the agreed plan for the day of operations, informing ATC units and aircraft operators about the ATFCM measures affecting European airspace from one to six days ahead of operations.

The tactical phase updates the plan for the day of operations according to real-time traffic demand where the NMOC monitors the situation and continuously optimizes capacity. Delays are minimized by providing aircraft affected by changes with alternative solutions on the day of operations.

The post-operational phase is related to operational process improvement by comparing planned and measured outcomes covering all ATFCM domains and units. Operational processes are measured in order to develop best practices and/or analyze lessons learned after the day of operations.

In this paper, we focus on the pre-tactical phase. This phase has to solve the very important problem of determining how the available air traffic controllers (ATCos) are assigned to each open sector to cover a sectorization structure (established in the strategic phase) for a specified amount of time. This assignment has to comply with a number of strong constraints accounting for the ATCo working conditions.

The sectorization changes throughout the day depend on aircraft traffic. More sectors are opened if the air traffic volume increases. This steps up the demand for ATCos who can only handle a limited amount of traffic.

This is a timetabling and scheduling problem. Timetabling and scheduling problems are combinatorial problems, which, on the grounds of size and complexity, cannot be solved by exact methods. For examples of other timetabling and scheduling problem-solving approaches, see [2,3].

ATCo scheduling software has already been developed within the ATM field [4]. These tools have both strengths and weaknesses [5]. Hardly any of this software has been reported in detail, sometimes because they are in-house tools. Three ATCo scheduling problem codifications were reported alongside three optimization techniques [6]. Another solution [7] is composed of a hybrid technique combining propositional satisfiability problem solving [8] and hill climbing.

A simplified version of the ATCo work shift scheduling problem for Spanish airports was solved by minimizing the number of ATCos required to cover a given airspace sectoring in compliance with Spanish ATCo working conditions [9]. The search process employs regular expressions to check solution feasibility. The solutions that output an optimal number of ATCos are used as the starting point for another optimization process targeting balanced ATCo workloads. This simplified version of the problem analyses straightforward scenarios and accounts for a core with only one type of sector for a 24-h period. Consequently, there are no constraints on ATCo distribution across sectors.

Cores including two sector types (en-route and approach sectors) and accounting for ATCos with different operating credentials were considered in [10]. This proposal focuses on the optimization of only one shift in accordance with a previously specified number of ATCos to cover a specified airspace sector configuration. This proposal adopts a multi-objective approach, accounting for ATCo work and rest periods, positions and workload distribution, the number of control center changes, and the solution structure. It proposes a three-phase problem-solving methodology. In the first phase, a template-based heuristic was used to identify unfeasible solutions. In the second phase, a number of independent simulated annealing (SA) metaheuristic runs were conducted to arrive at feasible solutions using regular expressions to check compliance with ATCo working conditions. In the third phase, simulated annealing was conducted by multiple independent runs to optimize the objective functions of the original feasible solutions again taking into account ATCo working conditions. This optimization process took into account the ordinal information on objective importance using the rank-order centroid function to transform a multiple into a single optimization problem.

In this paper, we consider the same multi-objective problem as [10], albeit using an adaptation of variable neighborhood search (VNS) rather than SA in the three-phase problem-solving methodology. Four representative and complex instances of the problem corresponding to different airspace sectorings provided by the Spanish ATM Research, Development and Innovation Reference Center (CRIDA) are now used to compare the performance of both metaheuristics in the three-phase problem-solving methodology. Moreover, the use of regular expressions to verify the ATCo labor conditions (constraints) is compared against implementation in the code in terms of execution times.

The paper is structured as follows. Section 2 describes the ATCo work-shift scheduling problem. Section 3 describes the proposed problem-solving methodology. Section 3.1 presents a template-based heuristic to identify unfeasible solutions. Then, some notions of VNS and its adaptation to the ATCo work-shift scheduling problem are provided in Section 3.2. Finally, we describe the second and third phases of the methodology aimed at reaching a feasible and an optimal solution in Sections 3.3 and 3.4, respectively. In Section 4, four real instances are used to illustrate the proposed methodology and to compare the performance of SA against the proposed adaptation of VNS and analyze the use of regular expressions. Finally, some conclusions are provided in Section 6.

2. Problem Description

There are limits on the amount of traffic that human ATCos can handle. Therefore, air traffic conditions the number of ATCos required, as airspace sectors are created and reduced to deal with demand, resulting in varying numbers of ATCos. The sectorization of the airspace according to estimated traffic for a specified period can be defined in advance and is denoted as airspace sector configuration.

A core is composed of a set of sectors, and any one sector may belong to several cores. A control center may be responsible for managing one or more cores. Each core should be solved separately, unless there are sectors belonging to more than one core. In this case, ATCos should be simultaneously assigned to the respective cores.

There are two types of sectors: approach and en-route sectors. Depending on airport procedures, approach sectors are generally five to 10 nautical miles (9 to 18 km) from the airport, whereas en-route sectors are usually further way.

Two ATCos with different roles operate each sector. The executive ATCo communicates with aircraft, instructing pilots on how to avoid each other, whereas the planner ATCo foresees possible conflicts between aircrafts which he or she reports to the executive ATCo. ATCos are accredited to operate a particular sector and categorized as PTD or CON ATCos. A PTD ATCo can operate en-route and approach sectors, whereas a CON ATCo can only operate en-route sectors. Figure 1 is an example of an airspace sectorization for the Barcelona eastern route in Spain. Each interval is associated with a configuration (3C, 4A, 6A, ...), where the number represents the number of open sectors and the letter refers to the sector configuration, i.e., there are two sectorizations with a different spatial distribution of the same number of sectors (5A and 5B in Figure 1).

Figure 1 shows one of the four examples used to illustrate our problem-solving methodology. The airspace is divided into three sectors (configuration 3C), after which one of the sectors is divided into a further two sectors. The result is configuration 4A, which is operational for one hour. The next configuration is 6A used for 40 min. See Figure 10 for further details.

Figure 1. Barcelona eastern route airspace sectoring.

A day (24 h) is divided into night (N), morning (M) and afternoon (A) periods, covered by five different ATC shifts: long morning (LMS) (5:40–14:00 h.), morning (MS) (6:20–14:00), afternoon (AS) (14:00–21:20), long afternoon (LAS) (14:00–22:20) and night (NS) (21:20–6:20). At certain times, ATCo shifts overlap: AS and LAS from 14:00 to 21:20; NS and LAS from 21:20 to 22:20, NS and LMS from 5:40 to 6:20, and MS and LMS from 6:20 to 14:00. NS ATCos are the only ATCos at work from 22:20 to 5:40.

On top of the division by shifts, ATCo working conditions also have to be taken into account. Royal Decree 1001/2010 and Law 9/2010, regulating the provision of air traffic services, stipulate these conditions, including constraints on minimum and maximum working and resting times, how long ATCos can spend in different positions, the maximum number of sectors that an ATCo can operate during a shift, etc. A list of ATCo working conditions is available in [10].

The ATCo work shift scheduling problem that we intend to solve should achieve the following objectives in accordance with a specified airspace sectorization and a specified number of ATCos with their respective accreditations:

- ATCo work and rest periods and position should respect the specified values.
- The number of control center changes should be reduced to the minimum.
- The solution structure should resemble the previous template-based solution for ease of understanding by control center staff with a view to potential manual changes.
- ATCo workload distribution should be balanced.

Experts from the Reference Center for Research, Development and Innovation in ATM (CRIDA, www.crida.es), a non-profit joint venture between ENAIRE, Spain's air navigation manager, the Universidad Politécnica de Madrid, and Ineco, a global infrastructure engineering and consultancy leader, ranked the above objectives by importance.

3. Problem-Solving Methodology

This section outlines a three-phase methodology to solve the stated ATCo work shift scheduling problem for a given airspace sectorization. Phase 1 sets out a heuristic which is used to construct ten different initial solutions by modifying rest period lengths using an optimized template. These initial solutions require more ATCos than are available and do not satisfy all working conditions.

In Phase 2, an algorithm based on variable neighborhood search (VNS) is applied to the infeasible solutions achieved in the first phase, in order to yield a feasible solution. The aim is to reduce the number of ATCos used to meet the number of available ATCos, while penalizing the number of times labour conditions are violated until a feasible solution is achieved.

A VNS-based algorithm is run again on the feasible solution output in Phase 2. This algorithm should optimize the objective functions. The objective functions represent the ATCo work and rest periods and positions, ATCo workload distribution, the number of control center changes, and the similarity of the solution structure to the previous template-based solution. The original multi-objective optimization problem is then transformed into a single weighted optimization problem taking into account the objectives ranked by importance by CRIDA experts. This ordinal information is used to specify the centroid-based weights. Solutions are represented by a matrix. The matrix columns represent time slots, and the rows ATCos. Time slots are equivalent to five minutes because five is the greatest common divisor for the applicable constraint times (e.g., ATCos have to work for at least 15 min in the new sector and a sector has to be open for at least 20 min). The Phase 1 heuristic establishes the number of rows, which will not necessarily be the same across the initial solutions.

Each matrix element (i, j) represents the state of ATCo i in time slot j. It is symbolized by three letters. The value 111 represents a resting ATCo, uppercase letters [A-Z] indicate that the ATCo is working as an executive operator, whereas lowercase letters [a-z] are used for planner positions.

Figure 2 illustrates solutions using colors to represent sectors. Rest periods are colored white. Figure 2 shows the solution for the airspace sector configuration illustrated in the Figure 1. This configuration is manned by 15 ATCos (number of rows).

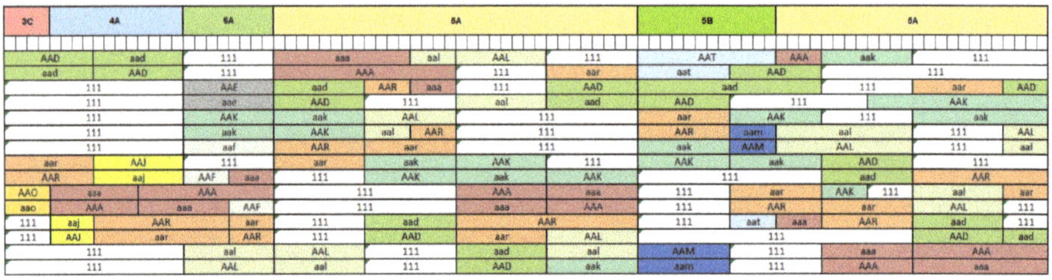

Figure 2. Example of solution representation.

3.1. Phase 1: A Heuristic for the Construction of Initial Solutions

We propose a heuristic to build a set of initial solutions with different rest periods. The Phase 2 algorithm then outputs a feasible solution based on the initial solutions output in Phase 1. The proposed heuristic is based on an optimized template (see Figure 3), with three ATCos covering a sector for 96 time slots (eight hours). ATCo positions (executive or planner) are not assigned until later.

ATCo 1	Working	Resting	Working	Working	Resting	Working	Working	...
ATCo 2	Resting	Working	Working	Resting	Working	Working	Resting	...
ATCo 3	Working	Working	Resting	Working	Working	Resting	Working	...

Figure 3. Shift template.

All work periods of are of equal length and twice as long as rest periods. The heuristic builds different, albeit similarly structured, initial solutions if the rest period duration varies. As the working conditions specify that each rest period should last at least 15 min (three slots) and the minimum and maximum work periods are six and twenty four slots, respectively, a rest period must be at least three and at most twelve slots long. Thus, the heuristic builds ten different initial solutions.

The steps of the heuristic are as follows. First, starting from an empty solution, the heuristic adds templates to cover the airspace sectoring. Note that more ATCos than available could be incorporated to the solution in this process. Then, ATCo positions (executive or planner) are allocated taking into account the number of open sectors in each time slot.

Next, we repair the solution to improve feasibility. To do this, we try to transfer a work period without the minimum length from one ATCo to another and extend a work period that does not have the minimum length using a work period from another ATCo.

Finally, we allocate available resources, i.e., we assign an available ATCo to each row in the built initial solution, taking into account the sectors in that row and the ATCo accreditations.

More details about the implementation of the heuristic are available in [10].

3.2. Variable Neighborhood Search and Its Adaptation to the ATC Work-Shift Scheduling Problem

The idea underlying variable neighborhood search (VNS) is to successively explore a set of predefined neighborhoods to find better solutions [11]. VNS explores a set of neighborhoods either at random or systematically in search of local optima. Conducting a local search of diverse neighborhoods potentially generates different local optima, where the global optimum is a local optimum for a given neighborhood. Different neighborhoods generate different landscapes.

The basic version of VNS is shown in Algorithm 1. However, other variants of this basic VNS, such as variable neighborhood descent, reduced variable neighborhood search and variable neighborhood decomposition search can be found in the literature. They depend on:

- The order in which the neighborhoods are used: forward VNS, which starts with $k = 1$ and increases k by one if no better solutions are found; otherwise set $k \leftarrow 1$; backward VNS, which starts with $k = k_{max}$ and decreases k by one if no better solutions are found, and extended version, which uses parameters k_{min} and k_{step}, sets $k \leftarrow k_{min}$ and increases k by k_{step} if no better solution is found.
- The acceptance of worse solutions. For instance, skewed VNS accepts if $f(x'') - \alpha d(x, x'') < f(x)$, being $d(x, x'')$ the distance between candidate solutions.

Algorithm 1 Basic VNS.

Require: N_k: set of neighborhood structures, $k = 1, ..., k_{max}$
Ensure: Best solution found.
  1: $k = 1$
  2: Generate an initial solution x
  3: **repeat**
  4:     Randomly generate $x' \in N_k(x)$
  5:     $x'' =$ Local-search(x', N_k)
  6:     **if** $(f(x'') < f(x))$ **then**
  7:         $x = x''$
  8:         $k = 1$
  9:     **else**
10:         $k = k + 1$
11:     **end if**
12: **until** stopping criterion

The variable neighborhood descent (VND), proposed by [12], changes the neighborhood deterministically. First, the set of neighborhoods and their order are determined, then the algorithm selects an initial solution, and it iterates until it has gone through all the neighborhoods. In each iteration, it performs a search process to reach a local optimum and evaluates if this local optimum is better than the previously derived best solution. If better, it becomes the starting solution for a new local search; otherwise, the algorithm moves on to the next neighborhood to be explored.

The reduced VNS, proposed by [13], is similar to basic VNS except that no iterative improvement procedure is applied. It explores different neighborhoods randomly and quickly reaches good quality solutions for large instances.

The variable neighborhood decomposition search, put forward by [13], generates subproblems by keeping all but k solutions components fixed, and applies local search only to the k "free" components.

In the biased VNS, reported by [13], once a good solution has been found after a good exploitation, it is usually necessary to move away from this good neighbor to try to find a better solution. This new search is usually very similar to a multi-start heuristic, but this method is usually too expensive in terms of efficiency. Therefore, the algorithm chooses to move towards the best neighbor depending on a distance function between the candidate solutions multiplied by a parametrizable value α.

Parallel VNS, described by [14], is a variant aimed at parallelizing the local search processes within VNS to derive and then compare several solutions resulting from starting points.

The VNS adaptation used in this paper is similar to VND in the sense that the algorithm restarts the search process once a solution is found that improves upon the previously best solution. However, whereas VND restarts the process using the same neighborhood definition, our version repeats the search process in all the previously used neighborhood definitions.

Besides, the basic VNS randomly generates a solution from the neighborhood under consideration and starts a local search from this solution, whereas our adaptation of VNS starts the search from the solution in the previous iteration since the local search process is not completely deterministic, rendering the previous step (randomly generate the initial solution) unnecessary.

The following four types of neighborhoods have been considered for the VNS adaptation proposed in this paper:

1. *First neighborhood.* It is based on a time slot exchange between two ATCos fulfilling the following constraints (see Figure 4):

    (a) The length of the time interval to be exchanged cannot be greater than 18 slots.

    (b) The first ATCo must be located in a row that is higher up than the second ATCo in the solution matrix.

    (c) The time interval for the first ATCo must correspond to a work period, whereas, for the second controller, it has to be a rest period.

(d) The exchange process will only be performed if the second ATCo was previously working in the same sector and position just before or after the exchanged time slot (see Figure 4) i.e., we are extending the work period for the second ATCo.

Figure 4. First neighborhood example.

2. *Second neighborhood.* It also consists of a time slot exchange between two ATCos, the only difference from the first neighborhood definition above being that the two time periods involved in the exchange process must be work periods (see Figure 5).

Figure 5. Second neighborhood example.

3. *Third neighborhood.* This neighborhood is similar to the first one except for the fact that the exchange process will be performed if the second ATCo was previously working in the same sector and position just before or after the exchange time slot (see Figure 6), i.e., it is not necessary to extend the work period for the second ATCo.

Figure 6. Third neighborhood example.

4. *Fourth neighborhood.* This neighborhood is similar to the second one except for the fact that the exchange process will be performed if the second ATCo was previously working in the same sector and position just before or after the exchange time slot (Figure 7), i.e., it is not necessary to extend the work period for the second ATCo.

Figure 7. Fourth neighborhood example.

Figure 8 describes how the four neighborhood definitions are used in our VNS adaptation for Phases 2 and 3, where *time_int* represents the time interval length to be exchanged, *neighborhood* identifies which of the neighborhood definitions under consideration is used, and *laps* refers to the times number the neighborhood definitions have been used.

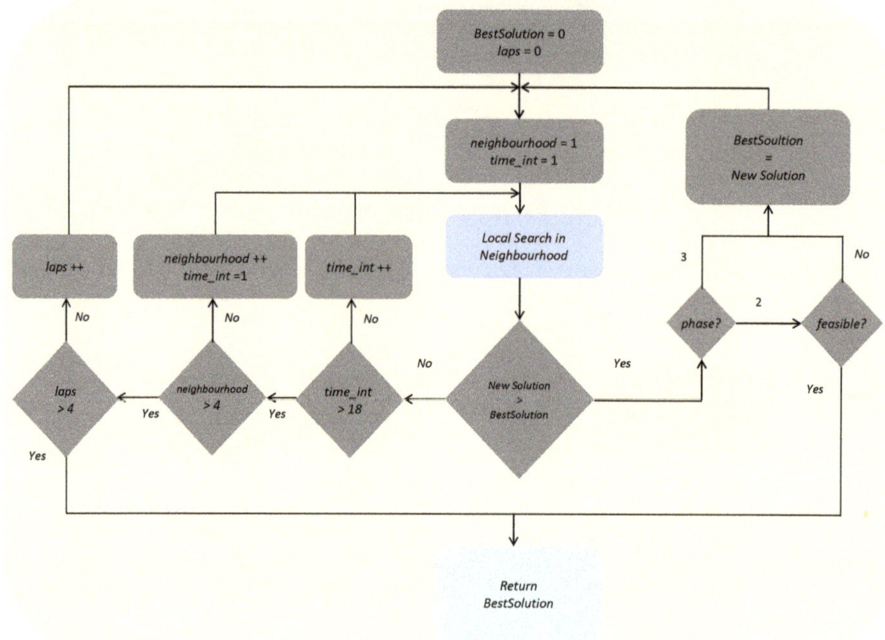

Figure 8. Adapted VNS algorithm for Phases 2 and 3.

The time interval length to be exchanged is initially set to one time slot ($time\_int = 1$) and we use the first neighborhood definition ($neighborhood = 1$). Then, if the local search outputs a new solution that is better than the current *BestSolution* in Phase 2, we check if the new solution is feasible. If it is, we output that solution and Phase 2 finishes. If the new solution is not feasible or we are performing Phase 3, then the new solution becomes the *BestSolution* and a new iteration is carried out.

If the solution achieved in the local search is worse than the current *BestSolution*, the time interval length is increased, and the local search is executed again. If the time interval length is 18 and the output solution is still worse than the current *BestSolution*, then we start using the second neighborhood definition and the time interval length is initialized ($time\_int = 1$). If we have used the four neighborhood definitions each with time interval lengths up to 18 and no solutions outperform the current *BestSolution*, then we use the first neighborhood definition again and repeat the process. This process ends when a feasible solution is reached (in Phase 2) or we have used the neighborhood definitions four times ($laps = 4$). In the second case, the current *BestSolution* is the optimal solution.

The local search consists of generating all the possible solutions within the neighborhood with a given time interval ($time\_int$). The generation process is parallelized and the respective solutions are stored in a list in the order in which they are generated. Four seeds are then established to initialize a search process within the list. We use the first seed, which identifies a solution in the list, and randomly move one position to the left or right along the list. If the fitness of the initial solution is not outperformed after exploring 25% of the solutions in the list, then we take the second seed and repeat the process. Once all four seeds have been used, the best solution visited in the process is returned.

Finally, note that we did not consider the same the neighborhood definition as in the SA used by [10] in the VNS adaptation proposed in this paper because the local searches we perform generate all the solutions in the respective neighborhood. This does not pose a problem for the four neighborhood definitions that we use in terms of cardinality. However, the number of solutions in a neighborhood based on the definition used in [10] is very high, which would lead to unaffordable computation times.

3.3. Phase 2: Deriving a Feasible Solution

In Phase 2, VNS is run on an initial solution output in Phase 1. VNS should reduce the number of ATCos covering all open sectors in a given airspace sector configuration to below the available number of ATCos, subject to the ATCo working condition constraints.

The solution matrix output by the Phase 2 search is reordered as follows: the ATCo with the smallest workload is placed at the top, followed by the ATCo with the second smallest workload, and so on. On the other hand, the ATCos with the largest workloads are placed at the bottom. This permutation tends to reduce the number of ATCos as the search progresses.

The fitness function considered in Phase 2 is:

$$max \ f = \begin{cases} w_1 h + w_2 g, & \text{if } c > n \\ g, & \text{if } c \geq n \end{cases} \tag{1}$$

where c is the number of ATCos for the current solution, n is the number of available ATCos, g is the penalty for all unsatisfied constraints (working conditions) and w_1 and w_2 represent the relative importance of the components h and g.

Function h should reduce the number of ATCos in the solution below:

$$h = \frac{1}{c^2} \sum_{i=1}^{c} ih_i, \tag{2}$$

where c is the number of ATCos and h_i is the number of work time slots for the i-th ATCo. The term $\frac{1}{c^2}$ means that a decrease in the number of ATCos improves the objective value enormously. Objective values are greater for solutions with a large workload and for ATCos with higher indexes (at the bottom of the solution matrix).

Looking at the objective function and the row permutation of the solutions in Phase 2, the search process appears to reallocate work periods from ATCos at the top of the solution matrix to ATCos in the bottom rows (with the largest workload). This augments the rest periods for the ATCos at the top of the matrix, thus decreasing the number of ATCos required to cover the airspace sector configuration.

Functions h and g must be normalized in order to correctly use the weighted fitness function, see [10].

Note that, in the four neighborhood definitions, condition (b) "The first ATCo must be located in the solution matrix in a row that is higher up than the second ATCo" is considered in the Phase 2 search process only when the number of ATCos in the solutions is greater than the available number of ATCos and is, otherwise, obviated.

Besides, the neighborhoods are initially used in the order in which they were listed in Section 3.2, but numerical experiments prove that the 4-1-3-2 order provides better solutions once the available number of ATCos is reached.

As already mentioned, VNS is conducted starting from an initial solution output in Phase 1. If a feasible solution is not reached in the search process, then we start again with another initial solution output in phase 1 (10 initial solutions were output in Phase 1). The Phase 3 search process starts from the respective feasible solution. This means that Phase 3 is only executed once. Note that a multi-start SA was proposed by [10], in which Phase 2 was executed for the 10 initial solutions derived from Phase 1, and Phase 3 was executed from each feasible solution reached in Phase 2.

3.4. Phase 3: Reaching the Optimal Solution

In Phase 3, VNS is run again, this time on the feasible solution output in Phase 2, in order to optimize the following four objective functions:

1. Objective 1: *Desirable ATCo work and rest periods, and positions.* This objective accounts for three equally important sub-objectives.

   (a) *The time ATCos remain in the same sector and working position (planner or executive) should be as close as possible to 45 min.*

   (b) *The optimal working time between breaks should be 90 min.*

   (c) *The percentage of ATCo working time in executive positions must be between 40% and 60% of the total working time (not including rest periods).*

2. Objective 2: *Similar solution structure to previous template-based solution.* Control center staff will find such a solution easier to understand, and this should facilitate any manual changes. To do this, we analyzed the template-based structure which was used as a benchmark. We concluded that rest and work periods for the same sector should be closely clustered.

3. Objective 3: *Minimization of the number of control center changes.* This objective can be achieved by reducing the number (not the duration) of rest periods.

4. Objective 4: *Balanced ATCo workloads.* ATCo workloads should be balanced in order to avoid high workloads for some, and low workloads for other, ATCs. This is achieved using the standard deviation of the work periods in the different solution matrix rows.

For the mathematical notation and the normalization process of the above objectives and sub-objectives, see [10].

We then transform the original multi-objective optimization problem into a single weighted optimization problem, whose weights, w_i, are derived using the rank-order centroid (ROC) method [15]. This single weighted optimization accounts for the ordinal information provided by CRIDA experts.

Finally, note that, in the four neighborhood definitions, condition (b) "The first ATCo must be located in the solution matrix in a row that is higher up than for the second ATCo" is not considered in the Phase 3 search process. Moreover, numerical experiments prove that the 2-1-3-4 neighborhood order provides better solutions. Note also that the infeasible solutions generated using the corresponding neighborhood definition in the Phase 3 local searches are not stored in the list.

4. Results: VNS and SA Performance Analysis

In this section, we illustrate the proposed methodology and compare the performance of VNS against SA in Phases 2 and 3 on the basis of four representative and complex instances of the problem corresponding to different airspace sectorings selected by CRIDA experts.

4.1. Illustrative Instances

Instance 1. Morning shift in Barcelona control center (10 different open sectors). Figure 9 shows the airspace sectoring of the Barcelona control center. It consists of a morning shift covering from 5:20 to 13:00 with 10 different open sectors:

- 3 open sectors (LECBLEGL, LECBLGU, LECBPPI) from 5:20 to 6:00,
- 4 open sectors (LECBLEGL, LECBLGU, LECBP1I, LECBPP2) from 6:00 to 7:40,
- 5 open sectors (LECBP1I, LECBPP2, LECBLVL, LECBLVS, LECBLVU) from 7:40 to 8:40,
- 6 open sectors (LECBPP2, LECBLVL, LECBLVS, LECBLVU, LECBP1L, LECBP1U) from 8:40 to 12:00, and
- 5 (LECBP1I, LECBPP2, LECBLVL, LECBLVS, LECBLVU) from 12:00 to 13:00. All the sectors and controllers involved belong to the *western route* core. The number of available controllers is 16, and their accreditation type is CON.

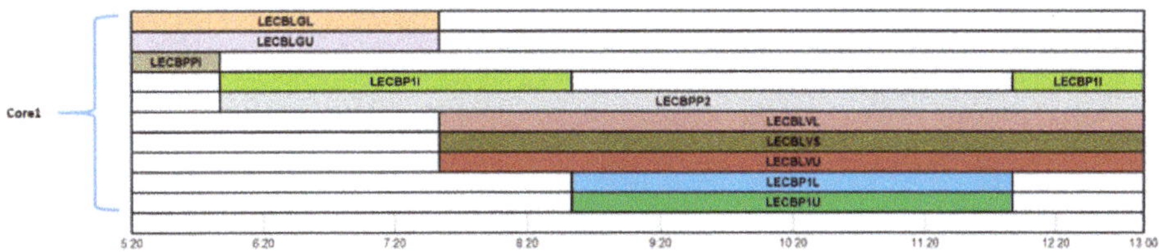

Figure 9. Sectorization of instance 1.

The problem posed by this instance is the percentage resting time constraint, where all ATCos must rest for 25% of the shift. If 16 ATCos cover 6 sectors (two templates of 3 sectors with 8 ATCos, which is optimized), the ATCos must work exactly 75% of the time. This implies that the work distribution slack is zero, and the working time for all ATCos is exactly equal, whereas all other constraints are met.

Instance 2. Morning shift in Barcelona control center (11 open sectors). Figure 10 shows the airspace sectoring of the Madrid control center. It consists of a morning shift, which covers from 5:20 to 13:00 with 11 open sectors: 3 open sectors from 5:20 to 8:40, 4 open sectors from 5:40 to 6:40, 6 open sectors from 6:40 to 7:20, 5 open sectors from 7:20 to 10:00, 5 open sectors from 10:00 to 11:00, and from 11:00 to 13:00. All the sectors and ATCos involved belong to the *eastern route* core. There are 15 available ATCos, whom are CON accredited.

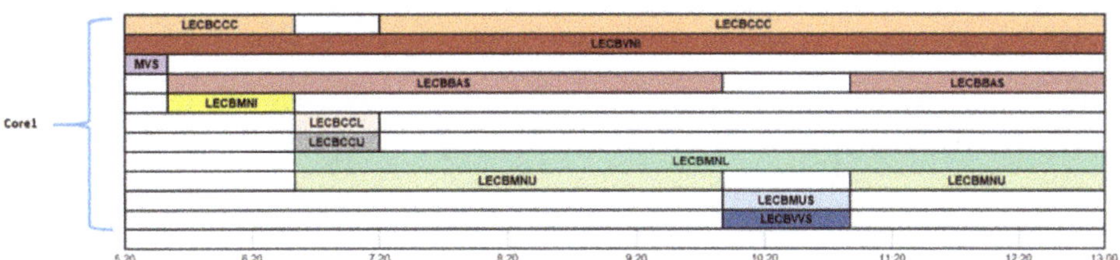

Figure 10. Sectorization of instance 2.

The problem posed by this instance is sector opening and closing. We must also cover 6 sectors opened for 40 min with only 15 controllers. Note that an additional ATCo would be necessary if these sectors were to remain open for longer.

Instance 3. Morning shift in Barcelona control center (9 open sectors). Figure 11 shows the airspace sectoring of the Barcelona control center for a morning shift, covering from 6:20 to 14:00 with 9 sectors open: 2 open sectors from 6:20 to 8:20, 4 sectors open from 8:20 to 10:20, 5 sectors open from 10:20 to 12:20, and 4, from 12:20 to 14:00. All the sectors and ATCos involved belong to the *eastern route* core. There are 14 available ATCos, all whom hold CON accreditation.

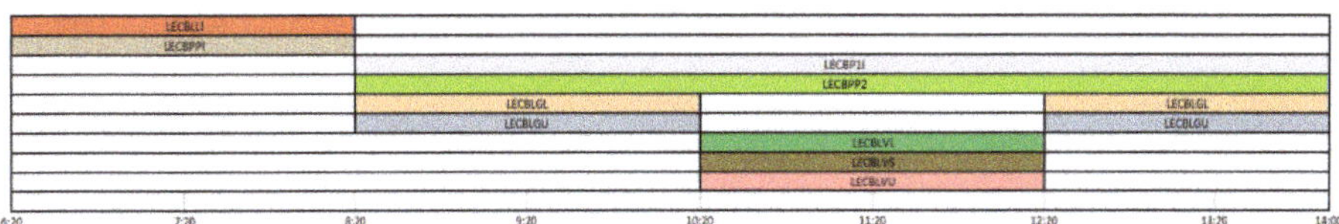

Figure 11. Sectorization of instance 3.

It is a relatively simple example, but there are quite a few changes of sectors that complicate the fulfillment of all constraints.

Instance 4. Afternoon shift in Barcelona control center (6 open sectors). Figure 12 shows the airspace sectoring of the Barcelona control center for an afternoon shift, covering from 14:00 to 21:20. There are 6 open sectors in the Barcelona *western route* core: 4 sectors open from 14:00 to 19:20, and 2 sectors open from 19:20 to 21:20; and 9 sectors open in the Barcelona *eastern route* core: 5 sectors open from 14:00 to 19:20, 4 sectors open from 19:20 to 21:00, and 2 sectors open from 21:00 to 21:20. The number of available ATCos is 28, whose accreditation type is CON, 14 belonging to the eastern and 14 to the western core.

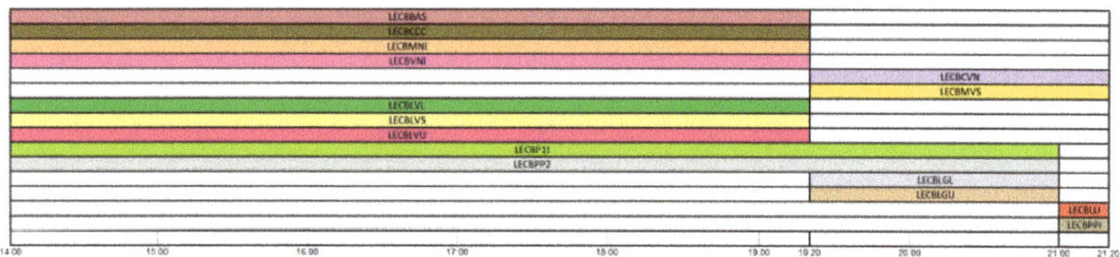

Figure 12. Sectorization of instance 4.

The main problem posed by this instance is its size. A large number of ATCos are needed to cover all sectors. In addition, there is a progressive closure of sectors, where ATCos are highly unlikely to comply with the minimum consecutive work constraint, among others.

Figures 13–15 illustrate an initial solution of this last instance, the corresponding feasible solution and the optimal solution reached using VNS in Phases 2 and 3, respectively. Looking at Figure 13, we find that optimized templates are used where three ATCos cover a sector for 96 time slots and the number of necessary ATCos (43) is greater than the number that are actually available ATCos (28). Figure 14 shows the feasible solution derived from Phase 2 using VNS. Now the number of ATCos matches 28, and all labor conditions are met. Finally, Figure 15 shows the optimal solution derived in Phase 3 using VNS. If we compare the initial feasible and the optimal solutions, it is clear that the structure of the optimal solution is like the previous template-based solution, where work and rest periods are more concentrated.

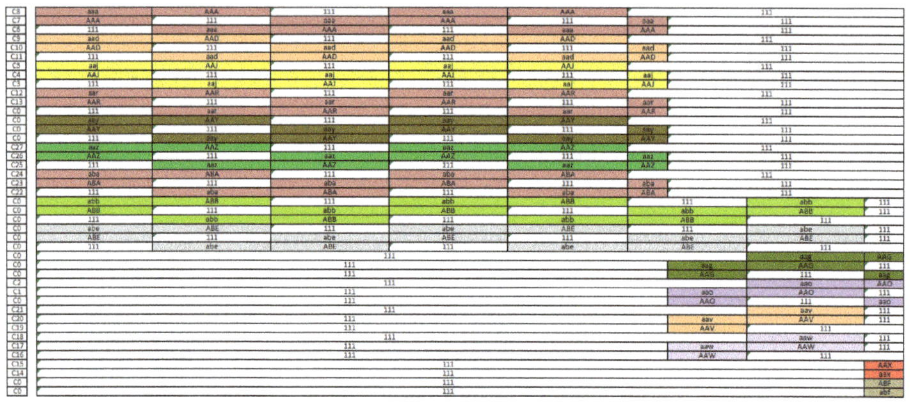

Figure 13. An initial solution in instance 4.

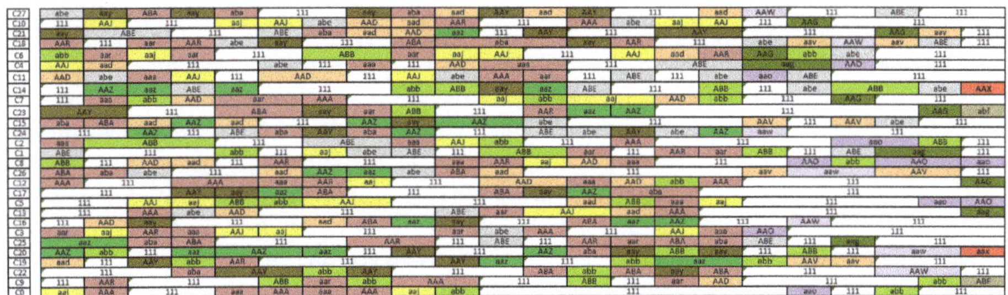

Figure 14. A feasible solution in instance 4.

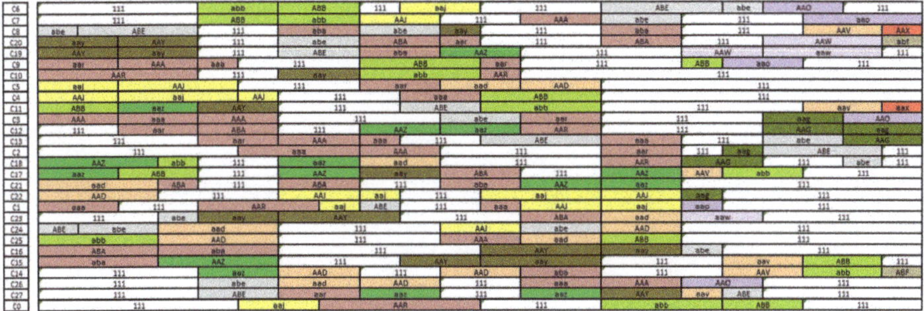

Figure 15. Optimal solution in instance 4.

Table 1 shows three out of the four original objectives, optimizing the ATCo work and rest periods and positions, the number of control center changes (rest periods), and the distribution of the ATCo workload.

Table 1. Objective functions in the initial and optimal solution in Instance 4.

	ATCo Work and Rest Periods and Positions			Objective 3	ATCo Workloads		
	Sub-Object. 1	Sub-Object. 2	Sub-Object. 3	Rest Periods	σ	min	max
Initial	$27.9, 44.9, 71.7\%$	$24.4, 24.4, 38.3\%$	$81.3, 85.4, 86.4\%$	122	87.3	17	293.5
VNS	$9.37, 11.45, 23.95\%$	$10.12, 12.65, 14.55\%$	$95, 98, 100\%$	160	58.13	115	330

The first column (sub-object. 1) shows the percentage of cases with a difference less than or equal to 10, 15 and 25 min, respectively, with respect to the goal of 45 min. The second column (sub-object. 2) shows the percentage of cases with a difference less than or equal to 15, 20 and 25 min, respectively, with respect to the goal of 90 min. The third column (sub-object. 1) shows the ATCo percentages whose differences are lower than or equal to 5%, 10% and 15%, respectively. The fourth column shows the number of rest periods, and, finally, the fifth column (ATCo workloads) lists the standard deviation, and the minimum and the maximum value of the ATCo workloads.

As expected, the optimal solution outperforms the initial one for all the objectives under consideration.

4.2. Computational Improvements

All the constraints with respect to the ATCo labor conditions must be checked to verify the feasibility of the analyzed solutions in the different iterations of the VNS execution. In Phase 2, this constraint verification is associated with the fitness function, since this phase is aimed at reaching a feasible solution, whereas the Phase 3 search process is carried out within the feasible region, on which ground we have to check the feasibility of the visited solutions.

Verifying constraints is very time consuming. Therefore, we have carried out a comparative analysis to check if this process is faster using regular expressions, which, apart from structuring the constraint modularly could lead to better computation times than implementing such constraints in the code.

Figure 16 shows the mean execution times throughout the iterations in the proposed VNS adaptation using both regular expressions and the code implementation for Instance 4. Contrary to what we had initially thought, the code implementation outperforms the use of regular expressions for higher numbers of iterations. This is due to the complexity of some of the constraints to be checked, requiring lot more than one number of regular expression.

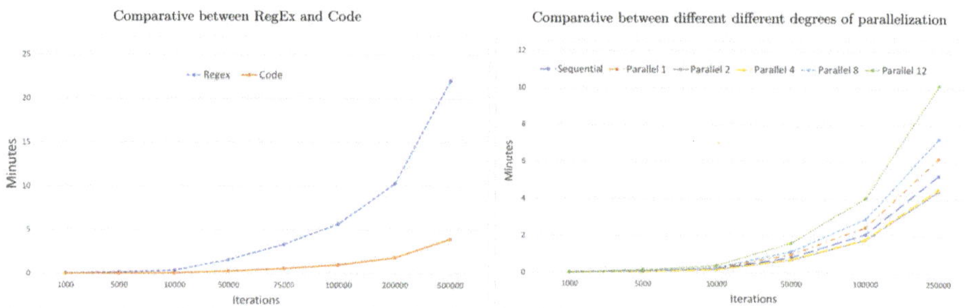

Figure 16. Computation improvements.

Additionally, the parallelization of the constraint verification process could improve computation times. Figure 16 shows the different execution times with and without different degrees of parallelization (up to 12 simultaneous threads), again for Instance 4. As expected, parallelization offers faster speeds with respect to sequential execution. We find that there are significant improvements as the number of threads increases, up to four threads.

These results are logical taking into account the computer that we used, which had a quad-core processor. Therefore, it works better when the number of threads used in the execution is also four. When more than four threads are used, they all share the CPU time and sometimes have to wait in a queue to be processed. These CPU inputs and outputs are time consuming and make the process less efficient. However, a CPU with a higher number of cores would take better advantage of more parallelized threads.

Similar results were output for the other three instances.

5. Discussion

In this section, we compare the performance of VNS against the multi-start SA used by [10] for the four complex instances under consideration, taking into account both the quality of the solution reached by means of the fitness function and the four original objectives and execution times.

Each instance was executed 10 times using a Intel(R) Xeon(R) E3-1240 PC with 3.50 GHz and 16 GB of RAM, running Windows 10.

Table 2 shows the minimum, mean and maximum values of the fitness function in the optimal solutions achieved by the VNS and SA. The mean values for SA are clearly higher than for VNS in all four instances under consideration. We can thus conclude that SA slightly outperforms VNS with respect to the quality of the solutions reached.

Table 2. Optimal fitness values in Phase 3 for SA and VNS.

	SA			VNS		
	Min	Mean	Max	Min	Mean	Max
Inst. 1	0.8637	**0.8689**	0.875	0.853	0.857	0.863
Inst. 2	0.8714	**0.8742**	0.8769	0.845	0.853	0.858
Inst. 3	0.8915	**0.8946**	0.8968	0.876	0.887	0.896
Inst. 4	0.8527	**0.8604**	0.867	0.854	0.855	0.856

Let us analyze in depth the optimal solutions derived by both metaheuristics in the four instances under consideration.

Tables 3–6 show the values of the initial solution and solutions reached by VNS and SA for the four instances under consideration, respectively, in terms of the four original objectives, optimizing the ATCo work and rest periods and positions, the similarity to the previous template-based solution, the number of control center changes (rest periods), and the distribution of the ATCo workload.

Note that Tables 3–6 provide mean values, and the respective best values for each column are highlighted in bold.

Table 3. Objective functions in the initial solution and solutions reached by SA and VNS for Instance 1.

	ATC Work and Rest Periods and Positions			Obj. 2	Obj. 3	ATCo Workloads		
	Sub-Object. 1	Sub-Object. 2	Sub-Object. 3	Similarity	Rest Periods	σ	Min	Max
Initial	26.7, 41.2, 70.8	18.2, 20.9, 30.8	87.3, 94.2, 94.5	**81.9**	51.2	67.1	35	292.5
SA	35.6, 35.6, 85.6	**31.4, 66.6, 66.6**	**100,100,100**	71.77	**51**	**20.45**	255	325
VNS	**47.9, 52, 86.7**	25.9, 61.1, 61.1	**100,100,100**	68.7	54	25.4	240	330

Table 4. Objective functions in the initial solution and solutions reached by SA and VNS for Instance 2.

	ATC Work and Rest Periods and Positions			Obj. 2	Obj. 3	ATCo Workloads		
	Sub-Object. 1	Sub-Object. 2	Sub-Object. 3	Similarity	Rest Periods	σ	Min	Max
Initial	24.3, 38, 62.5	16.4, 17.1, 25.4	61.9, 66.7, 67.3	**75.6**	59.3	116.6	15	319.5
SA	**65.2, 65.2, 80**	38.3, 38.3, 38.3	**100,100,100**	74.88	**46**	18.5	280	340
VNS	50, 63.1, 86.8	30.9, 41.8, 45.4	**100,100,100**	71	54	**16.3**	275	330

Table 5. Objective functions in the initial solution and solutions reached by SA and VNS for Instance 3.

	ATC Work and Rest Periods and Positions			Obj. 2	Obj. 3	ATCo Workloads		
	Sub-Object. 1	Sub-Object. 2	Sub-Object. 3	Similarity	Rest Periods	σ	Min	Max
Initial	25.3, 38.1, 67.6	16.5, 19.2, 25.2	85.1, 90.4, 94.1	**84.7**	43.5	60.8	70	238.5
SA	**75, 75, 88.1**	30.1, 30.1, 30.1	**100,100,100**	80.55	**38**	34.9	220	320
VNS	54.6, 63.9, **91.8**	**32.4, 40, 47.5**	**100,100,100**	79.6	42	**21.2**	210	280

Table 6. Objective functions in the initial solution and solutions reached by SA and VNS for Instance 4.

	ATC Work and Rest Periods and Positions			Obj. 2	Obj. 3	ATCo Workloads		
	Sub-Object. 1	Sub-Object. 2	Sub-Object. 3	Similarity	Rest Periods	σ	Min	Max
Initial	27.9, 44.9, 71.7	24.4, 24.4, 38.3	81.3, 85.4, 86.4	67.6	122	87.3	17	**293.5**
SA	56.6, 56.6, 76.6	40, 40, 40	**100,100,100**	**73.05**	77	32	**200**	320
VNS	**64.7, 74.6, 90**	**41.2, 53.9, 58.8**	84,88.7,88.7	68.6	193	**19.2**	212.5	290

Looking at sub-object. 1, we find that VNS outperforms SA for 8 out the 12 mean values shown in Tables 3–6; whereas SA is better than VS for sub-object. 2 in the first instance but not for instances 3 and 4.

As expected, the initial solution outperforms the solutions reached by SA and VNS in terms of similarity to previous template-based solution. However, SA outperforms VNS with respect to this objective in all four instances under consideration.

Regarding the number of rest periods, SA outperforms VNS for all four instances under consideration. For all four instances, the number of rest periods in VNS is [10.67, 17]% higher than for SA.

Finally, looking at the standard deviation representing the ATCo workload dispersion, we find that SA outperforms VNS for the first instance, whereas VNS is better for Instances 2, 3 and 4. In all four cases, the ATCo workload dispersions were very similar in both metaheuristics.

These tables were shown to CRIDA experts, who analyzed the quality of the solutions reached by VNS and SA for the four complex instances. They concluded that, although the solutions derived by SA slightly outperform those reached using VNS in terms of the fitness function, the quality of both solutions were very similar four the four instances under consideration and that the key factor was then the time it took to reach that solutions, i.e., the computation times.

Tables 7 and 8 show the minimum, mean and maximum computation times (in minutes) in Phases 2 and 3, respectively, for the original multi-start SA proposed by [10], an improved non multi-start SA using a constraint implementation in the code rather than regular expressions and with parallelization (4 threads), and the proposed adaptation of VNS (also using a constraint implementation in the code and with parallelization). The metaheuristic with the lowest computation time is highlighted in bold.

Table 7. Computation times (in minutes) in Phase 2.

	Original SA			Improved SA			VNS		
	Min	Mean	Max	Min	Mean	Max	Min	Mean	Max
Instance 1	206.21	254.71	333.24	5.79	**6.02**	6.25	8.73	9.26	10.15
Instance 2	207.34	249.59	308.14	16.34	17.64	19.76	4.28	**7.14**	11.29
Instance 3	111.7	128.96	157.88	3.64	6.75	12.92	4.11	**4.65**	5.50
Instance 4	394.69	507.25	619.82	48.88	**49.53**	50.48	100.71	111.85	122.18

Table 8. Computation times (in minutes) in Phase 3.

	Original SA			Improved SA			VNS		
	Min	Mean	Max	Min	Mean	Max	Min	Mean	Max
Instance 1	177.59	214.76	262.47	40.52	54.04	68.54	21.48	**31.12**	43.84
Instance 2	79.87	95.19	122.37	27.52	36.89	50.47	28.55	**30.68**	33.08-
Instance 3	103.19	139.24	210.34	24.67	25.33	26.46	16.01	**19.75**	24.68
Instance 4	802.16	958.25	1114.35	177.25	**247.87**	360.72	431.15	479.66	507.79

The first thing that we found is that the computation times for the improved SA are much lower than for the original SA in both phases and for the four instances under consideration, as was expected. Besides, computation times for Phase 3 are quite a lot higher than in Phase 2, accounting for the biggest share of the accumulated computation times shown in Table 9.

Table 9. Mean accumulated computation times (in minutes).

	Original SA	Improved SA	VNS
Instance 1	469.49	60.09	**40.38**
Instance 2	344.79	54.54	**37.81**
Instance 3	268.21	32.1	**24.40**
Instance 4	1465.52	**297.43**	591.52

If we focus on Phase 2, VNS and SA outperform each other in two out of the four instances. However, although the mean computation times for Instances 1 and 3 are similar, VNS clearly outperforms SA in Instance 2, whereas the opposite applies for Instance 4 (SA is more than 70 min faster than VNS). Note that Instance 4 involves a larger number of open sectors and required ATCos.

In Phase 3, VNS clearly outperforms SA in instance 1 but it is only slightly better for Instances 2 and 3. In Instance 4, SA clearly outperforms VNS with a difference of close to 200 min in the mean computation times.

Looking at the mean accumulated computation times in Table 9, we find that VNS clearly outperforms SA in Instances 1 and 2, they are similar for Instance 3, but SA is quite a lot better in Instance 4.

We have analyzed other sectorizations provided by CRIDA in order to verify whether or not the performance of VNS is sensitive to the instance dimension, as in the case of Instance 4, and we have confirmed this hypothesis: the improved SA outperforms VNS when the dimensionality (number of open sectors and required ATCos) is high in all cases, whereas VNS is clearly better in any other situation with a low and medium number of dimensions.

Note that although the mean computation time accumulated in Instance 4 is about 5 h (297.43 min) with the improved SA, which could be considered a high computation time, the complexity of this instance is the highest (9 simultaneously open sectors and 28 ATCos) that could materialize in Spanish airports. Consequently, the maximum computation time of this instance, 50.58 min + 360.72 min $\simeq$ 7 h could be considered, as an upper bound for the computation times. In most cases, solutions are reached in the less than an hour.

Taking into account that the pre-tactical phase takes place one to six days before the day of operations, the CRIDA experts considered this a good upper bound for computation times.

6. Conclusions

We have proposed a new methodology based on an adaptation of VNS for solving the ATCo work-shift scheduling problem. This problem involves covering a given airspace sectoring with a certain number of ATCos while satisfying a set of ATCo labor conditions according to Spanish regulations. The problem takes into account four objectives: ATCo work and rest periods and position should be as close as possible to fixed values, the solution structure should be similar to the previous template-based solution, the number of control center changes should be minimized, and the ATCo workload distribution should be balanced.

A comparative analysis proves that the constraint verification in search processes is faster with the implementation in the code of such constraints than using regular expressions. Besides, the parallelization of the constraint verification process has also been analyzed in terms of computation times.

Finally, the proposed methodology has been applied to four real complex scenarios selected by Spanish air navigation experts for the purposes of both methodology illustration and performance comparison against two versions of simulated annealing (SA).

Although the SA-derived solutions slightly outperform VNS solutions in terms of the fitness function, air navigation experts regard the quality of both solutions as very similar, where the key factor then is the time it takes to reach that solutions (computation times). VNS clearly outperforms SA in terms of computation times when the instance dimension (number of open sectors and ATCos required) is low or medium, but the improved version of SA is better for high dimensional instances.

Finally, although the CRIDA experts considered than the computation times for the most complex instances, such as instance 4, were good enough for the pre-tactical phase, which takes place one to six days before the day of operations, we propose as a future research line a further analysis to reduce these computation times together with the comparison of the considered metaheuristics with other in the literature.

Author Contributions: conceptualization, F.T., A.J.-M. and A.M.; methodology, F.T. and P.L.; software, F.T. and P.L.; validation, F.T., A.J.-M. and A.M.; formal analysis, F.T., A.J.-M., A.M. and P.L.; investigation, F.T., A.J.-M. and A.M.; writing–original draft preparation, F.T., A.J.-M. and A.M.; writing–review and editing, F.T., A.J.-M. and A.M.; visualization, F.T., A.J.-M. and A.M.; supervision, A.J.-M. and A.M.; project administration, A.J.-M. and A.M.; funding acquisition, A.J.-M. and A.M.

Funding: This research was funded by Spanish Ministry of Economy and Competitiveness projects grant number MTM2014-56949-C3-2-R and MTM2017-86875-C3-3-R.

Conflicts of Interest: The authors declare no conflict of interest.

References

1. EUROCONTROL. *Strategic, Pre-Tactical, Tactical and Post-Ops Air Traffic Flow and Capacity Management;* Technical Report; EUROCONTROL: Brussels, Belgium, 2017.
2. Telhada, J. Alternative MIP formulations for an integrated shift scheduling and task assignment problem. *Discret. Appl. Math.* **2014**, *164*, 328–343. [CrossRef]
3. Fonseca, G.H.; Santos, H.G.; Carrano, E.G. Integrating matheuristics and metaheuristics for timetabling. *Comput. Oper. Res.* **2016**, *74*, 108–117. [CrossRef]
4. EUROCONTROL. *Shiftwork Practices Study—ATM and Related Industries;* Technical Report; DAP/SAF-2006/56; EUROCONTROL: Brussels, Belgium, 2006.
5. EATCHIP Human Resources Team. *ATM Manpower Planning in Practice: Introduction to a Qualitative and Quantitative Staffing Methodology;* Technical Report; HUM.ET1.ST02.2000-REP-01; EUROCONTROL: Brussels, Belgium, 1998.
6. Stojadinovic, M. Air traffic controller shift scheduling by reduction to CSP, SAT and SAT-related Problems. In *Principles and Practice of Constraint Programming—20th International Conference (CP 2014);* Springer: Cham, Switzerland, 2015; pp. 886–902.
7. Stojadinovic, M. Hybrid of hill climbing and SAT solving for air traffic controller shift scheduling. *J. Inform. Technol. Appl.* **2015**, *5*, 81–87. [CrossRef]
8. Biere, A.; Heule, M.; van Maaren, H. *Handbook of Satisfiability;* Frontiers in Artificial Intelligence and Applications; IOS Press: Amsterdam, The Netherlands, 2009; Volume 185.
9. Tello, F.; Mateos, A.; Jiménez-Martín, A. ATC work shift scheduling using multistart simulated annealing and regular expressions. In Proceedings of the 2017 International Conference on Decision Support Systems Technology, Namur, Belgium, 3–6 July 2017; pp. 169–175.
10. Tello, F.; Mateos, A.; Jiménez-Martín, A. The air traffic controller work-shift scheduling problem in Spain from a multiobjective perspective: A metaheuristic and regular expression-based approach. *Math. Probl. Eng.* **2018**, *2018*, 15. [CrossRef]
11. Mladenovic, N.; Hansen, P. Variable neighborhood search. *Comput. Oper. Res.* **1997**, *24*, 1097–1100. [CrossRef]
12. Hertz, A.; Mittaz, M. A variable neighborhood descent algorithm for the undirected capacitated arc routing problem. *J. Algorithms* **2001**, *35*, 425–434. [CrossRef]
13. Hansen, P.; Mladenovic, N.; Perez-Britos, D. Variable neighborhood decomposition search. *J. Heuristics* **2001**, *7*, 39–47. [CrossRef]
14. Moreno, J.A.; Mladenovic, N. *Búsqueda por Entornos Variables para Planificación Logística;* Universidad de la Laguna: San Cristóbal de La Laguna, Spain, 2007.
15. Butler, J.; Olson, D.L. Comparison of centroid and simulation approaches for selection sensitivity analysis. *J. Multi-Criteria Decis. Mak.* **1999**, *8*, 146–161. [CrossRef]

Article

Multi-Attribute Decision-Making Methods as a Part of Mathematical Optimization

Irina Vinogradova

Department of Information Technologies, Vilnius Gediminas Technical University, 10223 Vilnius, Lithuania; irina.vinogradova@vgtu.lt; Tel.: +37-052-745-035

Received: 21 August 2019; Accepted: 26 September 2019; Published: 2 October 2019

Abstract: Optimization problems are relevant to various areas of human activity. In different cases, the problems are solved by applying appropriate optimization methods. A range of optimization problems has resulted in a number of different methods and algorithms for reaching solutions. One of the problems deals with the decision-making area, which is an optimal option selected from several options of comparison. Multi-Attribute Decision-Making (MADM) methods are widely applied for making the optimal solution, selecting a single option or ranking choices from the most to the least appropriate. This paper is aimed at providing MADM methods as a component of mathematics-based optimization. The theoretical part of the paper presents evaluation criteria of methods as the objective functions. To illustrate the idea, some of the most frequently used methods in practice—Simple Additive Weighting (SAW), Technique for Order of Preference by Similarity to Ideal Solution (TOPSIS), Complex Proportional Assessment Method (COPRAS), Multi-Objective Optimization by Ratio Analysis (MOORA) and Preference Ranking Organization Method for Enrichment Evaluation (PROMETHEE)—were chosen. These methods use a finite number of explicitly given alternatives. The research literature does not propose the best or most appropriate MADM method for dealing with a specific task. Thus, several techniques are frequently applied in parallel to make the right decision. Each method differs in the data processing, and therefore the results of MADM methods are obtained on different scales. The practical part of this paper demonstrates how to combine the results of several applied methods into a single value. This paper proposes a new approach for evaluating that involves merging the results of all applied MADM methods into a single value, taking into account the suitability of the methods for the task to be solved. Taken as a basis is the fact that if a method is more stable to a minor data change, the greater importance (weight) it has for the merged result. This paper proposes an algorithm for determining the stability of MADM methods by applying the statistical simulation method using a sequence of random numbers from the given distribution. This paper shows the different approaches to normalizing the results of MADM methods. For arranging negative values and making the scales of the results of the methods equal, Weitendorf's linear normalization and classical and author-proposed transformation techniques have been illustrated in this paper.

Keywords: optimization; decision-making; MADM; SAW; COPRAS; TOPSIS; PROMETHEE; MOORA; normalization; stability

---

1. Introduction

In a specific activity, a person consciously and intuitively seeks to find the best solutions to emerging problems or tasks. The action of making the best or most effective use of a situation or resource is called optimization. The Simple Additive Weighting (SAW), Technique for Order of Preference by Similarity to Ideal Solution (TOPSIS), Complex Proportional Assessment Method (COPRAS), Multi-Objective Optimization by Ratio Analysis (MOORA) and Preference Ranking Organization

Method for Enrichment Evaluation (PROMETHEE) methods applied in this paper have been described in different research papers as Multiple Criteria Decision Making (MCDM) [1–6], Multiple Attribute Decision Making (MADM) [7–9], Multiple Criteria Decision Analysis (MCDA) [10] or Multi-Attribute Decision Analysis (MADA) [11], and Multi-Criteria Analysis (MCA) [12,13]. Since this work is focused on decision-making and the number of alternatives are explicitly given and finite, the name MADM will be used to define the above-listed methods.

MADM methods are aimed at identifying the most satisfactory of several comparative alternatives or at ranking options according to their relevance in terms of the evaluated objective [14]. The methods are used for selecting the most satisfactory alternative/solution provided that there is no such alternative for which all criteria values are the best.

To solve an optimization problem with classical optimization methods, the function of its objective is fixed, establishing the set of objects to be optimized or the allowable area to be determined. The minimum or maximum values of the function are sought depending on the purpose of the problem being solved. The theoretical part of this work presents MADM methods as a component of mathematical optimization methods, and evaluation criteria for SAW, TOPSIS, COPRAS, MOORA and PROMETHEE methods appear as objective functions, which is a new form of presenting and interpreting methods. To illustrate the idea of this publication, some of the most widely applied MADM methods have been selected. The presented methodology can be transferred to other methods as well.

The judging matrix and the vector of criterion weights are the components of most of MADM methods. The judging matrix covers statistical data or the values of expert evaluation according to the criteria defining the objective [14]. Since the impact of criteria on the outcome of the problem to be solved is different, the significance (weights) of criteria is determined [15]. Criterion weights can be clarified directly or by employing certain weighting methods. The main idea of most of the used MADM methods is merging criterion values and their weights into a single evaluation characteristic (i.e., the summarized criterion of the method). Data on MADM methods are static, and their values do not vary in the problem-solving process.

Most of the assignments solved by people include problems that do not have sufficient numerical data or problems where the investigated objects are impossible to measure. In such cases, the judgment matrix is supplemented by the data obtained from the expert evaluation. Particular focus is switched to selecting experts in a particular field, considering their characteristics related to professional competence, work experience, scientific degree, research activity and the ability to address specific issues in the field given. MADM methods operate in numerical values, although the criteria themselves can be both quantitative and qualitative. The qualitative meanings of criteria, in some cases, facilitate expert evaluation that can be individual when the expert expresses individual opinions independently of other experts or shared and accepted in a group of professionals.

The research literature does not propose the best or most appropriate MADM method for dealing with a specific problem. This question is relevant, and thus there are many research papers focused on determining the stability of the method on the basis that any mathematical model or method can be applied in practice in the case that they remain stable with respect to the applied parameters [16]. A mathematical model is considered to be stable if a small change in the results is consistent with minor variations in the parameters for the model. Multiple MADM methods are applied in most complex decision-making tasks to ensure the accuracy of the final result. In the cases when several MADM methods are used for evaluation, it becomes unclear what results of which method are reliable. This paper proposes a new approach that helps the expert make the right decision. The core of the suggested approach is to apply several MADM methods and to determine the suitability/impact of the employed MADM methods on the problem solved (i.e., to clarify the stability of the method). The final result consists of the estimates of several methods taking into account the weight of the effect of each method.

The paper verifies the stability of multi-criteria methods when slightly changing data in the matrix of the initial solution (i.e., expert evaluations and weights of the vector, fixing recurrence frequency of the best alternative to the initial data). Previous papers of the author considered that the higher the number of imitations, the more accurate the evaluation of the stability of the multi-criteria method (i.e., the range of the varying result decreased). A sufficient number of recurrences was established when the result of evaluating MADM stability remained almost unchanged, because 10^5 times could be treated as an adequate number of estimations [17].

The practical part of the paper combines the results of several MADM methods into a single outcome and shows a few ways to normalize results obtained using MADM methods of different scales.

2. Literature Review

Analytical mathematical optimization problems were solved in as early as the 17th century. The first solution proposed investigating the problem of finding the minimum/maximum and was described by P. Fermat (XVII). Newton developed the method of fluxions. The technique was rediscovered and published in the paper "New Method for the Greatest and the Least" by G. W. Leibniz in 1684. Further, efforts exerted by Euler and Lagrange led to working out solutions to extreme tasks. In 1824, Fourier created the first algorithm for solving linear arithmetic constraints [18]. This algorithm made further advances in the field, such as the main duality theorem, the Farkas lemma, the Motzkin transfer theorem and others [19]. The traditionally employed model of optimization includes linear programming, sequential quadratic programming, nonlinear programming, and dynamic programming [20]. In 1939, the first formulation of the linear programming problem and the method for solving this problem were proposed by Leonid Kantorovich. In 1947, Danzig created the simplex method that was effectively used to solve linear programming problems [21]. Derivative-based stochastic optimization began with a seminal paper by Robbins and Monro (1951) that launched the entire field [22]. Richard Bellman developed the dynamic programming method in the 1950s [23].

Decision-making methods based on optimality were introduced by Pareto in 1896 and applied to a wide range of problems. The Multi-Objective Evolutionary Algorithm (MOEA) [24] is used to find the optimal Pareto solutions for specific problems [25]. Keeney and Raiffa [26] and Fishburn [27] introduced the Multi-Attribute Value Theory (MAVT), the Multi-Attribute Value Analysis (MAVA) and Multi-Attribute Utility Theory (MAUT) methods. Data envelopment analysis (DEA), introduced by Charnes et al., is a linear programming method for measuring the efficiency of multiple decision-making units by analysing the problems of multiple inputs and outputs [28].

Multiple criteria decision-making methods evolved from operations research theory by solving problems such as the development of computational and mathematical tools to support the subjective assessment of performance criteria by decision-makers [29]. MADM, as a discipline, has a relatively short history of approximately 30 years. Its role has increased significantly in different application areas along with the development of new methods and improved old methods in particular.

A work by Hwang and Yoon presented a plethora of methods for solving MADM problems [7]: Methods for Cardinal Preference of Attribute over Linear Assignment method [30], Simple Additive Weighting (SAW) method [31], Hierarchical Additive Weighting method, ELECTRE method, and Technique for Order of Preference by Similarity to Ideal Solution (TOPSIS) [7]. The most familiar and commonly used is the SAW method reflecting the idea of multi-criteria methods—merging criterion values and their weights into a single value [32].

Peng and Wang proposed the concept of hesitant uncertain linguistic Z-numbers (HULZNs) and presented the Multi-Criteria Group Decision-Making (MCGDM) method by integrating power operators employing the Vlse Kriterijumska Optimizacija I Kompromisno Resenje (VIKOR) [5] model. Peng and Wang merged the Multi-Objective Optimization by Ratio Analysis plus the Full Multiplicative From (MULTIMOORA) and power aggregation operators in order to create a comprehensive decision model for MCGDM problems with Z-numbers [33]. Outranking ELECTRE [34] and PROMETHEE [35] methods were described in the publication on multiple criteria decision analysis by Belton and Stewart in 2001 [10]. Opricovic and Tzeng conducted a comparative analysis of VIKOR and TOPSIS methods in 2004 [5,36].

New methods have recently emerged that are actively used in different fields of science: Weighted Aggregated Sum Product Assessment (WASPAS) [37], Complex Proportional Assessment Method (COPRAS) [38], Multi-Objective Optimization by Ratio Analysis (MOORA) [39], COPRAS grey (COPRAS-G), fuzzy additive ratio assessment (ARAS-F) [40], ARAS grey (ARAS-G) and MULTIMOORA (MOORA plus the full multiplicative form) [41,42], KEmeny Median Indicator Ranks Accordance (KEMIRA) [43], ARAS [44], and newest extensions of the ELECTRE [45] and PROMETHEE [46,47] methods. The examples of partial aggregation methods include Step-Wise Weight Assessment Ratio Analysis (SWARA) [48] and factor relationship (FARE).

Criterion weights are one of the components of MCMD methods and therefore have a strong impact on the final result [15]. For defining criterion weights, subjective evaluation is the most frequently applied technique when experts examine the significance of criteria, although objective and generalized estimates are known [49]. Weights can be set directly or using weighting methods such as Analytic Hierarchy Process (AHP) [50,51], Fuzzy Analytic Hierarchy Process (FAHP) [52,53], SWARA [54], Criterion Impact LOSs (CILOS) [55], Integrated Determination of Objective Criteria Weights (IDOCRIW) [14,56], etc. Recalculation of the weights of criteria under the Bayes theorem is proposed in the paper [56]. Regardless of the method, the principles of evaluation remain to take the position that the weight of the most important criterion is the highest. It was agreed that the sum of all weights should be equal to 1 [1]. Any measurement scale may be used for evaluations.

Based on a study by Sabaei et al., the most common decision management methods used in Scopus database publications are AHP, ELECTRE, and PROMETHEE [57]. The early 1990s witnessed the shift of focus toward methods that consider indifference and ensure the transparency of analysis processes [58]. An analogous study conducted by Mardani et al. aimed at determining the popularity of decision-making methods. The results showed that hybrid MADM and fuzzy MADM approaches (27.92%) were used more often than other methods. The most commonly used methods are AHP and fuzzy AHP [59] (24.87%), ELECTRE, fuzzy ELECTRE [60], MCDA and MCA (12.69%), and TOPSIS, fuzzy TOPSIS [61], PROMETHEE and fuzzy PROMETHEE [62] (5.08%) [1].

Mardani et al. carried out research and published the obtained material in the paper "Multiple Criteria Decision-Making Techniques and Their Applications," (i.e., a literature review for the period from 2000 to 2014 [2]). Another paper by Mardani et al. reviewed decision-making methods from the field of energy management for the period 1995–2015 [1].

The concept of sensitivity analysis in decision theory means the effective use and implementation of quantitative decision models, the purpose of which is to assess the stability of an optimal solution under changes in parameters, the impact of the lack of controllability of specific parameters and the need for the precise estimation of parameter values [63]. The first significant works on sensitivity analysis in the field of decision-making were done by Evans [63], who formulated the concepts of sensitivity analysis in linear programming to develop a formal approach applicable to classical decision-theoretic problems [64] and presented two simple computational procedures for sensitivity analysis of additive multi-attribute value models that yielded variations in attribute weights. Insua [65] developed a conceptual framework for sensitivity analysis in discrete multi-criteria decision-making, which allowed simultaneous variations in judgmental data and applied to many paradigms for decision analysis. Janssen [66] discussed the sensitivity of the rankings of alternatives to the overall uncertainty in scores, and priorities were analyzed using the Monte Carlo approach. Butler [67] presented a simulation approach allowing simultaneous changes in the weights and generating results that could be easily analyzed to provide insights into multi-criteria model recommendations statistically.

Wolters and Mareschal [68] proposed three novel types of sensitivity analysis focused on and elaborated for the PROMETHEE methods. Masuda [69] studied the sensitivity problems of the AHP method. In his work, he concentrated on how changes in the entire columns of the decision-making matrix might affect the values of the composite priorities of alternatives. Triantaphyllou [70] presented a methodology for performing a sensitivity analysis of the weights of decision criteria and identifying the performance values of the alternatives expressed in terms of decision criteria. The estimation of the effect/impact of uncertainty in the SAW method was performed by Podvezko [71], who determined the points of varying ranges of criterion weights of the investigated process, evaluated compatibility level and stability of expert opinions and assessed the effect of uncertainty on ranking comparable objects employing the imitation method. The impact of varying weights on the final result in the SAW method was studied by Zavadskas [72] and Memariani [73]. The influence of the elements of the decision matrix on the final ranking result was analyzed by Alinezhad [74]. The effect of the importance of criterion weights on the results of the TOPSIS method was studied by Yu [75] and Alinezhada [76]. Misra focused on a comparison of AHP, Decision-Making Trial and Evaluation Laboratory (DEMATEL), COPRAS, and TOPSIS methods [77]. Podvezko [32] compared SAW, TOPSIS and COPRAS methods. Moghassem [78] increased and decreased all criterion weights by 5%, 10%, 15%, and 20% in analyzing the sensitivity of TOPSIS and VIKOR. Hsu conducted the sensitivity analysis of TOPSIS by increasing and decreasing the top three weights by 10% [79].

3. MADM Methods as a Component of Mathematics-Based Optimization Techniques

To formulate the optimization problem, the paper presents a set of optimized elements and the measure of goodness of its elements (quality estimates).

The optimization problem takes the form of

$$\underset{x \in D}{opt} f(x), \tag{1}$$

where $f(x) : D \to Y$ is the objective function or criterion; D is the set or permissible area of the optimized objects; and *opt* is the minimum or maximum value of function $f(x)$.

The literature provides a number of different classifications of optimization problems. Typically, specific decision-making methods are created for each category of problems according to the characteristics of that particular class. Weights do not vary in SAW, TOPSIS, COPRAS, MOORA and PROMETHEE methods. Weights are determined using subjective or objective weighting methods. The number of comparable alternatives is finite in these methods.

MADM methods can be presented as a mathematical optimization problem as follows:

$$i_{opt}^{v}(r) = \arg\,max_i f^{v}(r, \omega), i = 1, \dots, n, \tag{2}$$

where v is the number of the MADM method. The merit of alternatives $i = 1, \dots, n$ is evaluated according to criteria $j = 1, \dots, m$, and the values are defined as $r = \left(r_{ij}\right)$. The influence of criteria on the evaluation result is different, and therefore the vector $\omega = (\omega_j), j = 1, \dots, m$, of the weights of criteria is determined, thus defining the importance of criteria.

3.1. SAW (Simple Additive Weighting) Method (v = 1)

$$i_{opt}^{1}(r) = \arg\,max_i \sum_{j=1}^{m} \left(\omega_j(\widetilde{r}_{ij})\right) \tag{3}$$

where the values of $\widetilde{r}_{ij}$ are normalized according to the formula:

$$\widetilde{r}_{ij} = \frac{r_{ij}}{\sum_{i=1}^{n} r_{ij}}. \tag{4}$$

When the values of criteria are multi-dimensional, they are transformed. The values of the maximized criteria are calculated according to the formula:

$$\overline{r}_{ij} = \frac{r_{ij}}{\max r_{ij}}. \tag{5}$$

Then, the highest value of $\overline{r}_{ij}$ is equal to 1. The value of minimized criteria r_i is correspondingly calculated according to the formula:

$$\overline{r}_{ij} = \frac{\min r_{ij}}{r_{ij}}. \tag{6}$$

Then, the lowest value of $\overline{r}_{ij}$ is equal to 1. For standard criteria, the principle of simple linear scalarization is applied.

3.2. TOPSIS (Technique for Order of Preference by Similarity to Ideal Solution) Method ($v = 2$)

$$i_{opt}^2(r) = \arg \max_i \frac{\sqrt{\sum_{j=1}^{m} (\omega_j(\widetilde{r}_{ij} - \widetilde{r}_j^-))^2}}{\sqrt{\sum_{j=1}^{m} (\omega_j(\widetilde{r}_{ij} - \widetilde{r}_j^+))^2} + \sqrt{\sum_{j=1}^{m} (\omega_j(\widetilde{r}_{ij} - \widetilde{r}_j^-))^2}}. \tag{7}$$

The method refers to vector data normalization:

$$\widetilde{r}_{ij} = \frac{r_{ij}}{\sqrt{\sum_{i=1}^{n} r_{ij}^2}}, \tag{8}$$

where $\widetilde{r}_{ij}$ is the normalized value of the jth criterion for the ith alternative.

The vector of the best R^+ value and the worst R^- value of criteria (ideal alternative) are calculated as

$$\begin{aligned}
R^+ &= \left\{\widetilde{r}_1^+, \widetilde{r}_2^+, \ldots, \widetilde{r}_m^+\right\} = \{(\max_i \widetilde{r}_{ij}/j \in J_1), (\min_i \widetilde{r}_{ij}/j \in J_2)\}, \\
R^- &= \left\{\widetilde{r}_1^-, \widetilde{r}_2^-, \ldots, \widetilde{r}_m^-\right\} = \{(\min_i \widetilde{r}_{ij}/j \in J_1), (\max_i \widetilde{r}_{ij}/j \in J_2)\},
\end{aligned} \tag{9}$$

where J_1 is a set of indices of the maximized criteria, J_2 is a set of indices of the minimized criteria, and $\widetilde{r}_j^- (\widetilde{r}_j^+)$ is the worst (best) value of the jth criterion.

The basic principle of the method is to find an alternative at the shortest overall distance from the best values of criteria and the maximum distance from the worst values. The method does not require the rearrangement of the minimized (maximized) criteria to the maximized (minimized) ones.

3.3. PROMETHEE (Preference Ranking Organization Method for Enrichment Evaluation) Method ($v = 3$)

$$\begin{aligned}
i_{opt}^3(r) &= \arg \max_i F_i = \arg \max_i\left(F_i^+ - F_i^-\right) = \\
&= \arg \max_i(\textstyle\sum_{g=1}^{n} \pi(A_i, A_g) - \sum_{g=1}^{n} \pi(A_g, A_i)) = \\
&= \arg \max_i(\textstyle\sum_{g=1}^{n} \sum_{j=1}^{m} \omega_j p_h(d_j(A_i, A_g)) - \sum_{g=1}^{n} \sum_{j=1}^{m} \omega_j p_h(d_j(A_g, A_i))),
\end{aligned} \tag{10}$$

where $i = 1, 2, \ldots, n$; $\sum_{j=1}^{m} \omega_j = 1$; $d_j(A_i, A_g) = r_{ij} - r_{gj}$ is the difference of alternatives A_i and A_g of inequality values r_{ij} and r_{gj} of the jth criterion R_j; and $p_h(d) = p_h(d_j(A_i, A_g))$ is the value of the hth priority function for the selected jth criterion.

The PROMETHEE method uses the basic ideas of other methods like combining the values of weights and normalized criteria into a single estimate (SAW method) and the pairwise comparison of criteria (AHP method). Instead of the normalized criteria values, the value of the priority function $p_h(d)$, $0 \leq p_h(d) \leq 1$ is used, and all possible pairs of alternatives for each of the criteria are compared with each other. A higher value of $p_h(d)$ corresponds to a better alternative; if the difference d is lower than the established critical value q, then $p_h(d) = 0$. If d is greater than the maximum limit s for the values of criteria, then $p_h(d) = 1$.

In practice, six ($h = 6$) functions of priorities $p_h(d)$ are applied [3,80].

The priority function of the usual criterion is equal to

$$p_1(d) = \begin{cases} 0, & \text{when } d \leq 0 \\ 1, & \text{when } d > 0. \end{cases} \tag{11}$$

The function chart is shown in Figure 1a.

The priority function of the U-shape criterion is equal to

$$p_2(d) = \begin{cases} 0, & \text{when } d \leq q \\ 1, & \text{when } d > q. \end{cases} \tag{12}$$

The function chart is shown in Figure 1b.

The priority function of the V-shape criterion (linear priority) is equal to

$$p_3(d) = \begin{cases} 0, & \text{when } d \leq 0 \\ \frac{d}{s}, & \text{when } 0 < d \leq s \\ 1, & \text{when } d > s. \end{cases} \tag{13}$$

The function chart is shown in Figure 1c.

The priority function of the level criterion is equal to

$$p_4(d) = \begin{cases} 0, & \text{when } d \leq q \\ 0.5, & \text{when } q < d \leq s \\ 1, & \text{when } d > s. \end{cases} \tag{14}$$

The function chart is shown in Figure 1d.

The priority function of the V-shape with indifference criterion is equal to

$$p_5(d) = \begin{cases} 0, & \text{when } d \leq q \\ \frac{d - q}{s - q}, & \text{when } q < d \leq s \\ 1, & \text{when } d > s. \end{cases} \tag{15}$$

The function chart is shown in Figure 1e.

The priority function of the Gaussian criterion is equal to

$$p_6(d) = \begin{cases} 0, & \text{when } d \leq 0 \\ 1 - \exp\left(-\frac{d^2}{2\sigma^2}\right), & \text{when } d > 0. \end{cases} \tag{16}$$

The function chart is shown in Figure 1f.

As mentioned above, PROMETHEE, similarly to the other multi-criteria decision methods, applies the idea of the SAW method instead of the normalized values $\widetilde{r}_{ij}$ of criteria and uses the values of the functions $p_h(d)$ of specifically selected priorities, where the argument d is the difference between the values of the criterion.

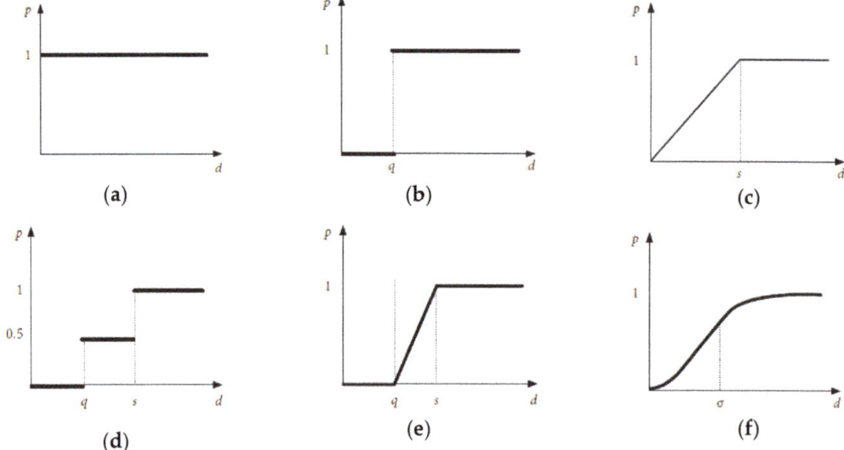

Figure 1. Function charts of criterion priorities: (**a**) function chart of the priorities of the usual criterion; (**b**) function chart of the priorities of the U-shape criterion; (**c**) function chart of the priorities of the V-shape criterion; (**d**) function chart of the priorities of the level criterion; (**e**) function chart of the priorities of the V-shape with indifference criterion; (**f**) function chart of the priorities of the Gaussian criterion.

3.4. COPRAS (Complex Proportional Assessment) Method (v = 4)

$$i^4_{opt}(r) = \arg\ max_i\left(\sum_{j}^{m} \omega_{+j}\widetilde{r}_{+ij} + \frac{\sum_{i=1}^{n}\sum_{j}^{m}\omega_{-j}\widetilde{r}_{-ij}}{\sum_{j}^{m}\omega_{-j}\widetilde{r}_{-ij}\ \sum_{i=1}^{n}\left(\sum_{j}^{m}\omega_{-j}\widetilde{r}_{-ij}\right)^{-1}}\right) \tag{17}$$

where $\omega_{+j}(\omega_{-j})$ are the maximized (minimized) weights of criteria; and $\widetilde{r}_{-ij}(\widetilde{r}_{+ij})$ are the normalized values of the minimized (maximized) criteria for each ith alternative. The values of the estimates of alternatives are normalized according to Equation (4).

The application of the COPRAS method separately assesses the effect of the minimized and maximized criteria on the result of the carried out evaluation [38,81].

3.5. MOORA (Multi-Objective Optimization on the Basis of Ratio Analysis) Method (v = 5)

$$i^5_{opt}(r) = \arg\ max_i\left(\sum_{j=1}^{g}\widetilde{r}_{ij} - \sum_{j=g+1}^{m}\widetilde{r}_{ij}\right). \tag{18}$$

For the value of $\widetilde{r}_{ij}$, vector normalization according to Equation (8) is applied. The initial version of the MOORA method did not take into account the importance of the criteria expressed in weights. The method calculation principle is the sum of the values of the minimized normalized criteria (from $g + 1$ to m) subtracted from the sum of the maximized normalized alternative criteria (from 1 to g). For developing the MOORA method, Brauers started using the weights of criteria [39]. The improved MOORA method is applied for calculations.

The presented methods have been selected as some of those most frequently applied in practice. Similarly, other familiar criteria such as VIKOR, ELECTRE, and others for evaluating the MADM method can be presented as objective functions.

4. Experimental Application of the Methodology Merging MADM Methods

The application of a few MADM methods may result in ranking the scale of evaluation results and reported findings, which is not a clear case of what decision should be made. Each method has an individual theoretical basis and logic, and therefore results in differences.

This chapter describes the methodology for merging the results of MADM methods and presents its practical application. The methodology proposes making calculations using several MADM methods and thus merging their results according to the importance of the method for the problem solved into a single value. SAW, COPRAS, TOPSIS, PROMETHEE and MOORA methods are used in the calculations.

To sum up the results of different methods into the single value, normalizing result data beforehand is required. Linear, classical, vector, logarithmic and other normalization techniques are known. Unlike other methods, the results received applying PROMETHEE are both positive and negative numbers. To transform the results of the PROMETHEE method and other MADM techniques to the uniform scale, PROMETHEE result data must be converted into positive values.

4.1. Methodology for Merging the Results of MADM Methods

The weight, representing the importance of the MADM method, is defined as Ω_ς. The result of the stability of a separate method is defined as S_ς and is expressed in percentage.

The weights of methods are normalized in the following way:

$$\Omega_\varsigma = \frac{S_\varsigma}{\sum_{\varsigma=1}^{v} S_\varsigma}, \quad \sum_{\varsigma=1}^{v} \Omega_\varsigma = 1. \tag{19}$$

The best alternative is established as

$$i_{opt}(\mu) = \arg\!\max \sum_{\varsigma=1}^{v} \Omega_\varsigma \cdot \mu_{i,\varsigma}. \tag{20}$$

where $\mu_{i,\varsigma}$ is the normalized result of the ςth MADM method of the ith alternative.

To merge the results of different methods into a single value, normalizing data on the obtained results is required beforehand. Linear, classical, vector, logarithmic and other normalization techniques are known. Unlike other methods, the results received applying PROMETHEE are both positive and negative numbers. To transform the results of the PROMETHEE and other MADM methods to the uniform scale, first, PROMETHEE result data must be converted into positive values.

For handling negative values and making the scales of the results of other methods equal, Wietendorf's [82] linear normalization rearranging data in the range of [0, 1] is suitable:

$$x_{tr} = \frac{x - x_{min}}{x_{max} - x_{min}}, \tag{21}$$

where x_{tr} is the normalized result of the method and $x_{tr} \in [0, 1]$, x is the initial obtained result of the method, x_{min} is the lowest value of the results of methods, and x_{max} is the highest value of the results of methods.

Another method for making data on MADM results equal in order to employ classical normalization [83] is as follows:

$$\widetilde{\mu}_{i\zeta} = \frac{\mu_{i,\zeta}}{\sum_{i=1}^{n} \mu_{i,\zeta}}. \tag{22}$$

Thus, the results of the PROMETHEE method are transformed into positive numbers beforehand. The transformed value of the evaluation result takes the form of $\widetilde{F}_i$, $i = 1, \ldots, n$. The results of F_i obtained applying the PROMETHEE method are sorted in ascending order. The lowest result of the transformed method is equal to $\widetilde{F}_1 = 1$. Other transformed values are calculated as follows:

$$\widetilde{F}_{i+1} = \widetilde{F}_i + F_{i+1} - F_i, \ i = 1, \ldots, n-1. \tag{23}$$

4.2. Algorithm for Defining MADM Stability

Any mathematical model or method can be applied in practice provided it is stable in terms of the applied parameters. The stability of MADM is verified by employing the statistical simulation method using a sequence of random numbers from the given distribution.

The algorithm for evaluating the stability of the MADM method is presented in Figure 2.

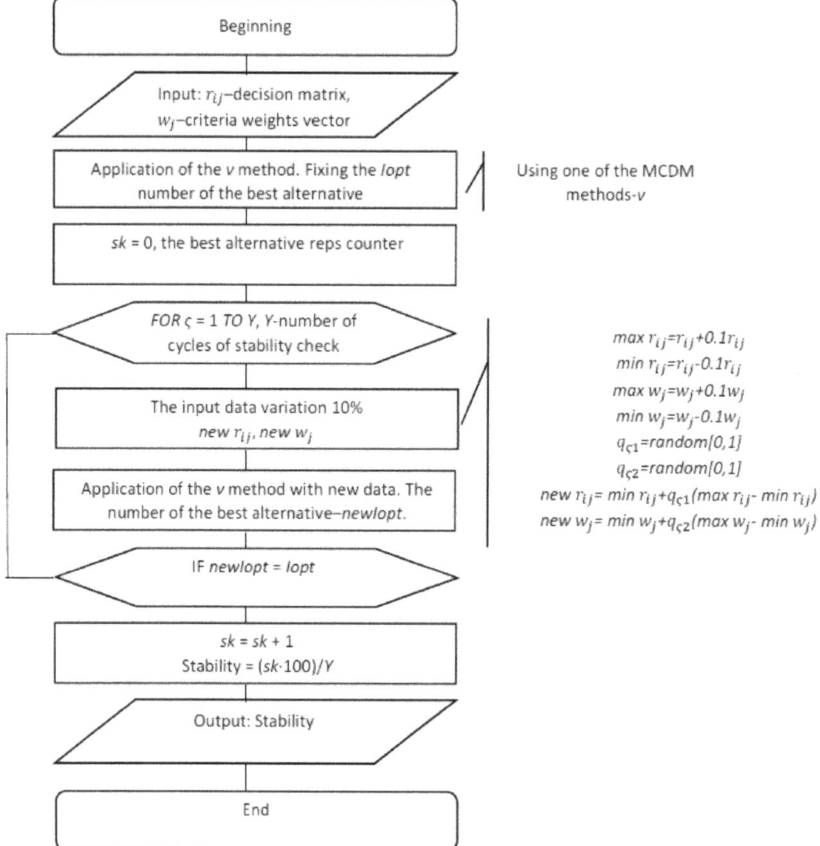

Figure 2. The algorithm for evaluating the stability of the Multi-Attribute Decision-Making (MADM) method.

MADM method v determines the best alternative i of the initial data and fixes the number of this alternative *Iopt*. Verifying the stability of multi-criteria methods brings slight changes in vector data in the initial judging matrix (i.e., expert evaluations r_{ij} and weights w_j). The calculation is made with the newly received values $new_{r_{ij}}$ and new_{w_j} using the MADM method, thus determining the number of the best alternative *newIopt*. The counter *sk* captures the amount of *newIopt* recurrence with the initial *Iopt*. As mentioned in the introduction, a sufficient number of cycles to evaluate the stability of the method to the nearest 0.1 was selected with $Y = 10^5$.

The stability coefficient that fixes the frequency of the recurrence of the best initial alternative is calculated by changing preliminary data. The method is more important for the result of the problem when the stability coefficient is higher.

When no information on the distribution of parameters for MADM methods is available, the uniform distribution is used for generating random values of $\overline{x}_\varsigma$ from the range $[\underline{X}, \overline{X}]$:

$$\overline{x}_\varsigma = \underline{X} + \widetilde{q}_\varsigma \cdot (\overline{X} - \underline{X}), \tag{24}$$

where $\widetilde{q}_\varsigma \in [0, 1]$.

The random values of alternate estimates and criterion weights are generated by slightly changing initial data r_{ij} and w_i by 10% when $\widetilde{q}_\varsigma \in [0, 1]$:

$$\begin{aligned} new_{r_{ij}} &= min\ r_{ij} + \widetilde{q}_\varsigma \cdot (max\ r_{ij} - minr_{ij}), \\ new_{w_i} &= min\ w_i + \widetilde{q}_\varsigma \cdot (max\ w_i - min\ w_i). \end{aligned} \tag{25}$$

The variation limits $[min\ r_{ij}, max\ r_{ij}]$ of alternative estimates r_{ij} are determined as

$$\begin{aligned} max\ r_{ij} &= r_{ij} + 0.1 \cdot r_{ij}, \\ min\ r_{ij} &= r_{ij} - 0.1 \cdot r_{ij}. \end{aligned} \tag{26}$$

Accordingly, the variation limits $[min\ w_i,\ max\ w_i]$ of criterion weights w_i are equal to

$$\begin{aligned} max\ w_i &= w_i + 0.1 \cdot w_i, \\ min\ w_i &= w_i - 0.1 \cdot w_i. \end{aligned} \tag{27}$$

By applying the algorithm for verifying the stability of the MADM method (Figure 2), the stability of all multi-criteria decision-making methods described in this paper is checked. The higher the frequency of the reoccurrence of the best alternative, the more stable the method. The proposed method considers the uncertainty of data on expert evaluation and therefore decreases the level of the subjectivity of the conducted evaluation. The evaluation carried out by applying multiple MADM methods allows selecting the result of the most stable method or merging the results of several methods into a single value.

4.3. Experimental Application of Merging the Results of MADM Methods

To illustrate the application of the method described in the paper, an example in which the estimates of alternatives differ slightly from each other has been chosen. The experts assessed the quality of the course units taught according to six criteria [17]. The descriptions of criteria, as well as the estimates of weights and course units, are given in Table 1. The mean of alternative estimates (i.e., course units), is in the range of [9.03, 9.34].

Table 1. Data on assessing course units.

w		Number of the Criterion	Alt. 1	Alt. 2	Alt. 3	Alt. 4	Alt. 5
0.27	max	1-clearly produced lecture material	9.00	9.00	10.00	8.75	9.20
0.11	max	2-arrangement of studies	9.00	9.50	8.00	10.00	7.75
0.33	max	3-competent teaching staff	9.75	9.40	9.25	9.75	10.00
0.17	max	4-relevance and practical benefits of the material	9.25	8.75	9.00	7.00	8.75
0.05	max	5-variety of techniques for presenting material	9.25	10.00	9.50	10.00	9.50
0.07	max	6-knowledge testing assignments	9.25	9.40	9.60	9.75	9.00
		Mean of estimates	9.25	9.34	9.27	9.21	9.03
		Ranking	3	1	2	4	5

Regarding the initial data (Table 1), the calculation has been conducted by applying the SAW (Equation (3)), TOPSIS (Equation (7)), MOORA (Equation (18)), COPRAS (Equation (17)) and PROMETHEE (Equation (10)) methods. Since all criteria are maximized in the problem solved (Table 1), the calculation of the SAW and COPRAS methods coincides [34]. Thus, only the SAW method will be mentioned below in the paper. The calculations of the PROMETHEE method used the function chart of the priority of the V-shape with indifference criterion (Equation (15)) with parameters q {0.25; 1.75; 0.3; 0.25; 0.2} and s {1.2; 2; 0.6; 1.5; 0.75}. The parameters q and s were not changed, testing the stability of the PROMETHEE method.

The final ranked results are presented in Figure 3. The best alternative is ranked 1, whereas the worst-rated alternative takes 5. Calculations revealed that the results of the methods differ: SAW method results {0.2022; 0.2001; 0.2020; 0.1958; 0.1999}, TOPSIS {0.6029; 0.5272; 0.6004; 0.3640; 0.5413; 3}, MOORA {4.2080; 4.1196; 4.2103; 3.9963; 4.1316}, PROMETHEE {0.2127; −0.3252; 0.0889; −0.2309; 0.2544} [84]. Therefore, it is not possible to unambiguously identify the best course unit from the results obtained.

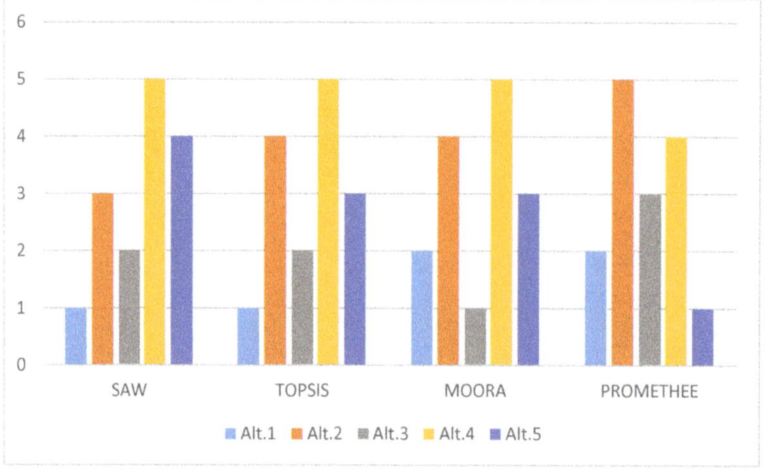

Figure 3. The results obtained using MADM methods. SAW: Simple Additive Weighting; TOPSIS: Technique for Order of Preference by Similarity to Ideal Solution; MOORA: Multi-Objective Optimization by Ratio Analysis; PROMETHEE: Preference Ranking Organization Method for Enrichment Evaluation.

According to the algorithm described above, the stability of the following methods has been determined: SAW 30.7%, TOPSIS 30.9%, MOORA 29.3% and PROMETHEE 26.8%.

The stability of all methods is low due to the similarity of the initial data. Even small variations in the initial data have changed ranking of the best alternative. Having applied Equation (19), the weights of methods are calculated: $\Omega_{SAW} = 0.2608$, $\Omega_{TOPSIS} = 0.2625$, $\Omega_{MOORA} = 0.249$, $\Omega_{PROMETHEE} = 0.2277$ (Figure 4). The weights of the methods are slightly different, and the most stable is the TOPSIS method.

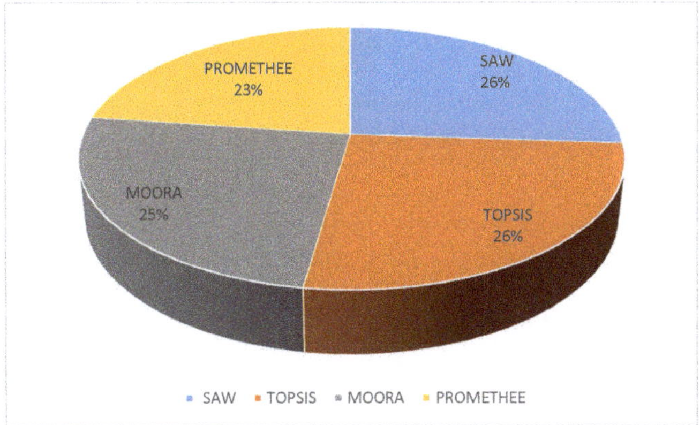

Figure 4. A comparison of stability determined by applying MADM methods.

In order to merge the results of all methods, their estimates need to be unified. Thus, the MADM results are normalized in the range of [0, 1] (Table 2). Wietendorf's [82] linear normalization is suitable for the results of different scales as well as for the negative values of the PROMETHEE method.

Table 2. Normalized MADM result in the range of [0, 1].

Methods	Alt. 1	Alt. 2	Alt. 3	Alt. 4	Alt. 5
SAW	1	0.66563	0.9609	0	0.6375
TOPSIS	1	0.68297	0.9895	0	0.7424
MOORA	0.98925	0.57617	1	0	0.6322
PROMETHEE	0.92816	0	0.7145	0.1626	1

Equation (20) is applied in summing up the estimates of the normalized methods considering their weights. The numerical results are presented in Figure 5. A comparison of the obtained results (Figure 5) with data provided in Table 1 shows changes in the findings. The weights of criteria had a significant impact on the result. Compared to the ranked results employing all methods, the merged MADM result matched with that determined by applying the TOPSIS method. The latter method had a higher weight (i.e., importance), in the problem solved.

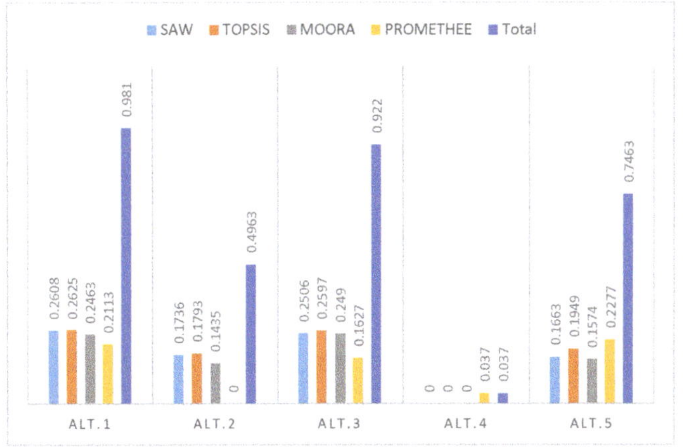

Figure 5. Merging the results of MADM methods following linear normalization.

Table 2 shows that Wietendorf's (Equation (21)) linear normalization has a disadvantage (i.e., zero estimates of alternatives). The weight of the method does not affect the worst-rated alternative as its result is normalized to the zero value.

When the results of two worst-rated alternative methods slightly differ from each other using different normalization, the result may change. Thus, no similar problems are encountered in finding the best alternative.

Another calculation method (i.e., technique for making values equal), involves classical normalization (Equation (22)) and pre-arranging the results of the PROMETHEE method using Equation (23). The transformed positive results of the PROMETHEE method are 1.5379, 1, 1.4141, 1.0943, and 1.5795. Table 3 shows the re-estimation of the methods using classical normalization [84]. The results of the MADM methods merged using Equation (20) are shown in Figure 6.

Table 3. Transformed MADM results applying classical normalization.

Methods	Alt. 1	Alt. 2	Alt. 3	Alt. 4	Alt. 5
SAW	0.2022	0.2001	0.2020	0.1958	0.1999
TOPSIS	0.2287	0.2000	0.2278	0.1381	0.2054
MOORA	0.2036	0.1993	0.2037	0.1934	0.1999
PROMETHEE	0.2321	0.1509	0.2134	0.1652	0.2384

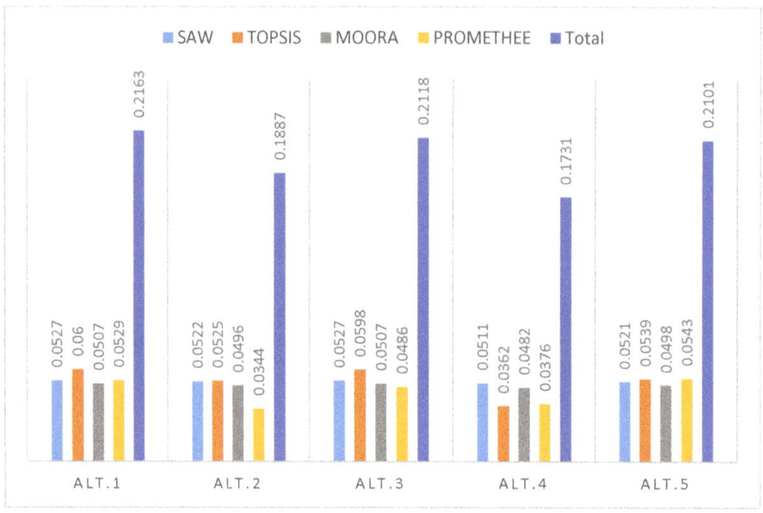

Figure 6. Merging the results of MADM methods following classical normalization.

The numerical results of the initial data (Table 1) and the merged results following classical (Figure 6) and linear (Figure 5) normalization are shown in Figure 7.

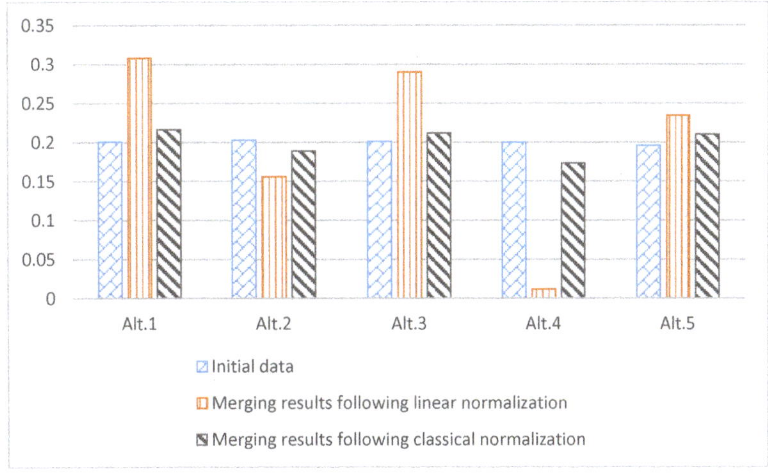

Figure 7. The results of evaluating alternatives, means of numerical values.

Before comparing the obtained information, the results were normalized so that the sum of all estimates of alternatives should be equal to one. The chart shows that the means of the estimates of the initial data differ slightly from each other. The merged results demonstrate that linear normalization leads to significant variations in the outcomes, which is clearly expressed in the evaluation of the fourth alternative. Differences in the results obtained following classical normalization are not significantly expressed in the chart.

The results expressed in ranks are shown in Figures 8 and 9. These charts indicate the mean ranks of the initial data, the ranks of the results of the merged MADM methods (following linear and classical normalization) and the means of the ranks of the results obtained by employing MADM methods. The best alternative is ranked 1, whereas the worst-rated alternative takes 5.

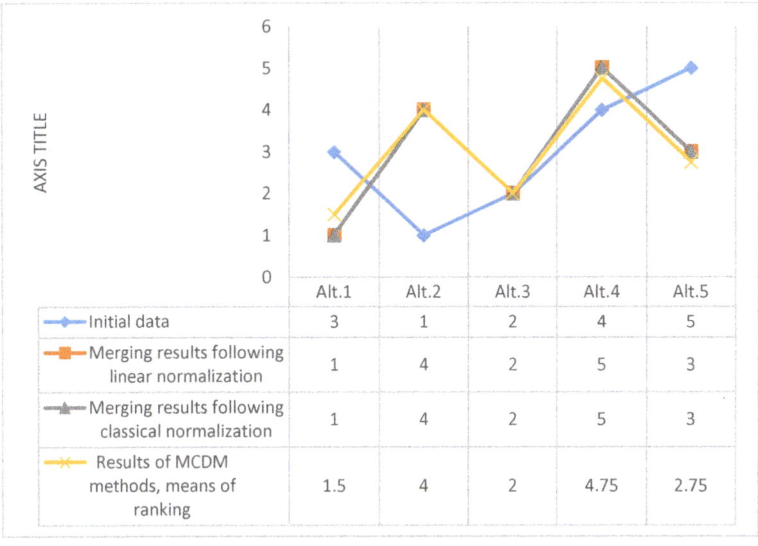

	Alt.1	Alt.2	Alt.3	Alt.4	Alt.5
Initial data	3	1	2	4	5
Merging results following linear normalization	1	4	2	5	3
Merging results following classical normalization	1	4	2	5	3
Results of MCDM methods, means of ranking	1.5	4	2	4.75	2.75

Figure 8. The results of evaluating alternatives, means of the ranked values.

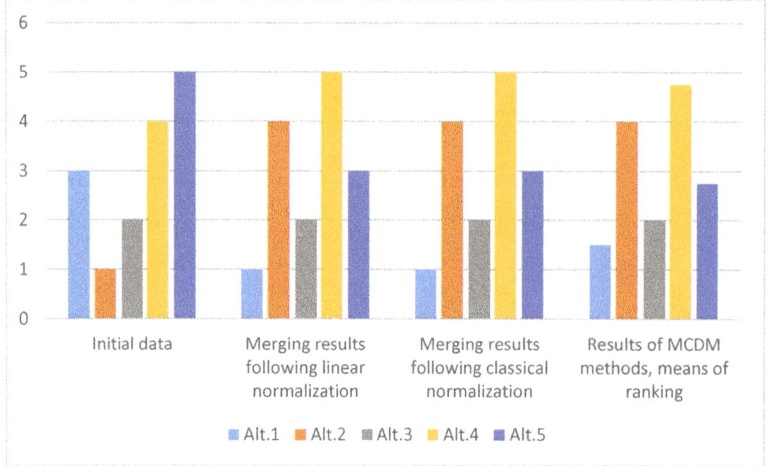

Figure 9. A comparison of the results of evaluating alternatives.

The results of the initial data differ from those achieved by evaluating the outcomes of the first and second alternatives. The merged results coincided following linear and classical normalization. The mean values of the results of MADM methods mainly coincided with the merged results of MADM methods. Since the values of the weights of MADM methods Ω are similar to each other (Figure 4), they did not have a significant effect on the final result. The average results of MADM ranks of the first and third alternatives may lead to different interpretations due to their estimates being equal to 1.5 and 2. The combined results have unequivocally identified the best alternative as Alt. 1.

Table 3 shows that the sum of the estimates for each alternative is equal to 1, which facilitates comparing them. A comparison of the ranked results provided in Figures 5 and 6 demonstrates that the employed methods of the linear and classical normalization of MADM results have determined all alternatives equally. For comparing the mean values of the initial estimates with the findings obtained using MADM methods, the ranking results have changed due to the effect of criterion weights.

5. Discussion and Conclusions

The paper has considered MADM methods as an integral part of the mathematical optimization theory. To illustrate the idea, some of the most applicable methods, SAW, TOPSIS, MOORA, PROMETHEE and COPRAS, have been preferred, and their evaluation criteria have been presented as objective functions, although this paper's methodology is not limited to the use of only these methods. Other MADM methods such as VIKOR, ELECTRE, Evaluation Based on Distance from Average Solution (EDAS), etc. can be similarly introduced as objective functions. The forthcoming papers of the author will focus on exploring more extensively the limitations to constraints on the variables of the above-listed and new MADM methods and will concentrate on the properties of the objective functions and their limitations.

The MADM methods introduced in this paper are employed for selecting the best alternative evaluated according to the established criteria. The purpose of classical optimization is analogous to MADM methods presented in the paper, which means finding an optimal solution from several or many possible options. The use of MADM makes sense in comparing alternatives that do not contain any dominant alternatives when considering all evaluation criteria. The data used in the presented MADM methods are not changed by searching for the optimal solution from all available ones. The decision matrix and the vector of criterion weights are static data, and the number of optional alternatives is finite.

Merging the results of the MADM methods in accordance with their importance showed their possibilities in evaluation. There is a large number of MADM methods, and therefore the literature does not provide unambiguous recommendations for the most appropriate one. Therefore, multiple MADM methods are frequently applied in practice. A methodology for merging the results of MADM methods was presented in this paper, based on summing up the normalized MADM results into a single value and considering the methods' stability.

The findings have demonstrated that weights have a significant influence on the result. In order to analyze the influence of the weights of criteria and methods on the obtained result, a problem example was presented in the practical part of the paper, and the averages of evaluating alternatives had little difference between them. Criterion weights have been found to significantly alter the primary outcomes. The established stability of the applied methods did not differ significantly: $\Omega_{SAW} = 0.2608$, $\Omega_{TOPSIS} = 0.2625$, $\Omega_{MOORA} = 0.249$, $\Omega_{PROMETHEE} = 0.2277$. Nevertheless, the influence of the weights of the methods on the result is noticeable. The ranked result obtained employing the TOPSIS method coincided with the ranked composite result, since the TOPSIS method had a greater influence of weight than the rest of the techniques had. The average results of MADM ranks of the first and third alternatives may lead to different interpretations due to their estimates being equal to 1.5 and 2. The combined results have unequivocally identified the best alternative.

Wietendorf's linear normalization is appropriate for rearranging the results of different scales as well as for the negative values of the PROMETHEE method. However, linear normalization has a disadvantage. Applying Wietendorf's linear normalization, the estimate of the worst alternative is converted into zero, and thus the weight of the influence of the method for determining the worst alternative has no effect on the combined result. The result data managed by applying classical normalization are convenient to be compared because the sum of all results is equal to one. In the case of classical normalization, the negative results of the methods require additional data transformation. The author of this paper proposes a method of transforming negative numbers. Hence, the normalization method had no influence on the final combined result in this task.

The article provides a method for verifying the stability of the MADM method, which ensures the validity of the evaluated result. The technique for validating the stability of the MADM method has a wide range of practical usability in different decision-making problems where evaluation is performed by employing several MADM methods. The proposed method considers the uncertainty of data on expert evaluation and therefore decreases the level of the subjectivity of the conducted evaluation. Further papers will focus more intensely on analyzing the sensitivity of fuzzy AHP methods by fluctuating the data and on investigating several algorithms of FAHP methods.

Funding: This research received no external funding.

Conflicts of Interest: The author declares no conflict of interest.

References

1. Mardani, A.; Zavadskas, E.K.; Khalifah, Z.; Zakuan, N.; Jusoh, A.; Nor, K.; Khoshnoudi, M. A review of multi-criteria decision-making applications to solve energy management problems: Two decades from 1995 to 2015. *Renew. Sustain. Energy Rev.* **2017**, *71*, 216–256. [CrossRef]
2. Mardani, A.; Jusoh, A.; Nor, K.; Khalifah, Z.; Zakwan, N.; Valipour, A. Multiple criteria decision-making techniques and their applications—A review of the literature from 2000 to 2014. *Econ. Res.* **2015**, *28*, 516–571. [CrossRef]
3. Brans, J.P.; Mareschal, B. PROMETHEE V: MCDM problems with segmentation constraints. *INFOR Inf. Syst. Oper. Res.* **1992**, *30*, 85–96. [CrossRef]
4. Liou, J.J.H.; Tzeng, G.-H. Comments on "Multiple criteria decision making (MCDM) methods in economics: An overview". *Technol. Econ. Dev. Econ.* **2012**, *18*, 672–695. [CrossRef]
5. Opricovic, S.; Tzeng, G.H. Compromise solution by MCDM methods: A comparative analysis of VIKOR and TOPSIS. *Eur. J. Oper. Res.* **2004**, *156*, 445–455. [CrossRef]

6. Ramani, T.L.; Quadrifoglio, L.; Zietsman, J. Accounting for nonlinearity in the MCDM approach for a transportation planning application. *IEEE Trans. Eng. Manag.* **2010**, *57*, 702–710. [CrossRef]

7. Hwang, C.L.; Yoon, K. *Multiple Attribute Decision Making: Methods and Applications a State-of-the-Art Survey*; Springer: Berlin/Heidelberg, Germany, 1981. [CrossRef]

8. Tzeng, G.H.; Huang, J.J. *Multiple Attribute Decision Making: Methods and Applications*; CRC Press: Boca Raton, FL, USA, 2011.

9. Chang, Y.H.; Yeh, C.H. Evaluating airline competitiveness using multiattribute decision making. *Omega* **2001**, *29*, 405–415. [CrossRef]

10. Belton, V.; Stewart, T.J. *Multiple Criteria Decision Analysis: An Integrated Approach*; Kluwer: Alphen aan den Rijn, The Nederlands, 2002. [CrossRef]

11. Malczewski, J.; Rinner, C. Multiattribute decision analysis methods. In *Multicriteria Decision Analysis in Geographic Information Science*; Springer: Berlin/Heidelberg, Germany, 2015; pp. 81–121. [CrossRef]

12. Yeh, C.H.; Deng, H.; Chang, Y.H. Fuzzy multicriteria analysis for performance evaluation of bus companies. *Eur. J. Oper. Res.* **2000**, *126*, 459–473. [CrossRef]

13. Mareschal, B.; Brans, J.P.; Vincke, P. PROMETHEE: A new family of outranking methods in multicriteria analysis. In Proceedings of the 10th Triennial Conference on Operational Research, Washington, DC, USA, 6–10 August 1984.

14. Zavadskas, E.K.; Podvezko, V. Integrated determination of objective criteria weights in MCDM. *Int. J. Inf. Technol. Decis. Mak.* **2016**, *15*, 267–283. [CrossRef]

15. Podvezko, V. Application of AHP technique. *J. Bus. Econ. Manag.* **2009**, *10*, 181–189. [CrossRef]

16. Taha, H.A. *Operations Research: An Introduction*, 6th ed.; Prentice Hall: Upper Saddle River, NJ, USA, 1997.

17. Vinogradova, I.; Kliukas, R. Methodology for evaluating the quality of distance learning courses in consecutive stages. *Proc. Soc. Behav. Sci.* **2015**, *191*, 1583–1589. [CrossRef]

18. Williams, H. Fourier's method of linear programming and its dual. *Am. Math. Mon.* **1986**, *93*, 681–695. [CrossRef]

19. Jorrand, P.; Sgurev, V. *Artificial Intelligence IV: Methodology, Systems, Applications*; Elsevier: Amsterdam, The Nederlands, 1990. [CrossRef]

20. Šipoš, M.; Klaić, Z.; Fekete, K.; Stojkov, M. Review of non-traditional optimization methods for allocation of distributed generation and energy storage in distribution system. *Teh. Vjesn.* **2018**, *25*, 294–301. [CrossRef]

21. Dantzig, G. Maximization of a linear function of variables subject to linear inequalities. In *Activity Analysis of Production and Allocation*; Koopmans, T.C., Ed.; Wiley: Hoboken, NJ, USA, 1951; pp. 339–347.

22. Robbins, H.; Monro, S. A stochastic approximation method. *Ann. Math. Statist.* **1951**, *22*, 400–407. [CrossRef]

23. Dreyfus, S.E. Richard Bellman on the birth of dynamic programming. *Oper. Res.* **2002**, *50*, 48–51. [CrossRef]

24. Tan, K.C.; Khor, E.F.; Lee, T.H. *Multiobjective Evolutionary Algorithms and Applications*; Springer: London, UK, 2005. [CrossRef]

25. Gunantara, N. A review of multi-objective optimization: Methods and its applications. *Elect. Electron. Eng.* **2018**, *5*, 1502242. [CrossRef]

26. Keeney, R.L.; Raiffa, H. *Decision Making with Multiple Objectives Preferences and Value Tradeoffs*; Wiley: New York, NY, USA, 1976.

27. Fishburn, P.C. Methods of estimating additive utilities. *Manag. Sci.* **1967**, *13*, 435–453. [CrossRef]

28. Charnes, A.W.; Cooper, W.W.; Rhodes, E. Measuring the efficiency of decision making units. *Eur. J. Oper. Res.* **1979**, *2*, 429–444. [CrossRef]

29. Zavadskas, E.K.; Turskis, Z.; Kildiene, S. State of art surveys of overviews on MCDM/MADM methods. *Technol. Econ. Dev. Econ.* **2014**, *20*, 165–179. [CrossRef]

30. Bernardo, J.J.; Blin, J.M. A programming model of consumer choice among multi-attributed brands. *J. Consum. Res.* **1977**, *4*, 111–118. [CrossRef]

31. MacCrimmon, K.R. *Decision Making among Multiple-Attribute Alternatives: A Survey and Consolidated Approach*; RAND Memorandum, RM-4823-ARPA; The RAND Corporation: Santa Monica, CA, USA, 1968.

32. Podvezko, V. The comparative analysis of MCDA methods SAW and COPRAS. *Econ. Eng. Decis.* **2011**, *22*, 134–146. [CrossRef]

33. Peng, H.; Wang, J.A. Multicriteria group decision-making method based on the normal cloud model with Zadeh's Z-numbers. *IEEE Trans. Fuzzy Syst.* **2018**, *26*, 3246–3260. [CrossRef]

34. Roy, B. The outranking approach and the foundations of ELECTRE methods. *Theory Decis.* **1991**, *31*, 49–73. [CrossRef]
35. Brans, J.P.; Vincke, P.; Mareschal, B. How to select and how to rank projects: The method. *Eur. J. Oper. Res.* **1986**, *24*, 228–238. [CrossRef]
36. Opricovic, S. *Multicriteria Optimization of Civil Engineering Systems*; University of Belgrade, Faculty of Civil Engineering: Belgrade, Srbija, 1998.
37. Zavadskas, E.K.; Turskis, Z.; Antucheviciene, J.; Zakarevicius, A. Optimization of weighted aggregated sum product assessment. *Electron. Elect. Eng.* **2012**, *6*, 3–6. [CrossRef]
38. Kildienė, S.; Kaklauskas, A.; Zavadskas, E. COPRAS based comparative analysis of the European country management capabilities within the construction sector in the time of crisis. *J. Bus. Econ. Manag.* **2011**, *12*, 417–434. [CrossRef]
39. Brauers, W.K.M.; Zavadskas, E.K. The MOORA method and its application to privatization in a transition economy. *Control. Cybern.* **2006**, *35*, 445–469.
40. Turskis, Z.; Keršulienė, V.; Vinogradova, I. A new fuzzy hybrid multi-criteria decision-making approach to solve personnel assessment problems. Case study: Director selection for estates and economy office. *Econ. Comput. Econ. Cybern. Stud. Res.* **2017**, *51*, 211–229.
41. Turskis, Z.; Zavadskas, E.K. A novel method for multiple criteria analysis: Grey additive ratio assessment (ARAS-G) method. *Informatica* **2010**, *21*, 597–610.
42. Turskis, Z.; Zavadskas, E.K. A new fuzzy additive ratio assessment method (ARAS-F). Case study: The analysis of fuzzy multiple criteria in order to select the logistic centers location. *Transport* **2010**, *25*, 423–432. [CrossRef]
43. Krylovas, A.; Zavadskas, E.K.; Kosareva, N.; Dadelo, S. New KEMIRA method for determining criteria priority and weights in solving MCDM problem. *Int. J. Inf. Technol. Decis. Mak.* **2014**, *13*, 1119–1133. [CrossRef]
44. Zavadskas, E.K.; Turskis, Z. A new additive ratio assessment (ARAS) method in multicriteria decision-making. *Technol. Econ. Dev. Econ.* **2010**, *16*, 159–172. [CrossRef]
45. Del Vasto-Terrientes, L.; Valls, A.; Slowinski, R.; Zielniewicz, P. ELECTRE-III-H: An outranking-based decision aiding method for hierarchically structured criteria. *Expert Syst. Appl.* **2015**, *42*, 4910–4926. [CrossRef]
46. Corrente, S.; Greco, S.; Slowinski, R. Multiple criteria hierarchy process with ELECTRE and PROMETHEE. *Omega* **2013**, *41*, 820–846. [CrossRef]
47. Ziemba, P. Towards strong sustainability management—A generalized PROSA method. *Sustainability* **2019**, *11*, 1555. [CrossRef]
48. Keršulienė, V.; Zavadskas, E.K.; Turskis, Z. Selection of rational dispute resolution method by applying new stepwise weight assessment ratio analysis (SWARA). *J. Bus. Econ. Manag.* **2010**, *11*, 243–258. [CrossRef]
49. Trinkūnienė, E.; Podvezko, V.; Zavadskas, E.K.; Jokšienė, I.; Vinogradova, I.; Trinkūnas, V. Evaluation of quality assurance in contractor contracts by multi-attribute decision-making methods. *Econ. Res. Ekon. Istraž.* **2017**, *30*, 1152–1180. [CrossRef]
50. Saaty, T.L. *The Analytic Hierarchy Process*; McGraw-Hill: New York, NY, USA, 1980.
51. Saaty, T.L. The analytic hierarchy and analytic network processes for the measurement of intangible criteria and for decision-making. In *Multiple Criteria Decision Analysis: State of the Art Surveys*; Greco, S., Ehrgott, M., Figueira, J., Eds.; Springer: Berlin, Germany, 2005; pp. 345–408. [CrossRef]
52. Kurilov, J.; Vinogradova, I. Improved fuzzy AHP methodology for evaluating quality of distance learning courses. *Int. J. Eng. Educ.* **2016**, *32*, 1618–1624. [CrossRef]
53. Kurilovas, E.; Vinogradova, I.; Kubilinskiene, S. New MCEQLS fuzzy AHP methodology for evaluating learning repositories: A tool for technological development of economy. *Technol. Econ. Dev. Econ.* **2016**, *22*, 142–155. [CrossRef]
54. Hashemkhani, Z.; Saparauskas, J. New application of SWARA method in prioritizing sustainability assessment indicators of energy system. *Inz. Ekon. Eng. Econ.* **2013**, *24*, 408–414. [CrossRef]
55. Čereška, A.; Zavadskas, E.K.; Bucinskas, V.; Podvezko, V.; Sutinys, E. Analysis of steel wire rope diagnostic data applying multi-criteria methods. *Appl. Sci.* **2018**, *8*, 260. [CrossRef]
56. Vinogradova, I.; Podvezko, V.; Zavadskas, E.K. The recalculation of the weights of criteria in MCDM methods using the Bayes approach. *Symmetry* **2018**, *10*, 205. [CrossRef]

57. Sabaei, D.; Erkoyuncu, J.; Roy, R. A review of multi-criteria decision making methods for enhanced maintenance delivery. *Proc. CIRP* **2015**, *37*, 30–35. [CrossRef]

58. Stewart, T.J. A critical survey on the status of multiple criteria decision making theory and practice. *Omega* **1992**, *20*, 569–586. [CrossRef]

59. Kim, J.; Kim, J. Optimal portfolio for LNG importation in Korea using a two-step portfolio model and a fuzzy analytic hierarchy process. *Energies* **2018**, *11*, 3049. [CrossRef]

60. Adeel, A.; Akram, M.; Ahmed, I.; Nazar, K. Novel m-polar fuzzy linguistic ELECTRE-I method for group decision-making. *Symmetry* **2019**, *11*, 471. [CrossRef]

61. Bae, H.J.; Kang, J.E.; Lim, Y.R. Assessing the health vulnerability caused by climate and air pollution in Korea using the fuzzy TOPSIS. *Sustainability* **2019**, *11*, 2894. [CrossRef]

62. Ziemba, P.; Becker, J. Analysis of the digital divide using fuzzy forecasting. *Symmetry* **2019**, *11*, 166. [CrossRef]

63. Evans, J.R. Sensitivity analysis in decision theory. *Decis. Sci.* **1984**, *15*, 239–247. [CrossRef]

64. Barron, H.; Schmidt, C.P. Sensitivity analysis of additive multiattribute value models. *Oper. Res.* **1988**, *36*, 122–127. [CrossRef]

65. Insua, D.R. *Sensitivity Analysis in Multi-Objective Decision Making*; Springer: Berlin/Heidelberg, Germany, 1990. [CrossRef]

66. Janssen, R. *Multiobjective Decision Support for Environmental Management*; Kluwer: Alphen aan den Rijn, The Nederlands, 1992. [CrossRef]

67. Butler, J.; Jia, J.; Dyer, J. Simulation techniques for the sensitivity analysis of multi-criteria decision models. *Eur. J. Oper. Res.* **1997**, *103*, 531–546. [CrossRef]

68. Wolters, W.T.M.; Mareschal, B. Novel types of sensitivity analysis for additive MCDM methods. *Eur. J. Oper. Res.* **1995**, *81*, 281–290. [CrossRef]

69. Masuda, T. Hierarchical sensitivity analysis of the priorities used in the analytic hierarchy process. *Int. J. Syst. Sci.* **1990**, *21*, 415–427. [CrossRef]

70. Triantaphyllou, E.; Sánchez, A. A sensitivity analysis approach for some deterministic multi-criteria decision-making methods. *Decis. Sci.* **1997**, *28*, 151–194. [CrossRef]

71. Podvezko, V. Multicriteria evaluation under uncertainty. *Bus. Theory Pract.* **2006**, *7*, 81–88. (In Lithuanian) [CrossRef]

72. Zavadskas, E.K.; Turskis, Z.; Dejus, T.; Viteikiene, M. Sensitivity analysis of a simple additive weight method. *Int. J. Manag. Decis. Mak.* **2007**, *8*, 555–574. [CrossRef]

73. Memariani, A.; Amini, A.; Alinezhad, A. Sensitivity analysis of simple additive weighting method (SAW): The results of change in the weight of one attribute on the final ranking of alternatives. *J. Ind. Eng.* **2009**, *4*, 13–18.

74. Alinezhad, A.; Sarrafha, K.; Amini, A. Sensitivity analysis of SAW technique: The impact of changing the decision making matrix elements on the final ranking of alternatives. *Iran. J. Oper. Res.* **2014**, *5*, 82–94.

75. Yu, V.F.; Hu, K.J. An integrated fuzzy multi-criteria approach for the performance evaluation of multiple manufacturing plants. *Comput. Ind. Eng.* **2010**, *58*, 269–277. [CrossRef]

76. Alinezhada, A.; Aminib, A. Sensitivity analysis of TOPSIS technique: The results of change in the weight of one attribute on the final ranking of alternatives. *J. Optim. Ind. Eng.* **2011**, *7*, 23–28.

77. Misra, S.K.; Ray, A. Comparative study on different multi-criteria decision making tools in software project selection scenario. *Int. J. Adv. Res. Comput. Sci.* **2012**, *3*, 172–178. [CrossRef]

78. Moghassem, A.R. Comparison among two analytical methods of multi-criteria decision making for appropriate spinning condition selection. *World Appl. Sci. J.* **2013**, *21*, 784–794.

79. Hsu, L.C.; Ou, S.L.; Ou, Y.C. A comprehensive performance evaluation and ranking methodology under a sustainable development perspective. *J. Bus. Econ. Manag.* **2015**, *16*, 74–92. [CrossRef]

80. Podvezko, V.; Podviezko, A. Dependence of multi-criteria evaluation result on choice of preference functions and their parameters. *Technol. Econ. Dev. Econ.* **2010**, *16*, 143–158. [CrossRef]

81. Zavadskas, E.K.; Kaklauskas, A.; Peldschus, F.; Turskis, Z. Multi-attribute assessment of road design solutions by using the COPRAS method. *Balt. J. Road Bridge Eng.* **2007**, *2*, 195–203.

82. Weitendorf, D. Beitrag Zur Optimierung Der Räumlichen Struktur Eines Gebäudes. Ph.D. Thesis, Hochschule für Architektur und Bauwesen, Weimar, Germany, 1976.

83. Zavadskas, E.K.; Turskis, Z. A new logarithmic normalization method in games theory. *Informatica* **2008**, *19*, 303–314.
84. Vinogradova, I. Integration of several MCDM results according to the importance of methods. *Liet. Matem. Rink. LMD Darb. B* **2016**, *57*, 77–82. (In Lithuanian)

MDPI

St. Alban-Anlage 66

4052 Basel

Switzerland

Tel. +41 61 683 77 34

Fax +41 61 302 89 18

www.mdpi.com

Mathematics Editorial Office

E-mail: mathematics@mdpi.com

www.mdpi.com/journal/mathematics